Basic Electrical Engineering

N K De

Formerly Professor
Department of Electrical Engineering
IIT Kharagpur
Visiting Professor
Narula Institute of Technology
Kolkata, West Bengal

Dipu Sarkar

Assistant Professor
Department of Electrical and Electronics Engineering
National Institute of Technology Nagaland

Universities Press

UNIVERSITIES PRESS (INDIA) PRIVATE LIMITED

Registered Office
3-6-747/1/A & 3-6-754/1, Himayatnagar
Hyderabad 500 029, Telangana, India
email: info@universitiespress.com, www.universitiespress.com

Distributed by
Orient Blackswan Private Limited

Registered Office
3-6-752 Himayatnagar, Hyderabad 500 029, Telangana, India

Other Offices
Bengaluru / Bhopal / Chennai / Ernakulam / Guwahati Hyderabad /
Jaipur / Kolkata / Lucknow / Mumbai / New Delhi / Noida / Patna

ISBN 978 81 7371 944 8

Typeset in TimeNew Roman at 10.6pt *by*
OSDATA, Hyderabad 500 029

Printed in India by
B.B. Press, Noida

Published by
Universities Press (India) Private Limited
3-6-747/1/A & 3-6-754/1, Himayatnagar
Hyderabad 500 029, Telangana, India

To

My wife, Sm. Sandhya De
and mother-in-law, Sm. Shefali Das

Nisit K. De

To

My father and mother,
Late Ranjit Sarkar and Sm. Rita Sarkar

Dipu Sarkar

Contents

Preface

This book has been developed as an introductory textbook on basic electrical engineering and follows the curriculum prescribed generally in the first year of engineering in colleges in India. It is intended to provide a firm foundation to the basic concepts of electrical circuits and systems. The material in the book can be considered in three parts—electric circuits (dc and ac), field parts (magnetic and electric), and electrical machines.

The first part consists of seven chapters. Chapter 1 gives an overview of the subject and also introduces the reader to electric circuits. The study of dc circuits and various theorems applicable to them is presented in Chapter 2, while in the next chapter (3), transients in R-L and R-C circuits, when fed from dc supply, are discussed. Before taking up the study of ac circuits, a brief chapter (4) on electrical quantities, including phasor algebra, is presented. In Chapters 5 and 6, the study of different types of circuits—series, parallel and series–parallel, fed from ac supply—are discussed in detail, along with phasor diagrams. In the last chapter (7) of this part, three-phase circuits, connected in star and delta, fed mainly from a balanced three-phase supply, is described with phasor diagrams along with the concept of power measurement in these circuits.

In the second part comprising three chapters, electromagnetism, including magnetic circuits, is taken up in detail in the first chapter (8). The various aspects of inductance are discussed in Chapter 9. The last chapter (10) of this part discusses electrostatics and capacitance.

In the third and last part of the book, the study of four types of electrical machines is taken up. Transformers, a widely used static ac machine, is discussed in Chapter 11, while in Chapter 12, the operation of dc machines, both as a generator and a motor is presented. All aspects are explained in detail. The study of three-phase induction motors — the work horse of industry — is taken up in Chapter 13. In Chapter 14, the synchronous machine which is a rotating ac machine and can be used both as a generator and a motor is discussed. This is followed by a brief chapter (15) on commonly used electrical measuring instruments, included in the syllabus of most Indian universities.

Both the authors have taken the help of many people in writing this book, whom they would like to acknowledge.

Firstly, they would like to thank the faculty members, including the head of the Electrical Engineering Department, Narula Institute of Technology, Kolkata, for their help and cooperation. The first author would also like to thank the senior and junior faculty members of the Electrical Engineering Department at IIT Kharagpur, where he was a faculty member from 1968–2007, for their assistance in developing this book. He also thanks Professor Gosai Das Roy and Professor Tapas K. Bhattacharya of the Electrical Engineering Department in particular, as they, along with the first author, worked together in preparing a web-based course on basic electrical engineering under the NPTEL Programme of the Government of India. Lastly, they would like to thank the publishers,

Universities Press, Hyderabad, for agreeing to publish this book and many others without whose help it would have been impossible to bring out the book in the present form.

Nisit K. De
Dipu Sarkar

CHAPTER 1

INTRODUCTION

In this introductory chapter, we will first learn the most basic law in electrical engineering—Ohm's law. It is used to find the current flowing through an element when it is connected to a dc supply. We will then define the terms, linear and non-linear resistors. We will determine the expressions for equivalent resistance when a number of them are connected in series or in parallel, or in a combination of both. The term 'conductance' is also introduced.

1.1 Ohm's Law

Ohm's law states that the current, I in a circuit (Fig. 1.1) is proportional to the applied voltage, V, if the resistance, R remains constant. Mathematically, it is written as,

$$R = \frac{V}{I}, \quad \text{or,} \quad V = I.R.$$

Fig. 1.1 *Ohm's law*

LINEAR AND NON-LINEAR RESISTORS

Materials that obey Ohm's law are called ohmic. If a graph is drawn with voltage, V on the x-axis, and current, I on the y-axis, the curve depicting an ohmic material is linear (Fig. 1.2) with its slope equal to the resistance, R, a constant. The current, I in the circuit can be obtained as

$$I = \frac{V}{R}.$$

As the graph is *linear*, if the applied voltage in a circuit, V is doubled, the current there is also doubled.

However, there are certain elements, such as a tungsten lamp or a diode, which exhibit *non-linear* characteristics between the applied voltage, V, and the current, I (Fig. 1.3(a)). This is because the slope of the graph, i.e., the resistance changes at different points, when the applied voltage is changed (Fig. 1.3(b)). The schematic of a diode is shown in Fig. 1.4.

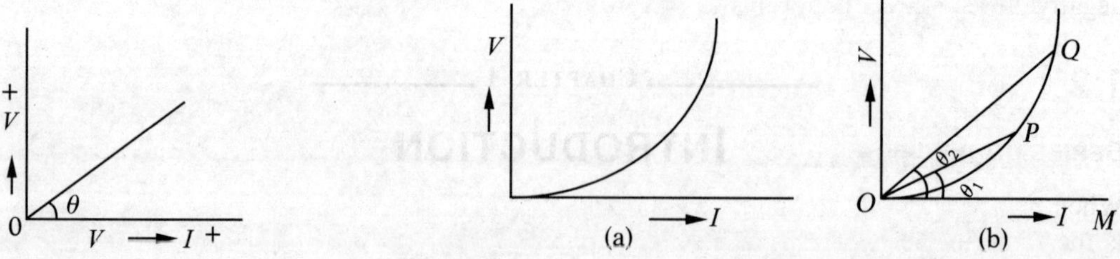

Fig. 1.2 *V – I graph* **Fig. 1.3** *V – I graph of diode*

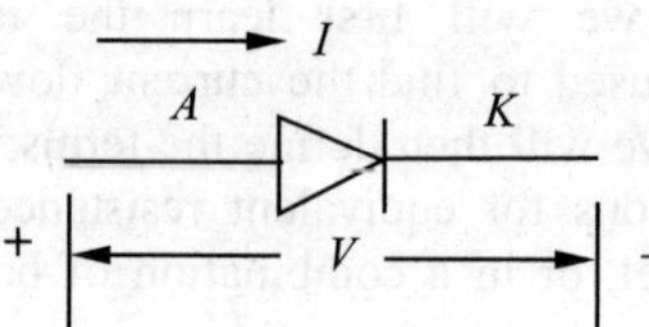

Fig. 1.4 *Schematic diagram of a diode*

BILATERAL AND UNILATERAL RESISTORS

Resistors can also be classified as bilateral and unilateral. In a *bilateral* element, assumed to be linear or to have constant resistance R, the $V - I$ characteristic is as shown in Fig. 1.5 obtained by extending the linear graph (Fig. 1.2) onto the third quadrant. If a positive voltage, $+V$ is applied across a resistor with resistance, R, the current also flows in the positive direction ($+I$). If negative voltage, $-V$ is applied across the same resistor, the current also flows in the negative direction ($-I$). Thus, in linear bilateral elements, the applied voltage only changes sign (with the current changing signs similarly); their magnitudes remain constant. Normally, a resistor is assumed to be bilateral—this assumption is used in Chapter 2, where dc circuits along with various theorems are presented in detail.

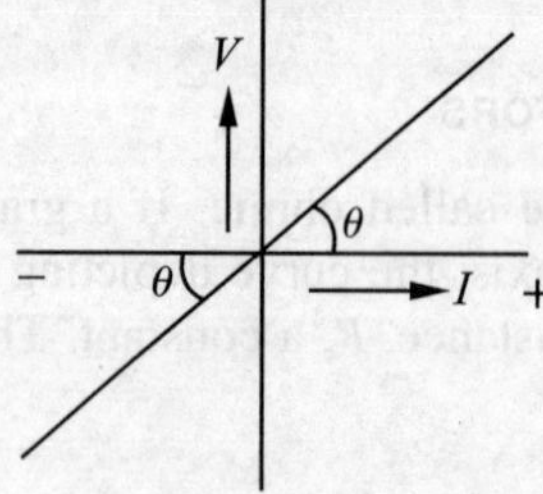

Fig. 1.5 *V – I graph of a bilateral element (R)*

In a *unilateral* element, the $V - I$ characteristic is of the type shown in Fig. 1.3(a), which is that of a diode. As the diagram indicates, when positive or forward voltage, $+V$ is applied across it, the current, I in the diode will be in the positive direction only, its value limited by the resistance, R (which is very low, nearly zero (0) in some cases) connected in series with it. When negative voltage, $-V$, is applied across it, little or no current will flow in the negative direction, as its resistance is very high, i.e., ∞.

1.2 Series and Parallel Connection of Resistors

SERIES CONNECTION

When two resistors, R_1 and R_2, are connected in series across a dc supply, V, the current, I, is the same in the two resistances (Fig. 1.6). If V_1 and V_2 are the voltage drops across the resistors, R_1 and R_2, respectively, then, using Ohm's law, we get,

$$V_1 = V_{AD} = I.R_1 \quad \text{and} \quad V_2 = V_{DB} = I.R_2.$$

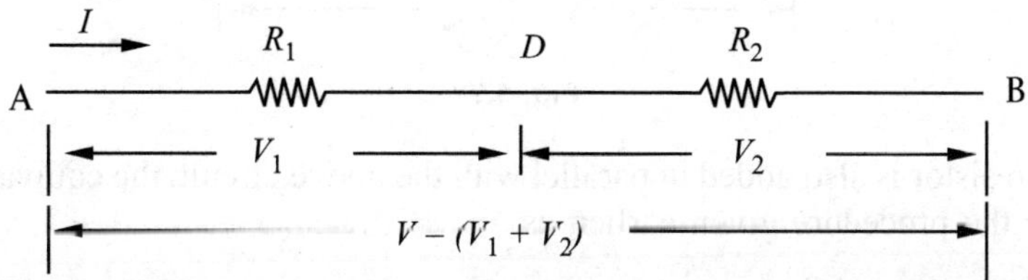

Fig. 1.6

The supply voltage is written as

$$V = V_{AB} = I \cdot R_{eq} = V_1 + V_2 = V_{AD} + V_{DB} = I \cdot (R_1 + R_2),$$

where, R_{eq} = equivalent resistance of the two resistors connected in series.

From the expression, $R_{eq} = (R_1 + R_2)$.

Now, if a third resistance is also added in series in the above circuit, the equivalent resistance is obtained by the procedure given earlier, as

$$R_{eq} = (R_1 + R_2 + R_3) = \sum_{i=1}^{3} R_i.$$

This can be easily extended to n number of resistances, R_i connected in series.

PARALLEL CONNECTION

If two resistors, R_1 and R_2, are connected in parallel across a DC supply, V, the voltage drops in the resistances are the same, i.e., V (Fig. 1.7). If I_1 and I_2 are the currents in the resistors, R_1 and R_2, respectively, then, using Ohm's law, we get,

$$I_1 = \frac{V}{R_1} \text{ and } I_2 = \frac{V}{R_2}.$$

The total current drawn from the supply voltage, V is given by,

$$I = \frac{V}{R_{eq}} = (I_1 + I_2) = \left(\frac{V}{R_1} + \frac{V}{R_2}\right) = V \cdot \left(\frac{1}{R_1} + \frac{1}{R_2}\right)$$

where, R_{eq} = equivalent resistance of the two resistors connected in parallel.

From the expression, $\frac{1}{R_{eq}} = \left(\frac{1}{R_1} + \frac{1}{R_2}\right)$, or, $R_{eq} = \frac{R_1 \cdot R_2}{(R_1 + R_2)}$.

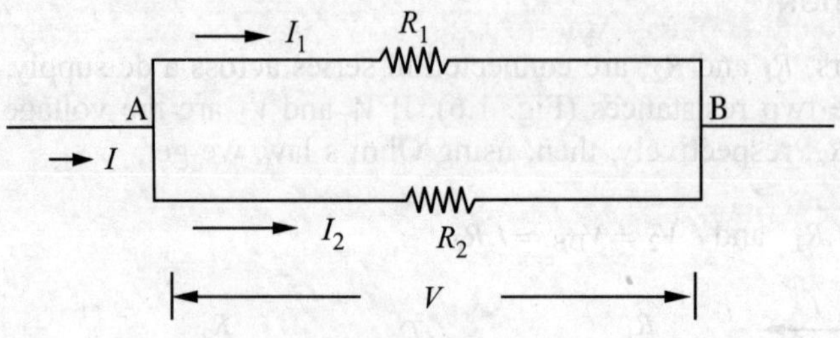

Fig. 1.7

Now, if a third resistor is also added in parallel with the above circuit, the equivalent resistance is obtained, by the procedure given earlier, as

$$\frac{1}{R_{eq}} = \left(\frac{1}{R_1} + \frac{1}{R_2} + \frac{1}{R_3}\right) = \sum_{i=1}^{3} \left(\frac{1}{R_i}\right).$$

This can be easily extended to n number of resistors, R_i connected in parallel.

The reciprocal of the resistance, R, is known as the conductance, G of the resistor, i.e., $G = \frac{1}{R}$. If this is used in the above equation, the expression for the equivalent conductance, G_{eq}, when three resistors are connected in parallel, is obtained as,

$$G_{eq} = \frac{1}{R_{eq}} = (G_1 + G_2 + G_3) = \sum_{i=1}^{3} G_i.$$

If two resistors are connected in parallel, then,

$$G_{eq} = \frac{1}{R_{eq}} = (G_1 + G_2).$$

Note that these expressions are the same as those obtained for the resistances of resistors connected in series, except that here, G is replaced by R.

This can also be easily extended to n number of resistors, R_i connected in parallel. Resistors can also be connected in series and parallel, in various combinations. The equivalent resistance between any two points can be found by using the above expressions.

Heating Effect of Current in a Resistor

When electric current is passed through a resistor, it generates heat due to the resistance it offers to the current flow. The work done in overcoming the resistance is generated as heat. If a constant current, I A, is passed through a resistor of resistance, $R\,\Omega$, the heat generated by the resistor, in time, t, is given by $I^2 \cdot R \cdot t$, the unit of which is joules (J). The time is measured in seconds (s), and the current is dc. As the rate of consumption of energy is uniform or constant, the power loss is given by energy consumed (for work done) in unit time. The power consumed (or loss) is given by $P = I^2.R$, its unit being watt (W). The phenomenon of electrical energy being converted into heat energy is used in immersion heaters and electric kettles to heat water. In the copper windings of electrical machines too, heat loss occurs. Here it is termed as copper loss.

We can also obtain another expression for power loss by using Ohm's law. The applied voltage or voltage drop (V) is

$$V = I.R.$$

The current can be written as

$$I = \frac{V}{R}.$$

If any of the two formulas are used, the power loss is found as

$$P = V \cdot I = \frac{V^2}{R}.$$

EXAMPLE 1.1 *Three resistors, R_1, R_2 and R_3 are connected in series across a dc supply of 12 V (Fig. 1.8). The value of the resistance, R_1 is 1 Ω, the voltage drop across R_2 is 4 V, and the power consumed in R_3 is 12 W. Find the values of resistances, R_2 and R_3.*

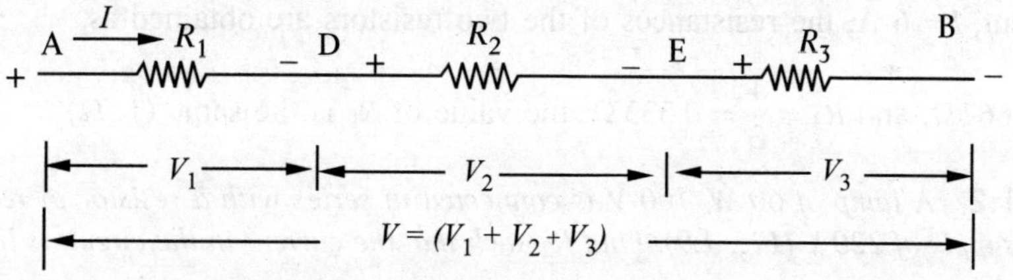

A I R_1 D R_2 E R_3 B

\+ − + − + −

V_1 V_2 V_3

$V = (V_1 + V_2 + V_3)$

Fig. 1.8

SOLUTION The circuit with three resistors, R_1, R_2 and R_3 connected in series is shown in Fig. 1.8.

Supply voltage $= V = V_{AB} = 12$ V.

$$R_1 = 1\ \Omega, V_2 = V_{DE} = 4\ \text{V}, P_3 = 12\ \text{W}.$$

If the current in the circuit is I, which is the same in all three resistances, the voltage drop across R_1,

$$V_1 = V_{AD} = I.R_1 = I.1 = I\ \text{V}.$$

Power consumed in R_3

$P_3 = V_3 \cdot I = I^2 \cdot R_3 = 12$ W.

The voltage drop across R_3

$$V_3 = V_{EB} = \frac{P_3}{I} = \frac{12}{I}.$$

Total voltage drop in the three resistors = supply voltage.

$$V = V_{AB} = (V_{AD} + V_{DE} + V_{EB}) = (V_1 + V_2 + V_3) = \left(I + 4 + \frac{12}{I}\right) = 12 \text{ V}.$$

Simplifying, we get,

$$I^2 - 8I + 12 = 0.$$

Solving it, the current

$$I = \frac{8}{2} \pm \sqrt{\left(\frac{8}{2}\right)^2 - 12} = 6 \text{ A or } 2 \text{ A}.$$

Using current, $I = 2$ A, the resistances of the two resistors are obtained as,

$$R_2 = \frac{V_2}{I} = \frac{4}{2} = 2\,\Omega, \text{ and } R_3 = \frac{P_3}{I^2} = \frac{12}{2^2} = 3\,\Omega \text{ , with } R_1 = 1\,\Omega.$$

Using current, $I = 6$ A, the resistances of the two resistors are obtained as,

$$R_2 = \frac{2}{3} = 0.667\,\Omega, \text{ and } R_3 = \frac{1}{3} = 0.333\,\Omega, \text{ the value of } R_1 \text{ is the same } (1\ \Omega).$$

EXAMPLE 1.2 *A lamp of 60 W, 100 V is connected in series with a resistor of resistance, R_2 across a dc supply of 220 V (Fig. 1.9). Find R_2 such that the current in the circuit is limited to the rating of the lamp.*

SOLUTION The circuit with two resistors, R_1 (lamp) and R_2 are connected in series as shown in Fig. 1.9.

Supply voltage $= V = V_{AB} = 220$ V.

The current in the circuit is I, which is the same in both resistors.

$$V_1 = V_{AD} = 100 \text{ V}, P_1 = V_1.I = 60 \text{ W}.$$

The current in the circuit, and also the lamp, $I = \dfrac{P_1}{V_1} = \dfrac{60}{100} = 0.6$ A.

The voltage drop across $R_2 = V_2 = V_{DB} = V - V_1 = V_{AB} - V_{AD} = I.R_2 = 220 - 100 = 120$ V.

The resistance $= R_2 = \dfrac{V_2}{I} = \dfrac{120.0}{0.6} = 200.0\,\Omega.$

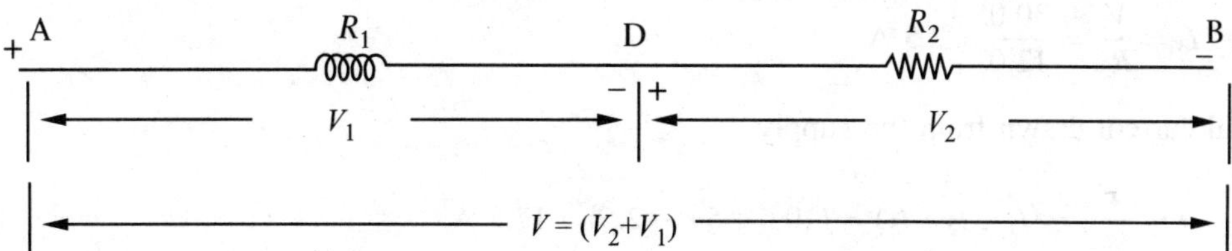

Fig. 1.9

EXAMPLE 1.3 *Calculate the total current and branch currents in a circuit with three resistors connected in parallel to a dc supply of 30 V (Fig. 1.10). The resistance values of the three resistors, R_1, R_2 and R_3, are 3 Ω, 6 Ω and 12 Ω, respectively. Also find the equivalent resistance. Determine the power consumed in the three resistors.*

SOLUTION The circuit with three resistors connected in parallel across a dc supply is shown in Fig. 1.10.

Supply voltage $= V = 30$ V.

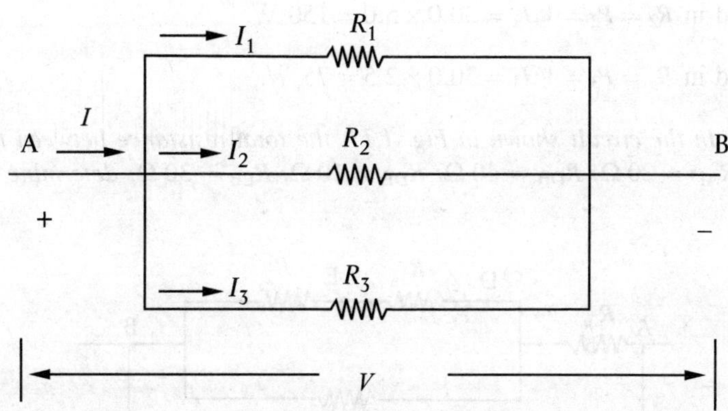

Fig. 1.10

The voltage drops across all three resistors are the same as they are when connected in parallel.

$$R_1 = 3\ \Omega, \quad R_2 = 6\ \Omega, \text{ and } R_3 = 12\ \Omega.$$

The current in the resistor, R_1

$$I_1 = \frac{V}{R_1} = \frac{30.0}{3.0} = 10.0 \text{ A}.$$

The other two branch currents are

$$I_2 = \frac{V}{R_2} = \frac{30.0}{6.0} = 5.0 \text{ A, and}$$

$$I_3 = \frac{V}{R_3} = \frac{30.0}{12.0} = 2.5 \text{ A}.$$

Total current drawn from the supply

$$I = \frac{V}{R_{eq}} = (I_1 + I_2 + I_3) = (10.0 + 5.0 + 2.5) = 17.5 \text{ A}.$$

The equivalent resistance $= R_{eq} = \frac{V}{I} = \frac{30.0}{17.5} = 1.7143\ \Omega.$

This resistance is also computed as

$$\frac{1}{R_{eq}} = \left(\frac{1}{R_1} + \frac{1}{R_2} + \frac{1}{R_3}\right) = \left(\frac{1}{3} + \frac{1}{6} + \frac{1}{12}\right) = \frac{7}{12} = 0.5833\ \Omega^{-1},$$

or, $\quad R_{eq} - \frac{12}{7} = 1.7143\ \Omega$

Power consumed in $R_1 = P_1 = V.I_1 = 30.0 \times 10.0 = 300$ W.

Power consumed in $R_2 = P_2 = V.I_2 = 30.0 \times 5.0 = 150$ W.

Power consumed in $R_3 = P_3 = V.I_3 = 30.0 \times 2.5 = 75$ W.

EXAMPLE 1.4 *In the circuit shown in Fig. 1.11, the total resistance between the terminals, A and B, is 24 Ω.* $R_{AD} = 20\ \Omega$; $R_{DB} = 40\ \Omega$; $R_{DE} = 10\ \Omega$; $R_{EB} = 30\ \Omega$; *determine the value of the resistance, R.*

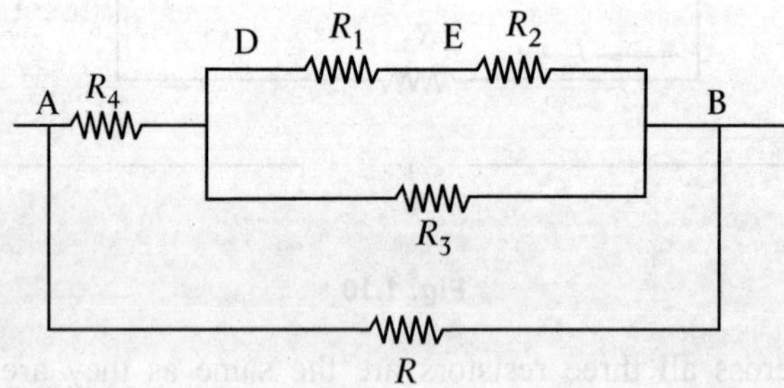

Fig. 1.11

SOLUTION The circuit is shown in Fig. 1.11.

$$R_{AD} = 20\ \Omega, \qquad R_{DB} = 40\ \Omega, \qquad R_{DE} = 10\ \Omega, \qquad R_{EB} = 30\ \Omega.$$

$$R_{DEB} = R_{DE} + R_{EB} = 10 + 30 = 40\ \Omega.$$

As the resistors, R_{DB} and R_{DEB} are connected in parallel, the equivalent resistance between D and B is

$$R'_{\text{DB}} = \frac{R_{\text{DB}} \cdot R_{\text{DEB}}}{(R_{\text{DB}} + R_{\text{DEB}})} = \frac{40 \times 40}{(40+40)} = 20\,\Omega.$$

This resistance is connected in series with R_{AD}.

The equivalent resistance is $R'_{\text{ADB}} = R_{AD} + R'_{\text{DB}} = 20 + 20 = 40\,\Omega$.

As this resistance is connected in parallel with $R_{\text{AB}} = R$, the equivalent resistance between A and B is

$$\frac{1}{R'_{\text{AB}}} = \left(\frac{1}{R'_{\text{ADB}}} + \frac{1}{R_{\text{AB}}}\right) = \left(\frac{1}{40} + \frac{1}{R}\right) = \frac{1}{24}\,\Omega^{-1}$$

or, $\frac{1}{R} = \left(\frac{1}{24} - \frac{1}{40}\right) = \frac{1}{60}\,\Omega^{-1}$ or, $R = 60\ \Omega$.

EXAMPLE 1.5 *Calculate the current drawn from a dc supply in the series–parallel circuit shown in Fig. 1.12.*

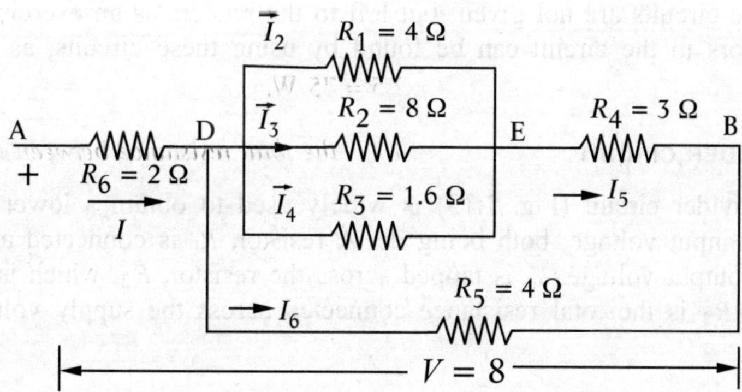

Fig. 1.12

SOLUTION The circuit is shown in Fig. 1.12.

$V_{\text{AB}} = 8$ V, $\quad R_{\text{AD}} = 2\,\Omega$, $\quad R_{\text{DB}} = 4\,\Omega$.

Three resistors of $4\,\Omega, 8\,\Omega$ and $1.6\,\Omega$, are connected in parallel across D and E.

The equivalent resistance between D and E is obtained as,

$$\frac{1}{R'_{\text{DE}}} = \left(\frac{1}{4} + \frac{1}{8} + \frac{1}{1.6}\right) = 1.0\,\Omega^{-1} \quad \text{or,} \quad R'_{\text{DE}} = 1\,\Omega.$$

A resistor, $R_{\text{EB}} = 3\,\Omega$ is connected in series with the above combination.

Total resistance across the path, DEB $= R'_{\text{DEB}} = R'_{\text{DE}} + R_{\text{EB}} = 1 + 3 = 4\,\Omega$.

This is connected in parallel with the resistor, $R_{DB} = 4\ \Omega$.

The equivalent resistance between D and B is

$$R'_{DB} = \frac{R'_{DEB} \cdot R_{DB}}{(R'_{DEB} + R_{DB})} = \frac{4 \times 4}{(4+4)} = 2\ \Omega.$$

A resistance, $R_{AD} = 2\ \Omega$ is connected in series with the above resistance, R'_{DB}, and also the dc supply.
The total resistance across the path, ADB, which is also the resistance connected in series with the dc supply, $V_{AB} = R'_{AB} = R_{AD} + R'_{DB} = 2 + 2 = 4\ \Omega$.

The current drawn from the dc supply

$$I_{AB} = \frac{V_{AB}}{R'_{AB}} = \frac{8}{4} = 2\ \text{A}.$$

Its direction is from A to B.

The intermediate circuits are not given, but left to the readers as an exercise. The currents in all the resistors in the circuit can be found by using these circuits, as we will see in Chapter 2.

POTENTIAL DIVIDER CIRCUIT

The potential divider circuit (Fig. 1.13) is widely used to obtain a lower output voltage from a constant input voltage, both being dc. A resistor, R_1 is connected across the input supply, V_i. The output voltage, V_o is tapped across the resistor, R_2, which is lower than R, where $R = R_1 + R_2$ is the total resistance connected across the supply voltage, V_i . Then mathematically,

$$V_o = \frac{R_2}{R} \cdot V_i.$$

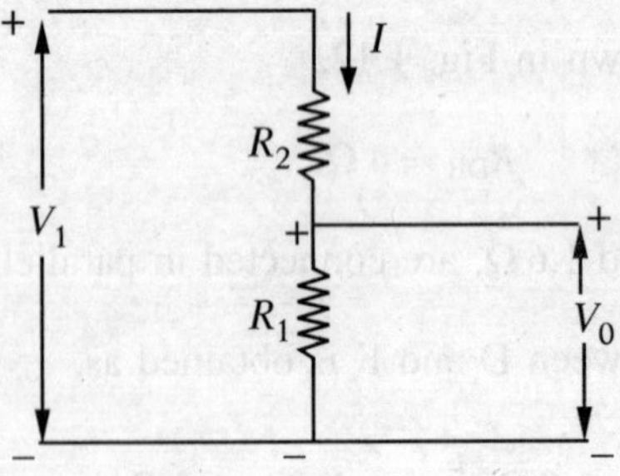

Fig. 1.13

The output voltage is lower that the input one, as $R_1 < R$. The current in the resistance, R is

$$I = \frac{V_i}{R}.$$

The output voltage is $V_o = I \cdot R_2$, from which the earlier expression can be obtained. As R_1 is varied, the output voltage also changes. This circuit is also called a potentiometer and is used in electrical measurement.

EXAMPLE 1.6 *Find the values of the resistances, R_1 and R_2 for the potentiometer or potential divider circuit (Fig. 1.14), so that the current, I is limited to 1 A, when $V_o = 20$ V. The input voltage, V_i is 100 V.*

SOLUTION The potentiometer or potential divider circuit is shown in Fig. 1.14.

Total resistance connected across the input voltage = $R = R_1 + R_2$.

The current in the resistance, $R = I = \frac{V_i}{R} = \frac{100}{R} = 1$ A.

Total resistance = $R = \frac{V_i}{I} = \frac{100}{1} = 100\,\Omega$.

Output voltage = $V_o = I \cdot R_2 = 1 \cdot R_2 = 20$ V.

Resistance, $R_2 = \frac{V_o}{I} = \frac{20}{1} = 20\,\Omega$.

Resistance, $R_1 = R - R_2 = 100 - 20 = 80\,\Omega$.

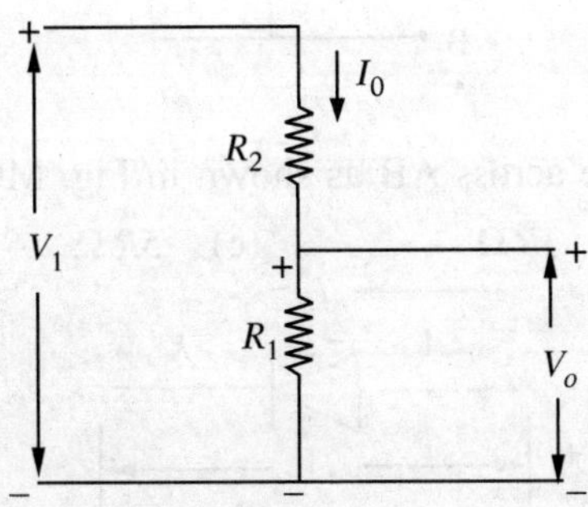

Fig. 1.14

Conclusion

In this introductory chapter, we introduced some basic points—Ohm's law, types of resistors, such as linear or constant and non-linear, and then, bilateral and unilateral. We also determined the equivalent resistance when a number of resistances are connected in series/parallel, and also in series–parallel. All these are used in the next chapter where we discuss dc circuits.

Multiple Choice Type Questions

Tick the correct option(s), out of the four choices provided in each question.

1.1 What is potential drop across a resistance dissipating 10 W of power when the current through the resistance is 2 A?

(a) 5 V (b) 6 V (c) 7 V (d) 2 V

1.2 For the circuit shown in Fig. MCQ 1.2, find the equivalent resistance across AB.

(a) 2 Ω (b) 3 Ω (c) 0.5 Ω (d) None of (a)–(c)

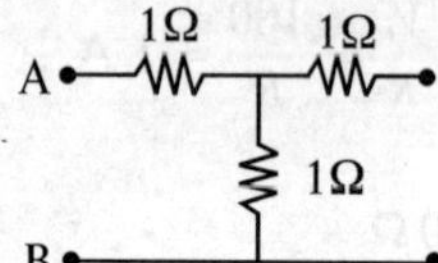

Fig. MCQ1.2

1.3 For the circuit shown in Fig. MCQ 1.3, find the equivalent resistance across AB.

(a) $\frac{2}{3}$ Ω (b) 3 Ω (c) 0.5 Ω (d) 1 Ω

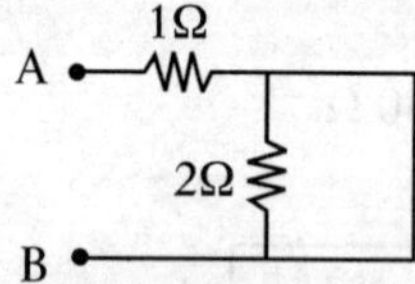

Fig. MCQ1.3

1.4 The equivalent resistance across AB as shown in Fig. MCQ 1.4 is

(a) $2R$ Ω (b) $4R$ Ω (c) $5R$ Ω (d) $6R$ Ω

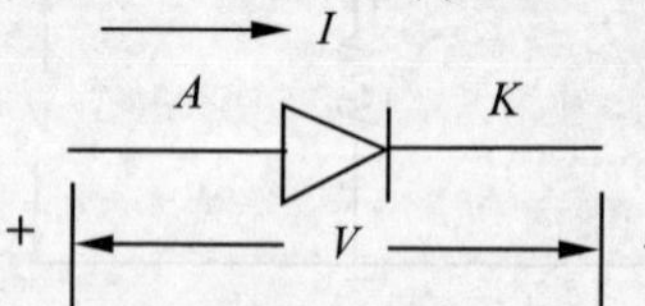

Fig. MCQ1.4

1.5 For the network shown in Fig. MCQ 1.5, find the reading of the voltmeter.

(a) 4 V (b) 6 V (c) 8 V (d) None of (a)–(c)

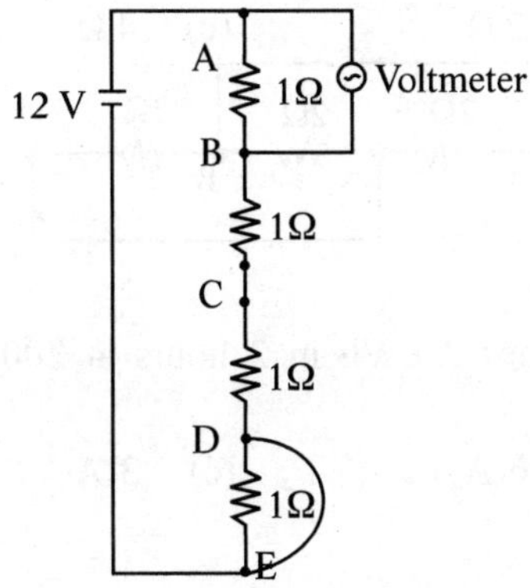

Fig. MCQ1.5

1.6 For the network shown in Fig. MCQ 1.6, find the voltage across AB.

(a) 10 V (b) 5 V (c) 7 V (d) None of (a)–(c)

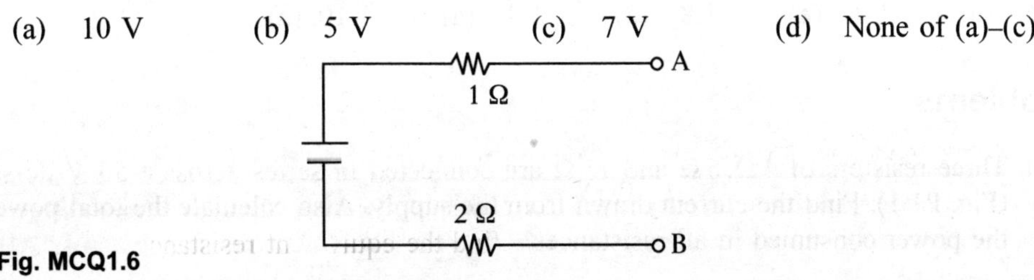

Fig. MCQ1.6

1.7 For the circuit shown in Fig. MCQ 1.7, find the voltage V_{FA}.

(a) – 6 V (b) 6 V (c) 7 V (d) 8 V

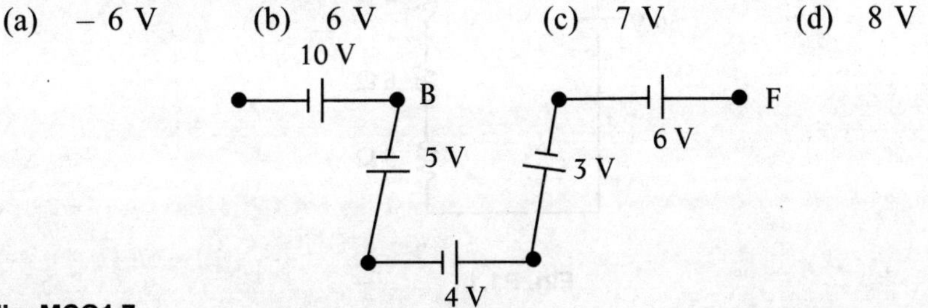

Fig. MCQ1.7

1.8 The equivalent resistance across A and B of the infinite ladder shown in Fig. MCQ 1.8 is

(a) $\frac{(1+\sqrt{5})}{2}R$ (b) $\frac{(1+\sqrt{3})}{2}R$ (c) $2R$ (d) None of (a)–(c)

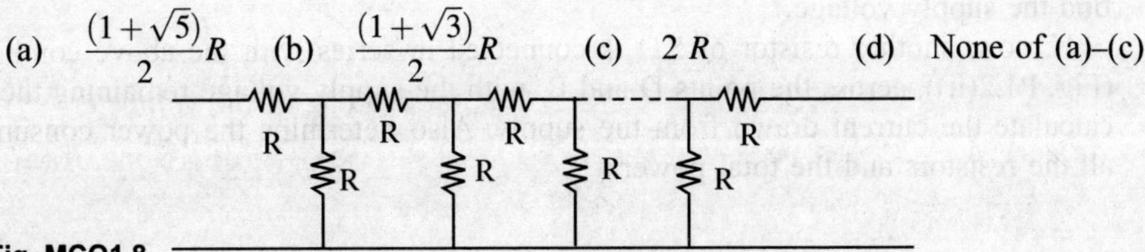

Fig. MCQ1.8

1.9 The equivalent resistance across AC for the circuit as shown in Fig. MCQ 1.9 is

(a) $\frac{2}{3}\,\Omega$ (b) $2\,\Omega$ (c) $4\,\Omega$ (d) $6\,\Omega$

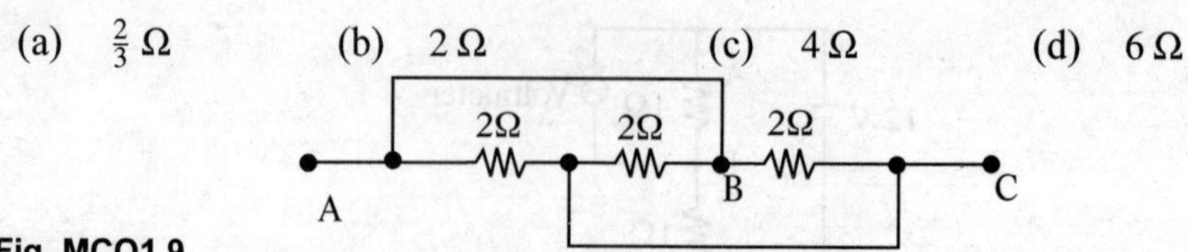

Fig. MCQ1.9

1.10 An electric heater consumes 2 kWh in 2 hours at 200 V. The current rating of the heating element is

(a) 5 A (b) 6 A (c) 3 A (d) None of (a)–(c)

Answers

1.1. (a)	1.2. (a)	1.3. (d)	1.4. (a)	1.5. (a)
1.6. (a)	1.7. (a)	1.8. (a)	1.9. (a)	1.10. (a)

Problems

1.1 Three resistors of $4\,\Omega, 6\,\Omega$ and $15\,\Omega$ are connected in series across a 50 V dc supply (Fig. P1.1). Find the current drawn from the supply. Also calculate the total power and the power consumed in all resistances.

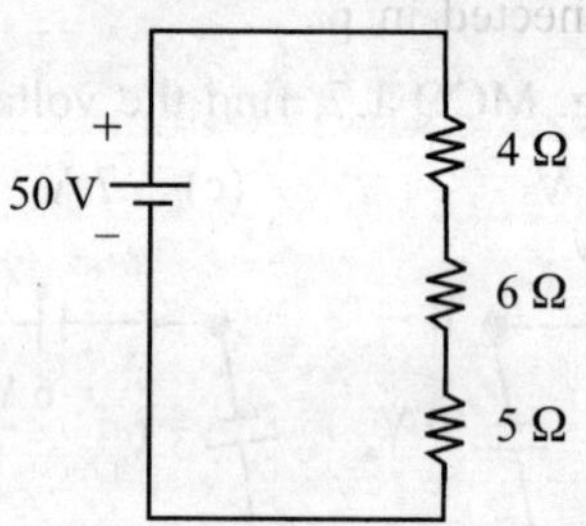

Fig. P1.1

1.2 Four resistors, R_1, R_2, R_3 and R_4, are connected in parallel across A and D, across a dc supply, and the equivalent resistance is $15\,\Omega$ (Fig. P1.2(i)). If the currents in the above resistors are 6, 3, 2 and 1, in that order, find the values of the four resistances. Also find the supply voltage.

If, now, another resistor of $5\,\Omega$ is connected in series with the above combination (Fig. P1.2(ii)), across the points D and B, with the supply voltage remaining the same, calculate the current drawn from the supply. Also determine the power consumed in all the resistors and the total power.

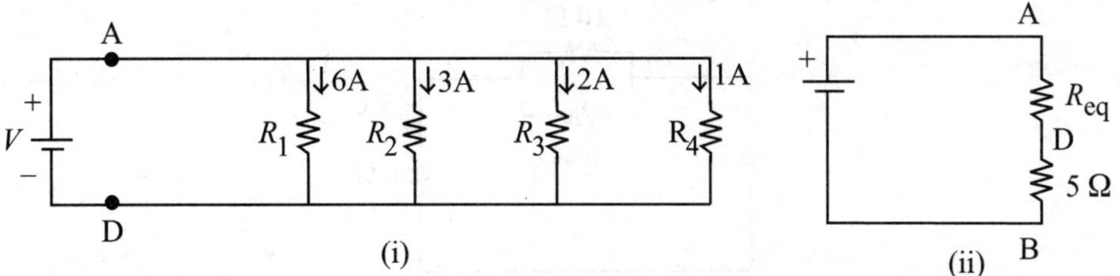

Fig. P1.2

1.3 Find the total resistance between the terminals, A and B, in the circuit shown in Fig. P1.3.

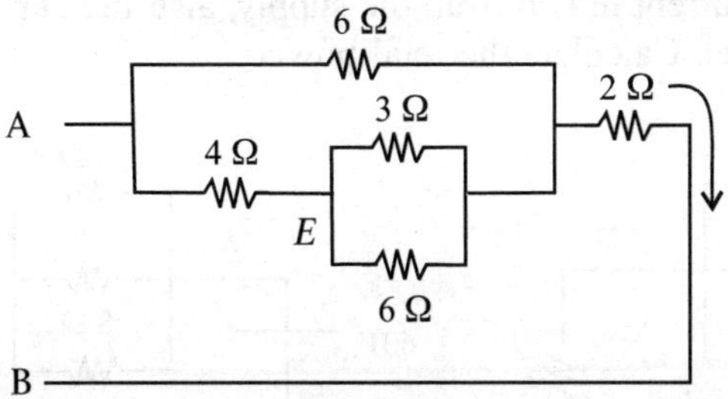

Fig. P1.3

1.4 A resistor of 10 Ω is connected in parallel with *R*, and the combination is in series with a resistor of 6 Ω (Fig. P1.4). When a dc supply is fed to the circuit, the voltage drop across the resistor of 6 Ω is 60 V, and the total power consumed in the circuit is 800 W. Find the value of the resistance, *R*. Also calculate the currents and the power consumed in the other two resistors.

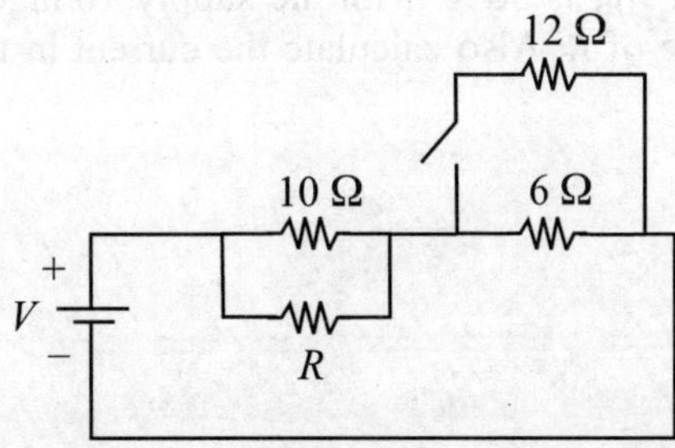

Fig. P1.4

If, now, another resistor of 12 Ω is connected in parallel with the resistor of 6 Ω, and the supply voltage remains the same, determine the total current drawn from the supply.

1.5 Two resistors of 10 Ω and 15 Ω are connected in parallel, and the combination is in series with another resistor of 8 Ω (Fig. P1.5). When a dc supply is fed to the circuit, the current in the resistor of 8 Ω is 5 A. Determine the voltage applied to the circuit. Also calculate the currents and the power consumed in the resistors.

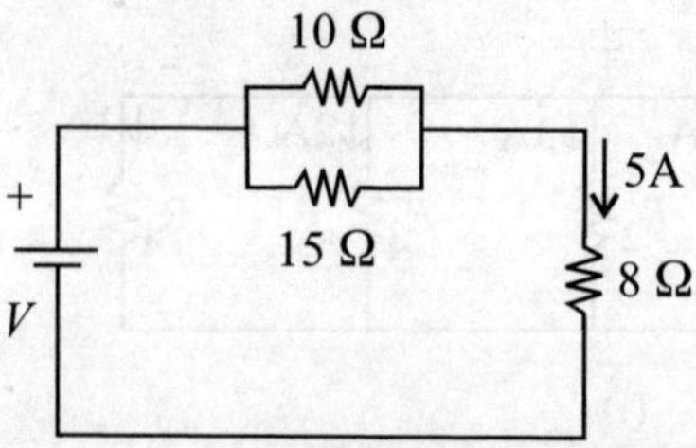

Fig. P1.5

1.6 Three resistors of 3 Ω, 8 Ω and 24 Ω are connected in parallel across A and D, and the combination is in series with another resistor of 4 Ω, across a 60 V dc supply (Fig. P1.6(i)). Find the current drawn from the supply, also the current in the three resistors connected in parallel. Calculate the total power

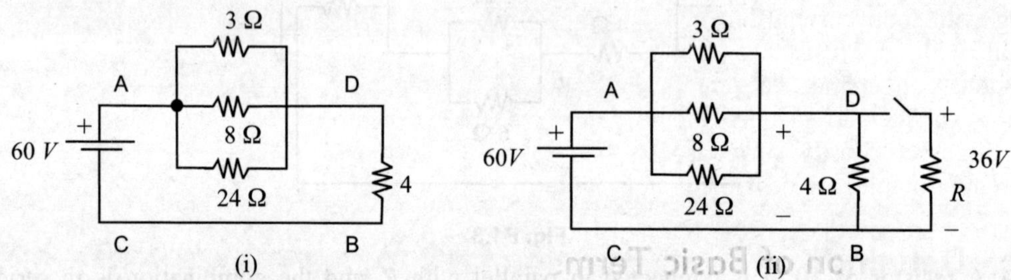

Fig. P1.6

If, now, a resistor (R) is connected in parallel with the resistor of 4 Ω across the points, D and B, and the voltage drop, V_{DB} is 36 V, with the supply voltage remaining the same (Fig. P1.6(ii)), determine the value of R. Also calculate the current in the two resistors connected across B and D.

CHAPTER 2

DC CIRCUITS

In this chapter, we will learn about dc circuits. We first define certain basic terms which are used extensively to describe dc circuits. Kirchoff's laws are described in detail, along with loop current and nodal voltage methods for calculating branch currents in dc circuits with only resistance elements. The circuit theorems—the superposition principle, the star–delta conversion, Thevenin's theorem—are then dealt with. Types of energy sources, such as voltage/current ones, along with independent/dependent types of these sources, are also discussed. Lastly, other aspects like Norton's theorem and maximum power transfer theorem/principle, are described, along with numerical examples.

2.1 Definition of Basic Terms Used in DC Circuits

In this section, some basic terms needed for the study of dc circuits, are described.

DC network: The dc network or circuit is the interconnection of circuit elements, such as resistors and energy sources, which are dc (where the flow of current is uni-directional). Resistors have been presented in detail in Chapter 1, and energy sources will be taken up later.

Network element: The circuit elements used to make up the network are called network elements. Examples of network elements are resistors which are linear and bilateral, i.e., obey Ohm's laws, inductances and capacitances (explained in detail in Chapters 5–7), which are also linear and used in ac circuits.

Network elements can be classified as linear or non-linear, and unilateral or bilateral (explained in Chapter 1).

They can also be classified as *active* or *passive*. Resistors, inductances and capacitances are *passive* in nature as they consume power in the circuits, when the circuits are fed from energy source(s). *Active* elements are those which deliver power to the passive circuit elements. They can be energy source(s), as described in Section 2.6.

Node: This term is used to denote a junction in an electric network or circuit where more than one branch (explained below) and at least two meet or intersect at that point. Three, four branches or more, may intersect at a node

Branch: A branch connects any two nodes. Normally, there is only one resistor in it. However, two or more resistors may be connected, either in series or parallel, which results in an equivalent resistance, or it may be assumed that there are two branches, especially when the resistances are connected in parallel. However, the resistors in a branch cannot be connected in series–parallel, as this would create another node. An energy source may also be connected in series or parallel with the resistor.

2.2 Kirchhoff's Laws

Kirchhoff's laws are two equations which deal with the current at the node and voltage for a loop in a circuit. They were described by Gustav Kirchhoff in 1835.

The *first law* states that the algebraic sum of the currents at a node is zero. The current entering the node is taken as positive, while the current leaving the node is taken as negative. Similar conditions apply for a current source, if any, present at the node. To take an example (Fig. 2.1), the equation for the currents at the node, P applying the law, is obtained as

$$I_1 + I_2 - I_3 + I_4 - I_5 = 0$$

or, $I_1 + I_2 + I_4 = I_3 + I_5$

From the second equation, it can be stated that the currents entering the node, P is equal to the currents leaving the node. In this example, there is no current source connected at the node.

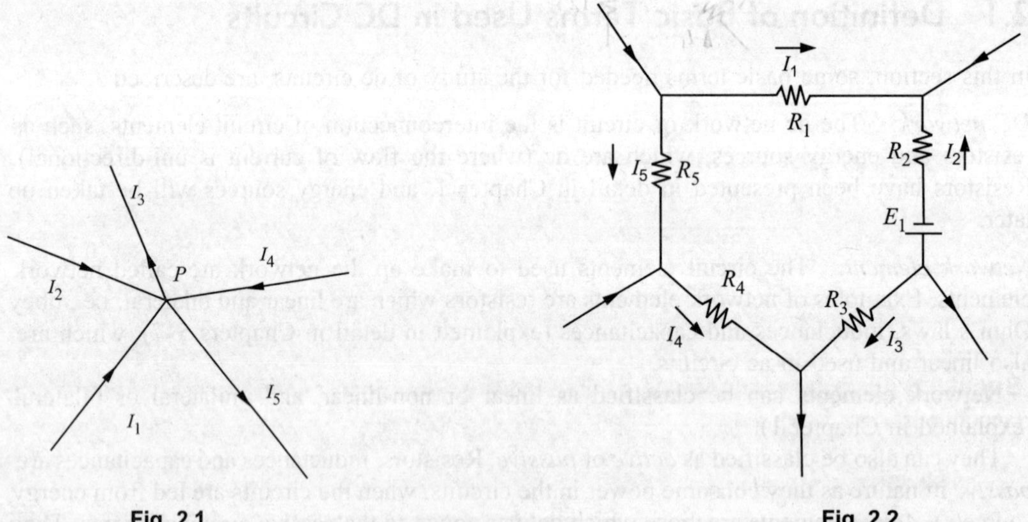

Fig. 2.1

Fig. 2.2

The *second law* states that the algebraic sum of the voltage drops, including voltage sources, if any, in the loop, is zero. Voltage drop is taken as positive, while voltage rise is taken as negative. The same convention applies for voltage sources, if any, in the loop. To take an example (Fig. 2.2), the equation for the voltages in the loop, applying the law, is obtained as

$$I_1R_1 - I_2R_2 - E_1 + I_3R_3 - I_4R_4 - I_5R_5 = 0$$

or, $I_1R_1 - I_2R_2 + I_3R_3 = E_1 + I_2R_2 + I_4R_4 + I_5R_5.$

From the second equation, it can be stated that the total voltage drop in the loop is equal to the total voltage rise in it. This includes any voltage source present in the loop. Please note the direction for the voltage source, E_1 in the loop.

2.3 Loop Current Method

In the loop current method, a system of equations with the current in each loop as the unknown variables is set up and solved. In a circuit with resistive elements and voltage/current sources, the current in a loop is taken in the clockwise direction. The current in a branch which is mutual between two loops is taken as the difference of the two loop currents, using Kirchhoff's current (first) law.

EXAMPLE 2.1 *In the circuit shown in Fig. 2.3, determine the current in the load resistor,* $R_L = 10\ \Omega$.

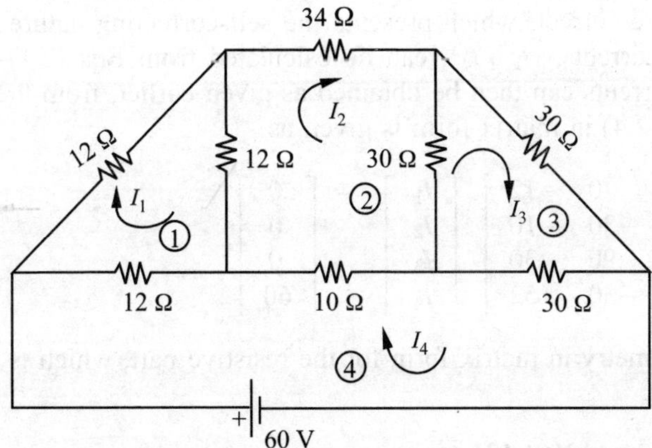

Fig. 2.3

SOLUTION For the circuit (Fig. 2.3), four (4) loops are taken. Applying Kirchhoff's voltage (second) law for Loop 1, the voltage equation is

$$12I_1 + 12(I_1 - I_2) + 12(I_1 - I_4) = 0$$

or, $(12 + 12 + 12)I_1 - 12I_2 - 12I_4 = 0$

or, $36I_1 - 12I_2 - 12I_4 = 0.$ (2.1)

The branch current in the 12 Ω resistor, between Loops 1 and 2, is the difference between the two loop currents, i.e., $(I_1 - I_2)$, as taken earlier. It may be noted that the self resistance in a loop is the sum of the resistances. The voltage drops in all mutual resistors between the loops are negative. No voltage source is there in Loop 1.

For Loop 2, the voltage equation is

$$12(I_2 - I_1) + 34I_2 + 30(I_2 - I_3) + 10(I_2 - I_4) = 0$$

or, $-12I_1 + 86I_2 - 30I_3 - 10I_4 = 0.$ (2.2)

It may be noted that the mutual terms in the equations are the same, which means that the equations are symmetrical. The branch current between Loops 2 and 1 is opposite to the one used in Eq. (2.1), as all loop currents are taken in the clockwise direction (shown in Fig. 2.3).

For Loops 3 and 4, the voltage equations are

$$-30I_2 + 90I_3 - 30I_4 = 0. \quad (2.3)$$

$$-12I_1 - 10I_2 - 30I_3 + 52I_4 = 60. \quad (2.4)$$

It may be noted that there is one voltage source in Loop 4 with a voltage rise in the clockwise direction. Hence, it is taken as positive on the RHS of Eq. (2.4). If all equations (Eqs (2.1)–(2.4)) are put together, there is symmetry, as described earlier. This shows that the voltage equations are correct, which presents the self-correcting nature of the loop current method. The loop currents, (I_1 – I_4), can be calculated from Eqs (2.1)–(2.4) as described here. The branch currents can then be obtained as given earlier, from the loop currents.

Equations (2.1)–(2.4) in matrix form is given as

$$\begin{bmatrix} 36 & -12 & 0 & -12 \\ -12 & 86 & -30 & -10 \\ 0 & -30 & 90 & -30 \\ -12 & -10 & -30 & 52 \end{bmatrix} \begin{bmatrix} I_1 \\ I_2 \\ I_3 \\ I_4 \end{bmatrix} = \begin{bmatrix} 0 \\ 0 \\ 0 \\ 60 \end{bmatrix}$$

Please note the symmetry in matrix form for the resistive part, which is a check in the loop current method.

From Eq. (2.1), $36I_1 = 12I_2 + 12I_4$

or, $3I_1 = I_2 + I_4.$ (2.5)

From Eq. (2.3), $90I_3 = 30I_2 + 30I_4$

or, $3I_3 = I_2 + I_4.$ (2.6)

From Eqs (2.5–2.6), $I_1 = I_3 = (I_2 + I_4)/3$ (2.7)

Using Eqs (2.6–2.7) in Eq. (2.2),

$$-14(I_2 + I_4) + 86I_2 - 10I_4 = 0$$

or, $72I_2 - 24I_4 = 0$

or, $3I_2 = I_4.$ (2.8)

Using Eqs (2.6–2.8) in Eq. (2.4),

$$-14(I_2+I_4)-10I_2+52I_4=60$$

or, $-24I_2+38I_4=60$

or, $-8I_4+38I_4=30I_4=60$

or, $I_4=60/30=2.0$ A.

Using Eq. (2.8), $I_2=2/3=0.667$ A.

Using Eq. (2.7), $I_1=I_3=(2.0+0.667)/3=0.889=8/9$ A.

So, the branch current in the 10 Ω resistor, between Loops 4 and 2,

$$I=I_4-I_2=2.0-0.667=1.333=4/3\text{ A}.$$

2.4 Nodal Voltage Method

The nodal voltage method is used for calculating node voltages. In a circuit with resistive elements and voltage/current sources, one node is assumed to be the reference (earth) node with zero voltage. The other node voltages are taken with respect to this reference node. In this case, Kirchhoff's current (first) law is used to find the current equation of that node with respect to the node voltages. The branch voltage between two nodes is the difference between the two node voltages. The branch current is the above voltage divided by the value of the resistance connected across that branch.

EXAMPLE 2.2 *In the circuit shown in Fig. 2.4, determine the current in the load resistance,* $R_L=6\ \Omega$.

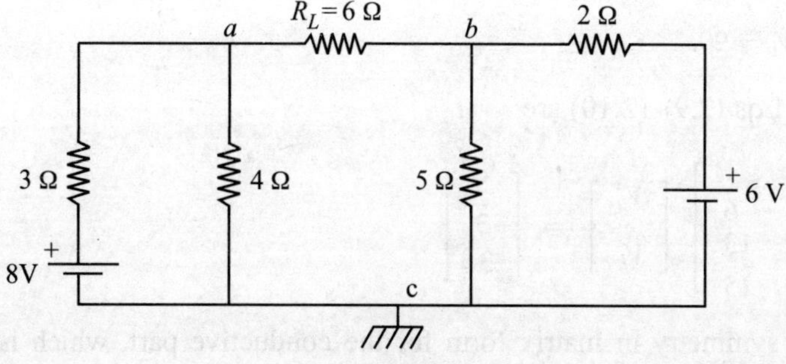

Fig. 2.4

SOLUTION For the circuit (Fig. 2.4), two (2) nodes, a and b, are used. The third node, c is the reference (earth).

Applying Kirchhoff's current law for node a, the current equation is,

$$\frac{V_a-8}{3}+\frac{V_a}{4}+\frac{V_a-V_b}{6}=0$$

or, $$V_a\left(\frac{1}{3}+\frac{1}{4}+\frac{1}{6}\right)-\frac{V_b}{6}=\frac{8}{3}$$

or, $$\frac{3}{4}V_a-\frac{1}{6}V_b=\frac{8}{3}. \tag{2.9}$$

The voltage at node a is V_a (it is not denoted as V_{ac}) as the voltage at node *c* is zero (earthed). The self conductance at node a is the sum of the conductance (reciprocal of resistance) of the resistors connected to that node. The mutual conductance is the conductance (reciprocal of the resistance) of the resistor connected between the two nodes, a and b. This is the same as the resistance terms in the voltage equations obtained using the loop current method.

Simplifying (i.e., multiplying by 12) the above equation, we get

$$9V_a-2V_b=32 \tag{2.9a}$$

For node b, the current equation is

$$\frac{V_b-6}{2}+\frac{V_b}{5}+\frac{V_b-V_a}{6}=0$$

or, $$-\frac{V_a}{6}+V_b\left(\frac{1}{2}+\frac{1}{5}+\frac{1}{6}\right)=\frac{6}{2}=3$$

or, $$-\frac{1}{6}V_a+\frac{13}{15}V_b=3. \tag{2.10}$$

It may be noted that the mutual conductance terms being the same, the current equations are symmetrical, as is the case in the loop current method.

Simplifying (i.e., multiplying by 30) the above equation, we get

$$-5V_a+26V_b=90. \tag{2.10a}$$

In matrix form, Eqs (2.9)–(2.10) are,

$$\begin{bmatrix}\frac{3}{4} & -\frac{1}{6}\\ -\frac{1}{6} & \frac{13}{15}\end{bmatrix}\begin{bmatrix}V_a\\ V_b\end{bmatrix}=\begin{bmatrix}\frac{8}{3}\\ 3\end{bmatrix}.$$

Please note the symmetry in matrix form for the conductive part, which is a check in the nodal current method.

Solving Eqs (2.9a) and (2.10a), the node voltages are V_a = 4.518 V, and V_b = 4.33 V. The voltage drop in the branch, V_{ab} is the difference between the two node voltages, i.e.,

$$V_{ab}=(V_a-V_b)=(4.518-4.33)=0.188\text{ V}.$$

The branch current in the 6 Ω resistor between nodes a and b is

$$I_{ab}=V_{ab}/6=0.188/6=0.0313\text{ A}=31.3\text{ mA}.$$

The branch current in the 3 Ω resistor between nodes c and a is

$$I_{ca} = (8 - V_{ac})/3 = (8.0 - 4.518)/3.0 = 3.482/3 = 1.161 \text{ A}.$$

The branch current in the 2 Ω resistor between nodes c and b is

$$I_{cb} = (6 - V_{bc})/2 = (6.0 - 4.3)/2.0 = 1.67/2 = 0.835 \text{ A}.$$

In the two examples given above, only voltage sources are considered—not current source(s). Examples with both voltage and current sources in the circuits are given later.

2.5 Circuit Theorems

In the following section, we will introduce some circuit theorems, starting with the superposition principle.

2.5.1 Superposition Principle

The principle of superposition is applicable to linear circuits/systems. As the resistors used in dc circuits satisfy Ohm's law ($V = R.I$), they are linear and the superposition principle can be used on them. It must be noted that this statement is just a principle, not a theorem.

SUPERPOSITION PRINCIPLE: STATEMENT

In a circuit with more than one voltage/current sources, the current/voltage drop in a particular branch is the algebraic sum of the currents/voltage drops in that branch with only one source (voltage/current) present; the other sources are replaced by their internal resistances.

EXAMPLE 2.3 *In the circuit shown in Fig. 2.4, determine the current in the load resistance, $R_L = 6\ \Omega$.*

SOLUTION This is the same circuit used in EXAMPLE 2.2. To apply the superposition principle, only one voltage source is taken at a time, for example, let us take the 8 V voltage on the left first (Fig. 2.5(a)); the other voltage source is short-circuited (a short circuit is one where current is allowed to flow unhindered, i.e., its internal resistance is made zero—an ideal voltage source). The circuit is termed a ladder one (one configured only with a combination of parallel and series connections and thus its equivalent resistance is easy to derive) with only one voltage source. For an ideal current source, the internal resistance should be infinity. Hence, a current source is to be kept open-circuited.

The circuit is simplified starting from the right-hand side. The equivalent resistance, R_{bc} is a parallel combination of 5 Ω and 2 Ω. Therefore,

$$R_{bc} = \frac{(5 \times 2)}{(5 + 2)} = \frac{10}{7} = 1.43\ \Omega.$$

As the resistor of 6 Ω is in series with R_{bc}, the combined resistance, $R_{ac1} = 6.0 + 1.43 = 7.43\ \Omega$ (Fig. 2.5(b)). This resistance is in parallel with 4 Ω. The equivalent resistance is

$$R_{ac} = \frac{(4.0 \times 7.43)}{(4.0 + 7.43)} = \frac{29.714}{11.43} = 2.6\ \Omega.$$

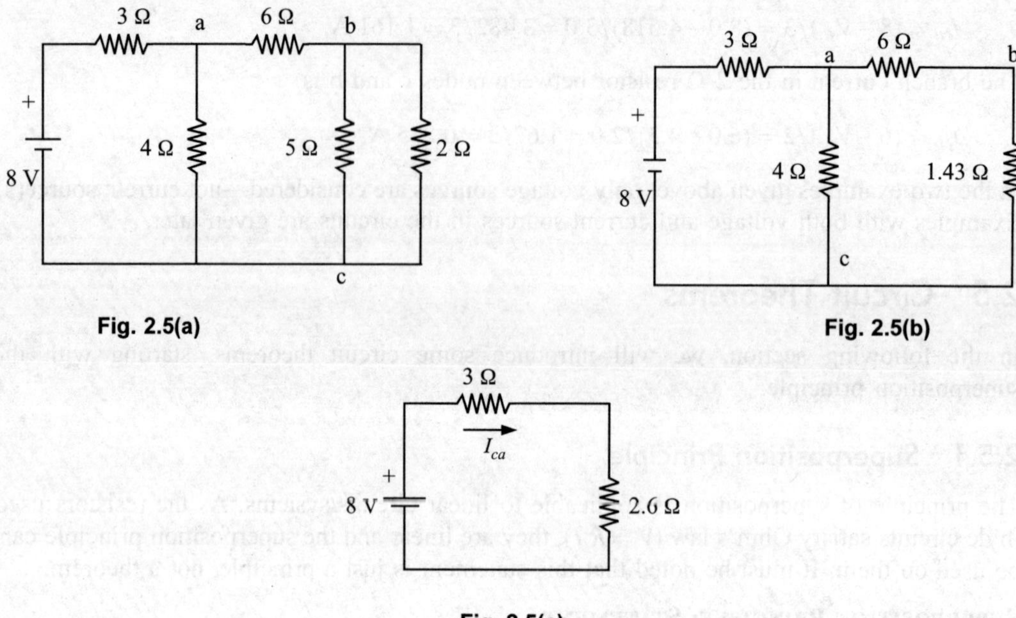

Fig. 2.5(a)

Fig. 2.5(b)

Fig. 2.5(c)

This resistance is in series with the resistor of 3 Ω (Fig. 2.5(c)). The current drawn from the 8 V source is

$$I_{ca} = \frac{8.0}{(3.0+2.6)} = \frac{8.0}{5.6} = 1.43 \text{ A}.$$

Now to find the branch currents under this condition, we go in the reverse way, and find the source current.

The current through the 6 Ω resistor is I_{abc} (Fig. 2.5(b)) $= \left(\frac{4.0}{11.43} \times 1.43\right) = 0.5$ A.

The current through the 4 Ω resistor is I_{ac} (Fig. 2.5(b)) $= 1.43 - 0.5 = 0.93$ A.

The current through the 2 Ω resistor is I_{bc} (Fig. 2.5(b)) $= \left(\frac{5.0}{7.0} \times 0.5\right) = 0.357$ A.

Now, consider the second voltage source of 6 V, with the first voltage source short-circuited (Fig. 2.5(d)). The circuit is simplified starting from the left-hand side. The equivalent resistance, R_{ac} is a parallel combination of 3 Ω and 4 Ω.

So, $$R_{ac} = \frac{(3\times 4)}{(3+4)} = \frac{12}{7} = 1.7143\ \Omega.$$

As the resistance of 6 Ω is in series with R_{ac}, the combined resistance, R_{bc1} is

$$R_{bc1} = 6.0 + 1.7143 = 7.7143\ \Omega \text{ (Fig. 2.5(e))}.$$

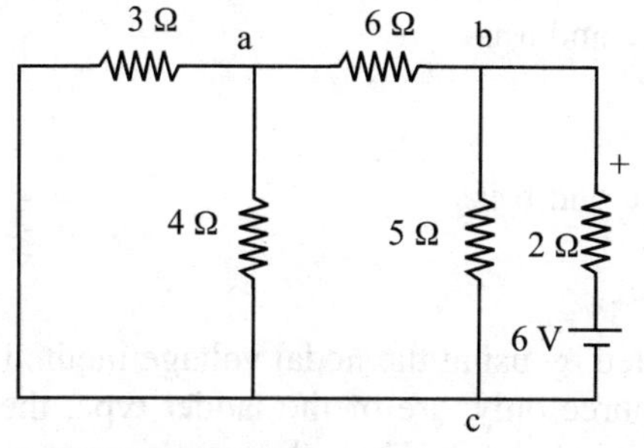

Fig. 2.5(d)

Fig. 2.5(e)

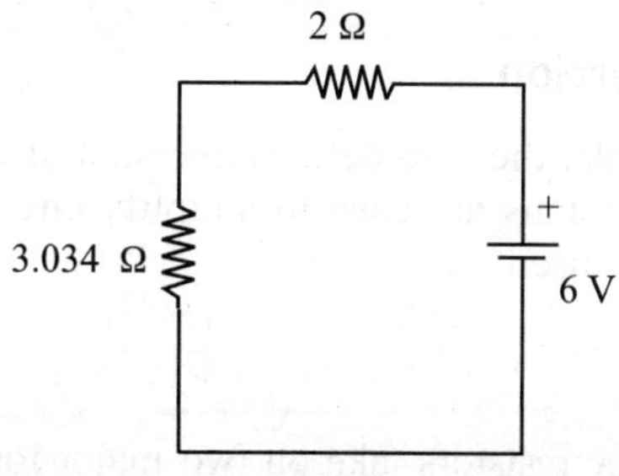

Fig. 2.5(f)

This resistance is in parallel with 5 Ω. The equivalent resistance is

$$R_{bc} = \frac{(5.0 \times 7.7143)}{(5.0 + 7.7143)} = \frac{38.5714}{12.7143} = 3.034\ \Omega.$$

This resistance is in series with the resistor of 2 Ω (Fig. 2.5(f)). The current drawn from the 6 V source is

$$I_{cb} = \frac{6.0}{(2.0 + 3.034)} = \frac{6.0}{5.034} = 1.192\ \text{A}.$$

Now, as in the earlier case, to find the branch currents under this condition, we go in the reverse way, and find the source current. The current through the 6 Ω resistor is

$$I_{ba}\ \text{(Fig. 2.5(f))} = \left(\frac{5.0}{12.7143} \times 1.192\right) = 0.469\ \text{A}.$$

The current through the 3 Ω resistor is

$$I_{ac}\ \text{(Fig. 2.5(e))} = \left(\frac{4.0}{7.0} \times 0.469\right) = 0.268\ \text{A}.$$

Applying the principle of superposition, the branch currents are obtained in the following way. The branch current in the 6 Ω resistor between nodes a and b is

$$I_{ab} = 0.5 - 0.469 = 0.031\ \text{A} = 31\ \text{mA}.$$

The branch current in the 3 Ω resistor between nodes c and a is

$$I_{ca} = 1.43 - 0.268 = 1.162 \text{ A}.$$

The branch current in the 2 Ω resistor between nodes c and b is

$$I_{cb} = 1.192 - 0.357 = 0.835 \text{ A}.$$

These branch currents obtained are the same as calculated by using the nodal voltage method described earlier. As the circuits using one voltage source only, are of the ladder type, the currents can be obtained using the simple method presented above. Then, the total current in any branch is the algebraic sum of the above currents as described.

2.5.2 Star–Delta Conversion

Like the superposition principle, the star–delta conversion and the delta–star conversion are not theorems. The two conversions are used to simplify circuits; of the two, the delta star conversion is most commonly used.

DELTA–STAR CONVERSION

Circuits or networks with three resistors take on two major forms with names that represent the way in which the resistors are connected—a *star* connected network which has the symbol of the letter, *Y* (wye) (Fig. 2.6(b)) and a delta connected network which has the symbol of a triangle, Δ (delta) (Fig. 2.6(a)).

If a 3-phase, 3-wire supply or even a 3-phase load is connected in one type of configuration, it can be easily transformed or changed into an equivalent configuration of the other type by using either the star–delta transformation or the delta–star transformation process.

In Fig. 2.6(a), three resistors (R_{ab}, R_{bc} and R_{ca}) are connected in delta. In Fig. 2.6(b), the equivalent circuit of three resistance (R_A, R_B and R_C) are connected in star.

As the two circuits are equivalent, the resistance between A and B using star- and delta-connected circuits is

$$R_A + R_B = \frac{R_{ab}(R_{ca} + R_{bc})}{R_{ab} + R_{bc} + R_{ca}} = \frac{R_{ab}R_{ca} + R_{ab}R_{bc}}{R_{ab} + R_{bc} + R_{ca}}. \quad (2.11)$$

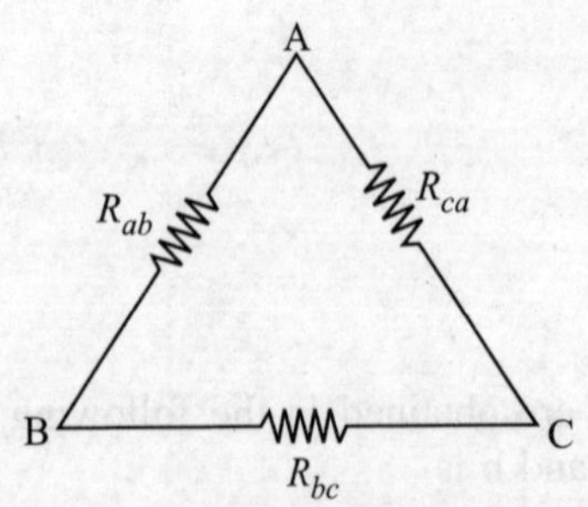

Fig. 2.6(a)

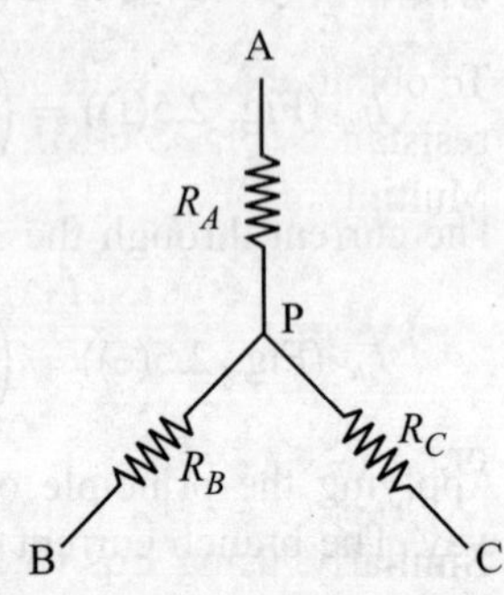

Fig. 2.6(b)

For the star-connected circuit, the resistance, R_A and R_B are connected in series between A and B. For the delta-connected circuit, the resistance, R_{ab} is in parallel with two resistances, R_{bc} and R_{ca} connected in series.

Similarly, the resistance between B and C is

$$R_B + R_C = \frac{R_{bc}(R_{ab} + R_{ca})}{R_{ab} + R_{bc} + R_{ca}} = \frac{R_{bc}R_{ab} + R_{bc}R_{ca}}{R_{ab} + R_{bc} + R_{ca}}. \tag{2.12}$$

The resistance between C and A is

$$R_C + R_A = \frac{R_{ca}(R_{bc} + R_{ab})}{R_{ab} + R_{bc} + R_{ca}} = \frac{R_{ca}R_{bc} + R_{ca}R_{ab}}{R_{ab} + R_{bc} + R_{ca}} \tag{2.13}$$

From Eqs (2.11)–(2.13), we get

$$2(R_A + R_B + R_C) = \frac{2(R_{ab}R_{bc} + R_{bc}R_{ca} + R_{ca}R_{ab})}{R_{ab} + R_{bc} + R_{ca}}$$

or, $$R_A + R_B + R_C = \frac{R_{ab}R_{bc} + R_{bc}R_{ca} + R_{ca}R_{ab}}{R_{ab} + R_{bc} + R_{ca}}. \tag{2.14}$$

Using Eqs (2.12) and (2.14),

$$R_A = \frac{R_{ab}R_{ca}}{R_{ab} + R_{bc} + R_{ca}}. \tag{2.15}$$

Similarly, using Eqs (2.14), (2.13) and (2.11),

$$R_B = \frac{R_{bc}R_{ab}}{R_{ab} + R_{bc} + R_{ca}}. \tag{2.16}$$

$$R_C = \frac{R_{ca}R_{bc}}{R_{ab} + R_{bc} + R_{ca}}. \tag{2.17}$$

The equivalent resistances connected in star, are obtained using delta-connected resistors. These formulas are widely used in solving problems related to dc circuits.

STAR–DELTA CONVERSION

To obtain the equivalent resistances connected in delta (Fig. 2.6(a)), in terms of star-connected resistance (Fig. 2.6(b)), the formulas obtained earlier, are used.

Multiplying Eq. (2.14) by R_{bc},

$$R_A R_{bc} = \frac{R_{ab}R_{bc}R_{ca}}{R_{ab} + R_{bc} + R_{ca}} = K \tag{2.18}$$

or, $$R_{bc} = K/R_A. \tag{2.19}$$

Similarly, using Eqs (2.13) and (2.15),

$$R_B R_{ca} = \frac{R_{ab} R_{bc} R_{ca}}{R_{ab} + R_{bc} + R_{ca}} = K \tag{2.20}$$

or, $$R_{ca} = K/R_B. \tag{2.21}$$

$$R_C R_{ab} = \frac{R_{ab} R_{bc} R_{ca}}{R_{ab} + R_{bc} + R_{ca}} = K \tag{2.22}$$

or, $$R_{ab} = K/R_C. \tag{2.23}$$

Using Eq. (2.19), Eq. (2.21) and Eq. (2.23) in Eq. (2.18),

$$R_A R_{bc} = \frac{(K/R_C)\,(K/R_B)\,(K/R_A)}{(K/R_C) + (K/R_B) + (K/R_A)} = K \tag{2.24}$$

Multiplying both the numerator and denominator by $(R_A.R_B.R_C)$ and simplifying,

$$R_A R_{bc} = \frac{K^2}{R_A R_B + R_B R_C + R_C R_A} = K$$

or, $$K = R_A R_B + R_B R_C + R_C R_A = R_A R_{bc} \tag{2.25}$$

or, $$R_{bc} = \frac{R_A R_B + R_B R_C + R_C R_A}{R_A} = R_B + R_C + \frac{R_B R_C}{R_A}. \tag{2.26}$$

Using Eq. (2.25) in Eq. (2.21) and Eq. (1.23),

$$R_{ca} = \frac{R_A R_B + R_B R_C + R_C R_A}{R_B} = R_C + R_A + \frac{R_C R_A}{R_B}. \tag{2.27}$$

$$R_{ab} = \frac{R_A R_B + R_B R_C + R_C R_A}{R_C} = R_A + R_B + \frac{R_A R_B}{R_C}. \tag{2.28}$$

The equivalent resistances connected in delta are obtained using star-connected resistors. These formulas are not much used in solving problems related to dc circuits, but may be used in certain special cases.

Before taking up an example, we need to look at how to use delta–star conversion as it decreases the number of loops in a circuit. For example, in EXAMPLE 2.1, where the current in the circuit was determined using the loop current method, four loops were used. However, there are two delta-connected resistors (Loop 1 and Loop 3), which can be converted into star-connected ones. Thus, the number of loops is decreased by two, leaving only two loops. As there is only one voltage source, the circuit is easily simplified. The author hopes that the students will be able to solve problems easily now. The problem can also be solved using Thevenin's theorem, which will be discussed next.

The example taken up here is that of an unbalanced bridge circuit, where star–delta conversion formula is used to solve the problem.

EXAMPLE 2.4 *In the circuit shown in Fig. 2.7(a), determine the current in the load resistance,* $R_L = 200\ \Omega$.

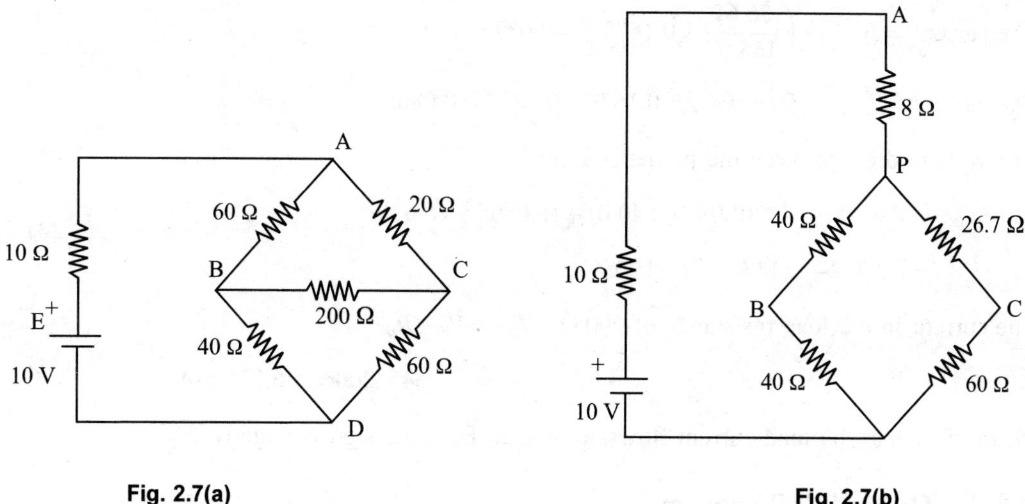

Fig. 2.7(a) **Fig. 2.7(b)**

SOLUTION Let us take the upper delta-connected circuit, ABC with three resistances (Fig. 2.7(a)), and convert it to a star-connected one.

The delta-connected resistances are

$$R_{ab} = 60\ \Omega, R_{bc} = 200\ \Omega \text{ and } R_{ca} = 40\ \Omega.$$

$$\sum R = R_{ab} + R_{bc} + R_{ca} = 60 + 200 + 40 = 300\ \Omega.$$

The formulas for the equivalent resistances to be connected in star (Fig. 2.7(b)) are

$$R_A = \frac{R_{ab}\,R_{ca}}{R_{ab} + R_{bc} + R_{ca}} = \frac{R_{ab}\,R_{ca}}{\sum R} = \frac{60 \times 40}{300} = 8\ \Omega.$$

R_A is also R_{AP}, where P is the star point.

$$R_{BP} = \frac{R_{bc}\,R_{ab}}{\sum R} = \frac{200 \times 60}{300} = 40\ \Omega.$$

$$R_{CP} = \frac{R_{ca}\,R_{bc}}{\sum R} = \frac{40 \times 200}{300} = \frac{80}{3} = 26.67\ \Omega.$$

Between the points P and D, two resistances, (40 + 40 = 80) Ω, and (26.67 + 60 = 86.67) Ω, are in parallel.

So, $$R_{PD} = \frac{(80.0 \times 86.67)}{(80.0 + 86.67)} = \frac{6{,}933.33}{166.67} = 41.6\ \Omega.$$

$$R_{ED} = R_{EA} + R_{AP} + R_{PD} = 10.0 + 8.0 + 41.6 = 59.6\ \Omega$$

The current drawn from a 10 V supply $= I_{EP} = \dfrac{10.0}{59.6} = 0.1678$ A.

The current $= I_{PBD} = \left(\frac{86.67}{166.67} \times 0.1678\right) = 0.0873$ A.

The current $= I_{PCD} = I_{EP} - I_{PBD} = 0.1678 - 0.0873 = 0.0805$. A

The voltage drop between the points B and C

$$V_{BC} = V_{BP} - V_{CP} = -(0.0873 \times 40.0) + (0.0805 \times 26.67)$$

$$= -3.492 + 2.147 = -1.345 \text{ V.}$$

The current in the load resistance of 200 Ω $= I_{BC} = V_{BC}/R_{BC}$

$$= -1.345/200 = -6.725 \text{ mA.}$$

Please note that the load current flows from C to B, as its sign is negative.

2.5.3 Thevenin's Theorem

This is the first theorem to be discussed here—and we shall do so with examples.

STATEMENT

Thevenin's theorem for linear electrical systems states that any number of voltage sources, current sources and resistors can be reduced to an equivalent voltage source and a resistor connected in series with the voltage source. Thevenin's equivalent circuit is usually calculated for an open circuit.

Consider a two-terminal linear network (circuit) with the load resistance R_L having at least one voltage/current source (Fig. 2.8(a)). Figure 2.8(a) is the equivalent circuit with a voltage source, which is measured/computed without the load resistance ($R_L = \infty$), i.e., open circuit, in series with the equivalent resistance of the circuit R_L, which is measured/computed with all voltage/current sources being replaced by their internal resistances.

The load current of the equivalent circuit is given by

$$I_L = \frac{V_{OC}}{R_{Th} + R_L}$$

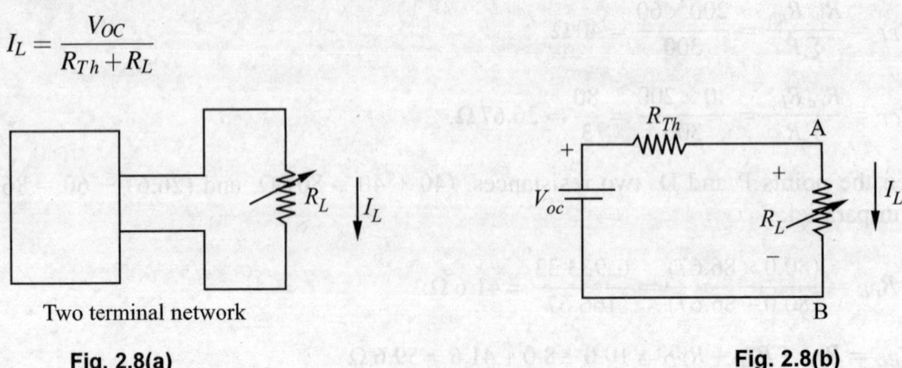

Fig. 2.8(a)

Fig. 2.8(b)

If the output of the network (circuit) is short-circuited, i.e., $R_L = 0$, the load current, which is now a short-circuit current, is

$$I_L = I_{SC} = \frac{V_{OC}}{R_{Th}}.$$

This shows that the equivalent resistance of the circuit is

$$R_{Th} = \frac{V_{OC}}{I_{SC}} .$$

This theorem is very useful because real components can be converted to ideal components (real and ideal components are explained in the next section). The properties of these ideal components are relatively easy to calculate.

EXAMPLES FOR THE APPLICATION OF THEVENIN'S THEOREM

Two examples are taken up here.

EXAMPLE 2.5 *In the circuit (Fig. 2.9(a)), which is the same as Fig. 2.3, determine the current in the load resistance,* $R_L = 10\,\Omega$.

SOLUTION As described earlier, there are two delta-connected circuits (Fig. 2.9(a)) with the resistances being equal. These can be converted to star equivalent ones. The formula for the equivalent resistances to be connected in star (Fig. 2.9(b)) is

$$R_A = \frac{R \cdot R}{(R+R+R)} = \frac{R}{3},$$

where R is the resistance connected in delta.

For the delta-connected circuit on the left (Fig. 2.9(a)), $R = 12\,\Omega$. Hence, the equivalent star-connected resistances are

$$R_A = R_B = R_C = \frac{12}{3} = 4\,\Omega.$$

For the delta-connected circuit on the right, $R = 30\,\Omega$. Hence, the equivalent star-connected resistances are,

$$R_D = R_E = R_F = \frac{30}{3} = 10\,\Omega.$$

The two star points are P and Q respectively.

Before proceeding further, please note the following: Earlier, there were four loops in the circuit (Fig. 2.9(a)). Now, in the equivalent circuit (Fig. 2.9(b))—after two delta-connected resistances are converted into star-connected equivalent ones—there are only two loops. The circuit can be further simplified, as there is one voltage source only. This is not described here, but can be easily solved to find the load current in the resistor of 10 Ω.

The circuit for finding the open-circuit voltage is shown in Fig. 2.9(c). The total resistance of the circuit is

$$R = 4 + (4 + 34 + 10) + 10 = 14 + 48 = 62\,\Omega.$$

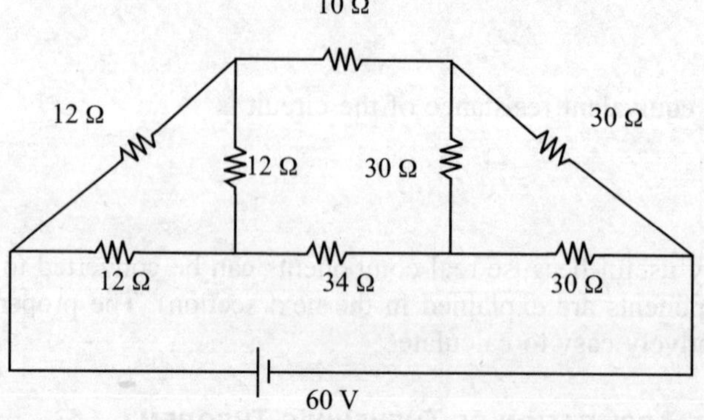

Fig. 2.9(a)

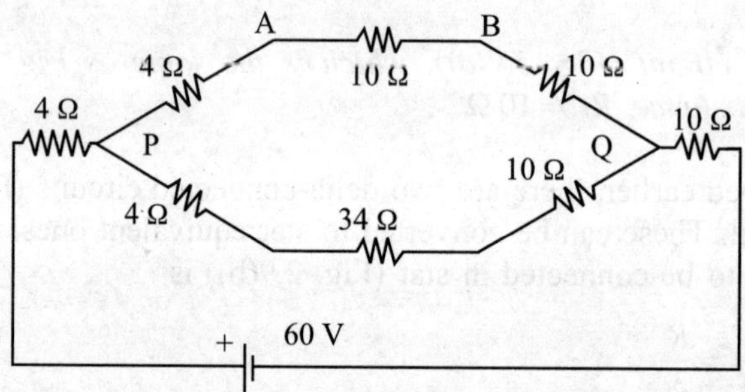

Fig. 2.9(b)

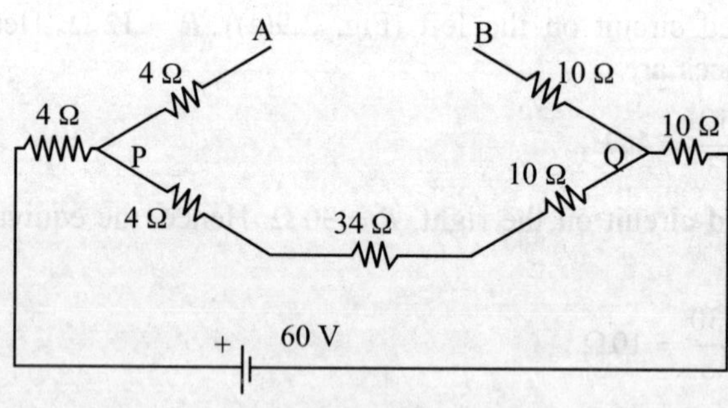

Fig. 2.9(c)

The current drawn from the source of 60 V is $\frac{60}{62} = 0.9677$A. The resistance between the points, P and Q is $R_{PQ} = 4 + 34 + 10 = 48\ \Omega$. Therefore, the open-circuit voltage is given by,

$$V_{OC} = V_{AB} = V_{PQ} = \left(\frac{60}{62} \times 48\right) = \frac{2,880}{62}\ \text{V}$$

With the voltage source short-circuited, as its internal resistance is zero, the circuit for computing equivalent or Thevenin's resistance is shown in Fig. 2.9(d). Between the points, P and Q, two resistors, 48 Ω and (4 + 10) = 14 Ω are in parallel. The resistance is,

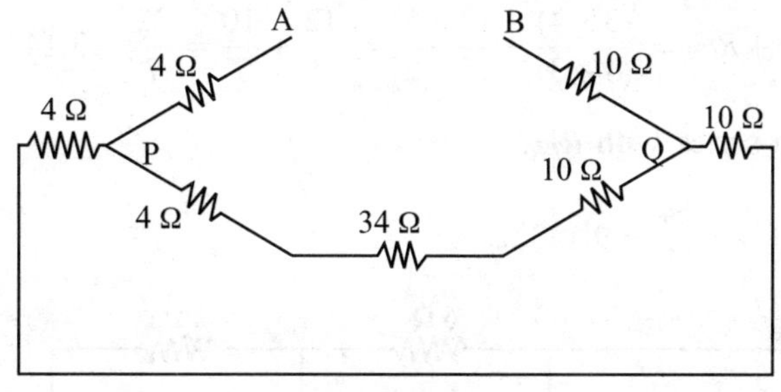

Fig. 2.9(d)

$$R_{PQ} = \frac{(14 \times 48)}{(14+48)} = \frac{672}{62}\ \Omega.$$

Between the points C and D, R_{PQ} is in series with a resistor of $(4+10) = 14\ \Omega$. The resistance is,

$$R_{Th} = R_{AB} = \frac{672}{62} + 14 = \frac{1,540}{62}\ \Omega.$$

$R_L = (10\ \Omega)$ is in series with R_{Th}.

So, $$R_{Th} + R_L = \frac{1540}{62} + 10 = \frac{2,160}{62}\ \Omega$$

The current through the load resistance is

$$I_L = \frac{V_{OC}}{R_{Th} + R_L} = \left(\frac{2,880}{62}\right) \times \left(\frac{62}{2.160}\right) = \frac{4}{3} = 1.333\ \text{A}.$$

The load current obtained is the same as determined earlier by the loop current method. However, this method is simple, if one understands the intricacies of Thevenin's method.

Thevenin's resistance can also be obtained by first finding the short-circuit current by shorting the load across A and B as seen in Fig. 2.9(c), and using the formula $R_{Th} = V_{OC}/I_{SC}$. This will be illustrated in the next example.

EXAMPLE 2.6 *In the circuit (Fig. 2.10(a)), determine the current in the load resistance,* $R_L = 6\ \Omega$.

SOLUTION The open-circuit condition with the load resistance removed is shown in Fig. 2.10(b).

The open-circuit voltage is

$$V_{OC} = V_{AB} = V_{AC} - V_{BC} = \frac{(8 \times 4)}{(4+3)} - \frac{(6 \times 5)}{(5+2)} = \frac{32}{7} - \frac{30}{7} = \frac{2}{7} = 0.286\ \text{V}$$

With two voltage sources short-circuited as their internal resistances are zero, the circuit for computing equivalent or Thevenin's resistance is shown in Fig. 2.10(c). Thevenin's or equivalent resistance is

$$R_{Th} = R_{AB} = R_{AC} + R_{CB} = \frac{(3 \times 4)}{(4+3)} + \frac{(2 \times 5)}{(5+2)} = \frac{12}{7} + \frac{10}{7} = \frac{22}{7} = 3.143\ \Omega.$$

R_L (= 6 Ω) is in series with R_{Th}.

So, $$R_{Th} + R_L = \frac{22}{7} + 6 = \frac{64}{7} = 9.143\ \Omega.$$

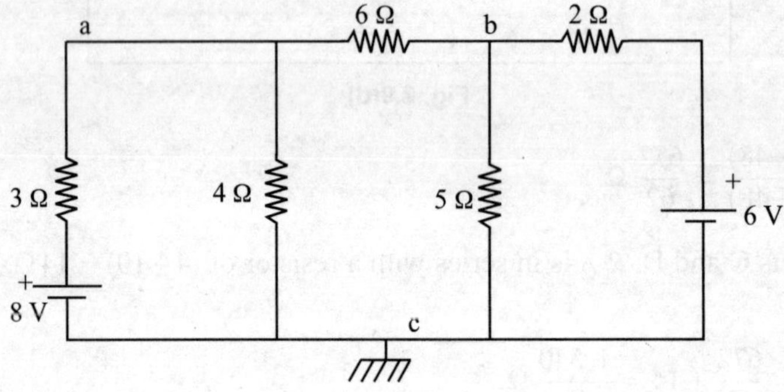

Fig. 2.10(a)

The current through the load resistance is

$$I_L = \frac{V_{OC}}{R_{Th} + R_L} = \left(\frac{2}{7}\right) \times \left(\frac{7}{64}\right) = \frac{1}{32} = 0.03125\ \text{A} = 31.25\ \text{mA}.$$

This current is the same as determined earlier by using the principle of superposition.

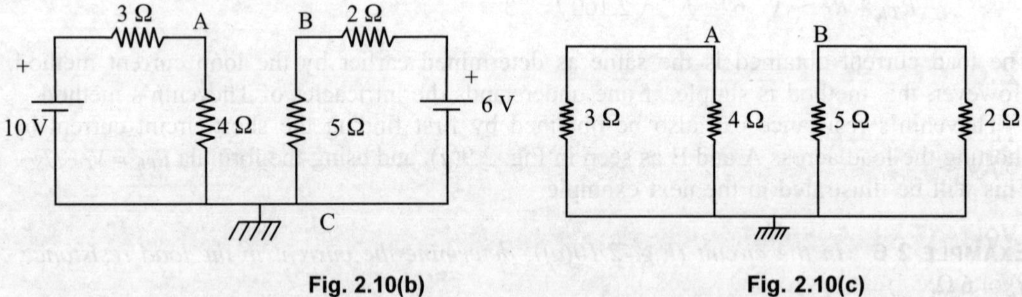

Fig. 2.10(b) **Fig. 2.10(c)**

An alternative method of finding Thevenin's resistance is presented below, where the short-circuit current is first determined. The load resistance is short-circuited by placing an ammeter of low resistance, instead of the load resistance. The voltage at point A is found using the nodal voltage method. Using Kirchoff's current (first) law at node A, the equation is

$$\frac{V-8}{3} + \frac{V}{4} + \frac{V}{5} + \frac{V-6}{2} = 0$$

or, $$V\left(\frac{1}{3}+\frac{1}{4}+\frac{1}{5}+\frac{1}{2}\right)=\frac{77}{60}\cdot V=\left(\frac{8}{3}+\frac{6}{2}\right)=\frac{34}{6}$$

or, $$V=\left(\frac{34}{6}\right)\times\left(\frac{60}{77}\right)=\frac{340}{77}\text{ V.}$$

The current drawn from the 8 V (LHS) source in the 3 Ω resistor

$$I_{CA}=\left(\frac{1}{3}\right)\times\left(8-\frac{340}{77}\right)=\left(\frac{1}{3}\right)\times\left(\frac{276}{77}\right)=\frac{92}{77}\text{ A.}$$

The current in the 4 Ω resistor

$$I_{AC}=\left(\frac{340}{77}\right)\times\left(\frac{1}{4}\right)=\frac{85}{77}\text{ A.}$$

The short current, using Kirchoff's current law at A

$$I_{SC}=I_{AB}=I_{CA}-I_{AC}=\frac{92}{77}-\frac{85}{77}=\frac{7}{77}=\frac{1}{11}=0.091\text{ A.}$$

Thevenin's resistance

$$R_{Th}=\frac{V_{OC}}{I_{SC}}=\left(\frac{2}{7}\right)\times 11=\frac{22}{7}=3.143\quad\Omega.$$

This value is the same as obtained earlier. More steps were needed in this problem. However, in certain cases, say in EXAMPLE 2.5, lesser steps may be needed.

2.6 Types of Energy Sources

There are different types of energy sources, which can be applied to a circuit consisting of only resistors.

VOLTAGE AND CURRENT SOURCES

Voltage/current sources are used in circuits (described in detail in the next chapter). These sources can be ideal or real. First, let us discuss ideal ones. In an *ideal voltage source*, of voltage, V_s, which can be a battery, the output voltage remains constant, irrespective of the current fed by the source to the load, or current fed to the source by the load. Same is the case with power drawn/absorbed by the voltage source. The $V-I$ characteristic, and its symbol, are shown in Fig. 2.11(a). The current is positive when the source supplies current to the load, while it is negative when current is fed to the source from the load. A two-terminal network is taken as the load, which is connected to the voltage source. Other sources are also present in the network.

In an *ideal current source*, the output current, I_s, remains constant, irrespective of the voltage appearing across it, which depends upon the circuit to which it is connected. This voltage is positive when the current is fed by the source to the load, while it is negative when

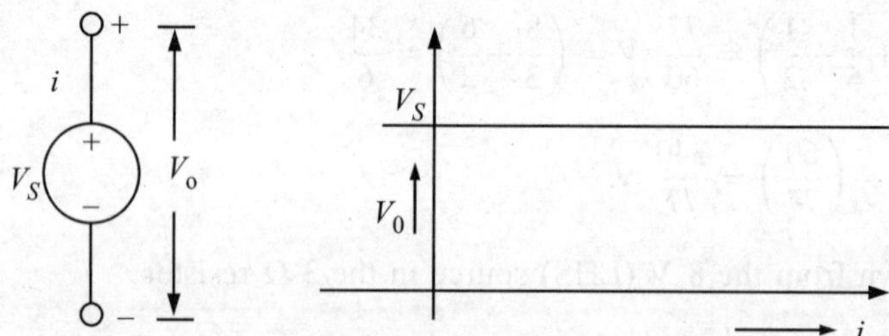

Fig. 2.11(a) *V–I characteristic and symbol of an ideal voltage source*

current is fed to the source by the load. The power drawn/absorbed by the current source depends upon the developed voltage across it. The $V - I$ characteristic, and its symbol, is shown in Fig. 2.12. As given earlier, a two-terminal network is taken as load, which is connected to the current source. Other sources are also present in the network.

For an ideal current source, the output current remains the same or is constant, irrespective of the voltage developed across it, which depends upon the circuit to which it is connected. The voltage developed in the source depends upon the same circuit. The sign of the voltage may be +ve or –ve. The power drawn/absorbed by the current source depends upon the developed voltage across it.

A *real (non-ideal) voltage source* can be considered as a voltage source, of voltage, V_s, connected in series with its internal or source resistance, r_s, which is normally small (Fig. 2.11(b)). In this case, the output voltage, V_o depends upon the current fed by the source to the load, or the current fed to the source by the load. Mathematically, it is found as, $V_o = V_s - (I.r_s)$, where I is the current in the voltage source. The $V - I$ characteristic is also shown in Fig. 2.11(b). As stated earlier, the current is positive when the source supplies current to the load, while it is negative when current is fed to the source from the load. A two-terminal network is taken as load, which is connected to the voltage source. Other sources are also present in the network. Therefore, the output voltage, V_o decreases from its initial value, V_s, when the current, I is drawn from the source, and is positive. If the current, I, is fed to the source, the output voltage, V_o increases from its initial value, V_s. This is shown in Fig. 2.11(b). If there is no current, i.e., $I = 0$, then the output voltage, V_o is equal to the source voltage, V_s. From the above, it can be observed that the source or internal resistance of an ideal voltage source is very low or zero (0), i.e. the resistance is short-circuited.

A *real (non-ideal) current source* is one in which an ideal current source, I_s, is connected in parallel with a resistance, r_s (Fig. 2.12). As contrasted with an ideal current source, the voltage drop across the real current source depends upon the current drawn from it. When no current is drawn from this source, the current, I is 0 (zero), and the voltage drop across the current source is $V_0 = I_s.r_s$. For any other case, the voltage drop across this source is, $V_0 = (I_s - I).r_s$, I being the current drawn or fed to the real current source. The $V - I$ characteristic is also shown in Fig. 2.12. It can be observed that the $V - I$ characteristics of the two types of real sources, viz, voltage and current are the same. A real voltage source can be converted to a real current source, and vice-versa (explained in Section. 2.7). Equating the two expressions, we get, $V_s = I_s r_s$. Now, it can be stated that the resistance connected in parallel with an ideal current is very high or infinity (∞), i.e., it is open-circuited.

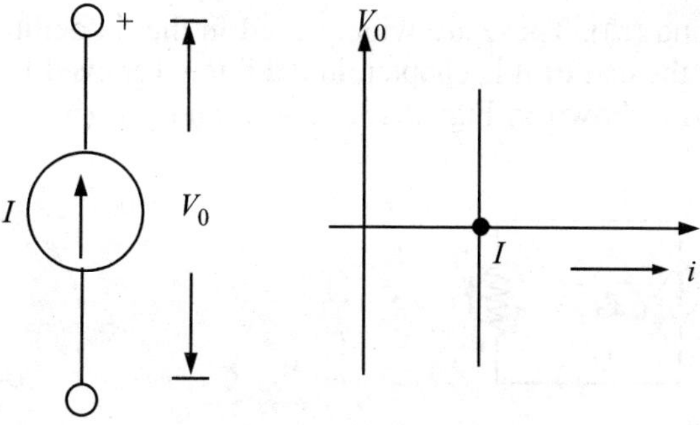

Fig. 2.11(b) *V-I characteristics and symbol of a non-ideal voltage source*

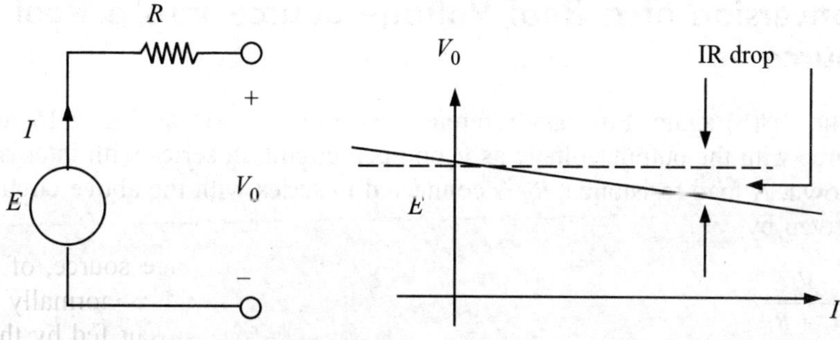

Fig. 2.12 *V-I characteristics and symbol of an non-ideal current source*

INDEPENDENT AND DEPENDENT SOURCES

All the sources of voltage/current types, described in the earlier section, can be termed as *independent sources*, as in these cases, the source voltage or current, as the case may be, is constant. It does not depend on any other parameter.

Fig. 2.13(a) *A voltage dependent voltage source* **Fig. 2.13(b)** *A current dependent voltage source*

In *dependent sources*, which may be of the voltage/current type, the voltage/current is related to the current or voltage in other branches of the circuit. Figure 2.13(a) illustrates a dependent source where the output of the voltage source depends upon the voltage in other parts (branches) of the circuit. There can be cases where the output voltage is related to the current in the other branches of the circuit (Fig. 2.13(b)). Similarly, the current in the current source may depend on the voltage or current in another branch of the circuit, as given for dependent voltage (Fig. 2.14(a) and (b)). These are widely used in the modelling of electronic circuits. Examples are given at the end of this chapter, in Additional Solved Problems. Different types of dependent sources are shown in Figs 2.13, 2.14(a) and 2.14(b).

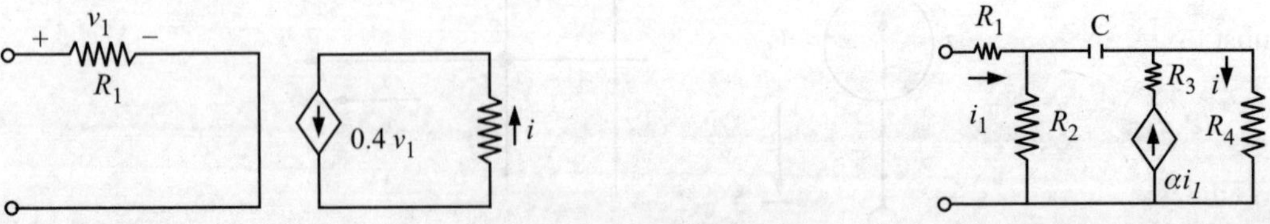

Fig. 2.14(a) *A voltage dependent current source* **Fig. 2.14(b)** *A current dependent current source*

2.7 Conversion of a Real Voltage Source into a Real Current Source

Consider Fig. 2.8(b) again. For easy reference, we bring it here as Fig. 2.15, where a real voltage source with the output voltage as V_S on open circuit, in series with internal resistance as R_S, is shown. A load resistance, R_L is connected in series with the above circuit. The load current is given by

$$I_L = \frac{V_s}{R_s + R_L}.$$

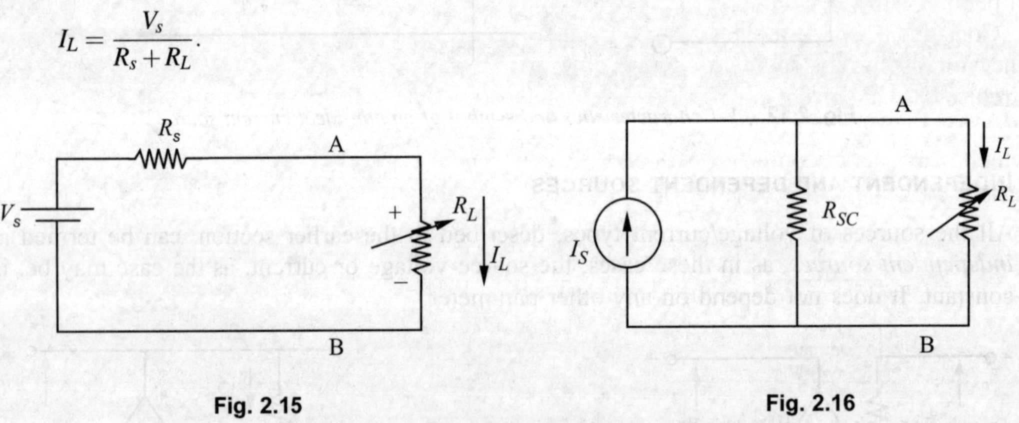

Fig. 2.15 **Fig. 2.16**

As shown earlier, if the output is short-circuited ($R_L = 0$), the short-circuit current is

$$I_L = I_{SC} = \frac{V_S}{R_S}.$$

A real current source is shown in Fig. 2.16. The current is the same as $I_S = I_{SC}$. Please note that a resistance is also connected in parallel with the current source. The value of the resistance is the same, i.e., R_S. The load resistance, R_L is connected in parallel.

So, the load current is $I_L = \left(\dfrac{R_S}{R_s + R_L} \times I_S \right).$

It may be noted that the source current is the same as the short current, i.e.,

$$I_S = I_{SC} = \frac{V_S}{R_S}.$$

Substituting this value in the earlier equation, we get

$$I_L = \frac{V_s}{R_s + R_L}.$$

So, the real current source is equivalent to the real voltage source as described earlier. Conversely, a real current source can be converted to a real voltage source. This will be described in the next section, while presenting Norton's theorem.

2.8 Some Other Basic Theorems

In this section, Norton's theorem and the maximum power transfer theorem are presented.

NORTON'S THEOREM

Norton's theorem states that any number of voltage sources, current sources and resistors having two open ends can be simplified into an ideal current source and a resistor connected in parallel with the source.

Consider a two-terminal linear network (circuit) with load resistance R_L having at least one voltage/current source as shown in Fig. 2.17(a). Figure 2.17(b) shows the equivalent circuit with a current source I_S, which is measured/computed with the load resistance ($R_L = 0$), i.e., short-circuited, in parallel with the equivalent resistance $R_{EQ} = R_S$ of the circuit, which is measured/computed with all voltage/current sources being replaced by their internal resistances.

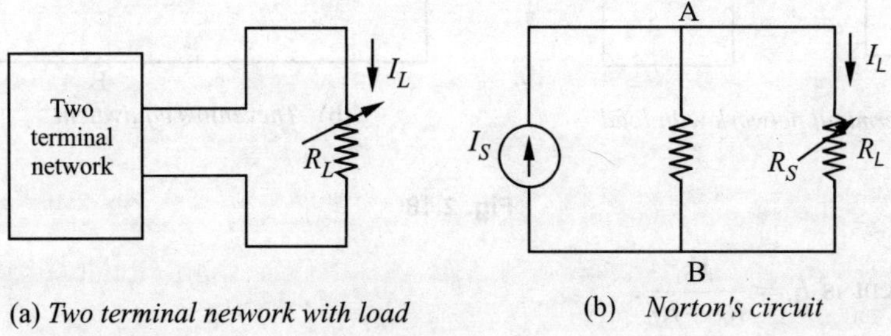

(a) *Two terminal network with load* (b) *Norton's circuit*

Fig. 2.17

The short-circuit current is the current in the current source. The equivalent resistance is the same as Thevenin's resistance, R_{Th} (or R_{EQ}, R_{SC}). The load current is given by

$$I_L = \left(\frac{R_s}{R_s + R_L} \times I_S\right).$$

If the output of the network (circuit) is short-circuited, i.e., $R_L = 0$, the load current, which is now a short-circuit current, is

$$I_L = I_S = I_{SC} = \frac{(I_S \cdot R_S)}{R_{EQ}}.$$

This circuit is equivalent to Thevenin's circuit shown in Fig. 2.8(b). So, the source current is

$$I_S = \frac{V_{Th}}{R_S},$$

or the open-circuit voltage is $V_{Th} = V_{OC} = (I_S.R_S)$. This is the same as described in the earlier section. In this case, a real current source is converted into a real voltage source.

Thus, though Norton's theorem and Thevenin's theorem are similar, there are some differences—Norton's theorem uses a current source, whereas Thevenin's theorem uses a voltage source. Thevenin's theorem uses a resistor in series, while Norton's theorem uses a resister set in parallel with the source.

2.8.1 Maximum Power Transfer Theorem

This is not a theorem in circuit analysis. However, it has its usefulness, as explained here. A two-terminal network is connected to a load resistance, R_L (Fig. 2.18(a)). To take Thevenin's equivalent of the network (Fig. 2.18(b)), a voltage source, V_S is in series with the equivalent resistance, R_S (can be taken as its internal resistance).

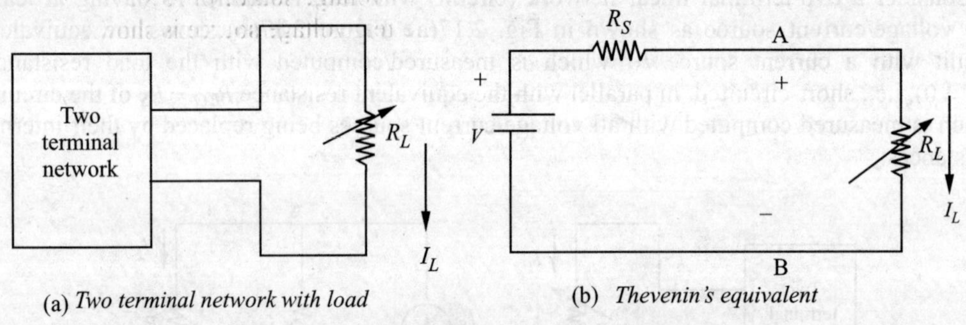

(a) *Two terminal network with load*　　(b) *Thevenin's equivalent*

Fig. 2.18

The load current is $I_L = \dfrac{V_S}{R_S + R_L}$.

The power consumed in the load $= P_L = (I_L)^2 R_L = \left(\dfrac{V_S}{R_S + R_L}\right)^2 R_L = \dfrac{(V_S)^2 R_L}{(R_S + R_L)^2}$.

To transfer maximum power to the load resistance, the condition is,

$$\frac{dP_L}{dR_L} = \frac{d}{dR_L}\left[(V_S)^2\left(\frac{R_L}{(R_S + R_L)^2}\right)\right] = 0.$$

Omitting the constant term, $(V_S)^2$, and simplifying,

$$\frac{dP_L}{dR_L} = \frac{1}{(R_S+R_L)^2} - \frac{2R_L}{(R_S+R_L)^3} = \frac{R_s+R_L-2R_L}{(R_S+R_L)^3} = \frac{R_s-R_L}{(R_S+R_L)^3} = 0$$

or, $R_s - R_L = 0$ or, $R_s = R_L$.

When the load resistance is equal to the source resistance, maximum power is transferred to the load resistance. The maximum power transferred to the load is

$$(P_L)_{\max} = \frac{(V_S)^2 R_s}{(2R_S)^2} = \frac{(V_S)^2}{4R_S}.$$

This is used in power amplifiers, such as radio, public address system etc., where a speaker is connected to the power amplifier via a transformer, to transfer maximum power to it.

ADDITIONAL PROBLEMS

EXAMPLE 2.7 *In the circuit shown in Fig. 2.19(a), determine the current in the load resistance, $R_L = 4\,\Omega$.*

SOLUTION FIRST METHOD: LOOP CURRENT METHOD

To simplify the circuit (Fig. 2.19(a)), the real current source of 6 A with a parallel resistance of 2 Ω, on the top of the RHS loop is converted to a real voltage source of $(6 \times 2 = 12)$ V in series with the same resistor of 2 Ω. The circuit with the real voltage source is shown in Fig. 2.19(b), which is taken as the modified one. There are three loops and also three nodes—A, B and C in the circuit (Fig. 2.19(b)). The node D is taken as the reference node. The loop currents are assumed to be in the clockwise direction.

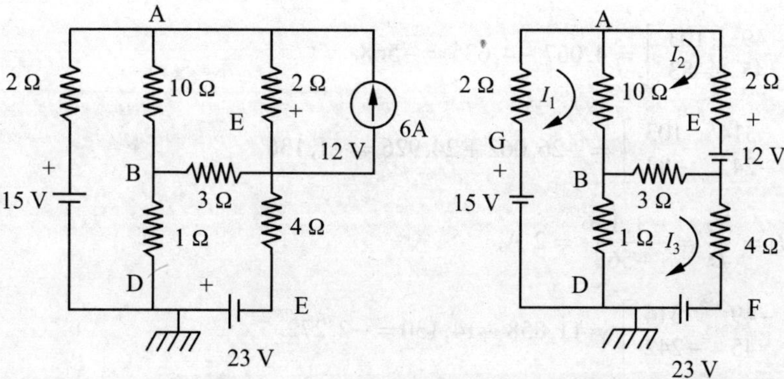

Fig. 2.19(a) **Fig. 2.19(b)**

Applying Kirchhoff's voltage (second) law in current loops, the voltage equations are,

$$(2+10+1)I_1 - 10I_2 - I_3 = 13I_1 - 10I_2 - I_3 = 15. \quad \text{(i)}$$

$$-10I_1 + (10+2+3)I_2 - 3I_3 = -10I_1 + 15I_2 - 3I_3 = -12. \quad \text{(ii)}$$

$$-I_1 - 3I_2 + (1+3+4))I_3 = -I_1 - 3I_2 + 8I_3 = 23. \qquad \text{(iii)}$$

In matrix form, the equations are

$$\begin{bmatrix} 13 & -10 & -1 \\ -10 & 15 & -3 \\ -1 & -3 & 8 \end{bmatrix} \begin{bmatrix} I_1 \\ I_2 \\ I_3 \end{bmatrix} = \begin{bmatrix} 1 \\ -12 \\ 23 \end{bmatrix}.$$

Note that the resistance matrix is symmetric.

The steps to solve the equations, by using the procedure of elimination, are given here in brief. The advantages, if any, will be given later.

The current, I_1 is eliminated from Eqs (i)–(iii). Multiplying Eq.(iii) by 13, and adding it with Eq.(i), we get,

$$-49I_2 + 103I_3 = 314. \qquad \text{(iv)}$$

Similarly, from Eqs (ii) and (iii), the following equation is obtained.

$$45I_2 - 83I_3 = -242. \qquad \text{(v)}$$

Writing in matrix form, Eqs (iv)–(v) are

$$\begin{bmatrix} -49 & 103 \\ 45 & -83 \end{bmatrix} \begin{bmatrix} I_2 \\ I_3 \end{bmatrix} = \begin{bmatrix} 314 \\ -242 \end{bmatrix}.$$

From the matrix, two equations are obtained from which the values of I_2 and I_3 are easily found as shown. The denominator and numerators needed to find the two currents are obtained as

$$D = \begin{bmatrix} -49 & 103 \\ 45 & -83 \end{bmatrix} = 4,067 - 4,635 = -568.$$

$$N_2 = \begin{bmatrix} 314 & 103 \\ -242 & -83 \end{bmatrix} = -26,062 + 24,926 = -1,136.$$

$$\text{The current } I_2 = \frac{N_2}{D} = \frac{-1,136}{-568} = 2 \text{ A}.$$

$$N_3 = \begin{bmatrix} -49 & 314 \\ 45 & -242 \end{bmatrix} = 11,858 - 14,130 = -2,272.$$

$$\text{The current, } I_3 = \frac{N_3}{D} = \frac{-2,272}{-568} = 4 \text{ A}.$$

Substituting these values in Eq.(iii), the current, $I_1 = -6 + 32 - 23 = 3$ A. The current in the load resistance (4 Ω) = 4 A.

Reverting to the original circuit (Fig. 2.19(a)), the current in the 2 Ω resistor (I_{AC}) in parallel with the current source of 6 A, which is on the top of the RHS loop, by applying Kirchhoff's current law at the node A = $I_{AC} = 6.0 + 2.0 = 8.0$ A.

The voltage drop, V_{AC}, across the above current source $= 2 \times 8.0 = 16.0$ V.

This voltage in the modified circuit (Fig. 2.19(b)) is found as $V_{AC} = [12 + (2 \times 2)] = 16.0$ V, which is the same as that obtained earlier. The node voltages are found as

$$V_A = 15 - (2 \cdot I_1) = 15 - (2 \times 3) = 9 \text{ V}.$$

$$V_C = -23 + (4 \cdot I_3) = -23 + (4 \times 4) = -7 \text{ V}.$$

The current in the branch ($R = 10\ \Omega$) is $(I_1 - I_2) = (3 - 2) = 1$ A. This current is assumed to be positive in the downward direction.

$$V_A - V_B = (10 \times 1) = 10 \text{ A}.$$

So, $V_B = (9 - 10) = -1$ V.

SECOND METHOD: NODAL VOLTAGE METHOD

The problem can also be solved using the nodal voltage method. Applying Kirchhoff's current (first) law at three nodes, A, B and C, the current equations are

$$\left(\frac{1}{2}+\frac{1}{10}+\frac{1}{2}\right)V_A - \frac{1}{10}V_B - \frac{1}{2}V_C = \frac{11}{10}V_A - \frac{1}{10}V_B - \frac{1}{2}V_C = \frac{15}{2} + \frac{12}{2} = \frac{27}{2}.$$

$$-\frac{1}{10}V_A - \left(\frac{1}{10}+\frac{1}{3}+1\right)V_B - \frac{1}{3}V_C = -\frac{1}{10}V_A + \frac{43}{30}V_B - \frac{1}{3}V_C = 0.$$

$$-\frac{1}{2}V_A - \frac{1}{3}V_B - \left(\frac{1}{2}+\frac{1}{3}+\frac{1}{4}\right)V_C - \frac{1}{2}V_A - \frac{1}{3}V_B + \frac{13}{12}V_C = -\left(\frac{23}{4}+\frac{12}{2}\right) = -\frac{47}{4}.$$

In matrix form, the equations are

$$\begin{bmatrix} \frac{11}{10} & -\frac{1}{10} & -\frac{1}{2} \\ -\frac{1}{10} & \frac{43}{30} & -\frac{1}{3} \\ -\frac{1}{2} & -\frac{1}{3} & \frac{13}{12} \end{bmatrix} \begin{bmatrix} V_A \\ V_B \\ V_C \end{bmatrix} = \begin{bmatrix} \frac{27}{2} \\ 0 \\ \frac{47}{4} \end{bmatrix}.$$

Note that the conductance matrix is symmetric. Solving the equations, we get

$$V_A = 9 \text{ V}, V_B = -1 \text{ V, and } V_C = -7 \text{ V}.$$

$$V_{AC} = V_A - V_C = (9+7) = 16 \text{ V}.$$

The current in the load resistance (4 Ω)

$$I_3 = \frac{23-7}{4} = 4 = 4 \text{ A}.$$

The other currents are

$$I_1 = \frac{15-9}{2} = 3 \text{ A, and } I_2 = \frac{16-12}{2} = 2 \text{ A}.$$

These currents are the same as those obtained earlier.

THIRD METHOD: THEVENIN'S THEOREM

The problem can also be solved using Thevenin's theorem. For finding the open-circuit voltage, the circuit (Fig. 2.19(b)) is shown in Fig. 2.19(c), with the load resistance of 4 Ω removed. There are two current loops here.

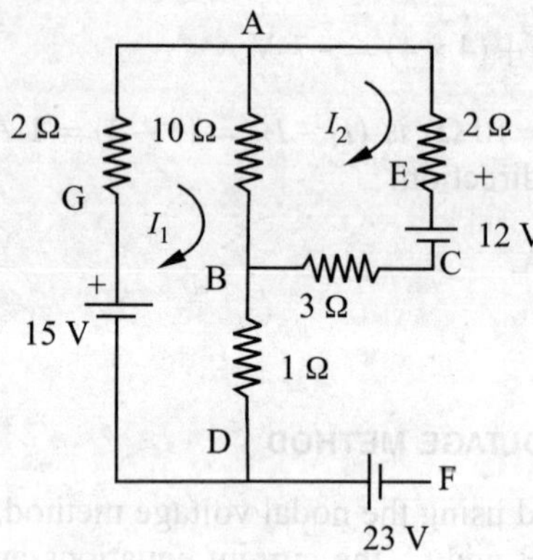

Fig. 2.19(c)

Applying Kirchhoff's voltage law in the current loops, the voltage equations are

$$13I_1 - 10I_2 = 15.$$

$$-10I_1 + 15I_2 = -12.$$

Solving, we get $I_1 = 1.1053$ A and $I_2 = -0.06316$ A.

Open-circuit voltage $= V_{OC} = V_{EC} = V_{ED} + V_{DB} + V_{BC}$

$$= 23.0 + (1 \times 1.1053) - (3 \times 0.06316) \approx 23.0 + 1.1053 - 0.1895 = 23.916 \text{ V}.$$

To determine Thevenin's or equivalent resistance, all voltage sources in Fig. 2.19(b) are short-circuited (Fig. 2.19(d)), as none of them have an internal resistance. Three delta-connected resistances (LHS) are converted into star-equivalent ones (Fig. 2.19(e)). The delta-connected resistances are $R_{da} = 2\ \Omega, R_{ab} = 10\ \Omega$ and $R_{bd} = 1\ \Omega$.

$$\sum R = R_{da} + R_{ab} + R_{bd} = 2 + 10 + 1 = 13\ \Omega.$$

The formulas for the equivalent resistances to be connected in star (Fig. 2.19(d)) are

$$R_D = \frac{R_{da}\,R_{db}}{R_{da} + R_{ab} + R_{bd}} = \frac{R_{da}\,R_{db}}{\sum R} = \frac{2 \times 1}{130} = \frac{2}{13} = 0.154 \quad \Omega.$$

R_D is also R_{DP}, where P is the star point.

$$R_{AP} = \frac{R_{ab}\,R_{ad}}{\sum R} = \frac{10 \times 2}{13} = \frac{20}{13} = 1.54\ \Omega.$$

$$R_{BP} = \frac{R_{ba}R_{bd}}{\Sigma R} = \frac{10\times 1}{13} = \frac{10}{13} = 0.77\ \Omega.$$

Between the points P and C, two resistances, $(2.0 + 1.54 = 3.54\ \Omega$, and $3.0 + 0.77 = 3.77\ \Omega)$, are in parallel.

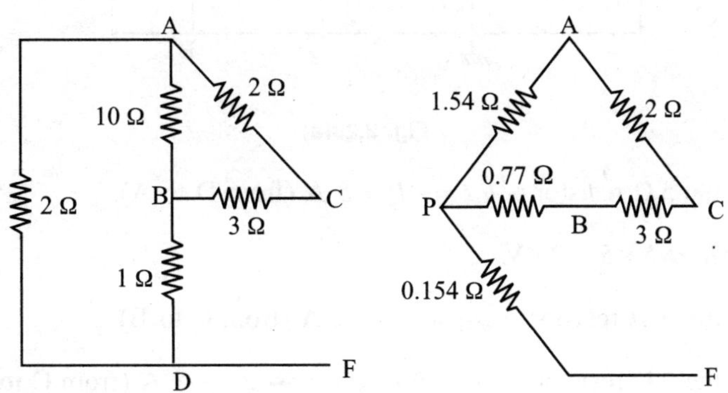

Fig. 2.19(d) **Fig. 2.19(e)**

So, $R_{PC} = \frac{3.54\times 3.77}{3.54+3.77} = \frac{3.54\times 3.77}{7.31} = 1.825\ \Omega.$

$$R_{Th} = R_{DC} = R_{DP} + R_{PC} = 0.154 + 1.825 = 1.979 \approx 1.98\ \Omega.$$

The load current is

$$I_L = I_{DC} = \frac{V_{OC}}{R_{Th}+R_L} = \frac{23.916}{1.979+4.0} = \frac{23.916}{5.979} = 4.0\text{ A}.$$

This current is the same as obtained earlier by the other two methods.

EXAMPLE 2.8 *Determine all branch currents in the circuit shown in Fig. 2.20(a). Determine the power input/ output to the current and voltage sources.*

SOLUTION **FIRST METHOD: LOOP CURRENT METHOD**

There are three loops in the circuit (Fig. 2.20(a)). However, the current in the second loop is $I_2 = -5$ A, as it is in the anti-clockwise direction. For the other two loops, applying Kirchhoff's voltage law, the equations are

$$5I_1 - 4I_2 = 5; \qquad -2I_2 + 5I_3 = 0.$$

From the first equation, $5I_1 = 5 - (4\times 5) = -15$ or, $I_{CA} = I_1 = -15/5 = -3$ A.

From the second equation, $5I_3 = -(2\times 5) = -10$ or, $I_{BC} = I_3 = -10/5 = -2$ A.

Current through the 1 Ω resistor $= I_{CA} = -3$ A (from A to C).

Current through the 4 Ω resistor $= I_{AC} = I_1 - I_2 = -3 + 5 = 2$ A (from A to C).

Voltage drop $= V_{AC} = 4\times 2 = 8$ V.

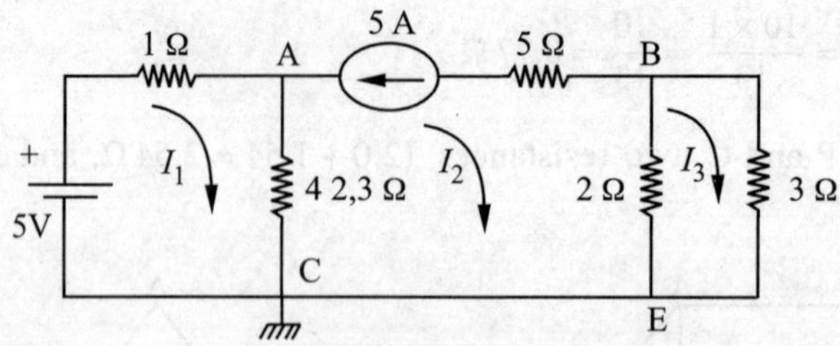

Fig. 2.20(a)

Current through the 5 Ω resistor $= I_{DA} = -I_2 = 5$ A (from D to A).

Voltage drop $= V_{DA} = 5 \times 5 = 25$ V.

Current through the 3 Ω resistor $= I_{BC} = I_3 = -2$ A (from C to B).

Current through the 2 Ω resistor $= I_{BC} = I_2 - I_3 = -5 + 2 = -3$ A (from C to B).

Voltage drop $= V_{BC} = -2 \times 3 = -6$ V.

The current through the 2 Ω resistor, is found as

$$I_{BC} = I_2 \times \left(\frac{3}{3+2}\right) = \left(-5 \times \frac{3}{5}\right) = -3 \text{ A (from C to B).}$$

Power drawn by the voltage source of 5 V $= 5 \times I_1 = -5 \times 3 = -15$ W (i.e., power fed by the circuit).

Voltage drop across current source of 5 A.

$$= V_{BD} = V_{BC} + V_{CA} + V_{AD} = -6 - 8 - 25 = -39 \text{ V.}$$

Power drawn by the current source of 5 A is $= I_2 \times V_{BD} = 5 \times 39 = 195$ W.

Total power fed by two sources to the circuit $= 195 - 15 = 180$ W.

Total power in all resistances $= [(1 \times 3^2) + (4 \times 2^2) + (5 \times 5^2) + (2 \times 3^2) + (3 \times 2^2)]$

$$= (9 + 16 + 125 + 18 + 12) = 180 \text{ W,}$$

which is equal to the total power fed by the two sources to the circuit.

SECOND METHOD: NODAL VOLTAGE METHOD

There are two nodes, A and B in the circuit (Fig. 2.20(a)).

Applying Kirchhoff's current law at two nodes, the equations are,

$$V_A\left(1 + \frac{1}{4}\right) - \frac{5}{1} = 5 \text{ or, } \left(\frac{5}{4}\right) V_A = 5 + 5 = 10.$$

$$V_A = 10 \times \frac{4}{5} = 8 \text{ V.}$$

$$\left(\frac{1}{2} + \frac{1}{3}\right) V_B = \left(\frac{5}{6}\right) V_B = -5$$

or, $$V_B = -5 \times \frac{6}{5} = -6 \text{ V.}$$

The voltages are the same as obtained earlier. Other steps are not shown here.

THIRD METHOD: THEVENIN'S THEOREM

This theorem is used to find only the current in the load resistance of 3 Ω (Fig. 2.20(a)). It can be applied to solve some problems in a simple way, as in this case, where both modified circuits are simple. Please note that the nodes E and C are equipotential, i.e., both these points are earthed.

To find the open-circuit voltage, the modified circuit with the load resistance of 3 Ω omitted, is as shown in Fig. 2.20(b). The current through the resistor of 2 Ω is 5 A, the current source current. Therefore, the open-circuit voltage is $V_{OC} = V_{EB} = 2 \times 5 = 10$ V. Other parts of the circuit need not be looked at.

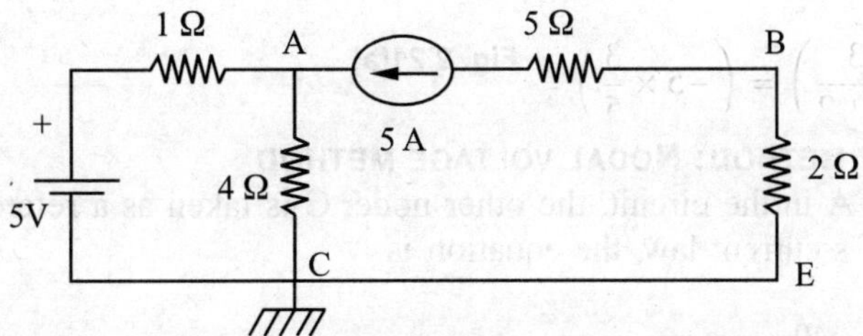

Fig. 2.20(b)

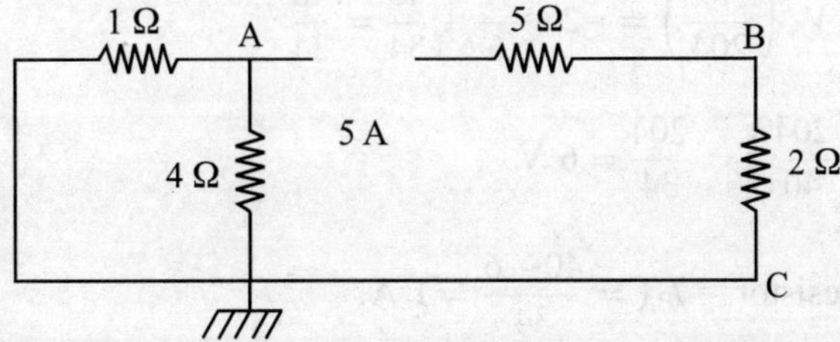

Fig. 2.20(c)

For finding Thevenin's or equivalent resistance, the modified circuit is shown in Fig. 2.20(c), with the voltage source short circuited, and the current source open circuited. The reasons for these are explained earlier. The equivalent resistance is $R_{Th} = R_{EB} = 2$ Ω. Please note that the circuit between the nodes B and D is open. As in the earlier case, other parts of the circuit need not be looked at.

The load current is $I_L = I_{EB} = \dfrac{V_{OC}}{R_{Th}+R_L} = \dfrac{10}{2+3} = 2$ A.

This value is the same as found earlier. The problem can also be solved using the *principle of superposition.*

EXAMPLE 2.9 *Find the voltage across the resistance of 20 Ω in the circuit shown in Fig. 2.21(a).*

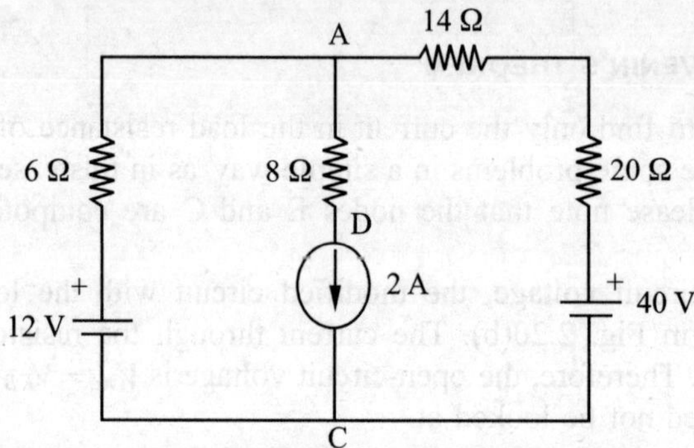

Fig. 2.21(a)

SOLUTION FIRST METHOD: NODAL VOLTAGE METHOD

There is one node, A in the circuit, the other node, C is taken as a reference one.

Using Kirchhoff's current law, the equation is

$$\frac{V_A - 12}{6} + \frac{V_A - 40}{20+14} = -2$$

or, $$V_A\left(\frac{1}{6}+\frac{1}{34}\right) = V_A\left(\frac{40}{204}\right) = -2 + \frac{12}{6} + \frac{40}{34} = \frac{40}{34}$$

or, $$V_A = \left(\frac{40}{34}\right) \times \left(\frac{204}{40}\right) = \frac{204}{34} = 6 \text{ V.}$$

Current in the 20 Ω resistor $= I_{BA} = \dfrac{40-6}{34} = 1$ A.

Current in the 6 Ω resistor $= I_{CA} = \dfrac{12-6}{6} = 1$ A.

Voltage drop in the 8 Ω resistor $= V_{AD} = 2 \times 8 = 16$ V.

Voltage drop across the current source of 2 A $= V_{DC} = V_{DA} + V_{AC} = -16 + 6 = -10$ V.

SECOND METHOD: LOOP CURRENT METHOD

There are two loops in the circuit (Fig. 2.21(b)). The loop currents are assumed to be in the clockwise direction. However, one current source of 2 A is there between two loops. The equation for this case is

$$I_1 - I_2 = 2.$$

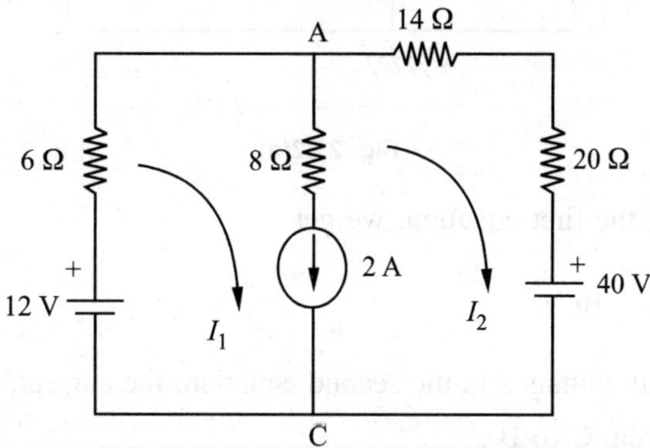

Fig. 2.21(b)

As we cannot take any of the conventional loops here, we take the total loop.

The voltage equation is

$$6I_1 + 34I_2 = 12 - 40 = -28.$$

Solving the two equations, we get

$$I_1 = 1 \text{ A and } I_2 = -1 \text{ A}.$$

These two values agree with those obtained by the first method. This problem can also be solved using the *principle of superposition*.

EXAMPLE 2.10 *In the circuit (Fig. 2.22(a)), find the current delivered by the current source, such that the node voltage at point A is zero.*

SOLUTION FIRST METHOD: NODAL VOLTAGE METHOD

There are two nodes, A and B in the circuit (Fig. 2.22(a)). The voltage at node A is given as zero.

Applying Kirchhoff's current law at the two nodes, the equations are

$$V_A\left(\frac{1}{4}+\frac{1}{6}+\frac{1}{4}\right)-\frac{V_B}{4}=\frac{2\,V_A}{3}-\frac{V_B}{4}=\frac{16}{4}=4.$$

$$-\frac{V_A}{4}+V_B\left(\frac{1}{4}+\frac{1}{8}\right)=-\frac{V_A}{4}+\frac{3\,V_B}{8}=-I.$$

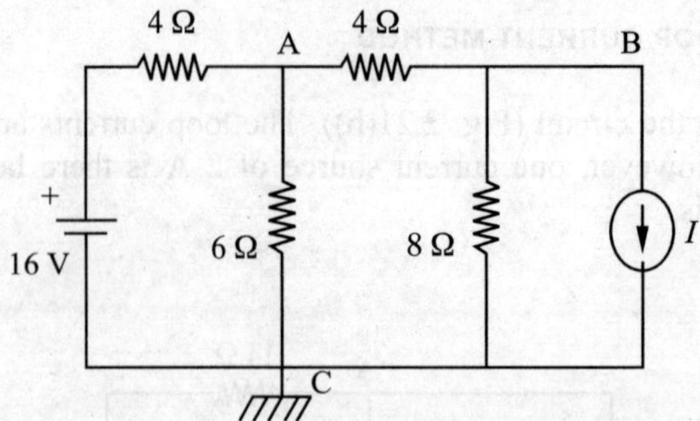

Fig. 2.22(a)

Substituting $V_A = 0$, in the first equation, we get

$$\frac{V_B}{4} = -4 \text{ or, } V_B = -16 \text{ V.}$$

Again, substituting both voltages in the second equation, the current, I in the current source is $I = \frac{3 \times 16}{8} = 6$ A (from C to B).

The current, $I_1 = I_{CA} = \frac{16}{4} = 4$ A.

The current, $I_2 = I_{AB} = \frac{16}{4} = 4$ A.

SECOND METHOD: LOOP CURRENT METHOD

There are three loops in the circuit (Fig. 2.22 (b)). The current in the third loop is $I_3 = I$.

Applying Kirchhoff's voltage law, the voltage equations are

$$(4+6)I_1 - 6I_2 = 10I_1 - 6I_2 = 16.$$

$$-6I_1 + (6+4+8)I_2 - 8I_3 = -6I_1 + 18I_2 - 8I_3 = 0.$$

The voltage at node A is

$$16 - 4I_1 = 0 \quad \text{or,} \quad I_1 = \frac{16}{4} = 4 \text{ A.}$$

Substituting $I_1 = 4$ A in the first equation, we get

$$I_2 = \frac{(4 \times 10) - 16}{6} = \frac{24}{6} = 4 \text{ A.}$$

Again, substituting both currents in the second equation, the current, I in the current source is

$$I = I_3 = \frac{-(6 \times 4) + (18 \times 4)}{8} = \frac{48}{8} = 6 \text{ A. (from C to B).}$$

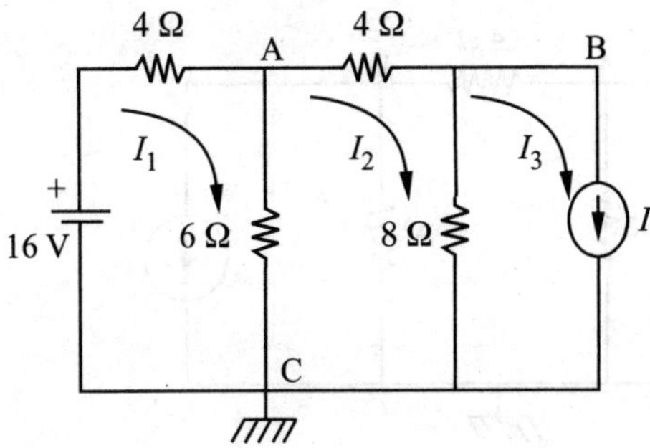

Fig. 2.22(b)

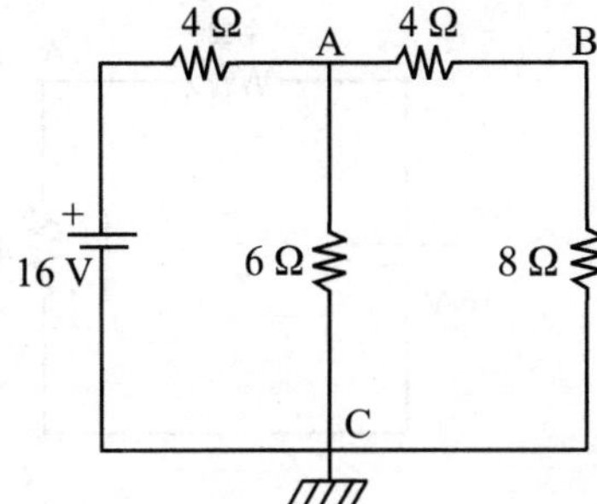

Fig. 2.22(c)

$$V_B = -(4 \times 4) = -16 \text{ V}.$$

This value is the same as the one obtained by the first method.

THIRD METHOD: PRINCIPLE OF SUPERPOSITION

The procedure is not quite simple in this case. There are two sources in the circuit (Fig. 2.22(a))—a voltage source and an ideal current source.

Firstly, let us take the voltage source (Fig. 2.22(c)), with the current source open circuited, as its internal resistance is very high (∞). Starting from RHS, the resistor, $R_{AC} = 4 + 8 = 12\ \Omega$. This resistor is in parallel with another resistor of 6 Ω (Fig. 2.22(d)). So,

$$R_{AC} = \frac{(6 \times 12)}{(6 + 12)} = \frac{72}{18} = 4\ \Omega.$$

The current drawn from the voltage source

$$I_{CA} = \frac{16}{(4 + 4)} = 2 \text{ A}.$$

The current in the resistor of 6 Ω (Fig. 2.22(c)),

$$I_{AC} = 2 \times \frac{12}{18} = \frac{4}{3} = 1.333 \text{ A}.$$

The voltages, $V_A = 6 \times \frac{4}{3} = 8$ V, and $V_B = 8 \times \frac{2}{3} = \frac{16}{3} = 5.333$ V.

Now, the current source (Fig. 2.22(e)) is taken up with the voltage source short circuited, as its internal resistance is zero. Again, starting from LHS, the resistance,

$$R_{AC} = \frac{(4 \times 6)}{(4 + 6)} = \frac{24}{10} = 2.4\ \Omega.$$

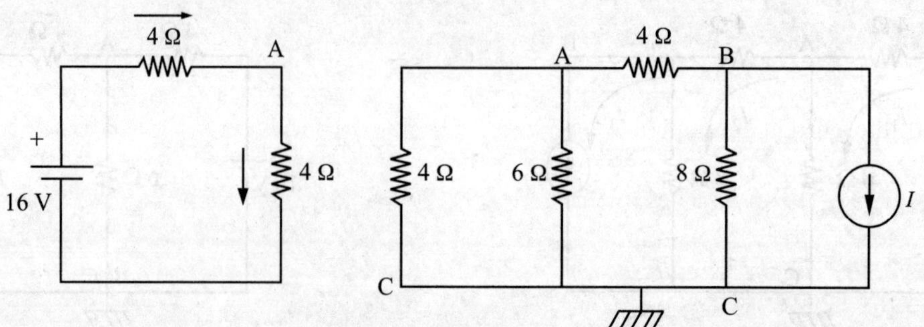

Fig. 2.22(d)

Fig. 2.22(e)

This is in series with the resistor of 4 Ω. So, $R_{AC} = (4+2.4) = 6.4\ \Omega$ (Fig. 2.22(f)). The current in the resistor of 8 Ω (Fig. 2.22(f)),

$$I_{CB} = I \times \frac{6.4}{(6.4+8.0)} = \frac{4}{9} \cdot I.$$

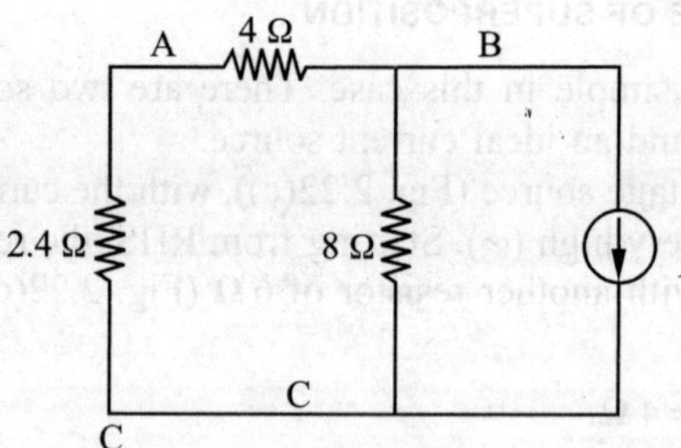

Fig. 2.22(f)

The current in the resistor of 4 Ω (Fig. 2.22(e)),

$$I_{CA} = \left(1 - \frac{4}{9}\right) \times \left(\frac{6}{10}\right) \times I = \frac{30}{90} \cdot I = \frac{1}{3} \cdot I.$$

The voltages, $V_A = -4 \times \frac{1}{3} \times I = -\frac{4}{3} \cdot I$, and $V_B = -8 \times \frac{4}{9} \times I = -\frac{32}{9} \cdot I$.

Using the principle of superposition,

$$V_A = 8 - \left(\frac{4}{3}\right) \cdot I = 0, \text{ or, the current, } I = 8 \times \frac{3}{4} = 6 \text{ A}.$$

$$V_B = \frac{16}{3} - \left(\frac{32}{9} \cdot I\right) = \frac{16}{3} - \left(\frac{32}{9} \times 6\right) = \frac{16}{3} - \frac{64}{3} = -16 \text{ V}.$$

The current, $I_1 = I_{CA} = 2 + \frac{6}{3} = 4$ A.

The current, $I_2 = I_{AB} = \frac{2}{3} + \left(\frac{5}{9} \times 6\right) = \left(\frac{2}{3} + \frac{10}{3}\right) = 4$ A.

As $V_A = 0$, the current through 6 Ω $= I_{AC} = 0$. I_{AC} is found as $I_{AC} = \frac{4}{3} - \left(\frac{2}{9} \cdot I\right) = 0$, from which $I = \left(\frac{4}{3} \times \frac{9}{2}\right) = 6$ A. These values are the same as found earlier.

EXAMPLE 2.11 *In the circuit shown in Fig. 2.23(a), determine the current in the load resistance,* $R_L = 200\ \Omega$.

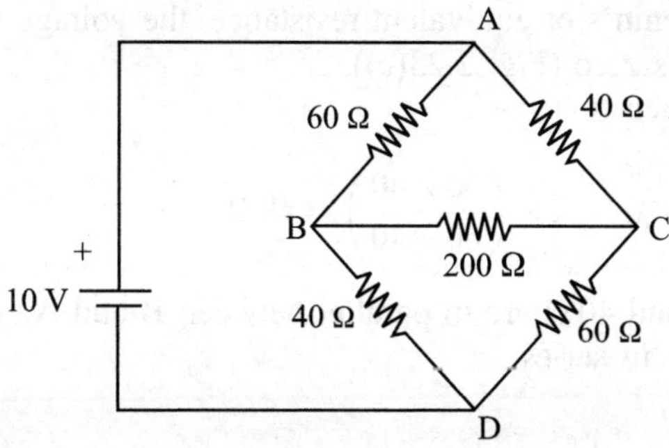

Fig. 2.23(a)

SOLUTION Please note that the circuit (Fig. 2.23(a)) is the same as the one in Fig. 2.7(a) with the internal resistance of the dc source, 10 Ω omitted from the circuit. This makes the problem simpler to solve using Thevenin's theorem.

Let us find the open-circuit voltage, V_{BC} (Fig. 2.23(b)), with the load resistance removed. Two resistances of $(60 + 40 = 100)\ \Omega$ and $(40 + 60 = 100)\ \Omega$ are in parallel between the points, A and D. Hence, the resistance is

$$R_{AD} = \frac{(100 \times 100)}{(100 + 100)} = \frac{10,000}{200} = 50\ \Omega$$

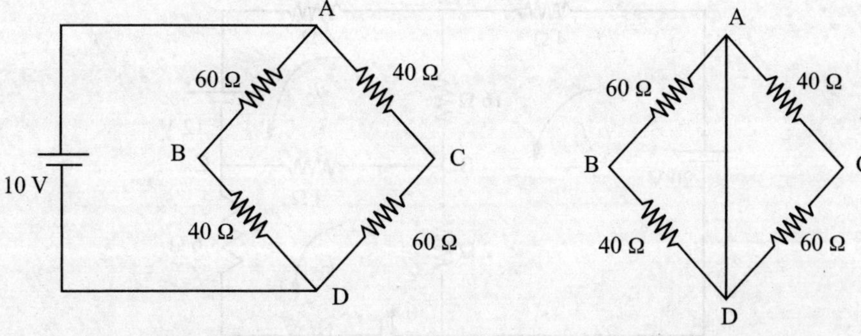

Fig. 2.23(b) **Fig. 2.23(c)**

The current drawn from the source $= \frac{10}{50} = 0.2$ A.

The current through two parallel paths is the same, and is $I = 0.2 \times \frac{100}{200} = 0.1$ A.

This current in parallel paths can be simply calculated as

$$I = \frac{10}{(60+40)} = \frac{10}{100} = 0.1 \text{ A}.$$

The open-circuit voltage $= V_{BC} = V_{BA} + V_{AC} = 0.1 \times (-60+40) = -2.0$ V.

For determining Thevenin's or equivalent resistance, the voltage source is short-circuited, as its internal resistance is zero (Fig. 2.23(c)).

The equivalent resistance is

$$R_{EQ} = R_{BC} = R_{BA} + R_{AC} = 2 \times \left(\frac{60 \times 40}{60+40}\right) = 48\ \Omega,$$

as two resistors of 60 Ω and 40 Ω are in parallel between B and A, and the same is between A and C, though they are in series.

The load current is

$$I_L = I_{BC} = \frac{V_{OC}}{R_{EQ}+R_L} = -\frac{2}{48+200} = -\frac{2}{248} = -\frac{1}{124} \approx -8.0 \quad \text{mA}.$$

Please note that the load current flows from C to B, as its sign is negative.
If the same circuit (Fig. 2.7(a)) has been used in this case, the delta–star conversion formula is needed to find the equivalent or Thevenin's resistance. For finding the open-circuit voltage, the same procedure is followed.

EXAMPLE 2.12 *Determine the current in the resistor of 2 Ω, in the circuit shown in Fig. 2.24(a). Also find the power input/output to the current and voltage sources.*

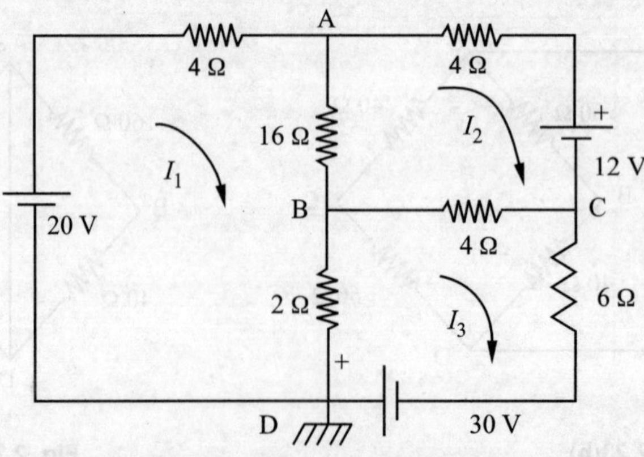

Fig. 2.24(a)

SOLUTION Please note that there are two real (not ideal) current sources (LHS and RHS) in the circuit (Fig. 2.24(a)). These have to be converted into real voltage sources. The procedure is briefly explained. For the current source (LHS), the source voltage is $(5 \times 4) = 20$ V, which is in series with the resistor of 4 Ω. Similarly, for the current source (RHS), the source voltage is $(3 \times 4) = 12$ V, which is in series with the resistor of 4 Ω.

FIRST METHOD: LOOP CURRENT METHOD

There are three loops in the modified circuit (Fig. 2.24(b)). Applying Kirchhoff's voltage law, the equations are

$$(4+16+2)I_1 - 16I_2 - 2I_3 = 22I_1 - 16I_2 - 2I_3 = 20.$$

$$-16I_1 + 24I_2 - 4I_3 = -12.$$

$$-2I_1 - 4I_2 + 12I_3 = 30.$$

In matrix form, the equations are

$$\begin{bmatrix} 22 & -16 & -2 \\ -16 & 24 & -4 \\ -2 & -4 & 12 \end{bmatrix} \begin{bmatrix} I_1 \\ I_2 \\ I_3 \end{bmatrix} = \begin{bmatrix} 20 \\ -12 \\ 30 \end{bmatrix}$$

Note that the resistance matrix is symmetric.

Solving the above equations, the loop currents are

$$I_1 = I_{DA} = 2.5 \text{ A}, \qquad I_2 = I_{AC} = 1.75 \text{ A}, \qquad \text{and } I_3 = I_{CD} = 3.5 \text{ A}.$$

These currents are either drawn from, or fed to the voltage sources (batteries).

The current in the resistor of 2 Ω $= I_{DB} = I_3 - I_1 = I_{CD} - I_{DA} = 3.5 - 2.5 = 1.0$ A.

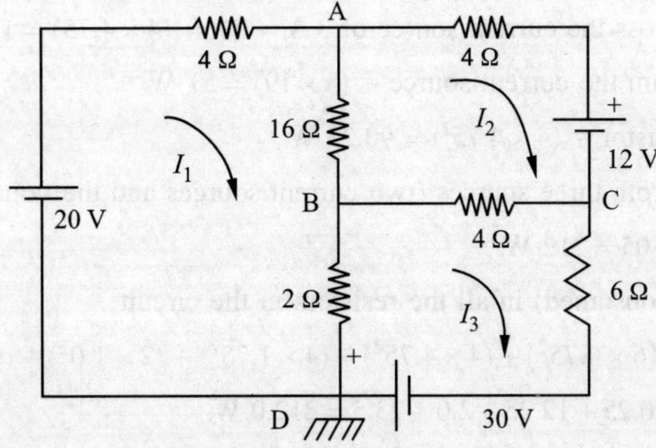

Fig. 2.24(b)

Other two branch currents are

$$I_{AB} = I_1 - I_2 = I_{DA} - I_{AC} = 2.5 - 1.75 = 0.75 \text{ A (in the 16 } \Omega \text{ resistor).}$$

$$I_{BC} = I_3 - I_2 = I_{CD} - I_{AC} = 3.5 - 1.75 = 1.75 \text{ A (in the 4 } \Omega \text{ resistor).}$$

The sum of the three branch currents, taken algebraically (with sign), is zero, thereby satisfying Kirchhoff's current law at node B.

Now, let us compute the power drawn from, or fed to the voltage sources in the circuit (Fig. 2.24(b)). The power drawn from, or fed to two current sources in the equivalent circuit (Fig. 2.24(a)) will be computed later.

The power drawn from the 20 V supply $= (20 \times 2.5) = 50$ W.

The power fed to the 12 V supply $= (12 \times 1.75) = 21$ W.

The power drawn from the 30 V supply $= (30 \times 3.5) = 105$ W.

Total power drawn from the three voltage sources $= 50 - 21 + 105 = 134$ W.

The power loss (or consumed) in all resistances in the circuit

$$= (4 \times 2.5^2) + (16 \times 0.75^2) + (4 \times 1.75^2) + (4 \times 1.75^2) + (2 \times 1.0^2) + (6 \times 3.5^2)$$

$$= 25.0 + 9.0 + 12.25 + 12.25 + 2.0 + 73.5 = 134.0 \text{ W},$$

which is the same as the power drawn from three voltage sources. Reverting to the circuit shown in Fig. 2.24(a)),

the current in the resistor of $4\,\Omega = (5.0 - 2.5) = 2.5$ A.

The voltage drop across the current source of 5 A $= V_{EC} = (4 \times 2.5) = 10$ V.

The power drawn from the current source $= (5 \times 10) = 50$ W.

The power in the resistor $= (4 \times 2.5^2) = 25$ W.

The current in the resistor of $4\,\Omega = (3.0 + 1.75) = 4.75$ A.

The voltage drop across the current source of 3 A $= V_{EC} = (4 \times 4.75) = 19$ V.

The power drawn from the current source $= (3 \times 19) = 57$ W.

The power in the resistor $= (4 \times 4.75^2) = 90.25$ W.

Total power drawn from three sources (two current sources and the voltage one)

$$= 50 + 57 + 105 = 212 \text{ W}.$$

The power loss (or consumed) in all the resistors in the circuit

$$= (4 \times 2.5^2) + (16 \times 0.75^2) + (4 \times 4.75^2) + (4 \times 1.75^2) + (2 \times 1.0^2) + (6 \times 3.5^2)$$

$$= 25.0 + 9.0 + 90.25 + 12.25 + 2.0 + 73.5 = 212.0 \text{ W},$$

which is the same as the power drawn from the three sources as stated above.

SECOND METHOD: NODAL VOLTAGE METHOD

There are three nodes—A, B and C, in the circuit (Fig. 2.24(c)). The same equations in a different form can be obtained from the circuit (Fig. 2.24(a)).

Applying Kirchhoff's current law at the nodes, the equations are

$$\frac{V_A-20}{4}+\frac{V_A-V_B}{16}+\frac{V_A-V_C-12}{4}=V_A\left(\frac{1}{4}+\frac{1}{16}+\frac{1}{4}\right)-\frac{V_B}{16}-\frac{V_C}{4}$$

$$=\frac{20+12}{4}=\frac{32}{4}=8$$

or, $\frac{9}{16}V_A-\frac{V_B}{16}-\frac{V_C}{4}=8$

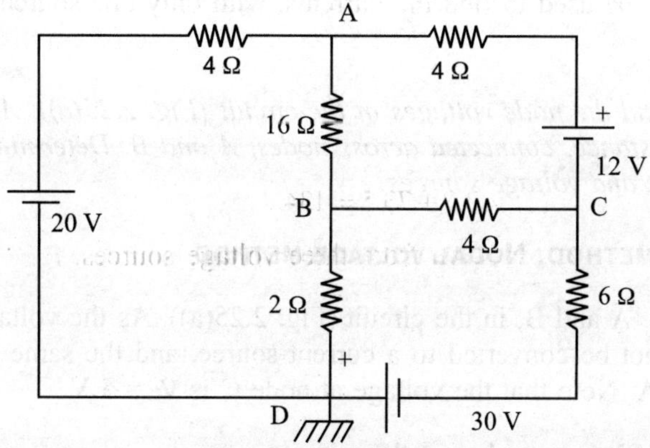

Fig. 2.24(c)

or, $9V_A-V_B-4V_C=128.$

$$-\frac{V_A}{16}+V_B\left(\frac{1}{2}+\frac{1}{16}+\frac{1}{4}\right)-\frac{V_C}{4}=-\frac{V_A}{16}+\frac{13}{16}V_B-\frac{V_C}{4}=0$$

or, $-V_A+13V_B-4V_C=0.$

$$-\frac{V_A}{4}-\frac{V_B}{4}+V_C\left(\frac{1}{4}+\frac{1}{4}+\frac{1}{6}\right)=-\frac{V_A}{4}-\frac{V_B}{4}+\frac{2}{3}V_C=-\left(\frac{12}{4}+\frac{30}{6}\right)=-(3+5)=-8.$$

or, $-3V_A-3V_B+8V_C=-96.$

In matrix form, the equations are

$$\begin{bmatrix} \frac{9}{16} & -\frac{1}{16} & -\frac{1}{4} \\ -\frac{1}{16} & \frac{13}{16} & -\frac{1}{4} \\ -\frac{1}{4} & -\frac{1}{4} & \frac{2}{3} \end{bmatrix}\begin{bmatrix} V_A \\ V_B \\ V_C \end{bmatrix}=\begin{bmatrix} 8 \\ 0 \\ -8 \end{bmatrix}.$$

Note that the conductance matrix is symmetric.

Solving the above equations, the node voltages are

$V_A = 10$ V, $V_B = -2$ V, and $V_C = -9$ V.

The loop currents are $I_1 = I_{DA} = (20 - 10)/4 = 2.5$ A,

$I_2 = I_{AC} = (10 + 9 - 12)/4 = 1.75$ A, and $I_3 = I_{CD} = (30 - 9)/6 = 3.5$ A.

These currents are the same as obtained earlier. Other branch currents can also be computed.

Also please note that in the two methods used here, the currents or the voltages, as the case may be, are computed using determinants of the order 3.

The problem can also be solved using the *principle of superposition*. Note that there are *three* sources, either current or voltage, in the circuits. In some cases, delta–star conversion formula may have to be used to find the currents, with only one source (voltage/current, as the case may be).

EXAMPLE 2.13 *Find the node voltages in the circuit (Fig. 2.25(a)). Also find the current through the 5 Ω resistance, connected across nodes, A and B. Determine the power input/output to the current and voltage sources*

SOLUTION FIRST METHOD: NODAL VOLTAGE METHOD

There are two nodes, A and B, in the circuit (Fig. 2.25(a)). As the voltage source of 5 V is an ideal one, it cannot be converted to a current source, and the same is true for the ideal current source of 1 A. Note that the voltage at node C is $V_C = 5$ V.

Applying Kirchhoff's current law at the nodes,

$$\frac{V_A - V_C}{5} + \frac{V_A - V_B}{7} = V_A\left(\frac{1}{5} + \frac{1}{7}\right) - \frac{V_B}{7} - \frac{5}{5} = 2$$

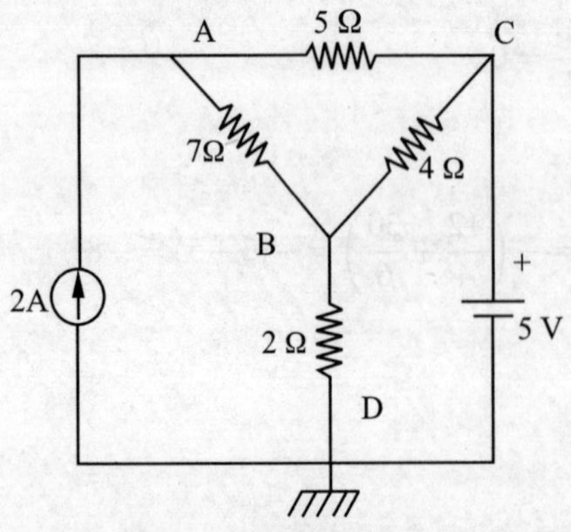

Fig. 2.25(a)

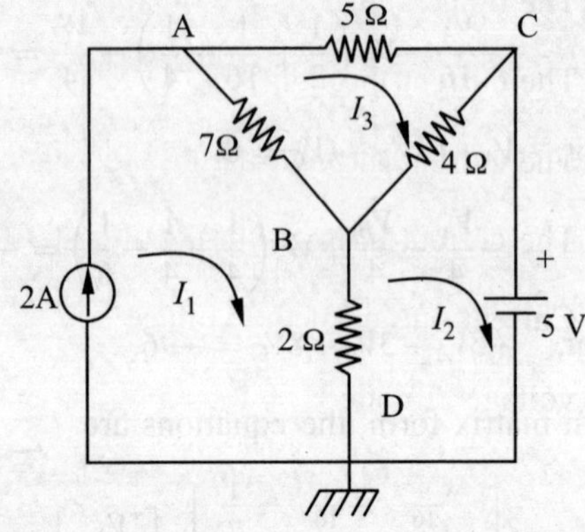

Fig. 2.25(b)

or, $\frac{12}{35}V_A - \frac{V_B}{7} = 2+1 = 3$

or, $12V_A - 5V_B = 105.$

$$-\frac{V_A}{7} + V_B\left(\frac{1}{7}+\frac{1}{2}+\frac{1}{4}\right) - \frac{V_C}{4} = 0$$

or, $-\frac{V_A}{7} + \frac{25}{28}V_B = \frac{5}{4}$

or, $-4V_A + 25V_B = 35.$

Solving the two equations, we get,

$V_A = 10$ V and $V_B = 3$ V.

Firstly, all loop currents, and then branch currents, can be easily obtained.

SECOND METHOD: LOOP CURRENT METHOD

There are three loops in the circuit (Fig. 2.25(b), of which the first one has a current source. So, $I_1 = 2$ A. Applying Kirchhoff's voltage in the other two loops, the voltage equations are

$-2I_1 + 6I_2 - 4I_3 = -5$ or, $6I_2 - 4I_3 = -5+4 = -1.$

$-7I_1 - 4I_2 + 16I_3 = 0$

or, $-4I_2 + 16I_3 = 14.$

Solving the two equations, we get,

$I_2 = 0.5$ A and $I_3 = 1.0$ A.

The current through the 5 Ω resistance $= I_3 = I_{AC} = 1.0$ A.

The branch currents are obtained as,

The current through the 7 Ω resistance $= I_{AB} = I_1 - I_3 = 2.0 - 1.0 = 1.0$ A.

The current through the 4 Ω resistance $= I_{CB} = I_3 - I_2 = 1.0 - 0.5 = 0.5$ A.

The current through the 2 Ω resistance $= I_{BD} = I_1 - I_2 = 2.0 - 0.5 = 1.5$ A.

Current output of the voltage source $= I_2 = I_{CD} = 0.5$ A.

voltage at node C $= V_C = 5$ V.

$V_{AC} = V_A - V_C = 5I_3 = 5 \times 1 = 5$ V.

voltage at node A $= V_A = V_{AC} + V_C = 5+5 = 10$ V.

Voltage at node B = $V_B = 2I_{BD} = 2.0 \times 1.5 = 3$ V.

The values are the same as obtained earlier.

Voltage at node C = $V_C = 5$ V.

$$V_{AC} = V_A - V_C = 5I_3 = 5 \times 1 = 5 \text{ V}.$$

Voltage at node A = $V_A = V_{AC} + V_C = 5 + 5 = 10$ V.

Voltage at node B = $V_B = 2I_{BD} = 2.0 \times 1.5 = 3$ V.

The values are the same as obtained earlier.

Now, let us compute the power drawn by the two sources, and also the power loss (or consumed) in all the resistors in the circuit.

The voltage drop across the current source = $V_A = 10$ V.

Power drawn from the current source = $10 \times 2 = 20$ W.

The current fed by the voltage source to the circuit = $-I_2 = I_{DC} = -0.5$ A.

Power drawn from voltage source = $-5.0 \times 0.5 = -2.5$ W.

Power drawn by the circuit from the two sources – the current source and the voltage one

$$= 20.0 - 2.5 = 17.5 \text{ W}$$

Power loss (or consumed) in all resistors in the circuit

$$= (5 \times 1.0^2) + (7 \times 1^2) + (4 \times 0.5^2) + (2 \times 1.5^2)$$

$$= 5.0 + 7.0 + 1.0 + 4.5 = 17.5 \text{ W},$$

which is equal to the power drawn from the two sources.

Now, this problem is solved using the delta–star conversion. The delta-connected circuit, ABC is converted into its star equivalent, with the star point assumed as P. The modified circuit is shown in Fig. 2.25(c).

$$\Sigma R = R_{AB} + R_{BC} + R_{CA} = 7 + 4 + 5 = 16\ \Omega.$$

$$R_{AP} = \frac{R_{AB} \cdot R_{CA}}{(R_{AB} + R_{BC} + R_{CA})} = \frac{7 \times 5}{\Sigma R} = \frac{35}{16} = 2.1875 \approx 2.19\ \Omega.$$

$$R_{BP} = \frac{R_{BC} \cdot R_{AB}}{\Sigma R} = \frac{4 \times 7}{16} = \frac{7}{4} = 1.75\ \Omega.$$

$$R_{CP} = \frac{R_{CA} \cdot R_{BC}}{\Sigma R} = \frac{5 \times 4}{16} = \frac{5}{4} = 1.25\ \Omega.$$

$$R_{PBD} = R_{BP} + R_{PD} = \frac{7}{4} + 2 = \frac{15}{4} = 3.75\ \Omega.$$

The currents in the circuit (Fig. 2.25(c)) are obtained using the principle of superposition. There are two sources—one current source and another voltage source.

First, let us take the current source of 2 A (Fig. 2.25(d)), with the voltage source short circuited. The current of 2 A in the resistance ($R_{AP} \approx 2.19\ \Omega$) drawn from the current source is divided into two parallel paths, PBD and PCD.

The total resistance in the path, PD $= R_{PD} = R_{PBD} + R_{PCD} = 3.75 + 1.25 = 5.0\ \Omega$

The current in $R_{PC} = 1.25\ \Omega$, and the battery path, i.e., the voltage source, is

$$I'_{PC} = I \cdot \frac{R_{PCD}}{R_{PD}} = 2.0 \times \frac{3.75}{5.0} = \frac{3}{2} = 1.5 \text{ A.}$$

The current in $R_{PBD} = 3.75\ \Omega$, is

$$I'_{PBD} = I - I'_{PC} = 2.0 - 1.5 = 0.5 = \frac{1}{2} \text{ A.}$$

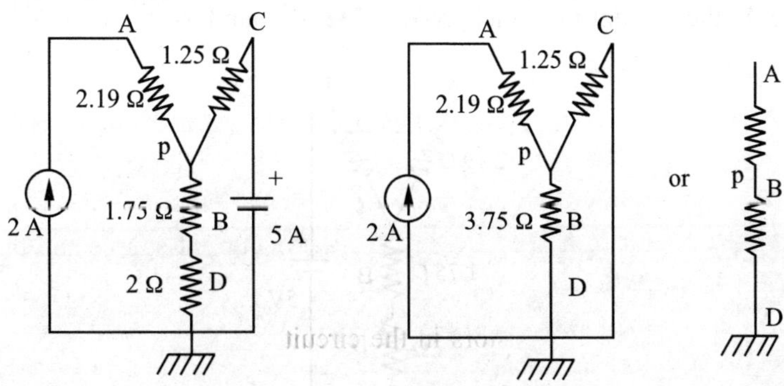

Fig. 2.25(c) Fig. 2.25(d)

Now, the voltage source of 5 V is there in the circuit, with the current source open circuited (Fig. 2.25(e)). The current in the path, CPBD, is

$$I''_{CP} = I''_{PBD} = \frac{V}{R_{PD}} = \frac{5}{5} = 1.0 \text{ A.}$$

As the current source is open circuited, the current in $R_{AP} \approx 2.19\ \Omega$, is $I''_{AP} = 0.0$ (A). Using the superposition principle, all currents in the modified circuit (Fig. 2.25(c)) are obtained as given here.

The current in $R_{AP} \approx 2.19\ \Omega$, is $I_{AP} = I'_{AP} + I''_{AP} = 2 + 0 = 2.0$ A, as $I = I'_{AP} = 2$ A.

The current in $R_{CP} = 1.25\ \Omega$, which is also fed to the battery or voltage source, is,

$$I_{PC} = I'_{PC} + I''_{PC} = 1.5 - 1.0 = 0.5 \text{ A.}$$

The current in $R_{PD} = 3.75\ \Omega$, is, $I_{PBD} = I'_{PBD} + I''_{BPD} = 0.5 + 1.0 = 1.5$ A.

Reverting to the original circuit (Fig. 2.25(a)), the currents in three branches, i.e., in the current source, which is the same (2 A), fed to the battery or voltage source, and in the resistance, R_{BD}, are 2.0 A, 0.5 A and 1.5 A respectively. Please note that the three currents satisfy Kirchhoff's current law at the node D. The other branch currents are obtained as follows.

To find the current in the branch $R_{AC} = 5\ \Omega$, the voltage drop across this branch is to be computed. For this, let us take the modified circuit (Fig. 2.25(e)).

$$V_{AC} = V_{AP} + V_{PC} = \left[\left(\frac{35}{16} \times 2\right) + \left(\frac{5}{4} \times \frac{1}{2}\right)\right] = \frac{35}{8} + \frac{5}{8} = 5\text{ V.}$$

$$\text{So,} \quad I_{AC} = \frac{V_{AC}}{R_{AC}} = \frac{5}{5} = 1.0\text{ A.}$$

The current in the other two branches can also be computed by this procedure. However, we will use Kirchhoff's current law at the nodes, A and C (or B). Applying Kirchhoff's current law at the node A, the current in branch, $R_{AB} = 7\ \Omega$, is found as

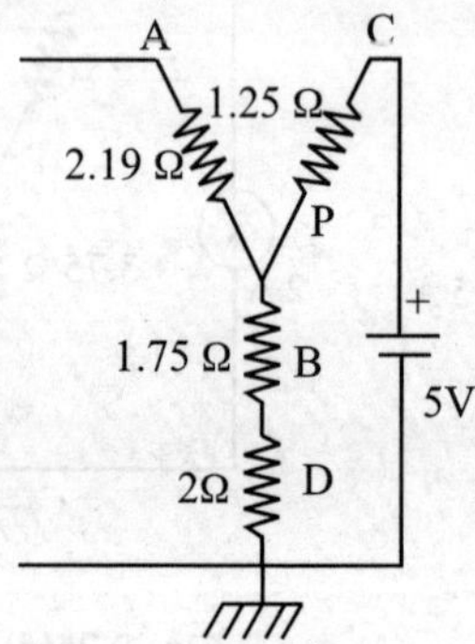

Fig. 2.25(e)

$$I_{AB} = I_{DA} - I_{AC} = 2 - 1 = 1.0\text{ A.}$$

Similarly, at the node C, the current in branch, $R_{BC} = 4\ \Omega$, is found as,

$$I_{CB} = I_{AC} - I_{CD} = 1.0 - 0.5 = 0.5\text{ A.}$$

The values of the current in the branches at the node B, can be easily checked to see if Kirchhoff's current law is satisfied there.

All currents are the same as obtained earlier.

EXAMPLE 2.14 *Find all node voltages and all branch currents in the circuit (Fig. 2.26). Also determine the power input/ output to the current sources.*

SOLUTION FIRST METHOD: NODAL VOLTAGE METHOD

There are three nodes—A, B and C, in the circuit (Fig. 2.26), with the node D taken as the reference node.

Applying Kirchhoff's current law, the current equations are,

$$(V_A - V_B)/1 + (V_A - V_C)/3 = [(4V_A)/3] - V_B - (V_C/3) = 3$$

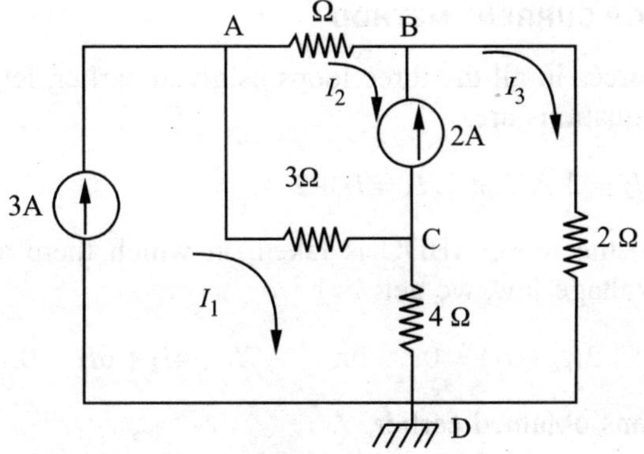

Fig. 2.26

or, $4V_A - 3V_B - V_C = 9.$

$$-V_A + [(3V_B)/2] = 2$$

or, $-2V_A + 3V_B = 4.$

$$-(V_A/3) + [(7V_C/12] = -2$$

or, $-4V_A + 7V_C = -24.$

Solving the equations, we get, $V_A = 6.7$ V, $V_B = 5.8$ V and $V_C = 0.4$ V.

There are three loops in the circuit (Fig. 2.26). The loop and also branch currents are obtained here. The loop currents are assumed to be in the clockwise direction.

The current through the resistor of 1 Ω

$$= I_2 = I_{AB} = \frac{V_{AB}}{R_{AB}} = \frac{V_A - V_B}{1} = (6.7 - 5.8) = 0.9 \text{ A.}$$

The current through the resistor of 2 Ω

$$= I_3 = I_{BD} = \frac{V_{BD}}{R_{BD}} = \frac{V_B}{2} = \frac{5.8}{2} = 2.9 \text{ A.}$$

The current through the resistor of 3 Ω

$$= I_{AC} = \frac{V_{AC}}{R_{AC}} = \frac{V_A - V_C}{3} = \frac{6.7 - 0.4}{3} = \frac{6.3}{3} = 2.1 \text{ A.}$$

The current through the resistor of 4 Ω

$$= I_{CD} = \frac{V_{CD}}{R_{CD}} = \frac{V_C}{4} = \frac{0.4}{4} = 0.1 \text{ A.}$$

It can easily be shown that Kirchhoff's current law at all three nodes is satisfied.

SECOND METHOD: LOOP CURRENT METHOD

As there are current sources in all the three loops as given earlier, let us find the currents in this way. The current equations are,

$$I_1 = 3\text{ A},\ I_3 - I_2 = 2\text{ A} \quad \text{or,} \quad I_3 = I_2 + 2.$$

For the other equation, the loop, ABDC is taken, in which there are no current sources. Applying Kirchhoff's voltage law, we get,

$$I_2 + 2I_3 + 4(I_3 - I_1) + 3(I_2 - I_1) = 0 \qquad \text{or,} \qquad -7I_1 + 4I_2 + 6I_3 = 0.$$

Substituting the equations obtained earlier,

$$-(7\times 3) + 4I_2 + 6(I_2 + 2) = 0$$

or, $10I_2 = 21 - 12 = 9$ or, $I_2 = 9/10 = 0.9$ A.

$$I_3 = 2.0 + 0.9 = 2.9\text{ A}.$$

The loop currents are the same as found earlier. The branch currents can be easily found from the loop currents. The node voltages can easily be obtained from loop and branch currents.

The voltage drop across the current source of 3 A $= V_A = 6.7$ V.

The power drawn from this source $= 3 \times 6.7 = 20.1$ W.

The voltage drop across the current source of 2 A $= V_{BC} = V_B - V_C = 5.8 - 0.4 = 5.4$ V.

The power drawn from this source $= 2 \times 5.4 = 10.8$ W.

Total power drawn from two current sources $= 20.1 + 10.8 = 30.9$ W.

The power loss in all resistances $= (1\times 0.9^2) + (2\times 2.9^2) + (3\times 2.1^2) + (4\times 0.1^2)$

$$= 0.81 + 16.82 + 13.23 + 0.04 = 30.9\text{ W},$$

which is the same as the power drawn by the two sources. The problem can also be solved by the *principle of superposition.*

EXAMPLE 2.15 *A circuit (Fig. 2.27) contains a voltage-dependent voltage source. Find the current supplied by the battery, and the power supplied by the voltage source.*

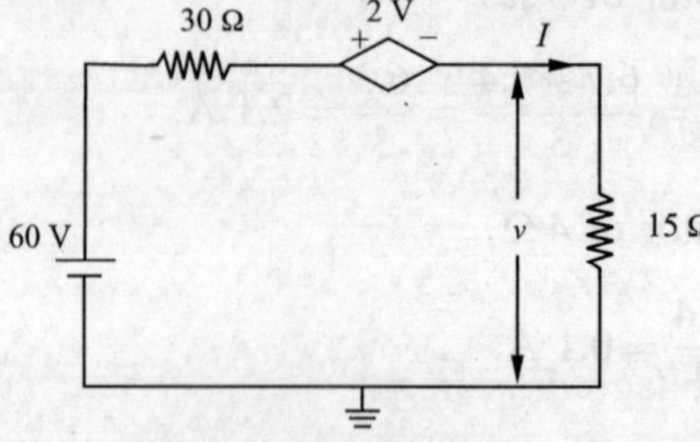

Fig. 2.27

SOLUTION Note that the voltage drop in the resistance of 15 Ω in the circuit (Fig. 2.27) is v (V).

Assuming that the current in the circuit is I in the clockwise direction, the voltage equation is $(30+15)I+2v=60$, where $v=15I$.

Substituting the second equation in the first one and simplifying, we get,

$$45I+2\times 15I=75I=60$$

or, $$I=\frac{60}{75}=0.8\text{ A}.$$

This is the current supplied by the battery.

Also note that the battery is an independent voltage source, while the other voltage source is voltage-dependent.

From the second equation, $v=15.0\times 0.8=12$ V.

The voltage of the voltage-dependent voltage source is $2\times 12=24$ V.

The power fed to it $=24.0\times 0.8=19.2$ W (Note its sign).

EXAMPLE 2.16 *Determine the voltage drop, v_2, across the current-controlled voltage source in the circuit shown in Fig. 2.28.*

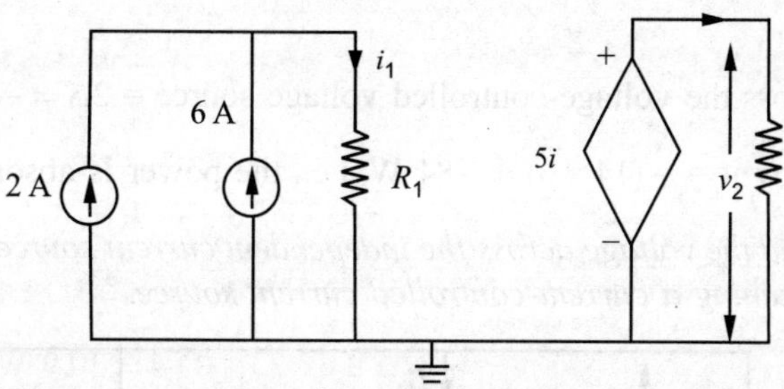

Fig. 2.28

SOLUTION Note that the current in the left-hand side of the circuit (Fig. 2.28) is i_1.

Applying Kirchhoff's current law at the node A, we get, $2+6=8=i_1$.

From the right-hand side of the circuit, the voltage drop across the current-controlled voltage source $=v_2=5i_1=5\times 8=40$ V.

EXAMPLE 2.17 *What is the power supplied by the voltage-controlled voltage source in the circuit (Fig. 2.29)?*

SOLUTION Note that the voltage drop in the resistance of 1 Ω in the circuit (Fig. 2.29) is $2v_0$ (V). Also there are two loops in the circuit, along with an independent current source of

1 A in loop 1 (right-hand side), such that $I_1 = -1$ A, assuming that the loop currents flow in the clockwise direction.

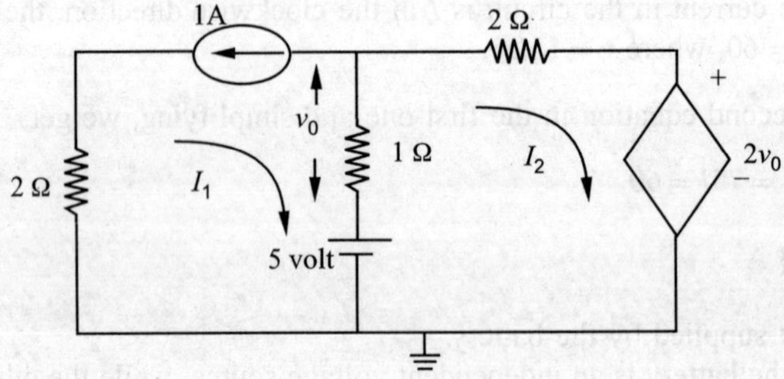

Fig. 2.29

Applying Kirchhoff's voltage law in loop 2 (LHS), we get,

$$-I_1 + (1+2)I_2 = 1 + 3I_2 = 5 - 2v_0, \text{ where } v_0 = (I_1 - I_2) = -(1 + I_2).$$

Substituting the second equation in the first one and simplifying, we get,

$$1 + 3I_2 = 5 + 2(1 + I_2) = 7 + 2I_2$$

or, $I_2 = 7 - 1 = 6$ A.

The voltage drop across the voltage-controlled voltage source $= 2v_0 = -2 \times (1+6) = -14$ V.

The power supplied by it $= -(14 \times 6) = -84$ W, i.e., the power is absorbed by the source.

EXAMPLE 2.18 *Find the voltage across the independent current source of 4 A in the circuit (Fig. 2.30) also containing a current-controlled current source.*

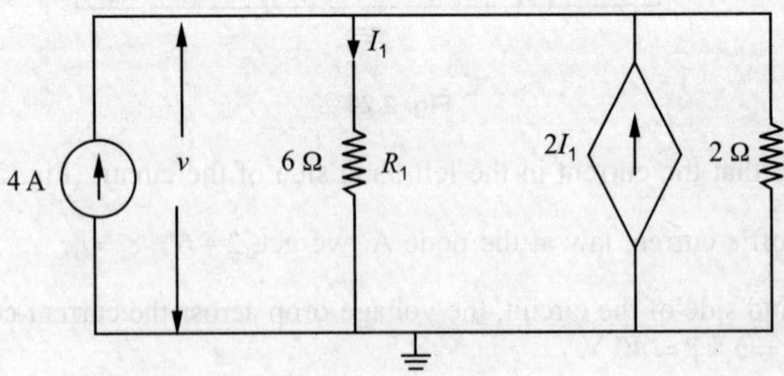

Fig. 2.30

SOLUTION Note that there is a current-controlled current source ($2I_1$) in the circuit (Fig. 2.30).

Applying Kirchhoff's current law at the node A, we get,

$$-I_1 + 2I_1 - \frac{v}{2} = I_1 - \frac{v}{2} = -4.$$

The current in the resistor of 6 Ω is $I_1 = \frac{v}{6}$.

Substituting the second equation in the first one, $\frac{v}{6} - \frac{v}{2} = -\frac{v}{3} = -4.$

The voltage across the independent current source of 4 A = v = 12 V.

The current in the current-controlled voltage source $= 2I_1 = \frac{2 \cdot v}{6} = \frac{v}{3} = 4$ A.

EXAMPLE 2.19 *Find the current in the resistor of 6 Ω in the circuit (Fig. 2.31) containing a voltage-controlled voltage source.*

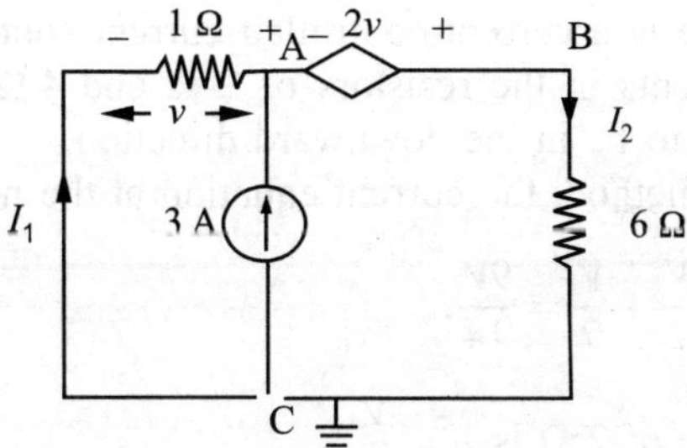

Fig. 2.31

SOLUTION Note that there is a voltage-controlled voltage source ($2v$) in the circuit (Fig. 2.31), where v is the voltage drop across the resistance of 1 Ω. Also there are two loops in the circuit, with an independent current source between the loops. The loop currents are assumed to be in the clockwise direction.

Applying Kirchoff's voltage law around two loops in the circuit, we get,

$$-v - 2v + 6I_2 = -3v + 6I_2 = 0$$

or, $I_2 = \frac{3v}{6} = \frac{v}{2} = 0.5 \cdot v.$

The current in the resistance of 1 Ω is $I_1 = -v$. Applying Kirchoff's current law at node A,

$$I_1 + 3 = I_2$$

or, $-v - 0.5v = -1.5v = \frac{3v}{2} = -3$

or, $v = -\frac{3 \cdot 2}{2} = -2.$

The current in the resistance of 6 Ω $= I_2 = -\frac{2}{2} = -1$ A, with the current being in the upward direction.

EXAMPLE 2.20 *Find the current in the resistor of 4 Ω in the circuit (Fig. 2.32) containing a current-controlled current source.*

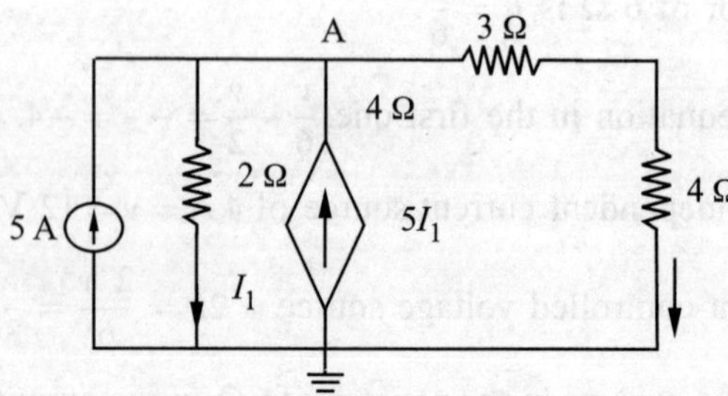

Fig. 2.32

SOLUTION Note that there is a current-controlled current source ($5I_1$) in the circuit (Fig. 2.32). The two branch currents in the resistors of 2 Ω and 4 Ω are I_1 and I_2 respectively. These currents are assumed to be in the downward direction.

Applying nodal voltage method, the current equation at the node is given by,

$$5+5I_1=\frac{V}{2}+\frac{V}{3+4}=\frac{V}{2}+\frac{V}{7}=\frac{9V}{14}.$$

The current in the resistance of 2 Ω is $I_1=\frac{V}{2}$.

Substituting the second equation in the first one, we get,

$$5+\frac{5V}{2}=\frac{9V}{14}.$$

or, $$\frac{5V}{2}-\frac{9V}{14}=\frac{13V}{7}=-5.$$

or, $$V=-\frac{35}{13}=-2.6923 \text{ V}.$$

The current in the resistor of 4 Ω is

$$I_2=\frac{V}{3+4}=\frac{V}{7}=-\frac{5}{13}=-0.3846 \text{ A},$$

with the current being in the upward direction.

EXAMPLE 2.21 *Find the loop currents in the circuit (Fig. 2.33) containing a current-controlled voltage source.*

SOLUTION Note that there is a current-controlled voltage source ($2I_1$) in the circuit (Fig. 2.33). There are two loops, with the currents assumed to flow in the clock-wise direction.

For the circuit shown in Fig. 2.33, find I_1 and I_2 using network analysis.

Mesh KVL (Kirchhoff's voltage law) equation for Loop 1 is $-5+2I_1+4(I_1-I_2)+2I_1=0$

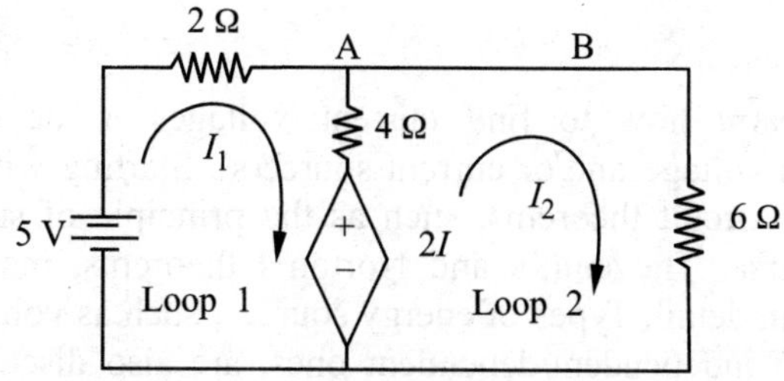

Fig. 2.33

or, $\quad 8I_1 - 4I_2 = 5.$ (i)

Mesh KVL equation for Loop 2 is: $4(I_2 - I_1) + 6I_2 - 2I_1$

or $\quad 10I_2 - 6I_1 = 0.$ (ii)

Now solving Eq. (i) and (ii), $I_1 = \frac{25}{28}$ A; $I_2 = \frac{15}{28}$ A.

Two loop currents are found as,

$$I_1 = \frac{25}{28} = 0.893 \text{ A and } I_2 = \frac{15}{28} = 0.536 \text{ A.}$$

EXAMPLE 2.22 *Find the current in the circuit (Fig. 2.34) containing a current-controlled voltage source.*

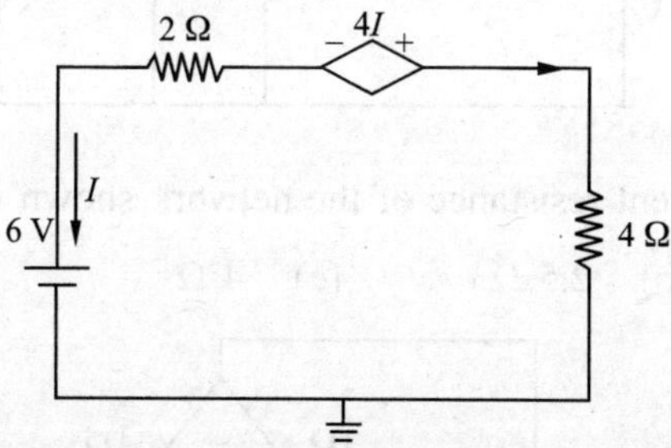

Fig. 2.34

SOLUTION Note that there is a current-controlled voltage source (4*I*) in the circuit (Fig. 2.34). There is only one loop, with the current assumed to flow in the clock-wise direction.

Applying Kirchhoff's voltage law in the loop, we get,

$$(2+4)I - 4I = 6I - 4I = 2I = 6, \text{ or } I = \frac{6}{2} = 3.0 \text{ A.}$$

The voltage drop across the current-controlled voltage source = $4 \times 3 = 12$ V, with the voltage being negative according to the polarity shown.

The voltage drop across the resistor of $2\ \Omega = 2 \times 3 = 6$ V.

The voltage drop across the resistor of $4\ \Omega = 4 \times 3 = 12$ V.

Conclusion

In this chapter, we learnt how to find currents/voltages in dc circuits with resistive elements only, fed from voltage and/or current source(s). Starting with the loop current and nodal voltage methods, circuit theorems, such as the principle of superposition, delta–star conversion and vice-versa, Thevenin's and Norton's theorems, maximum power transfer theorem, are explained in detail. Types of energy sources, such as voltage/currents ones (both ideal and real) and also independent/dependent ones, are also discussed. Solved problems are included for students to understand more thoroughly the various methods of finding currents/voltages discussed. Chapter 3 will deal with transients in R-L and R-C circuits, both when a dc voltage is switched on, and, when the circuits are short circuited, with initial inductor currents, and/or capacitor voltages/charges.

Multiple Choice Type Questions

Tick the correct option(s), out of the four choices provided in each question.

2.1 The current through the 5 Ω resistor of the network shown in Fig. MCQ2.1 is

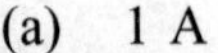

(a) 1 A (b) 2 A (c) 3 A (d) None of (a)–(c)

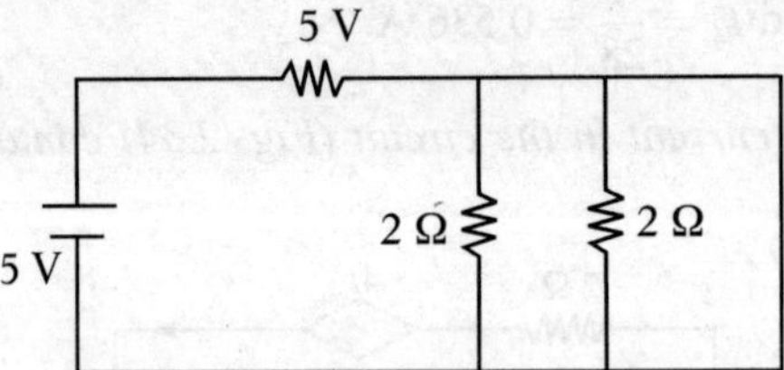

Fig. MCQ2.1

2.2 The Thevenin equivalent resistance of the network shown in Fig. MCQ2.2 is

(a) 1.5 Ω (b) 2.5 Ω (c) 4 Ω (d) None of (a)–(c)

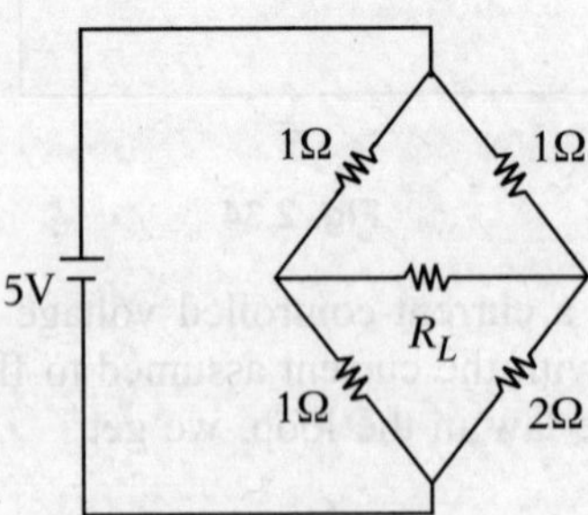

Fig. MCQ2.2

2.3 The Thevenin equivalent voltage of the network shown in Fig. MCQ2.3 is

(a) 33.35 V (b) 24 V (c) 36 V (d) None of (a)–(c)

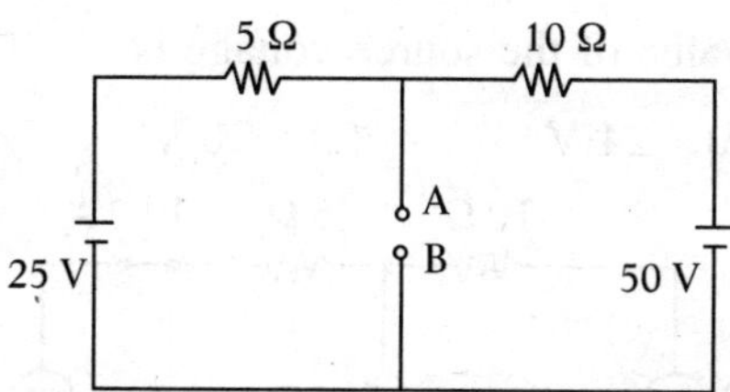

Fig. MCQ2.3

2.4 The V–I characteristic curve across the load terminal of a network is given by Fig. MCQ2.4.The Thevenin voltage and Norton current of the network is

(a) 4 V, 5 A (b) 5 V, 4 A (c) 4.5 V, 4.5 A (d) None of (a)–(c)

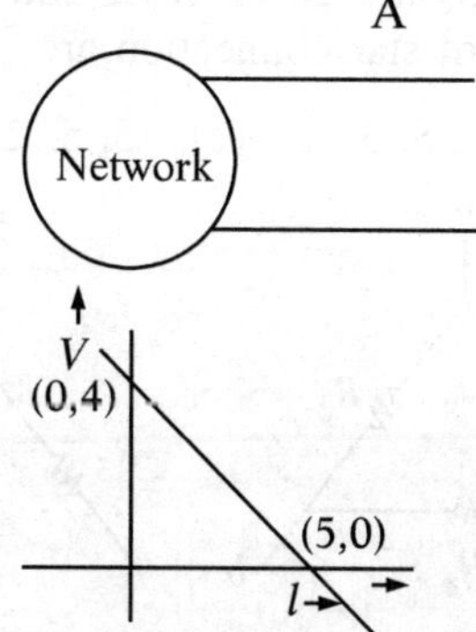

Fig. MCQ2.4

2.5 For the circuit shown in Fig. MCQ2.5, the value of R is

(a) 7.5 Ω (b) 5.0 Ω (c) 2.5 Ω (d) 10 Ω

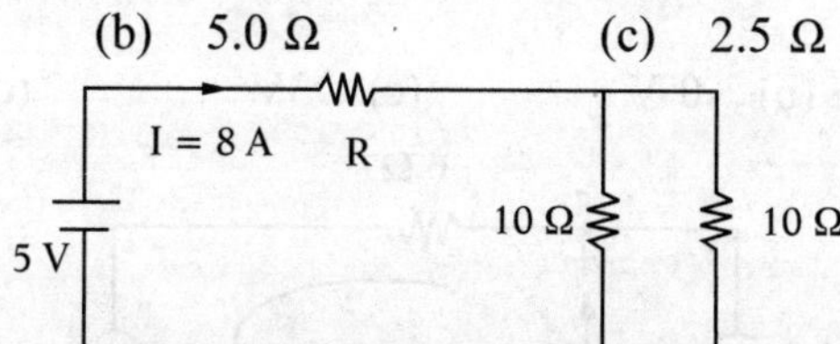

Fig. MCQ2.5

2.6 A segment of a circuit is shown in Fig. MCQ2.6. VR = 5 V, VC = $4 \sin 2t$. The voltage V_L is given by [GATE]

(a) $38 \cos 2t$ (b) $32 \sin 2t$ (c) $16 \sin 2t$ (d) $16 \cos 2t$

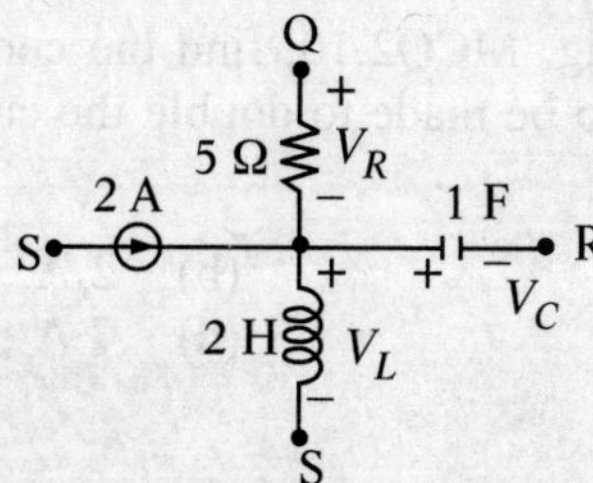

Fig. MCQ2.6

2.7 In Fig. MCQ2.7, the value of the source voltage is [GATE]

(a) 12 V (b) 24 V (c) 30 V (d) 44 V

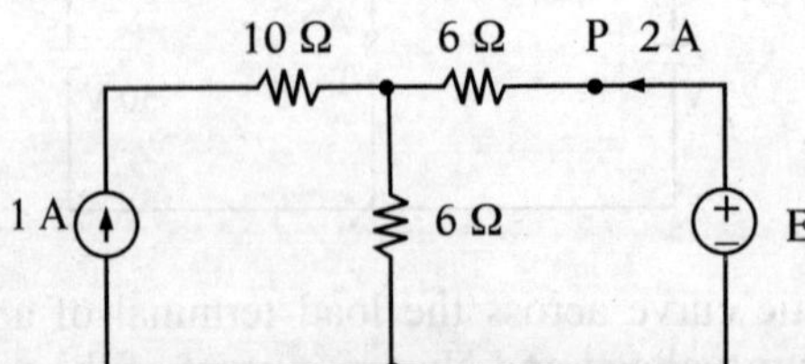

Fig. MCQ2.7

2.8 In Fig. MCQ2.8, R_a, R_b and R_c are 20 Ω, 10 Ω and 10 Ω respectively. The resistance R_1, R_2 and R_3 of an equivalent star connection are [GATE]

(a) 2.5, 5, 5 (b) 5, 2.5, 5 (c) 5, 5, 2.5 (d) 2.5, 5, 2.5

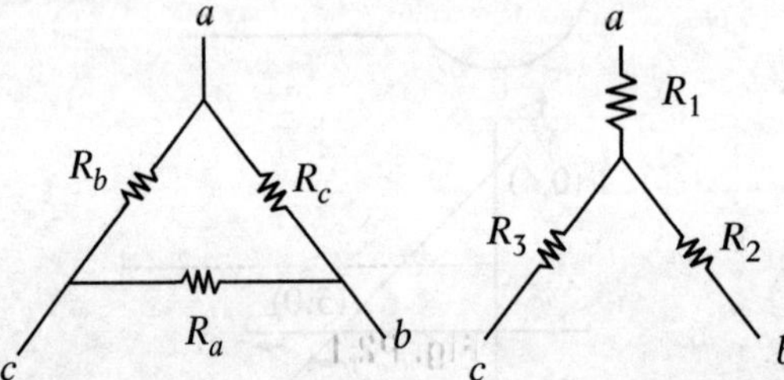

Fig. MCQ2.8

2.9 Assuming ideal elements in the circuits shown in Fig. MCQ2.9, the voltage V_{ab} will be

(a) 3 V (b) 0 V (c) 3V (d) 5 V

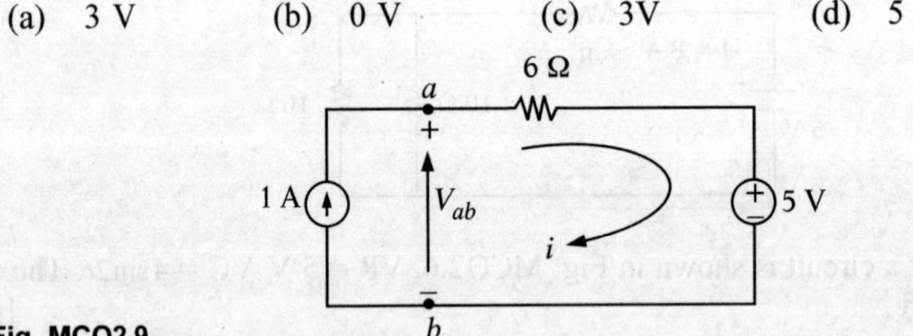

Fig. MCQ2.9 [GATE]

2.10 For the circuit shown in Fig. MCQ2.10, find the current through the 2 Ω resistance. Also identify the changes to be made to double the current through the 2 Ω resistance.

(a) 5 A; Put $V_s = 20$ V (b) 2 A; Put $V_s = 8$ V
(c) 5 A; Put $I_s = 10$ A (d) 7 A ; Put $I_s = 12$ A

[GATE]

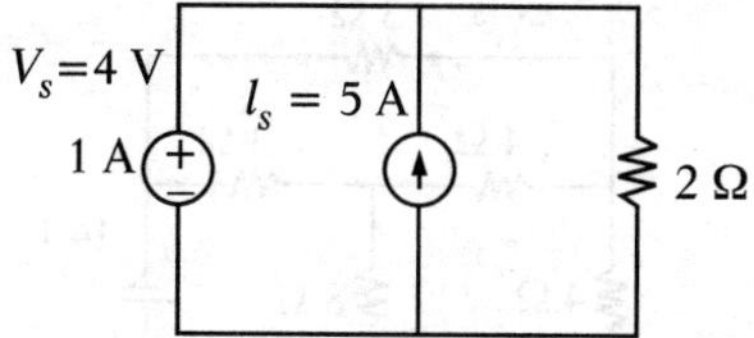

Fig. MCQ2.10

Answers

2.1. (a); 2.2. (a); 2.3. (a); 2.4. (a); 2.5. (a); 2.6. (b); 2.7. (c); 2.8. (a)
2.9. (a) 2.10. (b)

Problems

2.1 In the circuit shown in Fig. P2.1, find the current in the resistor of 5 Ω, which is connected across the points, B and C.

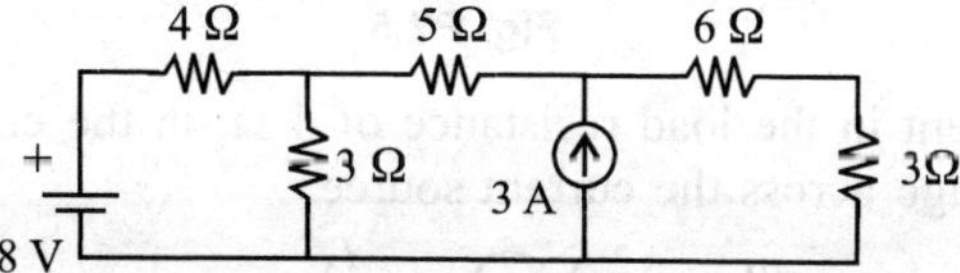

Fig. P2.1

2.2 Calculate the branch currents in the circuit shown in Fig. P2.2. Find also the power input/output to the voltage and/or current sources.

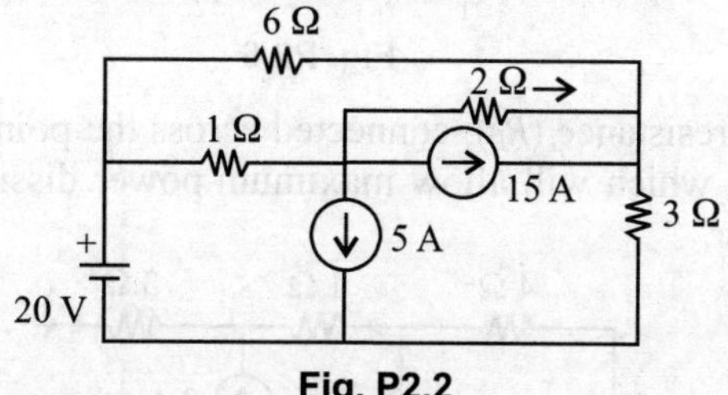

Fig. P2.2

2.3 In the circuit shown in Fig. P2.3, determine the current in the resistance of 3 Ω. Also find the voltage drop across the current source of 4 A.

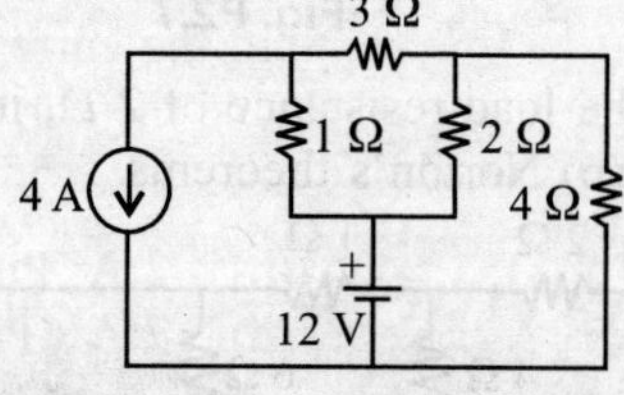

Fig. P2.3

2.4 Find the node voltages in the circuit shown in Fig. P2.4. Also, calculate the current in the load resistance of 3 Ω connected across the points, A and C, by Thevenin's theorem.

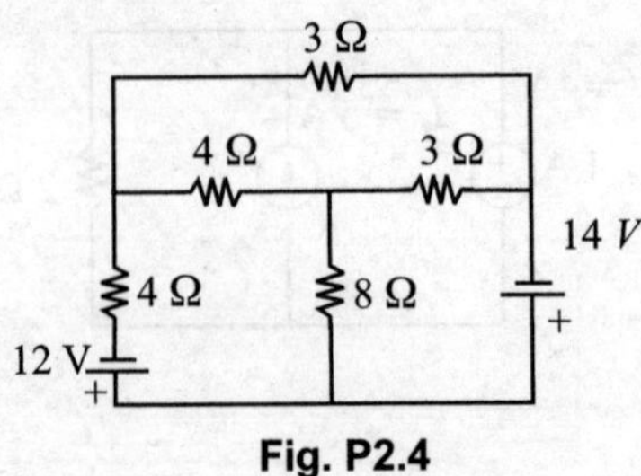

Fig. P2.4

2.5 In the circuit shown in Fig. P2.5, find the node voltages, and also the current in the resistor of 2 Ω.

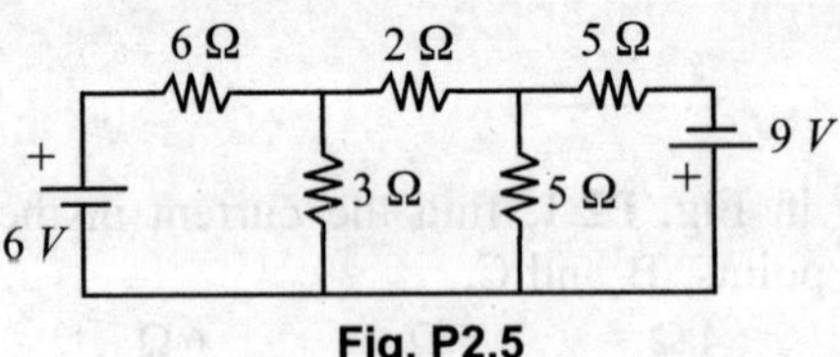

Fig. P2.5

2.6 Calculate the current in the load resistance of 4 Ω, in the circuit shown in Fig. P2.6. Find also the voltage across the current source.

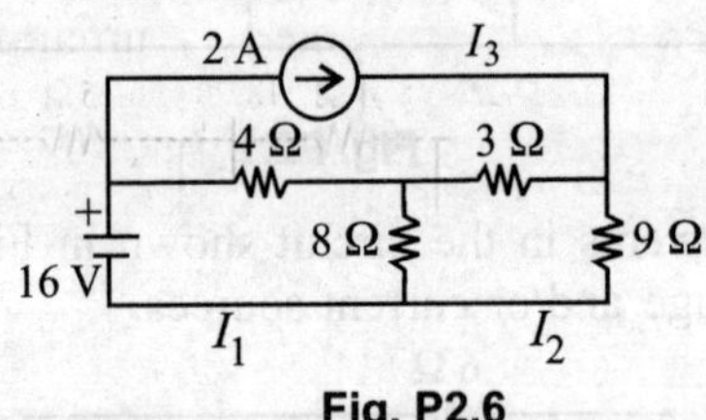

Fig. P2.6

2.7 Determine the load resistance (R_L), connected across the points, A and B in the circuit, shown in Fig. P2.7, which will allow maximum power dissipated in it. Also calculate the maximum power.

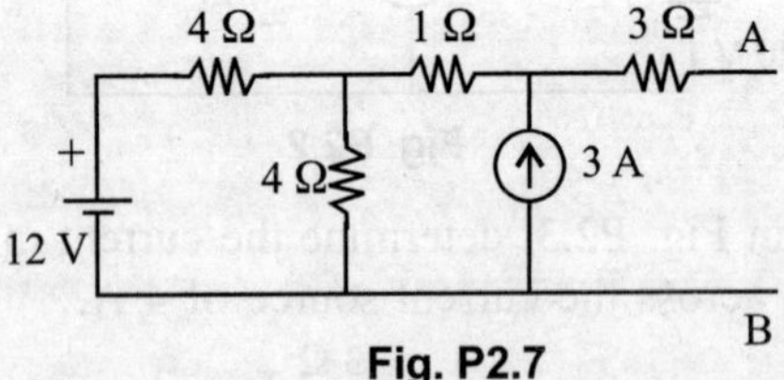

Fig. P2.7

2.8 Determine the current in the load resistance of 2 Ω, in the circuit shown in Fig. P2.8 using (a) Thevenin's, and (b) Norton's theorems.

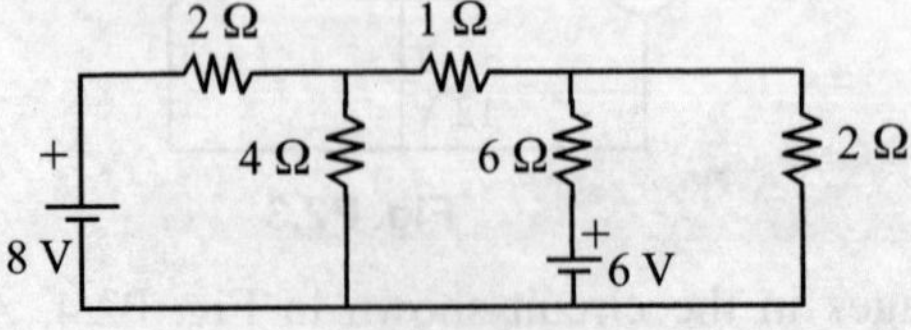

Fig. P2.8

2.9 In the circuit shown in Fig. P2.9, compute the current in the load resistance of 2 Ω, by Thevenin's theorem.

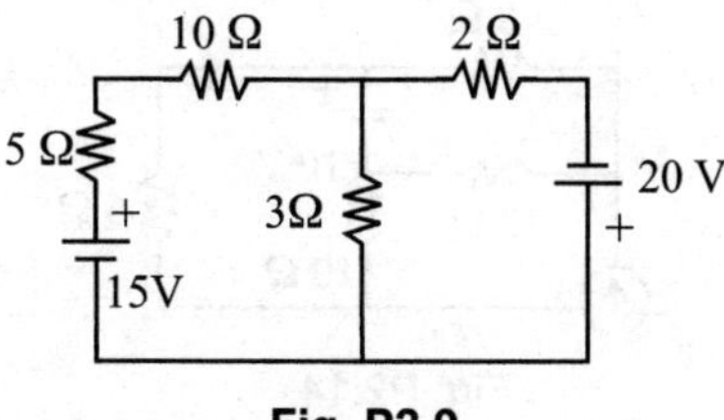

Fig. P2.9

2.10 Calculate the current in the load resistance of 10 Ω, in the circuit shown in Fig. P2.10 using Thevenin's theorem. Find also the node voltages.

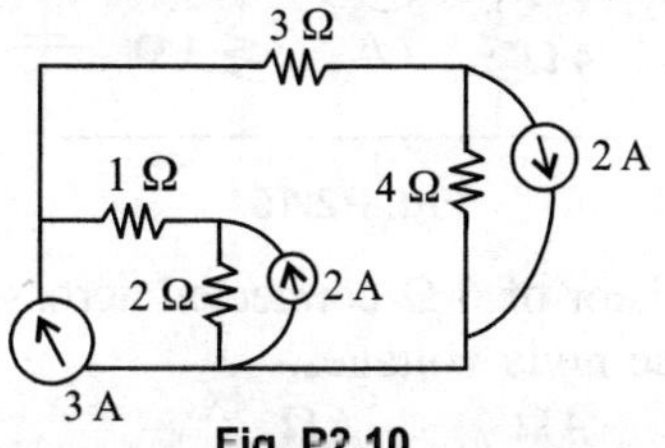

Fig. P2.10

2.11 In the circuit shown in Fig. P2.11, compute the current in the load resistance of 3 Ω.

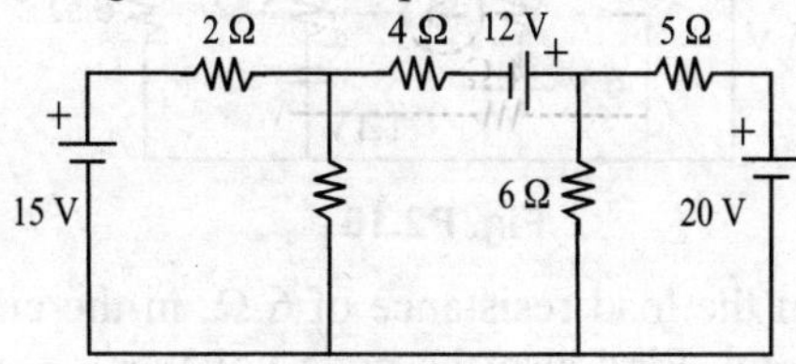

Fig. P2.11

2.12 Determine the current in the load resistance of 4 Ω connected across A and B, in the circuit shown in Fig. P2.12 using Thevenin's theorem. Also, find the node voltages.

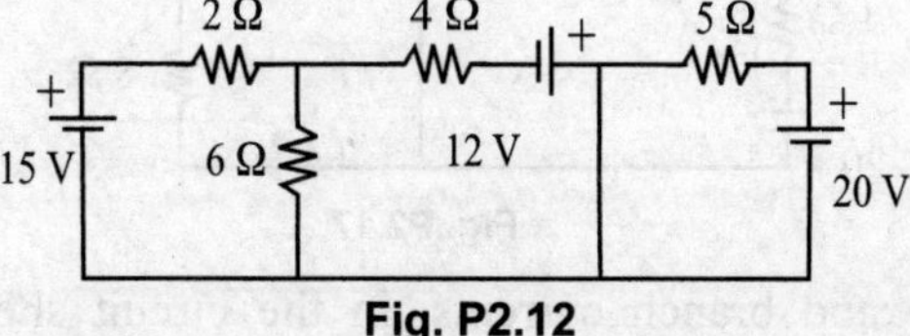

Fig. P2.12

2.13 In the circuit shown in Fig. P2.13, compute all node voltages and currents.

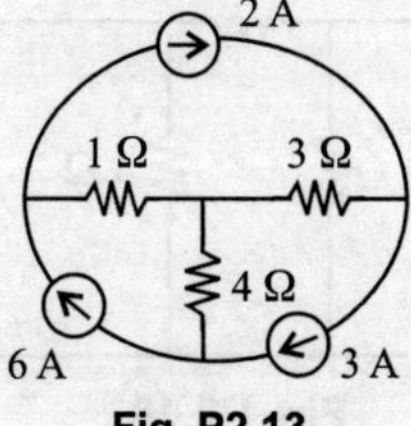

Fig. P2.13

2.14 Find the current in the resistor of 8 Ω, connected across C and D, in the circuit shown in Fig. P2.14. Determine also the power input/output to the voltage and/or current sources.

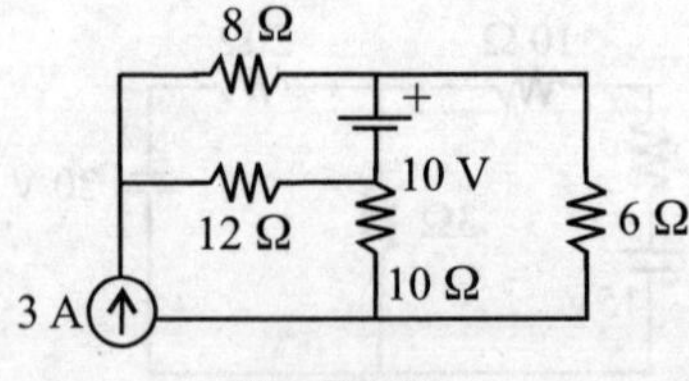

Fig. P2.14

2.15 In the circuit shown in Fig. P2.15, determine the node voltages.

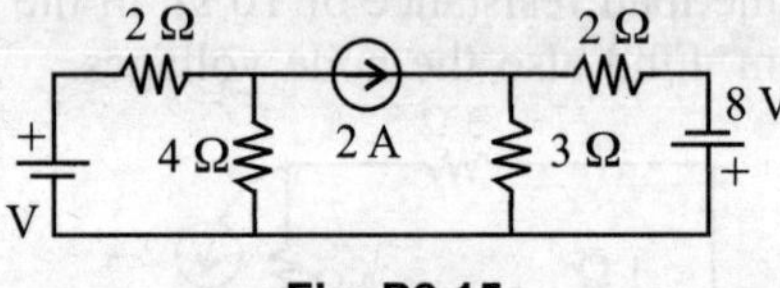

Fig. P2.15

2.16 Find the current in the resistor of 6 Ω connected across A and B, in the circuit shown in Fig. P2.16. Find also the node voltages.

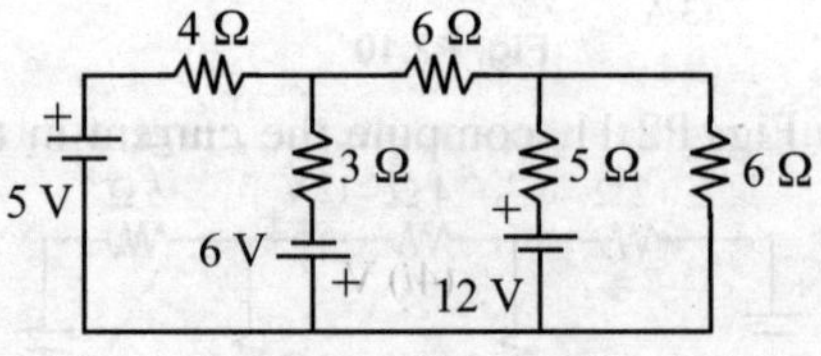

Fig. P2.16

2.17 Calculate the current in the load resistance of 6 Ω, in the circuit shown in Fig. P2.17, using Thevenin's theorem. Find also the node voltages.

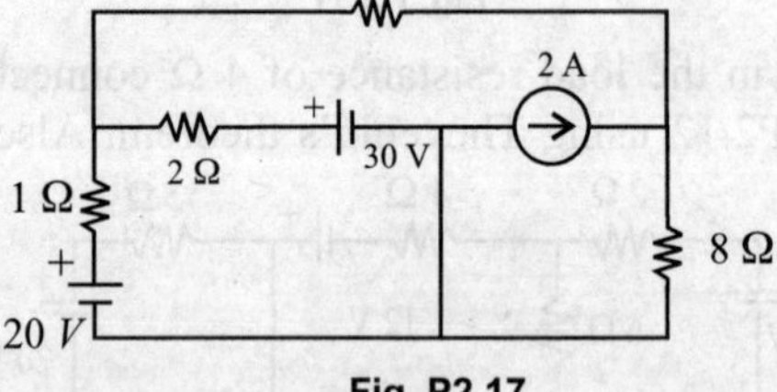

Fig. P2.17

2.18 Determine the loop and branch currents, in the circuit shown in Fig. P2.18. Use conventional methods, such as loop current/nodal voltage.

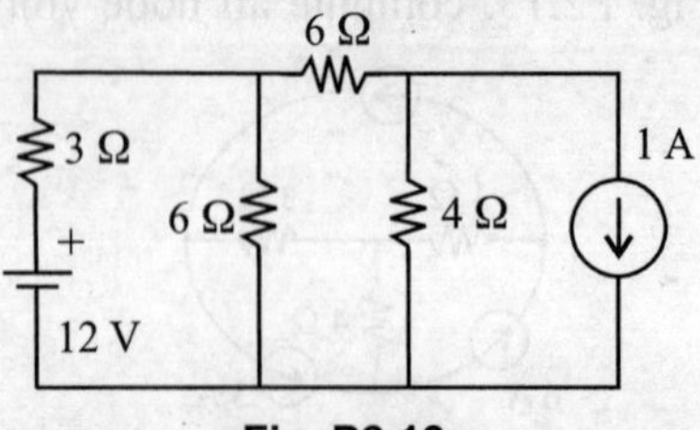

Fig. P2.18

2.19 Calculate the current, i_2, supplied by the independent voltage source in the circuit (Fig. P2.19), also containing a current-controlled voltage source.

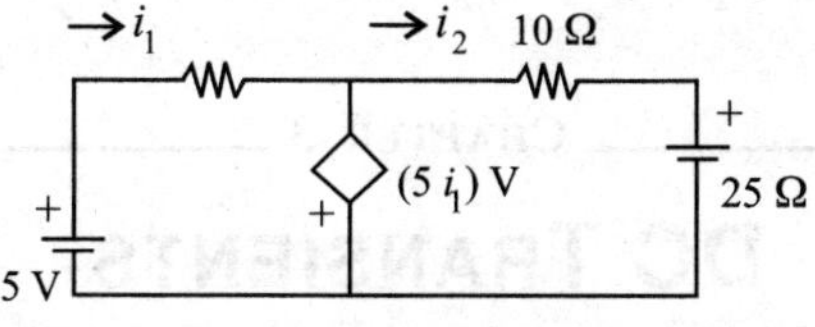

Fig. P2.19

2.20 Find the current in the resistor of 5 Ω in the circuit (Fig. P2.20), containing a current-controlled voltage source.

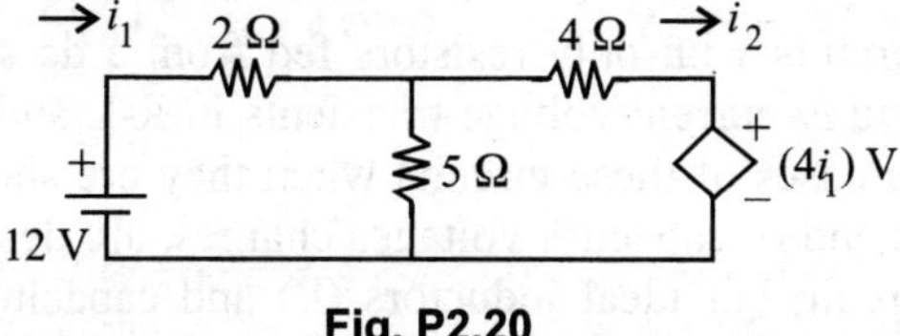

Fig. P2.20

2.21 Determine the values of the current, i and the voltage drops, v_1 and v_2, in the circuit (Fig. P2.21), containing a current-controlled voltage source. What is the voltage of the dependent source?

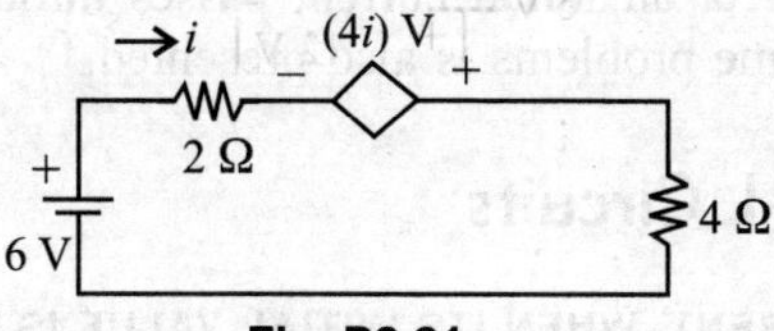

Fig. P2.21

2.22 Find the current (i) in the resistor of 2 Ω (LHS), and the voltage drop (v) across the resistor of 2 Ω (RHS), in the circuit (Fig. P2.22), containing a voltage-controlled current source.

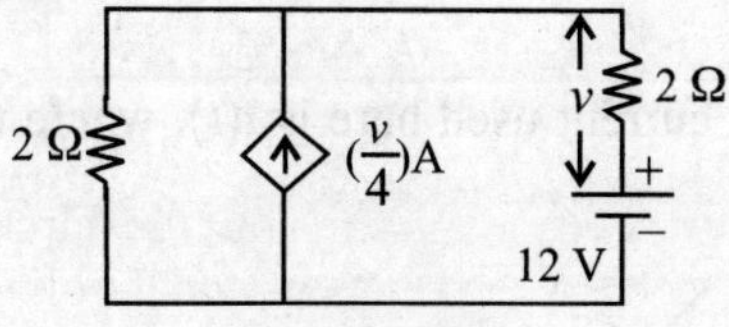

Fig. P2.22

CHAPTER 3

DC TRANSIENTS

In the previous chapter, circuits with only resistors fed from a dc supply was presented. In this chapter, we will first study current/voltage transients in R-L and R-C circuits when they are switched on. Then, the cases of these circuits when they are short circuited, with initial values of inductor currents and/or capacitor voltages/charges, are described. The resistors (R) consume power in any circuit, but ideal inductors (L) and capacitors (C) do not consume power, as there is no resistive element in them. However, energy is stored in the elements when L and C are fed from a dc supply. Normally, resistors are connected in series with L and/or C. Taking an inductive circuit (R-L) as an example, when the circuit is switched on to a dc supply, the current through the inductor, L increases for a certain time. Similarly, when an R-L circuit is short-circuited, an initial current passes through the inductor decreasing with time. The solution of some problems is also presented.

3.1 Transients in R-L Circuits

GROWTH OF INDUCTOR CURRENT, WHEN ITS INITIAL VALUE IS ZERO (0)

Figure 3.1 shows an R-L (R in series with L) circuit switched on to a dc supply, V. The equation for current in the circuit is

$$Ri + L\frac{di}{dt} = V$$

Please note that the symbol for current used here is $i(t)$, where the bracketed part is omitted.

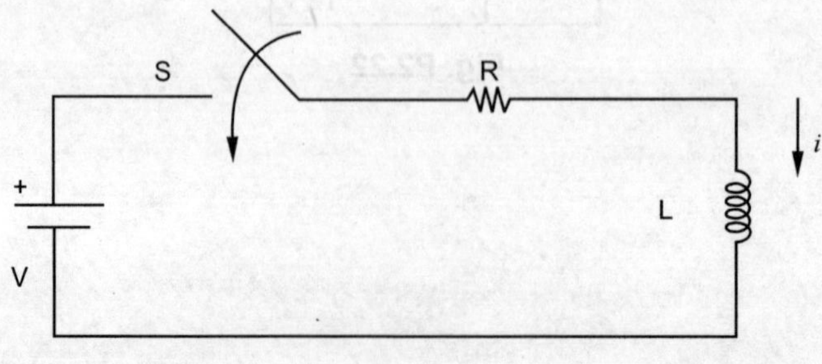

Fig. 3.1

$$L\frac{di}{dt} = V - Ri$$

or, $$\frac{di}{V - Ri} = \frac{dt}{L}.$$

Integrating the above equation with the initial condition that $i(+) = 0$, which means that no current is flowing in the inductor, we get

$$\int_0^i \frac{di}{V - Ri} = \int_0^t \frac{dt}{L}$$

or, $$\frac{\log_e (V - Ri)|_0^t}{R} = \frac{t}{L}$$

or, $$\log_e \left(\frac{V - Ri}{V}\right) = -\frac{Rt}{L}$$

Simplifying the above,

$$i(t) = \frac{V}{R}\left(1 - e^{-\frac{Rt}{L}}\right) = \frac{V}{R}\left(1 - e^{-\frac{t}{\tau}}\right).$$

The voltage drop across the inductor is

$$v_L(t) = \left(L \cdot \frac{V}{R} \cdot \frac{R}{L}\right) \cdot \frac{di}{dt} = V \cdot e^{-\frac{Rt}{L}} = V \cdot e^{-\frac{t}{\tau}}$$

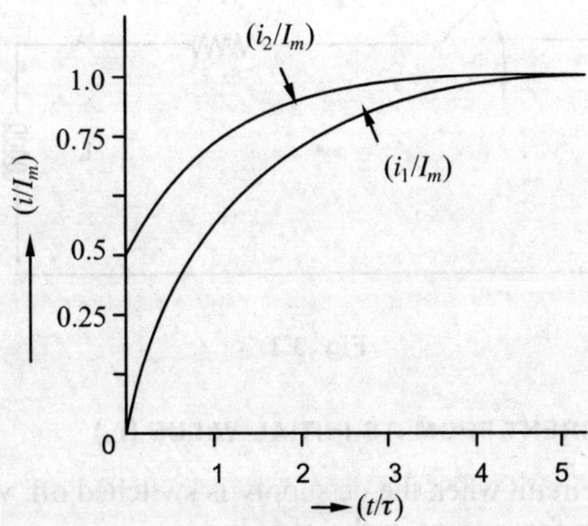

Fig. 3.2

The time constant (in seconds) of the R-L circuit is $\tau = \frac{L}{R}$. The two conditions for the current are $i(0) = 0$ at $t = 0$, and $i(\infty) = \frac{V}{R}$ at $t = \infty$ (steady state). The values of the current and time, in terms of their normalised values are given in Table 3.1. The graph of current vs time $(i-t)$ is plotted in Fig. 3.2. The following points may be noted. The steady state can be taken as occurring at $t = 4\tau$ (in seconds), at which time nearly 98% of the final value of the current is reached. The time axis is plotted in terms of the time constant of the circuit, τ. The current axis is in terms of the final, or maximum value of the current, i.e., $\frac{V}{R}$. Curve I (i_2) in Fig. 3.2 shows that the current grows from zero (initial value) to a nearly steady state (nearly 98% of the final value) in 4τ. It may be noted that the steady state (final) value of the current is maximum. The inductor is an energy storage element. The flux in it increases, as the current increases or grows, thus increasing the stored energy in its magnetic field.

Table 3.1 *Values of currents vs time (both normalised) plotted in Fig. 3.2 and Fig.3.4*

Sl.No.	Time (t/τ)	Currents		
		(i_1/I_m)	(i_2/I_m)	(i_3/I_m)
1	0.0	0.0	0.5	1.0
2	0.5	0.393	0.697	0.607
3	1.0	0.632	0.816	0.368
4	1.5	0.777	0.890	0.223
5	2.0	0.865	0.932	0.135
6	2.5	0.918	0.959	0.082
7	3.0	0.950	0.975	0.050
8	3.5	0.970	0.98	0.030
9	4.0	0.982	0.991	0.018

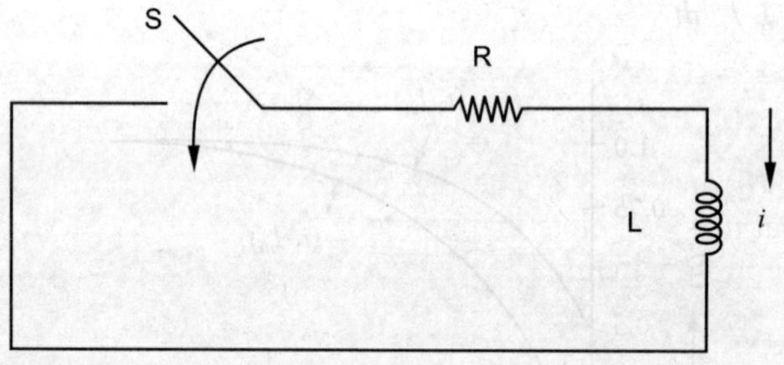

Fig. 3.3

DECAY OF INDUCTOR CURRENT FROM ITS INITIAL VALUE (I_0)

Figure 3.3 shows the R-L circuit when the dc supply is switched off, with the initial condition that $i(0) = I_0$. The equation for current in the circuit is

$$Ri + L\frac{di}{dt} = 0.$$

The steady state condition for the current is $i(\infty) = 0$. The current can be taken as

$$i(t) = A + Be^{-\frac{t}{\tau}}$$

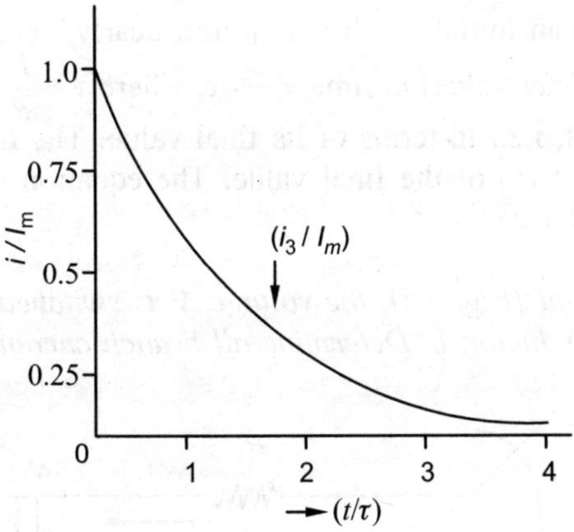

Fig. 3.4

At $t = 0$, $A + B = I_0$. At $t = \infty$, $A = 0$. Hence, $B = I_0 - A = I_0$. The current $i(t)$ takes the form

$$i(t) = I_0 e^{-\frac{t}{\tau}}$$

The graph of $i - t$ is plotted in Fig. 3.4. The current in the inductor decays (decreases) from I_0 to nearly zero (about 2% of the initial value) in 4τ, where $\tau = \frac{L}{R}$. The current axis is in terms of the initial or maximum value of the current in the inductor. The inductor is an energy storage element. If the current decreases or decays, the flux in it decreases, thus, decreasing the stored energy in its magnetic field.

GROWTH OF INDUCTOR CURRENT FROM ITS INITIAL VALUE (I_0)

Consider Curve II (i_2) in Fig. 3.2. The R-L circuit is switched on to a dc supply with the initial condition that $i(0) = I_0$. Note that normally, the initial current in the inductor is lower than the final value in the circuit. The equation for current in the circuit is

$$Ri + L\frac{di}{dt} = V.$$

The steady state condition for the current is $i(\infty) = \frac{V}{R}$, taken from the first case. The current can be taken as $i(t) = A + Be^{-\frac{t}{\tau}}$.

At $t = 0$, $A + B = I_0$. At $t = \infty$, $A = \frac{V}{R}$. So, $B = (I_0 - A) = \left(I_0 - \frac{V}{R}\right)$. The current, $i(t)$ is,

$$i(t) = \frac{V}{R}\left(1 - e^{-\frac{t}{\tau}}\right) + I_0\, e^{-\frac{t}{\tau}} = \frac{V}{R} + \left(I_0 - \frac{V}{R}\right) \cdot e^{-\frac{t}{\tau}}.$$

The current increases from an initial condition, I_0 to a nearly steady state (final) condition, $i(4\tau) = \frac{V}{R}$ (nearly 98% of final value) in time, $t = 4\tau$, where $\tau = \frac{L}{R}$. The current axis chosen is the same as given earlier, i.e., in terms of its final value. The initial value of the current in the inductor is taken as 50% of the final value. The equation shows both the terms for current growth as well as decay.

EXAMPLE 3.1 *In the circuit (Fig. 3.5), the voltage, V is switched on at t = 0. Initially, no current flows through the inductor, L. Determine all branch currents as a function of time.*

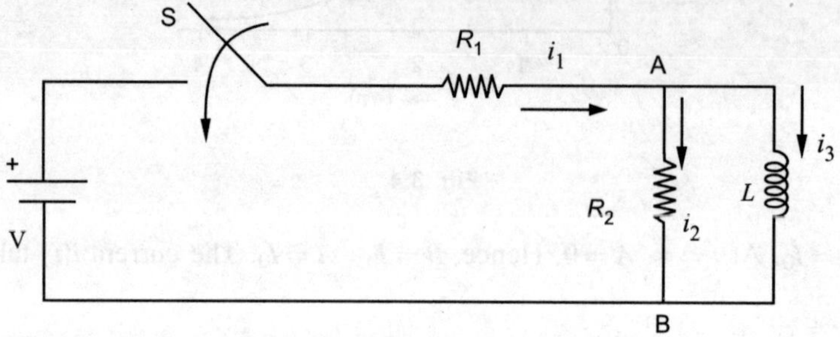

Fig. 3.5

SOLUTION *Thevenin's theorem* is applied to find the equivalent circuit across A and B (Fig. 3.6). The open-circuit voltage is

$$V_{OC} = V' = V\frac{R_2}{R_1 + R_2}$$

Thevenin's or equivalent resistance is obtained as

$$R_{Th} = R' = \frac{R_1 R_2}{R_1 + R_2}$$

The initial current at $t = 0$ in the inductor, L is assumed to be zero. The current equation in the load inductance, L is

$$R' i_3 + L \cdot \frac{di_3}{dt} = V'$$

The current is of the form

$$i_3(t) = \frac{V'}{R'}\left(1 - e^{-\frac{t}{\tau}}\right) = \frac{V'}{R'}\left(1 - e^{-\frac{Rt}{L}}\right).$$

$$\tau = \frac{L}{R'} = L\left(\frac{R_1 + R_2}{R_1 R_2}\right) = L\left(\frac{1}{R_1} + \frac{1}{R_2}\right), \quad a = \frac{1}{\tau} = \frac{1}{L}\left(\frac{R_1 R_2}{R_1 + R_2}\right).$$

$$\frac{V'}{R'} = \left(\frac{VR_2}{R_1 + R_2}\right) \cdot \left(\frac{R_1 + R_2}{R_1 R_2}\right) = \frac{V}{R_1}.$$

Hence, $i_3(t) = \frac{V}{R_1}\left(1 - e^{-\frac{t}{\tau}}\right) = \frac{V}{R_1}\left(1 - e^{-\frac{Rt}{L}}\right).$

The initial and final (steady state) conditions for i_3 are $i_3(0) = 0$, and $i_3(\infty) = \frac{V}{R_1}$.

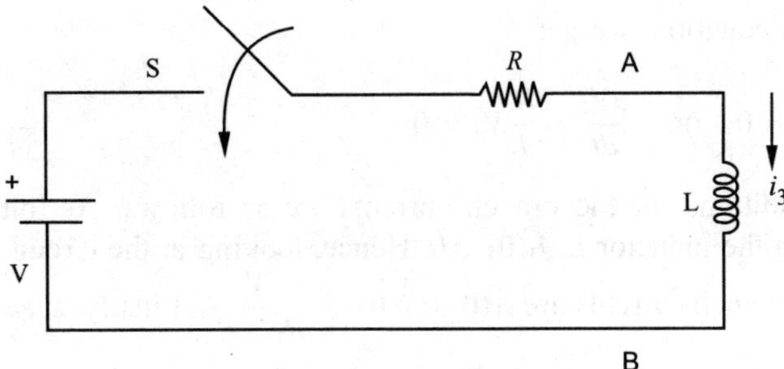

Fig. 3.6

Reverting to the circuit given in Fig. 3.5, the equation for i_2 is

$$R_2 i_2 = L\frac{di_3}{dt}$$

$$\text{or,}\quad i_2 = \frac{L}{R_2}\frac{di_3}{dt} = \left(\frac{LVR'}{R_1R_2L}\right)e^{-\frac{R't}{L}} = \left[\left(\frac{V}{R_1R_2}\right)\cdot\left(\frac{R_1R_2}{R_1+R_2}\right)\right]e^{-\frac{R't}{L}} = \left(\frac{V}{R_1+R_2}\right)e^{-\frac{t}{\tau}}.$$

The initial and final (steady state) conditions for i_2 are $i_2(0) = \frac{V}{R_1+R_2}$ and $i_2(\infty) = 0$. Applying Kirchhoff's current law at point B, the current, i_1 is

$$i_1 = i_2 + i_3 = \left(\frac{V}{R_1+R_2}\right)e^{-\frac{R't}{L}} + \frac{V}{R_1}\left(1 - e^{-R't/L}\right)$$

$$= \frac{V}{R_1} + \left(\frac{V}{R_1+R_2} - \frac{V}{R_1}\right)e^{-\frac{R't}{L}} = \frac{V}{R_1} - \left(\frac{V\cdot R_2}{R_1\cdot(R_1+R_2)}\right)e^{-\frac{R't}{L}}$$

The initial and final (steady state) conditions for i_1 are $i_1(0) = \frac{V}{R_1+R_2}$, and $i_1(\infty) = \frac{V}{R_1}$.

Nodal voltage method can be applied. Only a few steps are shown here. There is only one node, A in the circuit (Fig. 3.5). Before writing the current equation at the node, the following points may be noted.

$$v_L = V_{AB} = L\frac{di_3}{dt}, \quad \text{or} \quad i_3 = \frac{1}{L}\int v_L\,dt.$$

Applying Kirchhoff's current law at node A, the equation is,

$$V_A\left(\frac{1}{R_1}+\frac{1}{R_2}\right)+i_3=V_A\left(\frac{1}{R_1}+\frac{1}{R_2}\right)+\frac{1}{L}\int V_A\,dt=\frac{V_A}{R'}+\frac{1}{L}\int V_A\,dt=\frac{V}{R_1}$$

where, $\frac{1}{R'}=\left(\frac{1}{R_1}+\frac{1}{R_2}\right)$ or, $R'=\frac{R_1R_2}{R_1+R_2}$

Differentiating the equation, we get

$$\frac{1}{R'}\frac{dV_A}{dt}+\frac{V_A}{L}=0 \quad \text{or,} \quad \frac{dV_A}{dt}+\frac{R'}{L}V_A=0.$$

The boundary conditions for the branch currents are as follows. As initially no current is assumed to flow in the inductor L, $i_3(0)=0$. Hence, looking at the circuit (Fig. 3.5), initially (at $t=0$), the two branch currents are $i_1(0)=i_2(0)=\frac{V}{R_1+R_2}$. Finally, at steady state, $\frac{di_3}{dt}=0$. So, $v_L=V_A=0$. The current, $i_2(\infty)=0$. The two branch currents are $i_1(\infty)=i_3(\infty)=\frac{V}{R_1}$. The time constant of the circuit is,

$$\tau=L/R'=L\cdot\left(\frac{R_1+R_2}{R_1\cdot R_2}\right)=L\cdot\left(\frac{1}{R_1}+\frac{1}{R_2}\right).$$

From the above boundary conditions, $i_3(t)=\frac{V}{R_1}\left(1-e^{-\frac{t}{\tau}}\right)=\frac{V}{R_1}\left(1-e^{-\frac{R't}{L}}\right)$.

Similarly, the other two branch currents can be found. This current is the same as determined earlier.

EXAMPLE 3.2 *A coil having a self inductance of 5 mH and a resistor of 4 Ω, is connected to an 8 V dc supply. Find the current in the circuit (Fig. 3.7) as a function of time. Calculate how long it takes before it reaches 1.9 A. Also find the time constant of the circuit.*

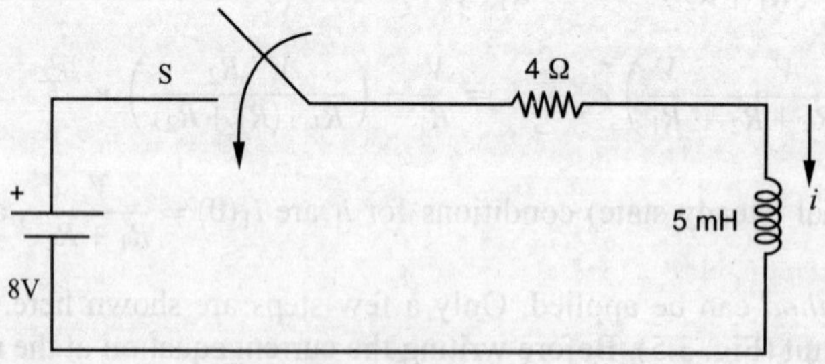

Fig. 3.7

SOLUTION The current in the circuit (Fig. 3.7) is given by

$$i(t)=\frac{V}{R}\left(1-e^{-\frac{R't}{L}}\right)=\frac{V}{R}\left(1-e^{-\frac{t}{\tau}}\right)$$

The initial current in the inductor, L is $i(0)=0$; $V=8$ V; $R=4\,\Omega$; $L=5$ mH.

Time constant $(\tau) = \frac{L}{R} = \frac{5}{4} = 1.25 \text{ ms} = 0.00125 \text{ s}.$

$$\frac{V}{R} = \frac{8}{4} = 2 \text{ A}; \; a = \frac{1}{\tau} = \frac{1}{1.25 \times 10^{-3}} = 800 \text{ s}^{-1}.$$

$$i(t) = 2.0 \cdot \left(1 - e^{-\frac{t}{0.00125}}\right) = 2.0 \cdot \left(1 - e^{-800t}\right) \text{ A}$$

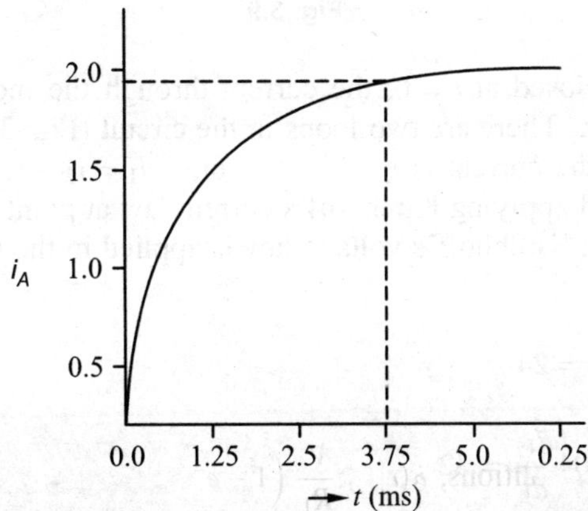

Fig. 3.8

The graph of current vs time $(i - t)$ is given in Fig. 3.8.

The values of t and $i(t)$ are given here.

Time, t in ms	0.625	1.25	1.875	2.5	3.75	5.0
Current, i in A	0.787	1.26	1.554	1.73	1.9	1.96

$i(T) = 1.9$ A or, $i(T) = 2.0\left(1 - e^{-800T}\right) = 1.9$ A.

$$T = -\frac{\log_e\left(1 - \frac{1.9}{2.0}\right)}{800} = \frac{\log_e 20.0}{800} \approx 3.745 \text{ ms.}$$

EXAMPLE 3.3 *In the circuit shown in Fig. 3.9, find the expression for the current $i_2(t)$, after the switch, S is closed at $t = 0$.*

SOLUTION $V = 24$ V, $R_1 = 12\,\Omega$, $R_2 = 8\,\Omega$, $L = 40$ mH, $I = 1$ A.

When the switch, S is open for a long time, the current through the inductor, L is $i_2(0^-) = 1$ A, as the current source of 1 A drives the current through it (Fig. 3.9).

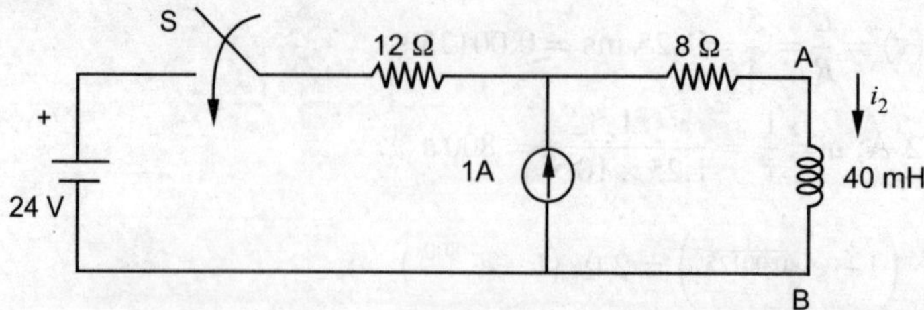

Fig. 3.9

When the switch, S is closed at $t = 0$, the current through the inductor remains the same, i.e., $i_2(0^+) = i_2(0^-) = 1$ A. There are two loops in the circuit (Fig. 3.9). In the mutual branch between the two loops, the current is $i_2 - i_1 = 1$ or, $i_1 = i_2 - 1$.

This can also be found applying Kirchhoff's current law at point B. As the mutual branch contains a current source, Kirchhoff's voltage law is applied in the loop, ACD. The equation is

$$12i_1 + 8i_2 + 0.04\frac{di_2}{dt} = 24$$

or, $$12(i_2 - 1) + 8i_2 + 0.04\frac{di_2}{dt} = 32$$

or, $$20i_2 + 0.04\frac{di_2}{dt} = 24 + 12 = 36.$$

Total resistance of the series circuit $= R' = R_1 + R_2 = 12 + 8 = 20\ \Omega$.

The time constant of the series circuit according to the above equation is

$$\tau = \frac{L}{R'} = \frac{0.04}{20.0} = 0.002 \text{ s} = 2 \text{ ms}, \ a = \frac{1}{\tau} = \frac{1.0}{0.002} = 500 \text{ s}^{-1}.$$

The current, i_2 is $i_2(t) = A + Be^{-500t}$.

Applying the initial condition for the current, i_2 (at $t = 0$),

$$i_2(0^+) = A + B = 1.$$

The final (steady state) condition for the current i_2 (at $t = \infty$), $i_2(\infty) = A = \frac{36}{20} = 1.8$.

So, $B = 1 - A = 1.0 - 1.8 = -0.8$.

The current, i_2 is $i_2(t) = (1.8 - 0.8e^{-500t})$ A.

The current, i_1 is $i_1(t) = i_1(t) - 1 = \left(1.8 - 0.8 \cdot e^{-500t}\right) - 1.0 = 0.8 \cdot \left(1 - e^{-500t}\right)$ A.

The values of the currents, i_1 and i_2 and time, t are given in the table. The graph of the currents vs time are plotted in Fig. 3.10.

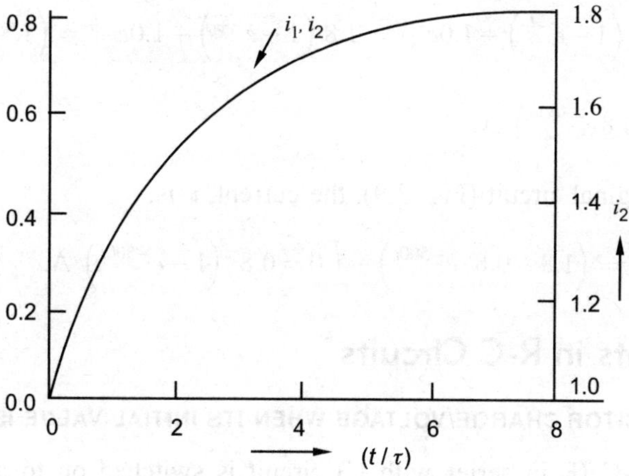

Fig. 3.10

Time, t in ms	1.0	2.0	3.0	4.0	6.0	8.0
Current, i_1 in A	0.315	0.506	0.62	0.69	0.76	0.785
Current, i_2 in A	1.315	1.506	1.62	1.69	1.76	1.785

The problem is now solved by finding Thevenin's equivalent circuit across the points, F and D (Fig. 3.11). Two circuits, being simple, are not shown here. To find the open-circuit voltage, we know that the current through the resistor of 12 Ω is $i_1 = -1$ A due to the presence of a current source in the circuit. The voltage drop across F and D is

$$V_{FD} = V_{BD} = 24 - 12i_1 = 24 + 12 = 36 \text{ V}.$$

Thevenin's or equivalent resistance $= R_{FD} = 12 + 8 = 20\,\Omega$.

Thevenin's equivalent circuit with the load inductance L, is shown in Fig. 3.11.

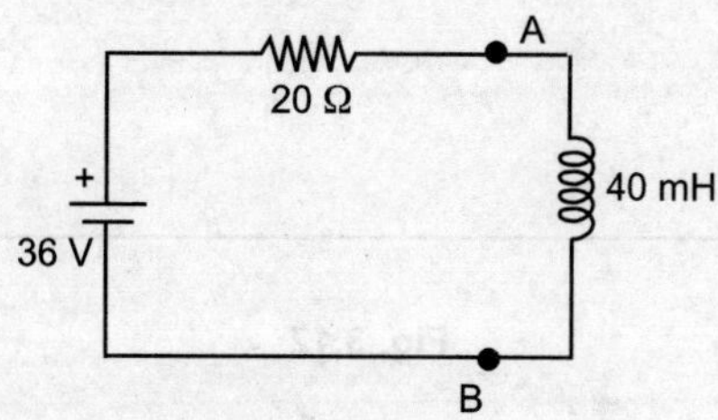

Fig. 3.11

The voltage equation is $20i_2 + 0.04\dfrac{di_2}{dt} = 36$.

The solution of the equation, with an initial current of 1 A as given earlier, is

$$i_2(t)=\left(\frac{36}{20}\right)\cdot\left(1-e^{-\frac{t}{\tau}}\right)+1.0e^{-\frac{t}{\tau}}=1.8\left(1-e^{-\frac{t}{\tau}}\right)+1.0e^{-\frac{t}{\tau}}=\left(1.8-0.8\,e^{-\frac{t}{0.002}}\right)$$

$$=\left(1.8-0.8\,e^{-500\,t}\right)\text{ A.}$$

Reverting to the original circuit (Fig. 3.9), the current, i_1 is,

$$i_1(t)=i_1(t)-1=\left(1.8-0.8\cdot e^{-500t}\right)-1.0=0.8\cdot\left(1-e^{-500t}\right)\text{ A.}$$

3.2 Transients in R-C Circuits

GROWTH OF CAPACITOR CHARGE/VOLTAGE WHEN ITS INITIAL VALUE IS ZERO (0)

In Fig. 3.12, the R-C (R in series with C) circuit is switched on to a dc supply, V. The equation for the current in the circuit is

$$Ri+\frac{1}{C}\int i\,dt=V.$$

The charge of the capacitor is given by $q(t)=\int_0^t i\,dt$, and the current is $i(t)=\dfrac{dq}{dt}$.

The capacitor is uncharged at $t=0$, i.e., $q(0)=0$. Substituting the two expressions, we get,

$$R\frac{dq}{dt}+\frac{q}{C}=V$$

or, $$\frac{dq}{dt}=\frac{1}{R}\left(V-\frac{q}{C}\right)=\frac{1}{RC}(CV-q)$$

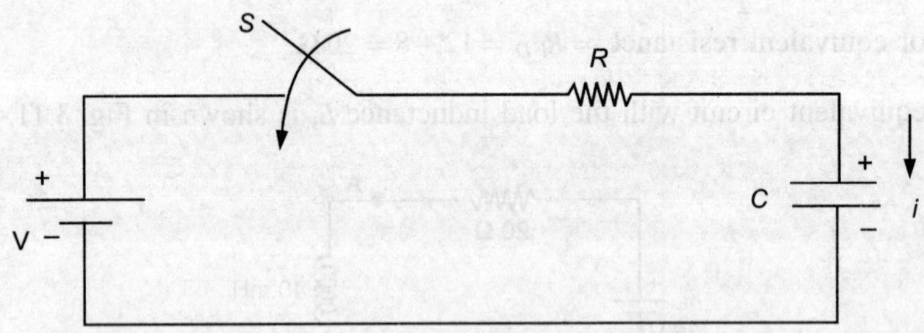

Fig. 3.12

or, $$\frac{dq}{CV-q}=\frac{dt}{RC}.$$

Integrating the above with the initial condition that $q(0^+)=0$.

$$\int_0^q\frac{dq}{CV-q}=\int_0^t\frac{dt}{RC}$$

or, $\quad -\log_e (CV - q)|_0^q = \frac{t}{RC}$

or, $\quad \log_e \left(\frac{CV - q}{CV}\right) = -\frac{t}{RC}.$

Simplifying the above,

$$q = CV\left(1 - e^{-\frac{t}{RC}}\right) = CV\left(1 - e^{-\frac{t}{\tau}}\right).$$

The capacitor voltage is $v_C = \frac{q}{C}$. So, $v_C = V\left(1 - e^{-\frac{t}{\tau}}\right)$.

The current in the capacitance is $i = \frac{dq}{dt}$. Therefore, the current is $i = \frac{V}{R} e^{-\frac{t}{\tau}}$.

The time constant (in seconds) of the R-C circuit is $\tau = R.C$. The two conditions for the charge are $q(0) = 0$ at $t = 0$, and $q(\infty) = C.V$ at $t = \infty$ (steady state). The graphs of charge and current vs time ($q - t$ and $i - t$) are plotted in Fig. 3.13. Values similar to that obtained in Table 3.1 are obtained here. The time taken to reach the steady state is $t = 4\tau$ (in seconds), at which nearly 98% of the final value of the charge is reached. The graph shows that the charge in the capacitance grows from zero to nearly the final value in 4τ. The capacitor voltage also follows the same nature. However, the current in it decreases or decays from an initial value $\frac{V}{R}$ to zero (nearly final value) at $t = 4\tau$. The charge/voltage axis is in terms of their final or maximum values. The current axis is in terms of its initial or maximum value. The time axis is same as given earlier. The capacitance is an energy storage element. The charge/voltage increases as the current flows in it, thus increasing the stored energy in its electric field.

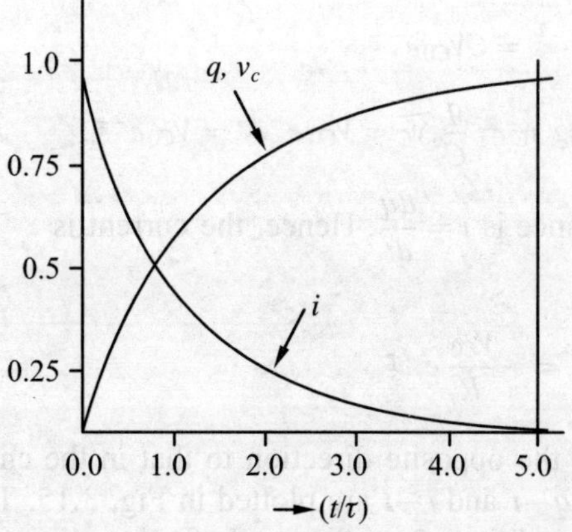

Fig. 3.13

DECAY OF CAPACITOR CHARGE/VOLTAGE FROM ITS INITIAL VALUE (Q_0)

In Fig. 3.14, the dc supply to the R-C circuit is switched off. The initial condition is $q(0) = Q_0 = C.V_{C0}$. The equation for current in the circuit is

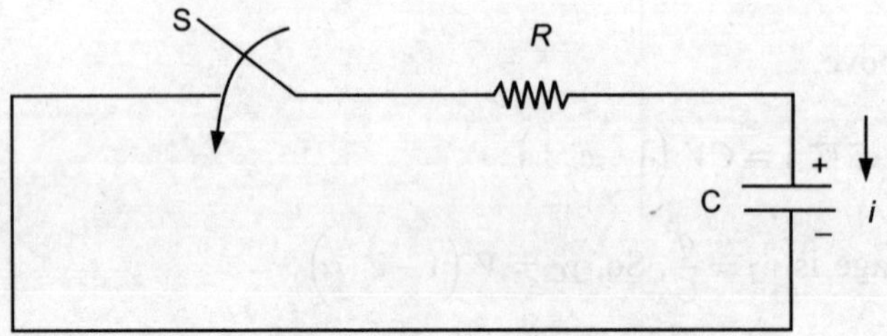

Fig. 3.14

$$Ri + \frac{1}{C}\int i\,dt = 0.$$

With $i = dq/dt$ for charge as given earlier, the equation takes the form

$$R\frac{dq}{dt} + \frac{q}{C} = 0.$$

The steady state condition for the charge is $q(\infty) = 0$. The charge can be taken as

$$q(t) = A + Be^{e^{-\frac{t}{RC}}} = A + Be^{-\frac{t}{\tau}}.$$

At $t = 0$, $A + B = Q_0$. At $t = \infty$, $A = 0$. So, $B = Q_0 - A = Q_0$. The charge, $q(t)$ is

$$q(t) = Q_0\, e^{-\frac{t}{RC}} = Q_0\, e^{-\frac{t}{\tau}} = CV_{C0}e^{-\frac{t}{\tau}}.$$

The capacitor voltage being $v_C = \dfrac{q}{C}$, $v_C = V_{C0}\, e^{-\frac{t}{RC}} = V_{C0}e^{-\frac{t}{\tau}}$.

The current in the capacitance is $i = \dfrac{dq}{dt}$. Hence, the current is

$$i = I_{C0} = -\frac{V_{C0}}{R}e^{-\dfrac{t}{RC}} = -\frac{V_{C0}}{R}e^{-\dfrac{t}{\tau}}$$

The capacitor current is in the opposite direction to that in the charging mode as it is being discharged. The graphs of $q-t$ and $i-t$ are plotted in Fig. 3.15. The charge in the capacitor decreases from the initial condition, Q_0 to the nearly final condition, $q(4\tau) = 0$, where $\tau = RC$. The capacitor voltage shows the same nature. The current, i decreases (decays) from I_{C0} to zero. The magnitude of the current in the capacitance decays (decreases) from I_{C0} to nearly zero in 4τ. The charge/voltage axis is in terms of initial or maximum values. The current axis is chosen in terms of its initial, or maximum value (in magnitude). The capacitor is an energy storage element. If the current decreases or decays, the charge in it decreases, thus, decreasing the stored energy in its electric field.

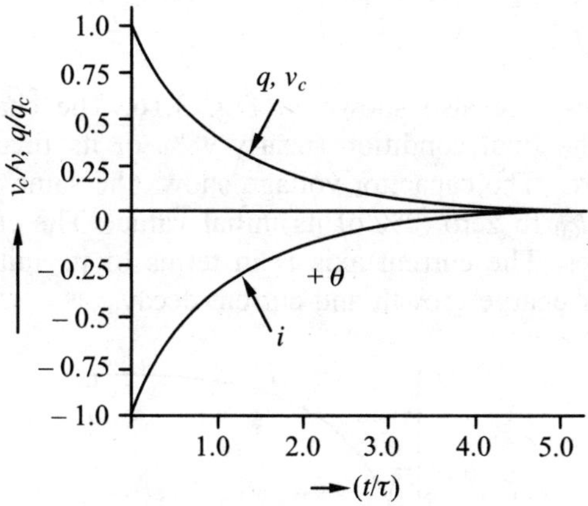

Fig. 3.15

GROWTH OF CAPACITOR CHARGE/VOLTAGE FROM ITS INITIAL VALUE (Q_0)

Consider the R-C circuit in Fig. 3.12 switched on to a dc supply, with the initial condition that $q(0) = Q_0 = C.V_{C0}$. It may be noted here that normally the initial charge of the capacitor is lower than its final value. The equation for current in the circuit is,

$$Ri + \frac{1}{C}\int i\,dt = V.$$

Converting this into the equation for charge as given earlier, it takes the form

$$R\frac{dq}{dt} + \frac{q}{C} = V.$$

The steady state condition for the charge is $q(\infty) = C.V$, when the capacitor is fully charged as given earlier. The charge is

$$q(t) = A + Be^{-\frac{t}{RC}} = A + Be^{-\frac{t}{\tau}}.$$

At $t = 0$, $A + B = Q_0$. At $t = \infty$, $A = C.V$. Hence, $B = Q_0 - A = Q_0 - CV$.

The charge $q(t)$ takes the form

$$q(t) = Q_0 e^{-\frac{t}{\tau}} + C \cdot V\left(1 - e^{-\frac{t}{\tau}}\right) = CV + (Q_0 - C \cdot V)e^{-\frac{t}{\tau}}.$$

The capacitor voltage being $v_C = \frac{q}{C}$,

$$v_C = V_{C0}e^{-\frac{t}{\tau}} + V\left(1 - e^{-\frac{t}{\tau}}\right) = V + (V_{C0} - V) \cdot e^{-\frac{t}{\tau}}.$$

The current in the capacitance is $i = \dfrac{dq}{dt}$. So, the current is

$$i = I_{C0} = \left(\frac{V - V_{C0}}{R}\right) e^{-\frac{t}{\tau}} = \left(\frac{V - V_{C0}}{R}\right) e^{-\frac{t}{RC}}$$

The plots of $q-t$ and $i(t)$ are also shown in Fig. 3.16. The charge increases from the initial condition, Q_0 to the final condition (nearly 98% of its final value), $q(4\tau) = CV$ in time, $t = 4\tau$, where $\tau = RC$. The capacitor voltage shows the same nature. The current, $i(t)$ decreases (decays) from I_{C0} to zero (2% of its initial value). The charge/ voltage axis is in terms of their final values. The current axis is in terms of its initial value. The equation shows both the terms, for charge growth and current decay.

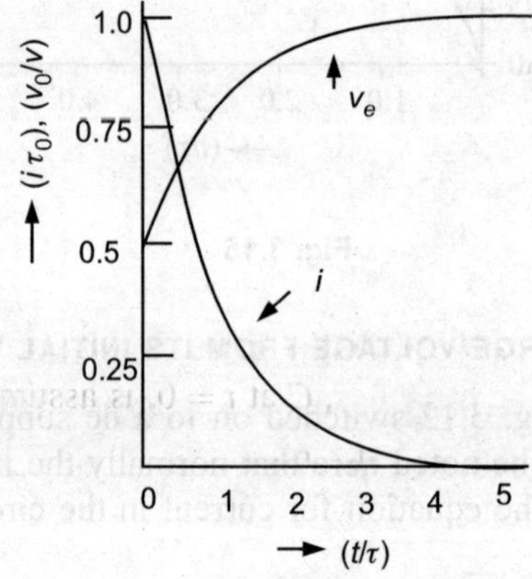

Fig. 3.16

EXAMPLE 3.4: *In the circuit (Fig. 3.17), the voltage, V is switched on at t = 0. Find all the branch currents as a function of time.*

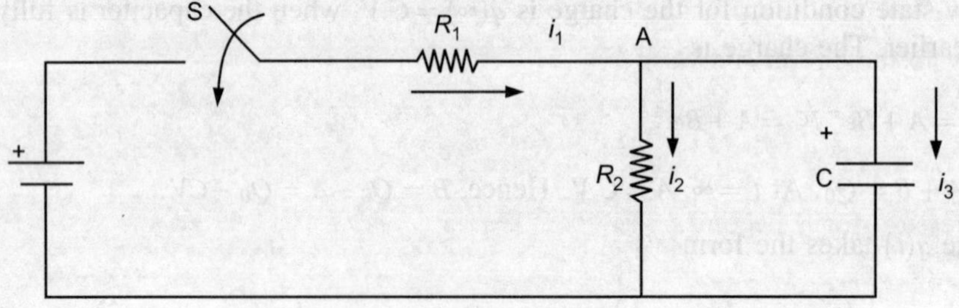

Fig. 3.17

SOLUTION Thevenin's theorem is applied to find the equivalent circuit across A and B (Fig. 3.18). As given earlier, the open circuit voltage is

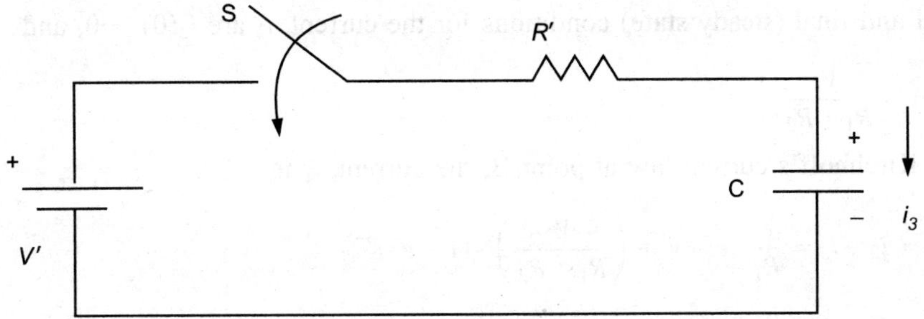

Fig. 3.18

$$V_{OC} = V' = V\,\frac{R_2}{R_1+R_2}.$$

Thevenin's or equivalent resistance is obtained as

$$R_{Th} = R' = \frac{R_1 R_2}{R_1+R_2}.$$

The initial charge/voltage in the capacitor, C at $t=0$, is assumed to be zero.

The current in the capacitance, C is of the form

$$i_3(t) = \frac{V'}{R'}e^{-\frac{t}{\tau}} = \frac{V}{R_1}e^{-\frac{t}{\tau}} = \frac{V}{R_1}e^{-\frac{t}{R'C}}.$$

The simplification procedure has been described in EXAMPLE 2.1.

$$\tau = R'C = C\left(\frac{R_1 R_2}{R_1+R_2}\right), \quad a = \frac{1}{\tau} = \frac{1}{C}\left(\frac{R_1+R_2}{R_1 R_2}\right) = \frac{1}{C}\left(\frac{1}{R_1}+\frac{1}{R_2}\right).$$

The initial and final (steady state) conditions for i_3 are $i_3(0) = \dfrac{V}{R_1}$, and $i_3(\infty) = 0$.

Reverting to the circuit shown in Fig. 3.17, the equation for i_2 is

$$R_2 i_2 = \frac{1}{C}\int i_3\,dt.$$

The expression for i_2 becomes

$$i_2(t) = \frac{1}{R_2 C}(1/R_2 C)\int_0^t i_3\,dt = \left(\frac{V}{R_1 R_2 C}\right)\int_0^t e^{-\frac{t}{R'C}}\,dt = \frac{V R' C}{R_1 R_2 C}\left(1-e^{-\frac{t}{R'C}}\right)$$

$$= \left(\frac{V}{R_1 R_2}\times\frac{R_1 R_2}{(R_1+R_2)}\right)\cdot\left(1-e^{-\frac{t}{R'C}}\right) = \left(\frac{V}{R_1+R_2}\right)\cdot\left(1-e^{-\frac{t}{R'C}}\right).$$

The initial and final (steady state) conditions for the current, i_2 are $i_2(0) = 0$, and

$$i_2(\infty) = \frac{V}{R_1 + R_2}.$$

Applying Kirchhoff's current law at point B, the current, i_1 is

$$i_1(t) = i_2 + i_3 = \frac{V}{R_1} \cdot e^{-\frac{t}{R'C}} + \left(\frac{V}{R_1 + R_2}\right) \cdot \left(1 - e^{-\frac{t}{R'C}}\right)$$

$$= \left(\frac{V}{R_1 + R_2}\right) + \left(\frac{V}{R_1} - \frac{V}{R_1 + R_2}\right) e^{-\frac{t}{R'C}}$$

$$= \frac{V}{R_1 + R_2} - \frac{V R_2}{R_1 (R_1 + R_2)} e^{-\frac{t}{R'C}}.$$

The initial and final (steady state) conditions for i_1 are $i_1(0) = \frac{V}{R_1}$, and $i_1(\infty) = \frac{V}{R_1 + R_2}$.

Nodal voltage method can also be applied, as in EXAMPLE 2.1. This is not included here.

EXAMPLE 3.5 *A series circuit (Fig. 3.19) consisting of a 2.5 kΩ resistor and a 4 μF capacitor is connected to a 300 V dc supply. A glow discharge tube is connected across the capacitor, and is triggered to strike every 20 ms. The capacitor is discharged completely in a negligibly short time when the tube strikes. Determine the voltage across the capacitor when the tube strikes, and find the rms value of the charging current.*

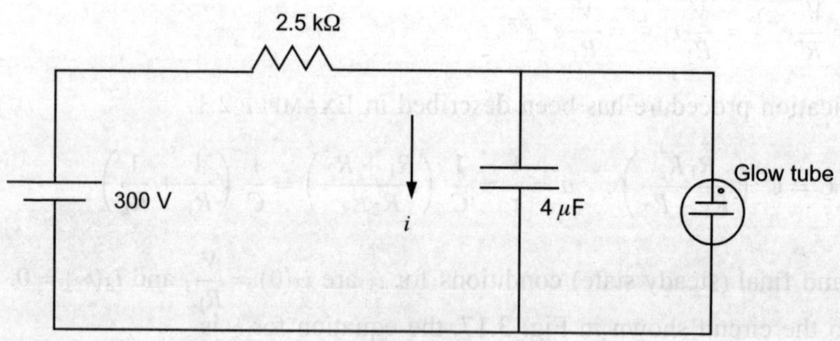

Fig. 3.19

SOLUTION The voltage equation for the circuit shown in Fig. 3.19 is

$$Ri + \frac{1}{C}\int i\, dt = V.$$

The equation in terms of the charge on the capacitor comes out as

$$R\frac{dq}{dt} + \frac{q}{C} = V.$$

Assuming that the capacitor is uncharged initially ($Q_0 = 0$), the solution for charge is of the form

$$q(t) = (C \cdot V) \cdot \left(1 - e^{-\frac{t}{\tau}}\right) = (C \cdot V) \cdot \left(1 - e^{-\frac{t}{RC}}\right).$$

$R = 2.5\ \text{k}\Omega\ = 2,500\ \Omega$; $C = 4\ \mu\text{F}$; $V = 300$ V;

Time constant $= \tau = R.C = (2.500 \times 4 \times 10^{-6}) = 0.01$ s $= 10$ ms.

Initial current through the capacitor $= I_0 = \dfrac{V}{R} = \dfrac{300}{2,500} = 0.12$ A $= 120$ mA.

The time period $= T = 20$ ms $= 0.02$ s.

The voltage across the capacitor after $T = 20$ ms

$$= v_C(T) = V.\left(1 - e^{-\frac{T}{\tau}}\right) = 300 \times \left(1 - e^{-\frac{20}{10}}\right) = 300.0 \times (1 - e^{-2}) = 259.4 \text{ V}.$$

The current (A) through the capacitor $= i_C(t) = \left(e^{-\frac{T}{\tau}}\right) = 0.12 \times \left(e^{-\frac{t}{0.01}}\right) = 0.12 \times (e^{-100t})$.

As the current repeats every 20 ms (T), the square of the rms value of the capacitor current

$$= (I_C)^2 = \frac{1}{T}\int_0^T \left[I_0\left(e^{-t/\tau}\right)\right]^2 .dt = \frac{1}{0.02} \times \int_0^{0.02} \left[0.12 \times (e^{-100t})\right]^2 .dt = \frac{(0.12)^2}{0.02} \times \int_0^{0.02} (e^{-200t}).dt$$

$$= \frac{(0.12)^2}{0.02 \times 200.0} \times (e^{-200t})|_{10.02}^{0} = (3.6 \times 10^{-3}) \times (1 - e^{-200 \times 0.02}) = (3.6 \times 10^{-3}) \times (1 - e^{-4})$$

$$= (3.6 \times 10^{-3} \times 0.9817) = 3.534 \times 10^{-3}.$$

The rms value of the capacitor current $= I_C = \sqrt{(3.534 \times 10^{-3})} = 0.05945$ A $= 59.45$ mA.

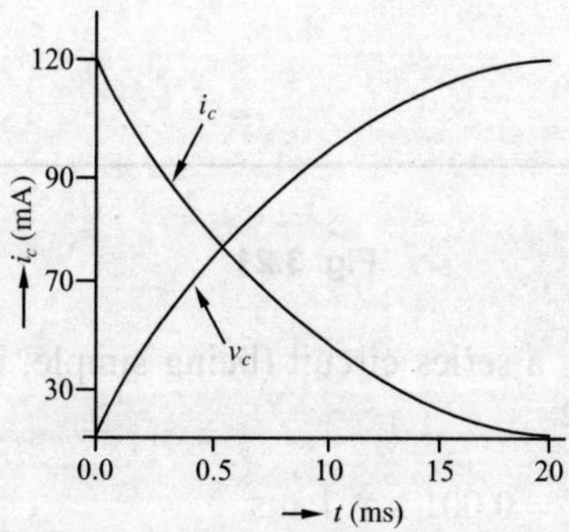

Fig. 3.20

The graph of the charging current,$(i_C - t)$ and the voltage across the capacitor, $(v_C - t)$ are given in Fig. 3.20. The time axis is taken in terms of the time constant of the circuit, τ. The values of $t, i_C(t)$ and $v_C(t)$ are given here.

Time, t in ms	5.0	10.0	15.0	20.0
Current, i_C in mA	72.8	44.15	26.78	16.24
Voltage, v_C in V	118.0	189.6	233.1	259.4

EXAMPLE 3.6 *In the circuit shown in Fig. 3.21, the switch, S is closed at $t = 0$. Find the current, $i_2(t)$ and the voltage across the capacitor, $v_C(t)$ for $t > 0$. Assume that the capacitor is uncharged at $t = 0$.*

SOLUTION $R_1 = 4\ \Omega$, $R_2 = 6\ \Omega$, $C = 100\ \mu F$, $I = 2$ A.

This problem is first solved by using the loop current method. There are two loops, of which the first one has a current source of 2 A. So, $i_1 = 2$ A. Applying Kirchhoff's voltage law, the voltage equation for the second loop is,

$$4(i_2 - i_1) + 6i_2 + \frac{1}{C}\cdot\int_0^t i_2\,dt = 4(i_2 - i_1) + 6i_2 + v_C = 0$$

or, $$10i_2 + \frac{1}{C}\cdot\int_0^t i_2\,dt = 10i_2 + v_C = 4 \times 2 = 8.$$

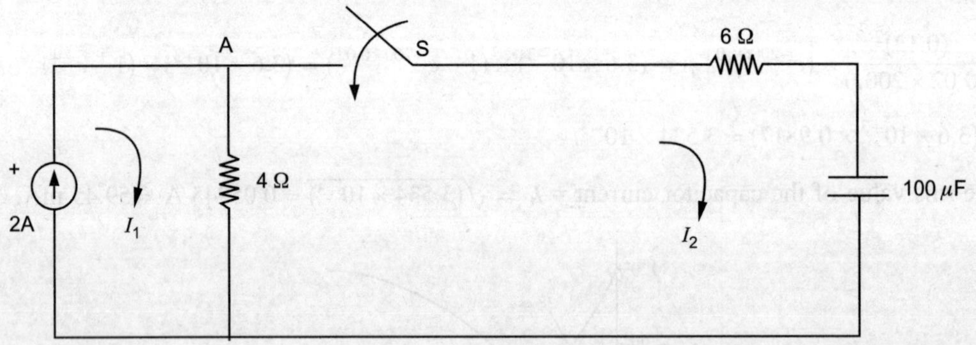

Fig. 3.21

The equation shows that this is a series circuit (being simple, it is not given) with $V = 8$ V, and $R' = R_1 + R_2 = 4 + 6 = 10\ \Omega$.

$$\tau = R'C = 10 \times 100 \times 10^{-6} = 0.001 \text{ s} = 1 \text{ ms}.$$

$$a = \frac{1}{\tau} = \frac{1.0}{0.001} = 1{,}000 \text{ s}^{-1}.$$

$$\frac{1}{C} = \frac{1}{100 \times 10^{-6}} = 10 \times 10^3 = 10{,}000.$$

The initial condition for the current, i_2 is given by $i_2(0) = \frac{V}{R'} = \frac{8}{10} = 0.8$ A, as the capacitor voltage is $v_C(0) = 0$.

The solution for the current, i_2 is given by

$$i_2 = \frac{V}{R'} \cdot e^{-\frac{t}{R'C}} = 0.8 \cdot e^{-\frac{t}{0.001}} = 0.8 \cdot e^{-1,000t} \text{ A.}$$

The current in the 4 Ω resistor is $i_3 = (i_1 - i_2) = \left(2.0 - 0.8 \cdot e^{-1,000t}\right)$ A.

The final (steady state) condition for the current, $i_2(\infty) = 0$, as the capacitor gets fully charged. The capacitor voltage, v_C is

$$v_C(t) = \left(\frac{1.0}{100 \times 10^{-6}}\right) \times 0.8 \cdot \int_0^t e^{-1,000\,t}\, dt$$

$$= \left(\frac{8,000}{1,000}\right) \cdot \left(1 - e^{-1,000\,t}\right) = 8.0 \cdot \left(1 - e^{-1,000\,t}\right) \text{ V.}$$

The real current source (Fig. 3.21) can be converted into a real voltage source to arrive at the same circuit (not given). Thevenin's theorem can also be applied to simplify the circuit which gives the same equivalent circuit.

The values of the currents, i_2 and i_3 and the capacitor voltage, v_C, are given in the table. The graphs of the currents, and capacitor voltage vs time, are plotted in Fig. 3.22.

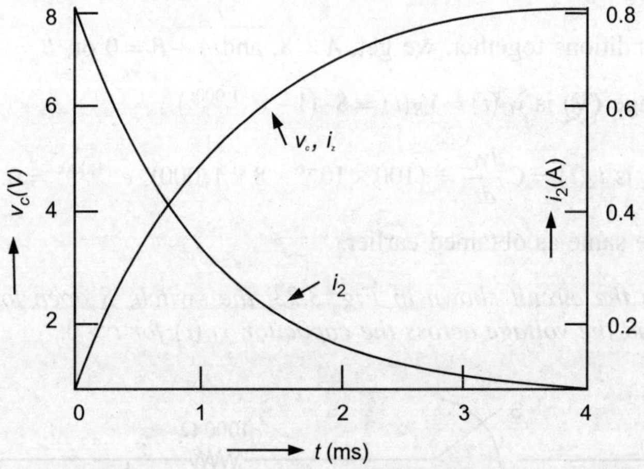

Fig. 3.22

Time, t in ms	0.5	1.0	1.5	2.0	3.0	4.0
Current, i_2 in mA	485.2	294.3	178.5	108.3	39.83	14.65
Current, i_3 in A	1.515	1.706	1.82	1.892	1.96	1.985
Voltage, v_C in V	3.148	5.057	6.215	6.917	7.602	7.853

This problem is also solved using the *nodal voltage method*. Two nodes, A and B are assumed in the circuit (Fig. 3.21). Applying Kirchhoff's current law at the nodes, the equations are,

$$V_A\left(\frac{1}{4}+\frac{1}{6}\right)-\frac{V_B}{6}=\frac{5V_A}{12}-\frac{V_B}{6}=2.$$

$$\frac{V_B-V_A}{6}=-i_2=-\frac{dq}{dt}=-C\frac{dv_C}{dt}=-C\frac{dV_B}{dt}.$$

From the first equation, we get

$$V_A=\frac{2\cdot V_B+24}{5}=0.4V_B+4.8.$$

Simplifying the second equation and substituting the above, we get

$$V_B-(0.4.V_B+4.8)=0.6V_B-4.8=0.6.(V_B-8)=-\left(6\times 100\times 10^{-6}\right)\cdot\frac{dV_B}{dt}$$

or, $$\frac{dV_B}{dt}+1{,}000\cdot V_B=8{,}000.$$

The initial value of $V_B=v_C$ is $V_B(0)=0$, as the capacitor is initially uncharged. The final (steady state) value of $V_B=v_C$ is $V_B(\infty)=8$ V, as $\frac{dV_B}{dt}=0$, at $t=\infty$, as the capacitor gets fully charged.

The solution of V_B is $V_B=A+Be^{-1{,}000t}$.

Putting the two conditions together, we get, $A=8$, and $A+B=0$ or, $B=-A=-8$.

The capacitor voltage (V) is $v_C(t)=V_B(t)=8\cdot\left(1-e^{-1{,}000t}\right)$.

The current (A), i_2 is $i_2(t)=C\frac{dv_C}{dt}=\left(100\times 10^{-6}\times 8\times 1{,}000\right)\cdot e^{-1{,}000t}=0.8e^{-1{,}000t}$

Both values are the same as obtained earlier.

EXAMPLE 3.7 *In the circuit shown in Fig. 3.23, the switch, S open for a long time, is closed at $t=0$. Find the voltage across the capacitor, $v_C(t)$ for $t>0$.*

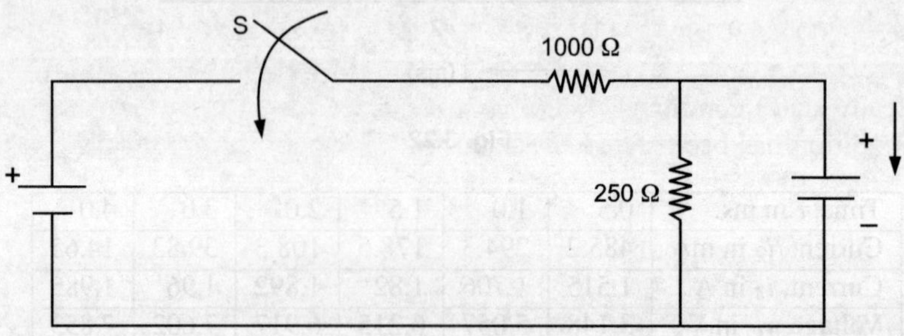

Fig. 3.23

SOLUTION $R_1 = 1\ \text{k}\Omega \quad R_2 = 250\ \Omega$, $C = 5\ \mu\text{F}$, $V = 100$ V.

As the switch, S is open for a long time (Fig. 3.23), the capacitor voltage is equal to the supply voltage, V, i.e., $V_C(0^+) = V = 100$ V. The circuit (Fig. 3.22) with switch, S closed, is converted into Thevenin's equivalent circuit across the points A and B (Fig. 3.24), with the load capacitor removed.

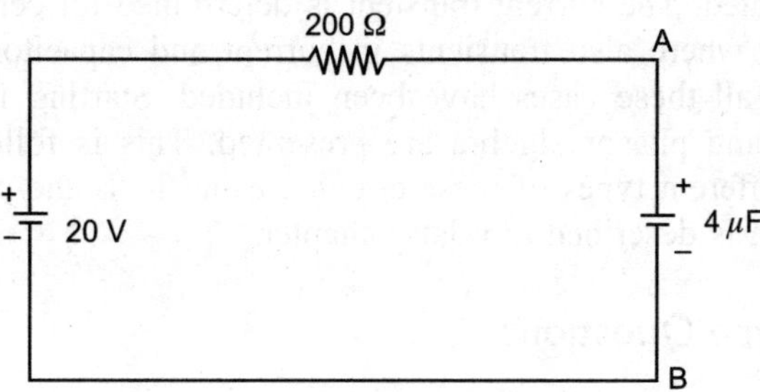

Fig. 3.24

The open circuit voltage is $V_{OC} = V_{AB} = V' = \dfrac{V \cdot R_2}{R_1 + R_2} = \dfrac{100 \times 250}{1{,}000 + 250} = 20$ V

Thevenin's or equivalent resistance is, $R_{Th} = R_{AB} = R' = \dfrac{R_1\,R_2}{R_1 + R_2} = \dfrac{1{,}000 \times 250}{1{,}000 + 250} = 200\ \Omega$

The voltage equation for the circuit (Fig. 3.24) is

$$R'i + (1/C)\int i\,dt = V'.$$

The equation in terms of the charge on the capacitor comes out as

$$R'\frac{dq}{dt} + \frac{q}{C} = V'.$$

The solution for charge ($q = C.v_C$) is as follows

$$q(t) = A + B\,e^{-\frac{t}{R'C}} = A + B\,e^{-\frac{t}{\tau}}.$$

Converting the above equation into capacitor voltage, $v_C = \dfrac{q}{C}$, we get

$$v_C(t) = A_1 + B_1\,e^{-t/\tau}.$$

$$\tau = R'.C = 200 \times 5 \times 10^{-6} = 1 \times 10^{-3} = 0.001\ \text{s} = 1\ \text{ms}.$$

The final (steady state) condition for the voltage across the capacitor is $V_C(\infty) = V' = 20$ V. The initial condition has been stated earlier. Using the equation derived at $t = \infty$, $A_1 = 20$ V. At $t = 0, A_1 + B_1 = 100$.

So, $\quad B_1 = 100 - 20 = 80$ V.

Substituting the above values, the equation for v_C is $v_C(t) = 20 + 80 \cdot e^{-t}$, where t is in ms. The graph of $(v_c - t)$ is not given here.

Conclusion

In this chapter, we study how to find current/voltage transients in dc circuits with resistors, inductance and/or capacitance, when the voltage and/or current source(s) are switched on. The expressions for the currents/voltages in these circuits are found when the circuits are short circuited, with initial values of inductor currents and/or capacitor voltages/charges. R-L circuits are first presented. The current transient is determined for certain cases. Then, R-C circuits are discussed, where also transients in current and capacitor voltages/charges are found. The graphs in all these cases have been included. Starting from the next chapter, alternating quantities and phasor algebra are presented. This is followed by the study of currents/voltages in different types of these circuits, using R, L and/or C, fed from a single phase ac supply, which is described in a later chapter.

Multiple Choice Type Questions

Tick the correct option(s), out of the four choices provided in each question.

3.1 In Fig. MCQ3.1, the capacitor initially has a charge of 10 Coulomb. The current in the circuit one second after the switch S is closed will be

(a) 14.7 A (b) 18.5 A (c) 40.0 A (d) 50.0 A

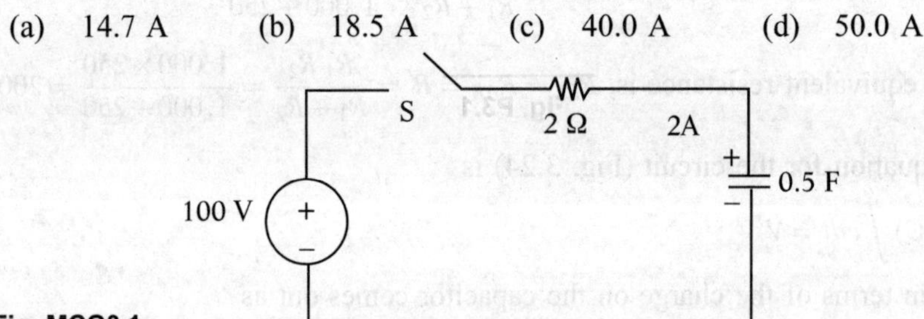

Fig. MCQ3.1

3.2 In Fig. MCQ3.2, the initial capacitor voltage is zero. The switch is closed at $t = 0$. The final steady state voltage across the capacitor is

(a) 20 V (b) 10 V (c) 5 V (d) 0

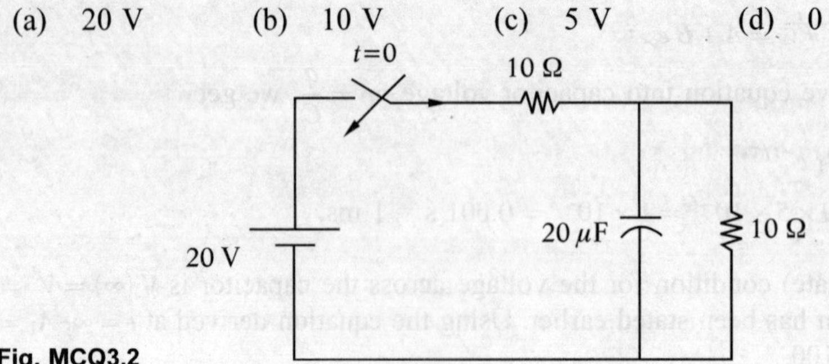

Fig. MCQ3.2

3.3 The circuit shown in Fig. MCQ3.3 is in steady state when the switch is closed at $t = 0$. Assuming that the inductance is ideal, the current through the inductor at $t = 0+$ equals

(a) 0 A (b) 0.5 A (c) 1 A (d) 2 A

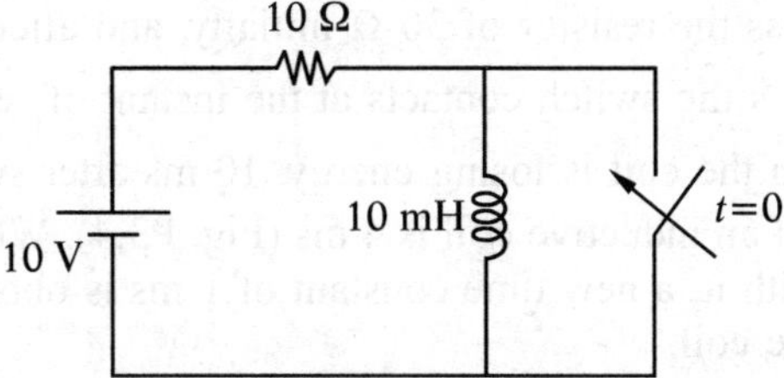

Fig. MCQ3.3

Answers

3.1. (a) 3.2. (b) 3.3. (a)

Problems

3.1 A coil, having a self inductance of 200 mH and a resistance of 10 Ω, is connected to a dc supply of 20 V (Fig. P3.1). Calculate how long it takes before the current reaches 1.8 A. Also find the time constant of the circuit.

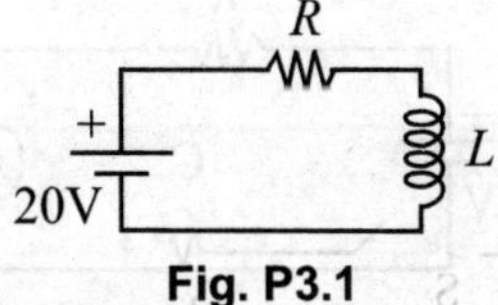

Fig. P3.1

3.2 The shunt field circuit of a 200 V dc motor has a resistance of 100 Ω. Find the value of the discharge resistor (Fig. P3.2) required:

(a) to limit the voltage between the field terminals to 400 V, and

(b) to limit the induced voltage to 400 V, when the field circuit is broken.

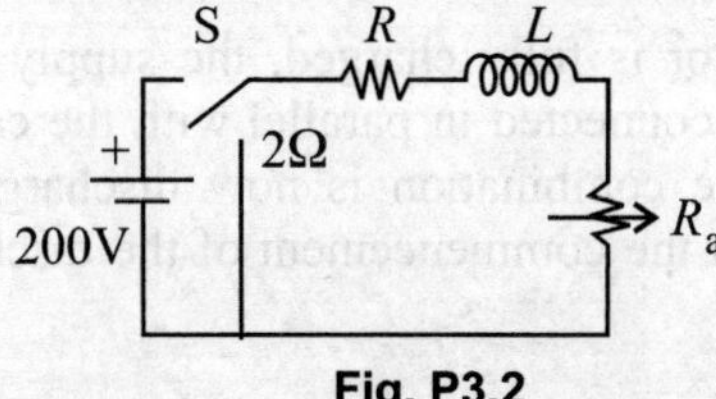

Fig. P3.2

3.3 A coil having an inductance of 0.5 H and a resistance of 10 Ω is connected in parallel with another resistance of 30 Ω, first to a 100 V dc supply, which is then suddenly switched off (Fig. P3.3). Calculate:

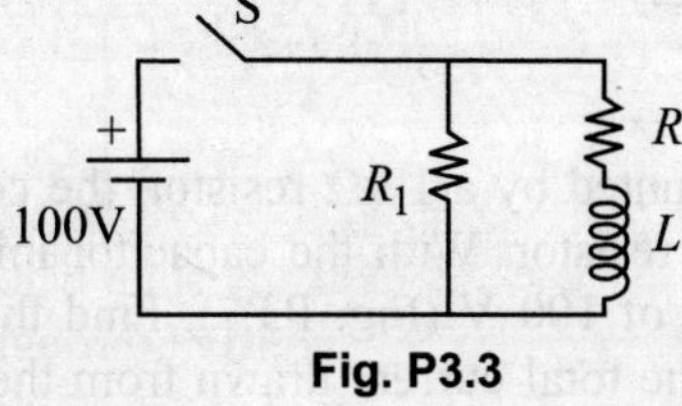

Fig. P3.3

(a) the initial rate of change of current after switching off the supply,

(b) the voltage across the resistor of 30 Ω initially, and after 10 ms,

(c) the voltage across the switch contacts at the instant of separation, and

(d) the rate at which the coil is losing energy, 10 ms after switching off.

3.4 The time constant of an inductive coil is 4 ms (Fig. P3.4). When a resistance of 30 Ω is inserted in series with it, a new time constant of 1 ms is obtained. Find the inductance and resistance of the coil.

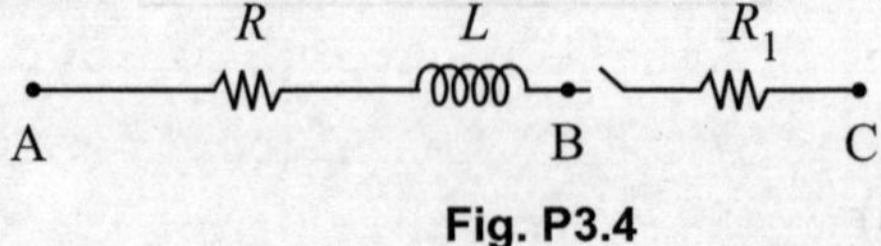

Fig. P3.4

3.5 A series circuit, consisting of a 2.5 kΩ resistor and a 4 μ F capacitor, is connected to a dc supply of 300 V. A glow discharge tube is connected across the capacitor (Fig. P3.5), and is triggered to strike every 20 ms. The capacitor is discharged completely in a negligibly short time when the tube strikes. Determine the voltage across the capacitor when the tube strikes. Also compute the rms value of the charging current.

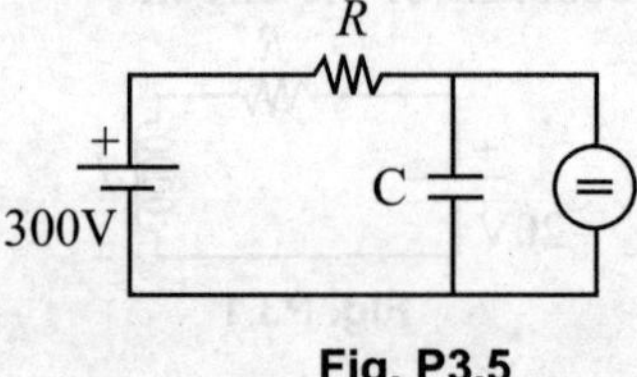

Fig. P3.5

3.6 A circuit consists of a capacitor, $C_1 = 20\ \mu$F, in series with a 10 kΩ resistor. It is switched on to a 150 V dc supply (Fig. P3.6(a)). Obtain an expression for the current. Also find (a) the time constant, and (b) the initial and final (steady state) values of the current.

When the above capacitor is fully charged, the supply is switched off, and another capacitor, $C_2 = 10\ \mu$F, is connected in parallel with the capacitor, with the total charge remaining the same. The combination is now discharged through the resistor (Fig. P3.6(b)). How long from the commencement of the discharge, will the voltage take to fall to 37 V?

Fig. P3.6(a) **Fig. P3.6(b)**

3.7 A 100 μF capacitor is shunted by a 1 kΩ resistor, the combination being connected in series with another 1 kΩ resistor. With the capacitor initially uncharged, the circuit is connected to a dc supply of 100 V (Fig. P3.7). Find the expression for the current in the capacitor, $i_C(t)$, and the total current drawn from the supply, $i_1(t)$.

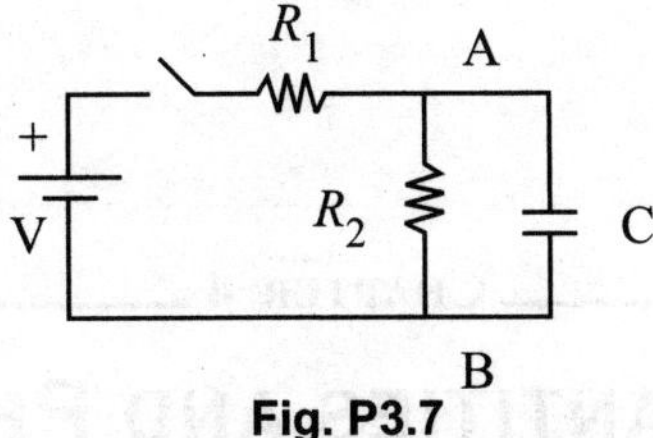

Fig. P3.7

3.8 The capacitor of 2 μF is uncharged in the circuit (Fig. P3.8). Determine the final voltage across the capacitor, after the switch has initially been in position 2 for 2 ms, and then in position 3 for 5 ms.

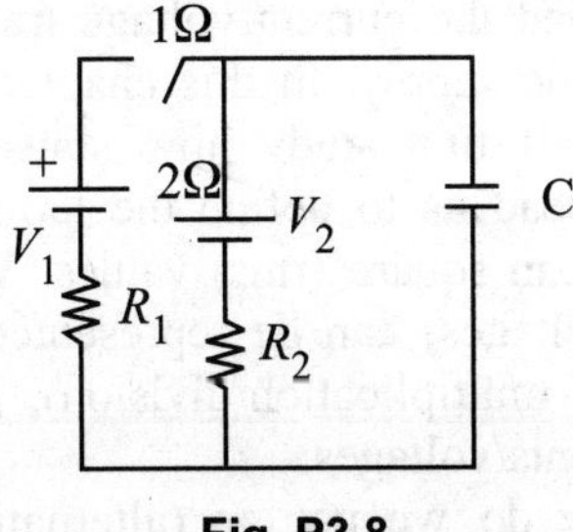

Fig. P3.8

CHAPTER 4

ALTERNATING QUANTITIES AND PHASOR ALGEBRA

In the previous chapter, we studied the current/voltage transients in R-L and R-C circuits when they are switched on to a dc supply. In this chapter, we will discuss two important aspects of ac waveforms. We will first study how sinusoidal voltage (ac) waveform is generated. This will necessarily lead us to define the terms used to describe time-varying waveforms—average and root mean square (rms) values. We will then see how sinusoidal alternating quantities (currents/voltages) can be represented by phasors. We will learn the algebra (addition/subtraction, and multiplication/division), including complex conjugate, of the phasor representation of currents/voltages

To begin our discussion, why do we use ac (alternating current) systems? Power is generated, transmitted, distributed and utilised these days in ac form, not dc. Transformers can operate only with ac supply. As the voltage required for generation, transmission, distribution, and utilisation by the consumer are all at different voltage levels, transformers are used in all cases. Most industrial and domestic drives are run by ac (induction) motors of the squirrel cage type (Fig. 4.1), which are both cheap and rugged (maintenance free). Only in special cases of bulk power transmission is a high voltage (HV) dc system used.

Fig. 4.1

4.1 Generation of Sinusoidal (ac) Waveform

The ac (sinusoidal) waveform is generated by an alternator. It is a simple rotating machine which transforms mechanical power to electrical power. Its operation follows the laws of electromagnetic induction.

When electricity is passed through a coil, it generates a magnetic field. One loop of wire is usually referred to as a turn or a winding, and a coil consists of one turn (single turn coil) or more turns (multi-turn coil). Let us consider a single turn coil rotating at a constant speed of ω rad/s about an axis perpendicular to a uniform magnetic field (Fig. 4.2). The maximum flux linkage, ϕ_m occurs when the coil is vertical. The flux linking the coil at an angular displacement θ is $\phi = \phi_m \cos\theta$. If time t is measured from the instant when the coil is vertical, then $\theta = \omega t$, so, $\phi = \phi_m \cos\omega t$.

The emf (e) induced in the coil is given by the law of electromagnetic induction, as $e = -\dfrac{d\phi}{dt}$.

So, $e = \omega\phi_m \sin\omega t = E_m \sin\omega t$, where $E_m = \omega\phi_m$.

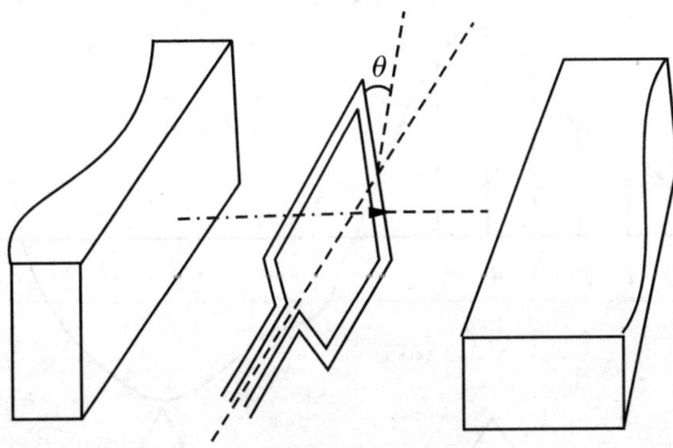

Fig. 4.2 *Generation of sinusoidal (ac) waveform by a single turn coil*

The induced voltage is therefore sinusoidal. In the alternators used in real life, the conductors usually remain stationary in the frame or stator and poles form the rotating element or rotor. The equations above apply to both cases. If the number of turns in the coil is T, then the induced or generated emf is $e = T\omega\,\phi_m \sin\omega t$.

For an alternator, if the number of poles of the rotor is P, speed of the rotor is N and frequency of the generated ac voltage is f, then

$$N = \frac{120f}{P}.$$

The above relation is also valid for any induction motor. The detailed mathematical explanation for the above expression is given in Section 13.1.

4.2 Average and RMS Values of Periodic Voltage or Current Waveform

The current waveform shown in Fig. 4.3(a) is periodic in nature, with time period as T. It is positive for the first half cycle and negative for the second. Let us divide the half time period $\left(\frac{T}{2}\right)$, into say 6 equal time intervals, each equal to ΔT. The average value of the current over half time period is computed as

$$I_{\text{av}} = \frac{(i_1 + i_2 + i_3 + \cdots + i_6) \cdot \Delta T}{6 \cdot \Delta T} = \frac{(i_1 + i_2 + i_3 + \cdots + i_6)}{6}.$$

Please note that the number of time intervals is $n = 6$.

The average value may be written as *area of the current curve over half cycle* divided by the *time period of the half cycle*, which is $\frac{T}{2}$.

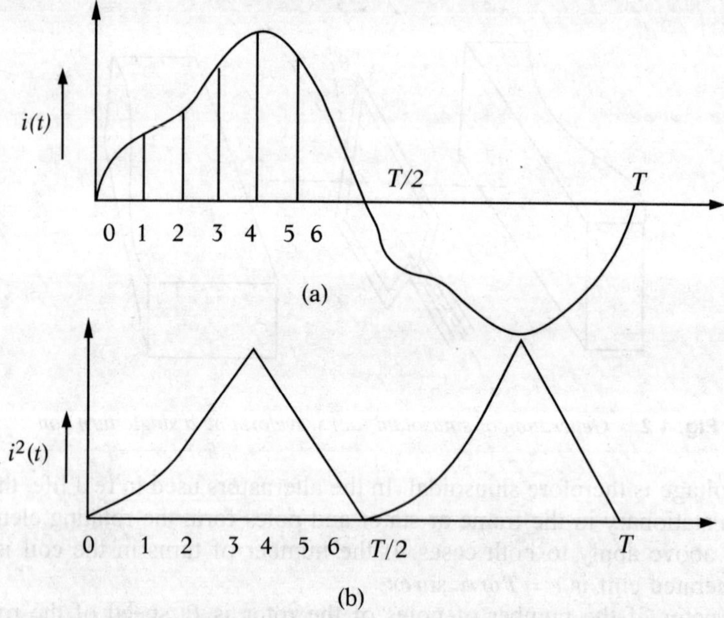

Fig. 4.3 *Waveforms of (a) current and (b) square of the current*

The average value can be expressed as

$$I_{\text{av}} = \frac{2}{T} \int_0^{\frac{T}{2}} i(t) \cdot dt,$$

when the current waveform is continuous.

The square of the current waveform is shown in Fig. 4.3(b). If this current in half time period is subdivided into 6 time intervals in a resistor of resistance R, the average heating during this period is proportional to

$$\frac{[(i_1)^2 + (i_2)^2 + (i_3)^2 + \cdots + (i_6)^2]}{6}.$$

Let I be the value of the direct current which produces the same heating in the resistor as produced by the above periodic waveform, with the half time period subdivided into n time intervals,

$$I^2 = \frac{[(i_1)^2 + (i_2)^2 + (i_3)^2 + \cdots + (i_n)^2] \cdot \Delta T}{n \cdot \Delta T} = \frac{[(i_1)^2 + (i_2)^2 + (i_3)^2 + \cdots (i_n)^2]}{n}.$$

The expression of I^2 may be written as the *the area of the square of the current curve over the half cycle* divided by the *time period of the half cycle,* which is $\frac{T}{2}$.

So, $\quad I = \sqrt{\frac{2}{T} \int_0^{\frac{T}{2}} i^2 \cdot dt}.$

This value is termed as the root mean square (RMS) or effective current.

AVERAGE AND RMS VALUES OF SINUSOIDAL VOLTAGE WAVEFORM

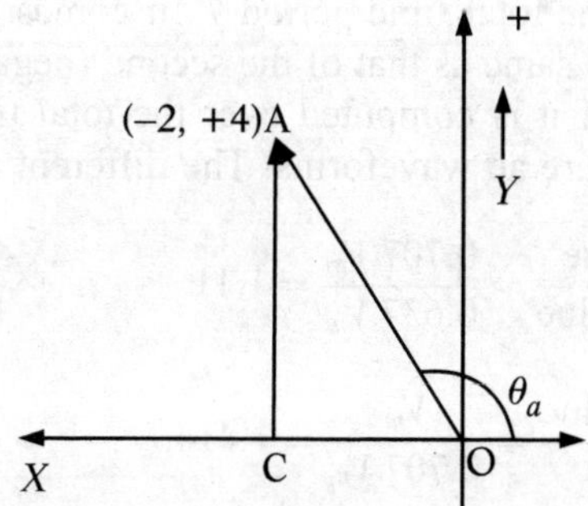

Fig. 4.4 *(a) Sinusoidal voltage waveform; (b) Square of sinusoidal voltage waveform*

The waveforms for the voltage, $v(t)$ and square of the voltage, $v^2(t)$ are shown in Figs 4.3(a) and 4.3(b) respectively.

Time period, $T = \frac{1}{f} = \frac{2\pi}{\omega}$, in radians ($\omega T = 2\pi$).

Half time period $= \frac{T}{2} = \frac{1}{2f} = \frac{\pi}{\omega}$; expressed in radians, $\frac{\omega T}{2} = \pi$.

$v(\theta) = V_m \sin\theta$ for $\pi \le \theta \le 0$.

In terms of time, $t = \frac{\theta}{\omega}$, it is written as $v(\theta) = V_m \sin \omega t$ for $\frac{\pi}{\omega} \le t \le 0$.

$$V_{\text{av}} = \frac{1}{\pi} \int_0^{\pi} v(\theta) \cdot d\theta = \frac{1}{\pi} \int_0^{\pi} V_m \sin\theta \cdot d\theta = \frac{V_m}{\pi} \cdot \cos\theta \Big|_{\pi}^{0} = \frac{2}{\pi} \cdot V_m = 0.637 \cdot V_m.$$

$$V = \sqrt{\frac{1}{\pi}\int_0^{\pi}[v(\theta)]^2 \cdot d\theta} = \sqrt{\frac{1}{\pi}\int_0^{\pi}(V_m \sin\theta)^2 \cdot d\theta} = \sqrt{\frac{(V_m)^2}{\pi}\int_0^{\pi}\frac{1}{2}(1-\cos 2\theta)\cdot d\theta}.$$

$$= \sqrt{\frac{(V_m)^2}{2\pi}\cdot\left(\theta - \frac{\sin 2\theta}{2}\right)\Bigg|_{\pi}^{0}} = \sqrt{\frac{(V_m)^2}{2\pi}\cdot\pi} = \frac{V_m}{\sqrt{2}} = 0.707\cdot V_m.$$

or, $\quad V_m = \sqrt{2}V.$

If time t, is used as a variable, instead of radians, θ, V_{av} can be obtained in this way.

$$V_{\text{av}} = \frac{\omega}{\pi}\int_0^{\frac{\pi}{\omega}} v(t)\cdot dt = \frac{\omega}{\pi}\int_0^{\pi} V_m \sin\omega t \cdot dt = \frac{\omega V_m}{\pi\omega}\cdot\cos(\omega t)\Bigg|_{\pi}^{0} = \frac{2}{\pi}\cdot V_m = 0.637\cdot V_m,$$

which is the same result as obtained before if the variable, t is changed to radians, $\theta(\theta = \omega t)$. The rms value, V can be determined in the same way. If the average value of the above waveform is computed over the total time period T, it comes out as *zero*, as the area of the first (positive) half cycle is the same as that of the second (negative) half cycle. However, the rms value remains the *same* if it is computed over the total time period. There are various ratios or factors used to measure ac waveforms. The different factors are defined below.

$$\text{Form factor} = \frac{\text{RMS value}}{\text{Average value}} = \frac{0.707.V_m}{0.637.V_m} = 1.11.$$

$$\text{Peak factor} = \frac{\text{Maximum value}}{\text{RMS value}} = \frac{V_m}{0.707.V_m} = 1.414.$$

Note: The rms value is always greater than the average value, except for a rectangular waveform, in which case, the heating effect remains constant and the average and the rms values are equal.

Examples

Examples of two waveforms which are periodic in nature are given here.

Saw-tooth Current Waveform

A saw-tooth (or triangular) current waveform is shown in Fig. 4.5. Its time period is T (s). Please note that the slope of the line is $m = \frac{I_m}{T}$ in V/s.

$$i(t) = I_m \cdot \frac{t}{T} = \frac{I_m}{T}\cdot t \text{ for } 0 < t < T.$$

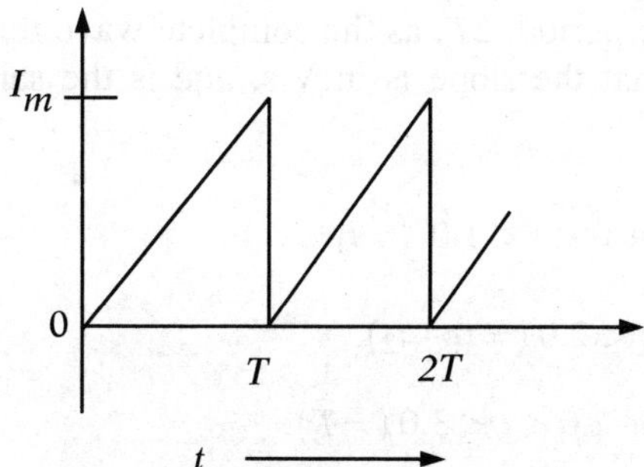

Fig. 4.5 *Triangular current waveform*

Average value of the current

$$= I_{\text{av}} = \frac{1}{T} \cdot \int_0^T i(t) \cdot dt = \left(\frac{I_m}{T^2}\right) \times \int_0^T t \cdot dt = \left(\frac{I_m}{T^2}\right) \times \left[\left(\frac{t^2}{2}\right)\Big|_0^T\right] = \frac{I_m \times T^2}{2 \times T^2} = \frac{I_m}{2} = 0.5 \cdot I_m.$$

Root mean square (rms) value of the voltage

$$V_{\text{rms}} = \sqrt{\frac{1}{T} \cdot \int_0^T i^2 dt} = \sqrt{\left(\frac{(I_m)^2}{T^3}\right) \times \left(\int_{0.0}^{1.0} t^2 \cdot dt\right)} = \sqrt{\left(\frac{(I_m)^2}{T^3}\right) \times \left[\left(\frac{t^3}{3}\right)\Big|_0^T\right]} = (I_m) \times \sqrt{\frac{T^3}{3 \times T^3}}$$

$$= \frac{I_m}{\sqrt{3}} = 0.577 \cdot I_m.$$

The average and rms values are in A, as maximum current, I_m is in A.

Form factor $= k_f = \dfrac{I_{\text{rms}}}{I_{\text{av}}} = \dfrac{0.577.I_m}{0.5.I_m} = 1.154.$

Trapezoidal Voltage Waveform

A trapezoidal voltage waveform is shown in Fig. 4.6. Its time period (T) is 3 s, which is divided into three parts.

Time period $= T = (t_1 + t_2 + t_3) = (1 + 1 + 1) = 3.0$ s.

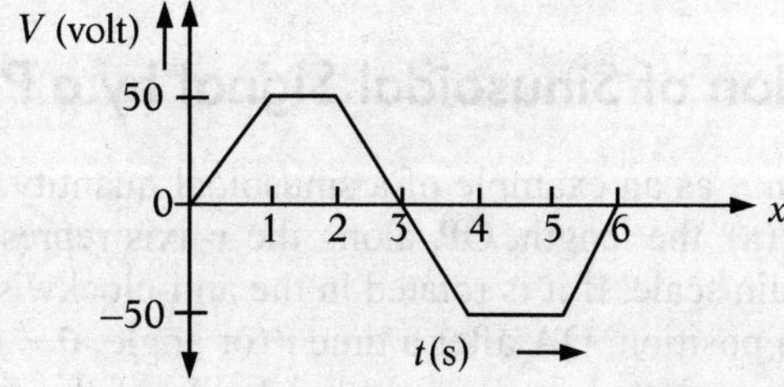

Fig. 4.6 *Trapezoidal voltage waveform*

This is half of the total time period, $2T$, as the complete wave form is symmetrical about the zero (t) axis. Please note that the slope is in V/s, and is the same in the first part and the third part.

$$(t) = \frac{50}{1} \cdot t = 50.t \text{ V for } 0 < t < 1.0\ (= t_1).$$

$$v(t) = 50 \text{ V for } 1.0 < t < 2.0\ (= t_1 + t_2).$$

$$v(t) = [50.(3 - t)] \text{ V for } 2.0 < t < 3.0\ (= T).$$

Average value of the voltage

$$= V_{av} = \frac{1}{T} \cdot \int_0^T v(t) \cdot dt = \left(\frac{50}{3}\right) \times \left[\int_{0.0}^{1.0} t \cdot dt + \int_{1.0}^{2.0} 1 \cdot dt + \int_{2.0}^{3.0} (3 - t) \cdot dt\right]$$

$$= \left(\frac{50}{3}\right) \times \left[\left(\frac{t^2}{2}\right)\Big|_0^1 + (t)\,|_1^2 + \left(\frac{(3-t)^2}{2}\right)\Big|_3^2\right] = \left(\frac{50}{3}\right) \times \left[\frac{1}{2} + (2-1) + \frac{1}{2}\right] = \frac{50 \times 2}{3} = \frac{100.0}{3}$$

$$= 33.33 \text{ V}.$$

Root mean square (rms) value of the voltage

$$= V_{rms} = \sqrt{\frac{1}{T} \cdot \int_0^T v^2 dt} = \left(\frac{50}{\sqrt{3}}\right) \times \sqrt{\left[\int_{0.0}^{1.0} t^2 \cdot dt + \int_{1.0}^{2.0} 1^2 \cdot dt + \int_{2.0}^{3.0} (3 - t)^2 \cdot dt\right]}$$

$$= \left(\frac{50}{\sqrt{3}}\right) \times \sqrt{\left[\left(\frac{t^3}{3}\right)\Big|_0^1 + (t)\,|_1^2 + \left(\frac{(3-t)^3}{3}\right)\Big|_3^2\right]} = \left(\frac{50}{\sqrt{3}}\right) \times \sqrt{\left[\frac{1}{3} + (2-1) + \frac{1}{3}\right]} = \frac{50 \times \sqrt{5}}{\sqrt{3 \times 3}}$$

$$= \quad 37.27 \text{ V}.$$

$$\text{Form factor} = k_f = \frac{V_{rms}}{V_{av}} = \frac{37.27}{33.33} = 1.118.$$

Similarly, average, and rms or effective values of periodic voltage or current waveforms can be computed.

4.3 Representation of Sinusoidal Signal by a Phasor

Let us take current, $i(t) \sin t$, as an example of a sinusoidal quantity which we will represent using phasors. In Fig. 4.7(a), the length, OP, along the x-axis represents the maximum value of the current I_m on a certain scale. If it is rotated in the anti-clockwise direction at an angular speed, ω, it will take up a position, OA after a time t (or angle, $\theta = \omega.t$ with the x-axis). The vertical projection of OA is plotted in the right-hand side of the figure with respect to the angle θ. It will generate a sine wave (Fig. 4.7(b)), as OA is at an angle, θ with the x-axis, as stated earlier. The vertical projection of OA along the y-axis is OC = AB = $i(\theta) = I_m \sin\theta$, which is the instantaneous value of the current at any time t or angle θ. The angle θ is in rad., i.e., $\theta = \omega t$. The angular speed, ω is in rad/s, i.e., $\omega = 2\pi f$, where f is the frequency in Hz or cycles/s. Thus,

$$i = I_m \sin\theta = I_m \sin(\omega t) = I_m \sin(2\pi f t)$$

OP represents the *phasor*, with respect to the above current, i. Phasors separate the dependencies on A, ω, and θ into three independent factors.

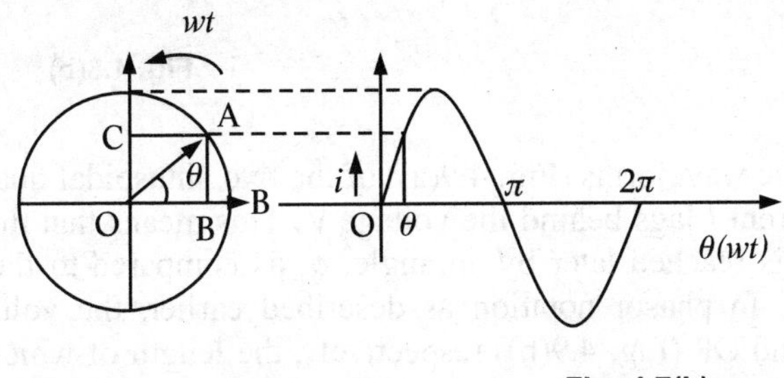

Fig. 4.7(a) **Fig. 4.7(b)**

The line, OP can be taken as the rms value, $I = \frac{I_m}{\sqrt{2}}$, instead of the maximum value, I_m. Then, the vertical projection of OA, in magnitude equal to OP, does not exactly represent the instantaneous value of I, but the current's value with the scale factor of $\frac{1}{\sqrt{2}} = 0.707$. The reason for the choice of phasor given here, will be explained later, and also in the next chapter.

GENERALISED CASE

Let us consider the current to be of the form,

$$i(t) = I_m \sin(\omega t + \alpha).$$

The phasor representation of this current is the line, OQ, at an angle, α, with the line OP along the x-axis (Fig. 4.8(a)). After time, t, OD will be at an angle $\theta = \omega t$ with OQ, which is at an angle $\theta + \alpha = \omega t + \alpha$, with the line OP along the x-axis. The sine wave is shown in Fig. 4.8(b). The vertical projection of OD along the y-axis gives the instantaneous value of the current,

$$i(t) = I_m \sin(\theta + \alpha) = I_m \sin(\omega t + \alpha) = \sqrt{2} I \sin(\omega t + \alpha),$$

where I_m and I are the maximum and rms values of the current respectively.

PHASOR REPRESENTATION OF VOLTAGE AND CURRENT

The voltage and current waveforms are given as,

$$v = \sqrt{2} V, \text{ where } \theta = \omega t \text{ and } i = \sqrt{2} I \sin(\theta - \phi).$$

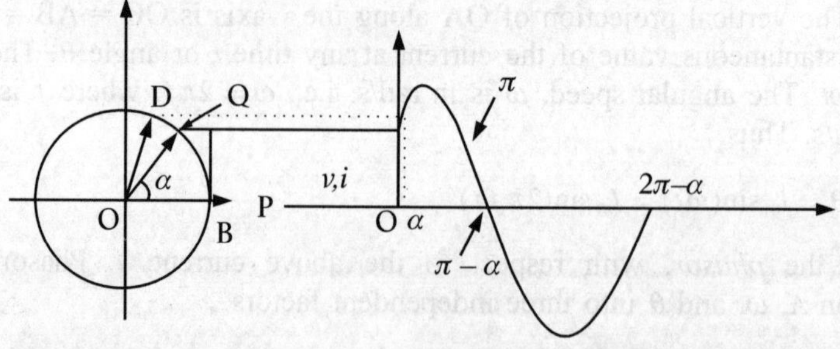

Fig. 4.8(a)

Fig. 4.8(b)

It can be seen from the waveforms (Fig. 4.9(a)) of the two sinusoidal quantities—voltage and current—that the current I lags behind the voltage V. This means that the positive maximum value of the current is reached later by an angle, ϕ, as compared to the positive maximum value of the current. In phasor notation as described earlier, the voltage and current are represented by OQ and OP (Fig. 4.9(b)) respectively, the length of which are proportional to voltage V, and current I, in different scales as applicable to each one. The voltage phasor, OP (V) leads the current phasor, OQ (I) by angle, ϕ, as the two phasors rotate in the anti-clockwise direction as stated earlier, where the angle, ϕ is also measured in the anti-clockwise direction.

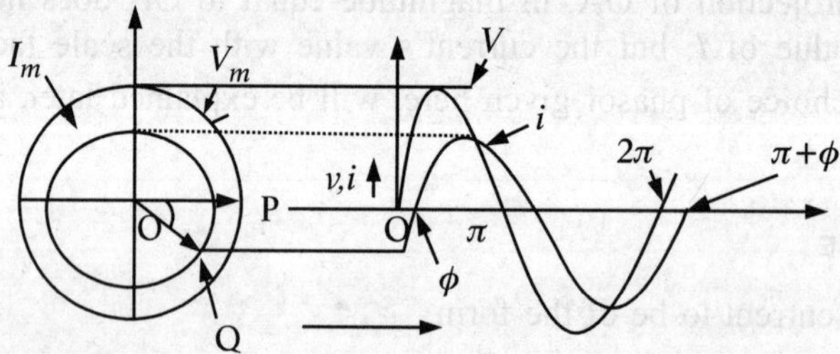

Fig. 4.9(a)

Fig. 4.9(b)

Mathematically, the two phasors can be represented in *polar form,* with the current phasor (I) taken as the reference, such that $I = I\angle 0°$, and $V = V\angle\phi°$.

In *Cartesian* or *rectangular form*, these are,

$$I = I\angle 0° = (I + j0), \text{ and } V = V\angle\phi = (V\cos\phi + jV\sin\phi),$$

where the symbol, j is given by $j = \sqrt{-1}$.

Of the two terms in each phasor, the first one is termed as *real* or its component is on the x-axis, while the second one is *imaginary* or its component is on the y-axis, as shown in Fig. 4.10. The angle, θ is in degree or rad.

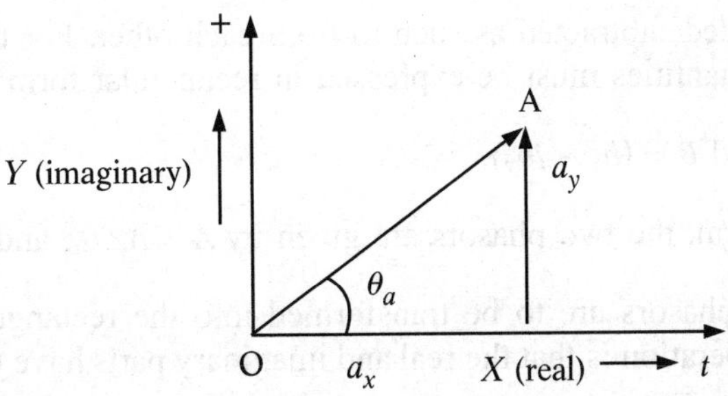

Fig. 4.10

4.4 Phasor Algebra

Before discussing mathematical operations, like addition/subtraction and multiplication /division, involving phasors, let us take a look at the two forms—polar and rectangular—by which a phasor or complex quantity is represented. It may be observed here that phasors are also taken as complex quantities, as given earlier.

Representation and Transformation of a Phasor

A phasor or a complex quantity in rectangular form (Fig. 4.10) can be represented as $A = (a_x + ja_y)$, where a_x and a_y are the real and imaginary parts of the phasor respectively. In polar form, it is expressed as

$$A = A\angle\theta_a = (A\cos\theta_a + jA\sin\theta_a),$$

where A and θ_a are the magnitude and phase angle of the phasor.

From the above two equations or expressions, the procedure or rule of transformation from polar to rectangular form is as given here.

$$a_x = A\cos\theta_a, \text{ and } a_y = A\sin\theta_a.$$

The rule for transformation from rectangular to polar form is,

$$A = \sqrt{(a_x)^2 + (a_y)^2} \text{ and } \theta_a = \tan^{-1}\left(\frac{a_y}{a_x}\right).$$

Examples of how this transformation works out using numerical values are given later.

Addition/Subtraction of Phasors

Addition/subtraction of phasors or complex quantities follow much of the same rules of addition/subtraction of scalar quantities, which may be positive or signed (decimal/fraction or fraction with an integer). For the two operations, the quantities must be either phasors or complex. An example of a phasor is voltage/current, and that of a complex quantity is impedance/admittance, which will be explained in the next chapter. A phasor and a complex quantity cannot be added/subtracted as such to/from each other. For the operations, the two phasors or complex quantities must be expressed in rectangular form as

$$A = (a_x + ja_y) \text{ and } B = (b_x + jb_y).$$

If they are in polar form, the two phasors are given by $A = A\angle\theta_a$, and $B = B\angle\theta_b$.

In this case, the two phasors are to be transformed into the rectangular form. The rule of addition/subtraction operation is that the real and imaginary parts have to be treated separately as,

$$C = A \pm B = [(a_x \pm ja_y) + (b_x \pm jb_y)] = (c_x + jc_y),$$

where $\quad c_x = (a_x \pm b_x)$, and $c_y = (a_y \pm b_y)$.

That is, for addition, real parts must be added together separately and imaginary parts must be added together separately. The same rule is also followed for subtraction. After the result is obtained in rectangular form, it can be transformed to a polar one.

It may be observed that the six values of a, b and c—parts of the two phasors and the resultant one, are all signed scalar quantities, though in the example, a and b are taken as positive, resulting in positive values of c. Also the phase angle, θ may lie in any of the four quadrants, though here the angles are in the first quadrant only. This rule for addition can be extended to three or more quantities.

Addition/subtraction operations can also be performed using the quantities as phasors in polar form (Fig. 4.11). The two phasors are A (OA) and B (OB). To find the sum C (OC), a line AC is drawn equal and parallel to OB. The line BC is equal and parallel to OA. Thus,

$$C = \text{OC} = \text{OA} + \text{AC} = \text{OA} + \text{OB} = A + B \text{ (as AC = OB).}$$

$$\text{Also,} \quad \text{OC} = \text{OB} + \text{BC} = \text{OB} + \text{OA} = B + A \text{ (as BC = OB).}$$

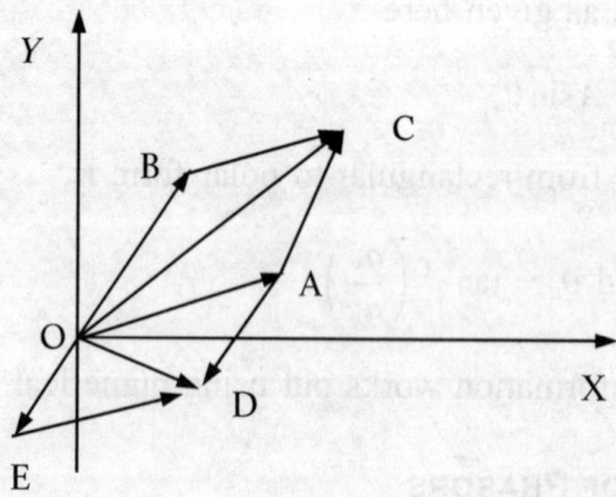

Fig. 4.11

The law of association can be applied to all operations, i.e., addition here, $(C = A + B = B + A)$, and later for multiplication.

To obtain the difference D (OD), a line AD is drawn equal and parallel to OB, but in the opposite direction to AC or OB. A line OE is also drawn equal to OB, but in the opposite direction to OB. Both AD and OE represent the phasor (– B). The line, ED is equal to OA. Thus,

$$D = \text{OD} = \text{OA} + \text{AD} = \text{OA} - \text{OB} = A - B \text{ (as AD = OB)}.$$

Also, $\text{OD} = \text{OE} + \text{ED} = -\text{OB} + \text{OA} = -B + A$ (as ED = – OB).

Here also the law of association is applied.

Examples using numerical values are given later.

MULTIPLICATION/DIVISION OF PHASORS

Firstly, we will describe the procedure for multiplication. We assume that the two phasors are available in polar form as $A = A\angle\theta_a$, and $B = B\angle\theta_b$.

Otherwise, they are to be transformed from the rectangular to the polar form. This must also be done if we need to divide phasors. Please note that a phasor is to be multiplied by a complex quantity only to obtain the resultant phasor. A phasor is not normally multiplied by another phasor, except in special cases. It may be divided by another phasor, to obtain a complex quantity.

A phasor can also be divided by a complex quantity to obtain the resultant phasor.

To find the magnitude of the product C, the two magnitudes of the phasors are to be multiplied, whereas for phase angle, the phase angles are to be added. Thus,

$$C = C\angle\theta_c = A \cdot B = A\angle\theta_a \cdot B\angle\theta_b = (A \cdot B)\angle(\theta_a + \theta_b),$$

where, $C = AB$, and $\theta_c = (\theta_a + \theta_b)$.

Please note that the same symbol, C is used for the product in this case. If the phasors are expressed in rectangular form as

$$A = A\angle\theta_a = (a_x + ja_y) \text{ and } B = B\angle\theta_b = (b_x + jb_y),$$

where, $A = \sqrt{(a_x)^2 + (a_y)^2}$, and $\theta_a = \tan^{-1}\left(\dfrac{a_y}{a_x}\right)$;

$$B = \sqrt{(b_x)^2 + (b_y)^2}, \text{ and } \theta_b = \tan^{-1}\left(\frac{b_y}{b_x}\right).$$

The resultant phasor is obtained as

$$C = C\angle\theta_c = (c_x + jc_y) = A \cdot B = [(a_x + ja_y) \cdot (b_x + jb_y)] = [(a_x b_x - a_y b_y) + j(a_x b_y + a_y b_x)].$$

Please note that $j^2 = -1$. The real and imaginary parts of C, which is in rectangular form, are given by,

$$c_x = (a_x b_x - a_y b_y), \quad \text{and} \quad c_y = (a_x b_y + a_y b_x)$$

The magnitude and phase angle of the resultant phasor, C are $\sqrt{(c_x)^2+(c_y)^2}$, and $\theta_c = \tan^{-1}\left(\frac{c_y}{c_x}\right)$ respectively.

Substituting the values as given above, we get

$$C = \sqrt{(a_xb_x - a_yb_y)^2 + (a_xb_y + a_yb_x)^2} = \sqrt{[(a_x)^2 + (a_y)^2]} \cdot \sqrt{[(b_x)^2 + (b_y)^2]} = A \cdot B, \text{ and}$$

$$\theta_c = \tan^{-1}\left(\frac{a_xb_y + a_yb_x}{a_xb_x - a_yb_y}\right).$$

The resultant for θ_c can be obtained by using the expression derived earlier

$$\theta_c = (\theta_a + \theta_b) = \left[\tan^{-1}\left(\frac{a_y}{a_x}\right) + \tan^{-1}\left(\frac{b_y}{b_x}\right)\right].$$

Substituting, $m = \frac{a_y}{a_x}$, and $n = \frac{b_y}{b_x}$, we get,

$$\theta_c = (\tan^{-1} m + \tan^{-1} n) = \tan^{-1}\frac{(m+n)}{(1 - m \cdot n)}.$$

Again substituting the expressions for m and n, and multiplying both the numerator and denominator by (a_xb_x), the relation as derived earlier, is obtained as

$$\theta_c = \tan^{-1}\left(\frac{a_xb_y + a_yb_x}{a_xb_x - a_yb_y}\right).$$

Let us now divide the phasor, A by another phasor, B, to obtain the resultant phasor, D. The magnitude is obtained by division of the magnitudes, and the phase is the difference of the two phase angles. Thus,

$$D = D\angle\theta_d = \frac{A}{B} = \frac{A\angle\theta_a}{B\angle\theta_b} = \left(\frac{A}{B}\right)\angle(\theta_a - \theta_b),$$

where $D = \frac{A}{B}$, and $\theta_d = \theta_a - \theta_b$.

Please note that the same symbol, D is used for the resultant phasor. The phasor, A is divided by another phasor, B, to obtain the resultant phasor, D. The phasors, A and B are in rectangular form. The phasor, D is

$$D = D\angle\theta_d = (d_x + jd_y) = \frac{A}{B} = \frac{(a_x + a_y)}{(b_x + jb_y)}.$$

To simplify D, i.e., to obtain the real and imaginary parts, both the numerator and denominator, are to be multiplied by the complex of B, so as to convert the denominator into the real value only. The complex conjugate of B is,

$$B^* = B\angle -\theta_b = (b_x - jb_y).$$

In the complex conjugate of the phasor, the sign of the imaginary part is negative. The phase angle is also negative.

$$D = (d_x + jd_y) = \frac{A}{B} = \frac{(a_x + ja_y)}{(b_x + jb_y)} = \frac{(a_x + ja_y) \times (b_x - jb_y)}{(b_x + jb_y) \times (b_x - jb_y)}$$

$$= \frac{[(a_x a_y + b_x b_y) + j(a_y b_x - a_x b_y)]}{[(b_x)^2 + (b_y)^2]}$$

$$= \left(\frac{(a_x a_y + b_x b_y)}{[(b_x)^2 + (b_y)^2]}\right) + j\left(\frac{(a_y b_x - a_x b_y)}{[(b_x)^2 + (b_y)^2]}\right).$$

The magnitude and phase angle of the resultant (phasor) are

$$D = \frac{\sqrt{[(a_x a_y + b_x b_y)^2 + (a_y b_x - a_x b_y)^2]}}{[(b_x)^2 + (b_y)^2]} = \frac{\sqrt{[(a_x)^2 + (a_y)^2]}}{\sqrt{[(b_x)^2 + (b_y)^2]}} = \frac{A}{B}, \text{ and}$$

$$\theta_d = \tan^{-1}\left(\frac{a_y b_x - a_x b_y}{a_x b_x + a_y b_y}\right).$$

The result for θ_d can be obtained by using the expression derived earlier as

$$\theta_d = (\theta_a - \theta_b) = \left[\tan^{-1}\left(\frac{a_y}{a_x}\right) - \tan^{-1}\left(\frac{b_y}{b_x}\right)\right].$$

The above expression is simplified by the procedure given earlier. Firstly, the substitutions, $m = \frac{a_y}{a_x}$, and $n = \frac{b_y}{b_x}$ are used. Then, we get,

$$\theta_d = (\tan^{-1} m - \tan^{-1} n) = \tan^{-1} \frac{(m-n)}{(1 + m \cdot n)}.$$

Substituting them back, and then, multiplying both the numerator and denominator by $(a_x.b_x)$, the relation as derived earlier, is obtained as

$$\theta_d = \tan^{-1}\left(\frac{a_y b_x - a_x b_y}{a_x b_x + a_y b_y}\right).$$

(a) Represent A in Fig. 4.12 in rectangular and polar form.

(b) If another vector B is included (Fig. 4.13), perform the operation of addition and subtraction on the vectors A and B.

(c) Perform the operation of multiplication and division on the vectors A and B.

(a) Representation of a phasor in rectangular and polar form.

The phasor, A in the rectangular form (Fig. 4.12) is

$$A = A\angle\theta_a = (A\cos\theta_a + jA\sin\theta_a) = (a_x + ja_y) = (-2 + j4),$$

where the real and imaginary parts are $a_x = -2$, $a_y = 4$.

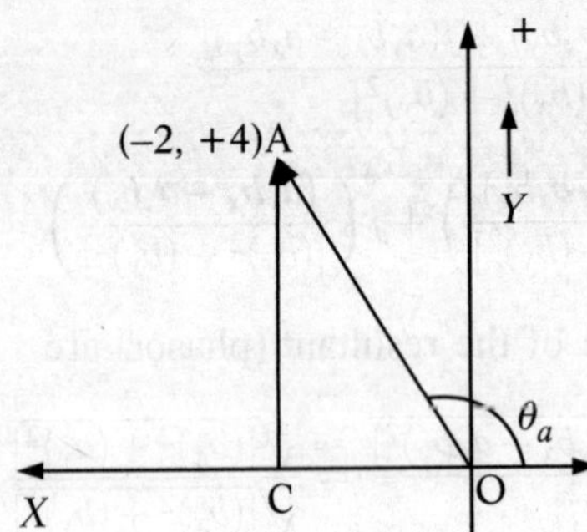

Fig. 4.12

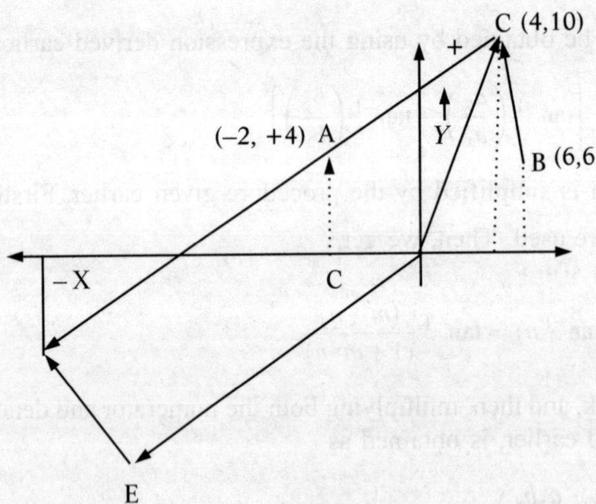

Fig. 4.13

To transform the phasor, A into the polar form, the magnitude and phase angle are

$$A = \sqrt{(a_x)^2 + (a_y)^2} = \sqrt{(-2)^2 + (4)^2} = 4.472, \text{ and}$$

$$\theta_a = \tan^{-1}\left(\frac{a_y}{a_x}\right) = \tan^{-1}\left(\frac{4}{-2}\right) = 116.565° = 2.034 \text{ rad}$$

Please note that θ_a is in the second quadrant, as the real part is negative, and the imaginary part is positive. Transforming the phasor, A into rectangular form, the real and imaginary parts are given by

$$a_x = A\cos\theta_a = 4.472 \times \cos 116.565^\circ = -2.0, \text{ and}$$

$$a_y = A\sin\theta_a = 4.472 \times \sin 116.565^\circ = 4.0.$$

(b) Addition/subtraction of phasors *A* and *B*

B in rectangular form is given below.

$$B = B\angle\theta_b = (B\cos\theta_b + jB\sin\theta_b) = (b_x + jb_y) = (6 + j6) = 8.845\angle 45^\circ,$$

The sum and difference of A and B are given by the phasors, C and D respectively (Fig. 4.12).

$$C = A + B = [(-2 + j4) + (6 + j6)] = [(-2 + 6) + j(4 + 6)] = (4 + j10) = 10.77\angle 68.2^\circ.$$

$$D = A - B = [(-2 + j4) - (6 + j6)] = [(-2 - 6) + j(4 - 6)] = -(8 + j2) = 8.246\angle -166^\circ.$$

The symbols, C and D, used here, are the same as given earlier. It may be noted that for addition and subtraction operations involving phasors, they should be represented in rectangular form as given above. If any one of the phasors is in polar form, it should be transformed into rectangular form. If the two phasors are both in polar form, the phasor diagram (the diagram must be drawn to scale), or the geometrical method can be used as shown in Fig. 4.13. The result obtained, using the diagram as shown are the same as obtained earlier.

$$C(\text{OC}) = 10.77, \angle\text{COE} = 68.2^\circ, \text{ and } D(\text{OD}) = 8.246, \angle\text{DOE} = 166^\circ.$$

(c) Multiplication/division of phasors *A* and *B*

The multiplication and division operations are performed on A and B in polar form. If any one of the phasors is in rectangular form, it may be transformed into polar form. Also note that the same symbols for the phasors are used here, as was used earlier. How to multiply and divide using the rectangular form of the phasor representation will be explained later.

The resultant phasor C, i.e., the product of the two phasors, A and B given in polar form, is

$$C = A \cdot B = [(4.472\angle 116.565^\circ) \times (8.485\angle 45^\circ)] = (4.472 \times 8.485)\angle(116.565^\circ + 45.0^\circ)$$

$$= 37.945\angle 161.565^\circ = (-36.0 + j12.0).$$

The resultant phasor C, i.e., the product of the two phasors, A and B in rectangular form, is

$$C = A \cdot B = [(-2 + j4) \times (6 + j6)] = [(-12 - 24) + j(24 - 12)] = (-36.0 + j12.0)$$

$$= 37.947\angle 161.565^\circ.$$

The result obtained is same as earlier.

The resultant phasor, D is obtained when the first phasor, A, is divided by the second phasor, B, both in polar form as

$$D = \frac{A}{B} = \frac{4.472\angle 116.565^\circ}{8.485\angle 45.0^\circ} = \left(\frac{4.472}{8.485}\right)\angle(116.565^\circ - 45.0^\circ)$$

$$= 0.527\angle 71.565^\circ = (0.167 + j0.5).$$

The above result can be calculated by the procedure described earlier, using the rectangular form of the two phasors as,

$$D = \frac{A}{B} = \frac{(-2+j4)}{(6+j6)} = \frac{(-2+j4)\times(6-j6)}{(6+j6)\times(6-j6)} = \frac{[(24-12)+j(24+12)]}{(6^2+6^2)} = \frac{(12.0+j36.0)}{72.0}$$

$$= (0.167 + j0.5) = 0.527\angle 71.565^\circ.$$

The result obtained is the same as earlier.

The simplification procedure applied to scalar quantities, using different elementary operations, which is well known, can be extended to phasor quantities. This will be used in the study of ac circuits to be presented in the next chapter.

Conclusion

In this chapter, we first studied how the sinusoidal voltage waveform was generated. We then defined average, and root mean square (rms) values of periodic voltage/current waveform, and other factors. Phasor representation of sinusoidal quantities (voltage/current), and its algebra—mathematical operations, such as addition/subtraction and multiplication/division, of phasors or complex quantities, including transformation of phasors from rectangular to polar form, and vice versa—have been discussed. The study of ac circuits, starting from series circuits, will be described in the next chapter.

Multiple Choice Type Questions

Tick the correct option(s), out of the four choices provided in each question.

4.1 The rms value of the periodic waveform given in Fig. MCQ4.1 is

(a) $2\sqrt{6}$ A (b) $6\sqrt{2}$ A (c) $\frac{4}{3}$ A (d) 1.5 A

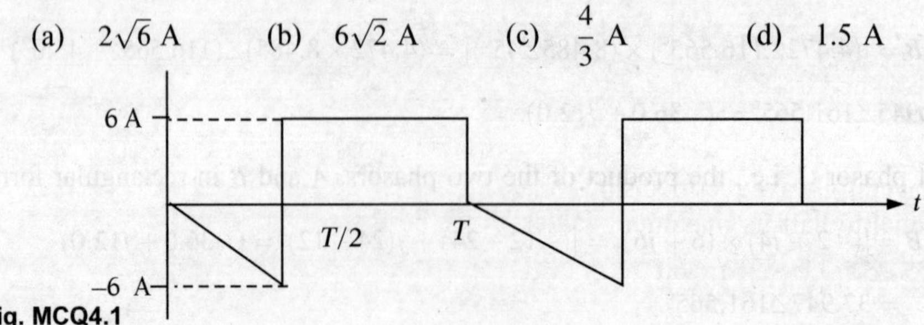

Fig. MCQ4.1

4.2 For the triangular waveform shown in Fig. MCQ4.2, the RMS value of the voltage is equal to

(a) $\sqrt{\frac{1}{6}}$ (b) $\sqrt{\frac{1}{3}}$ (c) $\frac{1}{3}$ (d) $\sqrt{\frac{2}{3}}$

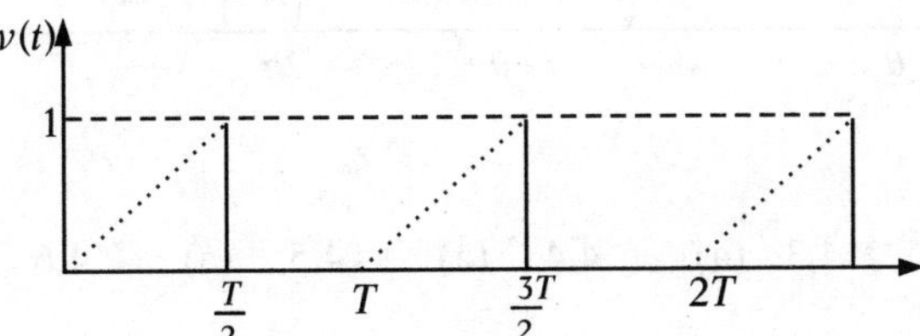

Fig. MCQ4.2

4.3 A periodic voltage waveform observed on an oscilloscope across a load is shown in Fig. MCQ4.3. A permanent magnet moving coil (PMMC) meter connected across the same load shows

(a) 4 V (b) 5 V (c) 8 V (d) 10 V

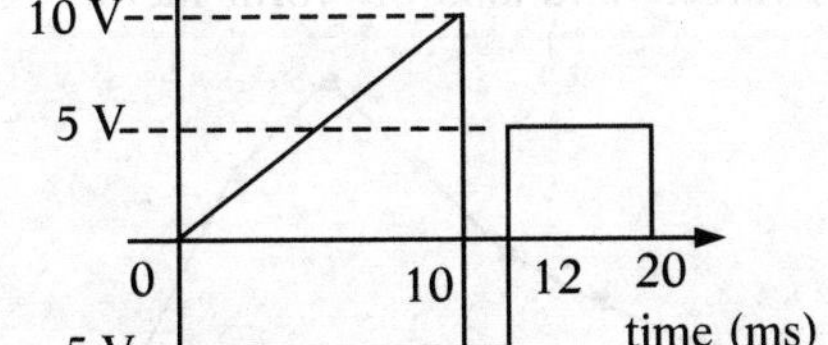

Fig. MCQ4.3

4.4 The rms value of the voltage $u(t) = 3 + 4\cos 3t$ is

(a) $\sqrt{17}$ V (b) 5 V (c) 7 V (d) $(3+2\sqrt{2})$ V

4.5 For the signal shown in Fig. MCQ4.5 The average value of the signal is

(a) $\frac{7}{10}I_m$ (b) $\frac{5}{17}I_m$ (c) $\frac{3}{10}I_m$ (d) $\frac{1}{10}I_m$

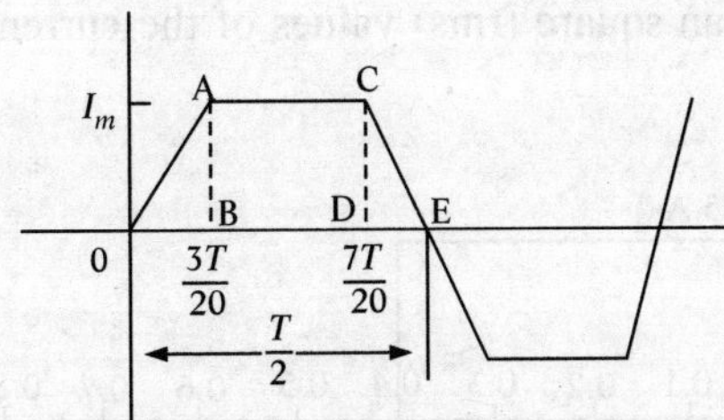

Fig. MCQ4.5

4.6 The delayed fullwave rectified sinusoidal signal shown in Fig. MCQ 4.6 has an average value equal to half its maximum value. Then, the delay angle θ is

(a) 55.25° (b) 60° (c) 30° (d) 40°

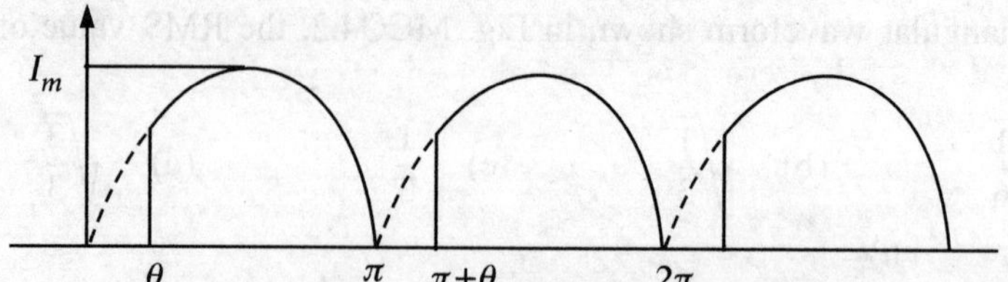

Fig. MCQ4.6

Answers

4.1. (d) 4.2. (b) 4.3. (a) 4.4. (a) 4.5. (a) 4.6. (a)

Problems

4.1 A voltage wave has the following values over one half cycle, both half cycles being symmetrical about the zero axis:

0, 30, 40, 45, 55, 80, 90, 56, 0

Plot the half cycle of the waveform in Fig. P4.1, and hence, determine the average and root mean square (rms) values, and also the form factor.

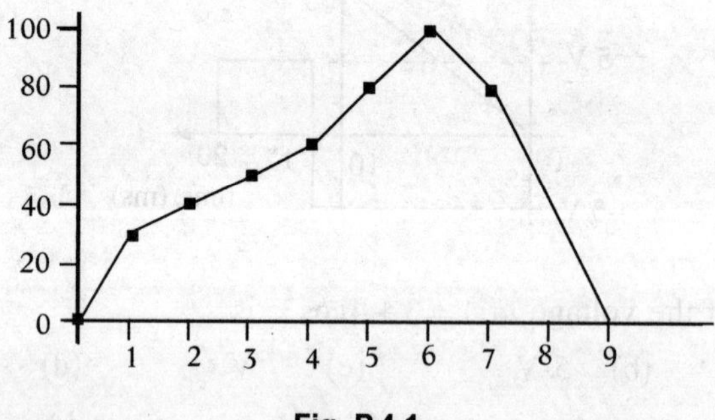

Fig. P.4.1

4.2 A constant current of 5 A flows for 0.04 s, and to complete the cycle, a constant current of 2 A flows in the opposite direction for 0.06 s (Fig. P4.2). Compute the average (or mean), and the root mean square (rms) values of the current waveform.

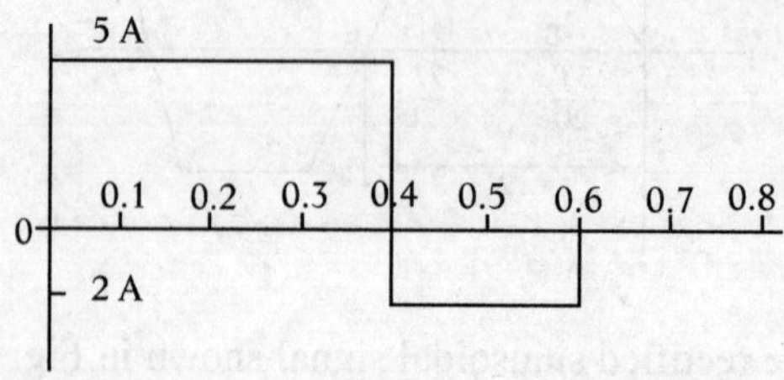

Fig. P.4.2

4.3 A trapezoidal waveform is given in Fig. P4.3 over one half cycle, both half cycles being symmetrical about the zero axis. The voltage increases linearly from 0 to 50 V in 1 s, remains constant at the same value of 50 V for 1 s, and then decreases linearly from 50 V to 0 in another 1 s. Find the average and rms values of the voltage waveform, and also the form factor.

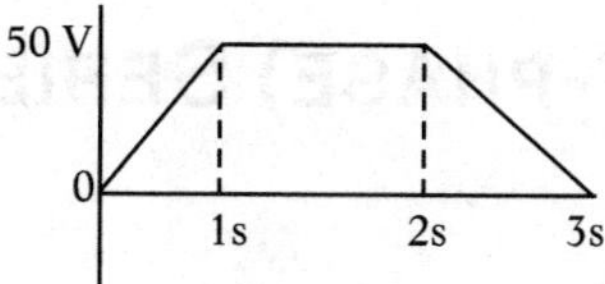

Fig. P.4.3

4.4 The waveform of an output current which consists of a part of the positive half cycle of a sine wave between the angles, α and $180°(\pi)$, is shown in Fig. P4.4, for one cycle. Note that it is zero in the negative half cycle. Find the average and rms values of the current for $\alpha = 30$.

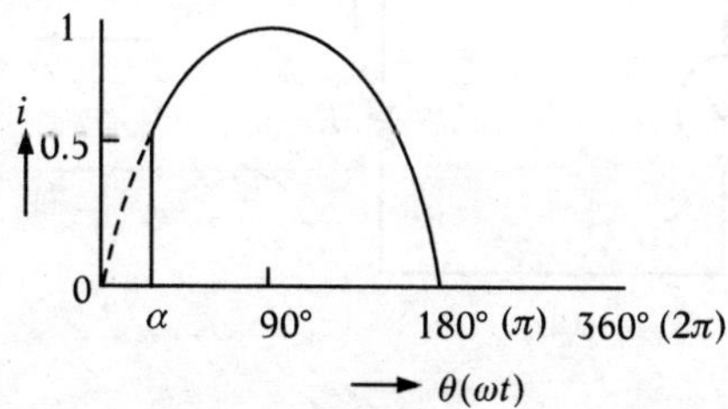

Fig. P.4.4

4.5 Transform the following complex quantities (from Cartesian or rectangular to polar form, and polar to Cartesian form, as the case may be):

(a) $(10 + j8)$ (b) $(-20 + j30)$

(c) $20\angle 70°$ (d) $30\angle -140°$

4.6 Perform the arithmetic operations (addition and/or subtraction) on the following complex quantities:

(a) $(20\angle 40° + 15\angle -30°)$ (b) $(25\angle 60° + 40\angle -15° + 35\angle 135°)$

(c) $(30\angle -120° - 10\angle 45°)$ (d) $(15\angle -40° - 55\angle 120° + 30\angle 70°)$

4.7 Perform the arithmetic operations (multiplication and/or division) on the following complex quantities:

(a) $(-12 + j16)(4 + j5)$ (b) $(3 + j9)(-4 + j6)(8 - j6)$

(c) $\dfrac{(5 - j8)}{(2 + j3)}$ (d) $\dfrac{(-10 + j15)}{(6 + j8)} \times (4 + j8)$

4.8 In a circuit, if two currents are in the form, $i_1 = 0.12 \sin(\omega t + 60°)$ and $i_2 = 0.08 \sin \omega t$, determine $(i_1 - i_2)$.

4.9 Three voltages are represented by

$v_1 = 40 \sin(314t + 30°)$, $v_2 = 50 \sin(314t + 60°)$ and $v_3 = 25 \sin(314t - 45°)$.

Calculate the resultant voltage and express it in the time domain form.

CHAPTER 5

AC (SINGLE-PHASE) SERIES CIRCUITS

In the previous chapter, we were introduced to sinusoidal voltage waveforms—how they are generated, and how they can be represented by phasors.

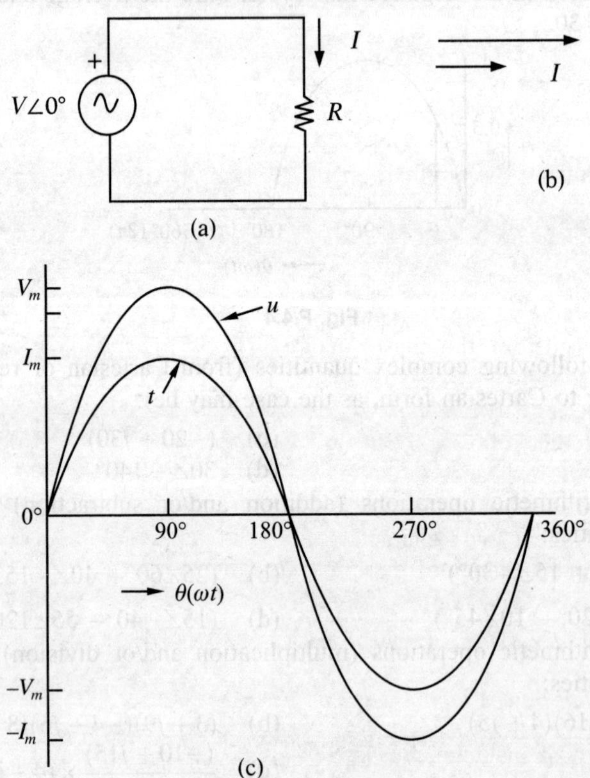

Fig. 5.1 *Circuit with resistance (R): (a) Circuit diagram; (b) phasor diagram; (c) waveforms (voltage/current)*

In this chapter, we will begin our study of ac current by discussing single-phase ac series circuits with the elements R, L and/or C. We will determine the current in the circuit when only one of the elements (R/L/C) is connected, when two elements (R and L or C) are connected, and all three elements (R, L and C), are connected in series. Impedance and reactance are explained using phasor diagrams. Lastly, we will study the resonance phenomenon in series R-L-C circuits fed from variable frequency ac (single-phase) supply.

5.1 Circuits using only a single element

5.1.1 Voltage, Current, Impedance, Phasor Diagram

PURELY RESISTIVE (R) CIRCUIT

The instantaneous value of a current i though a circuit (Fig. 5.1(a)) is given by,

$$i = \frac{v}{R} = \frac{V_m}{R}\sin\omega t = I_m \sin\omega t,$$

where I_m and V_m are the maximum values of current and voltage respectively. The rms value of current is given by,

$$I = \frac{I_m}{\sqrt{2}} = \frac{\frac{V_m}{\sqrt{2}}}{R} = \frac{V}{R}.$$

In phasor notation, using the applied voltage, V as reference,

$$V\angle 0^\circ = (V + j0), \text{ and } I\angle 0^\circ = (I + j0).$$

In this chapter on different types of series circuits fed from an ac supply, we will mostly take current, I as reference. In this case, both voltage and current are in phase ($\phi = 0^\circ$), which can be observed from the phasor diagram (Fig. 5.1(b)) with two (voltage and current) phasors, and also from the two waveforms (Fig. 5.1(c)).

In ac circuits, the term, *impedance* is defined as the ratio of voltage to current, similar to resistance in dc circuits following Ohm's law. The impedance, Z is a complex quantity. It consists of a real part, resistance R, and an imaginary part, reactance, X, which is zero (0) in this circuit, as there is no inductance or capacitance attached. All the components are taken as constant, having linear $V - I$ characteristics.

The impedance or resistance of the circuit is obtained as,

$$\frac{V\angle 0^\circ}{I\angle 0^\circ} = \frac{V}{I}\angle 0^\circ = Z\angle 0^\circ = (R + j0).$$

Power consumed and power factor

From the waveform of instantaneous power ($W = v.i$), shown in Fig. 5.1(c), for the above circuit, the average power is,

$$W = \frac{1}{2\pi}\int_0^{2\pi}(v\cdot i)\,d\theta = \frac{1}{2\pi}\int_0^{2\pi}[(\sqrt{2}V\sin\theta)\cdot\{\sqrt{2}I\sin(\theta)\}]\,d\theta$$

$$= \frac{1}{2\pi}\int_0^{2\pi} V\cdot I.2(\sin^2\theta)d\theta = \frac{1}{2\pi}V\cdot I\int_0^{2\pi}(1-\cos 2\theta)d\theta = VI.$$

Therefore, for a purely resistive (R) circuit, the power factor is unity (1), as $\phi = 0°$, and the average power is $V.I$.

PURELY INDUCTIVE (L) CIRCUIT

For the circuit in which an inductance is connected in series (Fig. 5.2(a)), the current i, is obtained by the procedure described before. As

$$v = L\frac{di}{dt} = V_m \sin \omega t = \sqrt{2}V \sin \omega t$$

or, $$di = \frac{\sqrt{2}V}{L}\sin(\omega t)dt.$$

Integrating, we get,

$$i = \int \frac{\sqrt{2}V}{L}\sin(\omega t)dt = -\frac{\sqrt{2}V}{\omega L}\cos(\omega t) = \frac{\sqrt{2}V}{\omega L}\sin(\omega t - 90°)$$

$$= \frac{\sqrt{2}V}{\omega L}\sin(\omega t - 90°) = I_m \sin(\omega t - 90°) = \sqrt{2}I \sin(\omega t - 90°).$$

It may be mentioned here that the current I, is the steady solution, neglecting the constant of integration. The rms value, I is

$$|I| = \frac{V}{\omega L} = \frac{V}{X_L}, \text{ its phase angle being } \phi = 90°.$$

So, $I\angle -90° = (0 - jI)$, while V is the same as given earlier, which is $V\angle 0° = (V + j0)$. The impedance of the circuit is obtained as,

$$Z\angle 90° = \frac{V\angle 0°}{I\angle -90°} = \frac{V}{I}\angle 90°.$$

Using phasor algebra (division) and the rectangular form, the impedance is given by,

$$Z\angle 90° = \frac{V}{-jI} = j\omega L = (0 + jX_L) = X_L\angle 90°,$$

where the inductive reactance in Ω is given by $X_L = \omega L = 2\pi f L$, with angular frequency in rad/s, $(\omega = 2\pi f)$.

Note that the current lags behind the voltage by $\phi = +90°$, which is taken as positive. This can be observed both from the phasor diagram (Fig. 5.2(b)), and the voltage and current waveforms (Fig. 5.2(c)). As the circuit has only inductive reactance $X_L = \omega L$ (positive, according to convention), and no resistance, the impedance, Z is only in the y-axis (imaginary).

An alternative method, using current, I, instead of voltage, V as reference, can also be used in this case, as given earlier, to obtain all the expressions of voltage, with its phase angle.

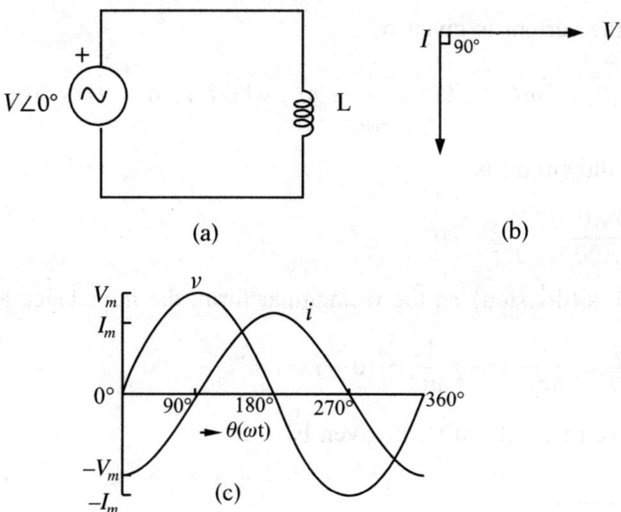

Fig. 5.2 *Circuit with inductance (L): (a) Circuit diagram; (b) phasor diagram; (c) waveforms (voltage/current)*

Power consumed and power factor

From the waveform of instantaneous power ($W = v.i$), shown in Fig. 5.2(c), for the above circuit, the average power is,

$$W = \frac{1}{2\pi}\int_0^{2\pi}(v \cdot i)\,d\theta = \frac{1}{2\pi}\int_0^{2\pi}[(\sqrt{2}V\sin\theta)\cdot\{\sqrt{2}I\sin(\theta - \frac{\pi}{2})\}]\,d\theta$$

$$= \frac{1}{2\pi}\int_0^{2\pi} V \cdot I.2(\sin\theta)(\cos\theta)d\theta = \frac{1}{2\pi}V \cdot I\int_0^{2\pi}(\sin 2\theta)d\theta = 0.$$

For the purely inductive (L) circuit, the power factor is zero (0), as $\phi = +90^\circ$, and the average power is zero.

Purely Capacitive (C) Circuit

For the circuit in which a capacitance is connected in series (Fig. 5.3(a)), the current i, is found as,

$$i = C\frac{dv}{dt}, \text{ where } v = V_m \sin\omega t = \sqrt{2}V\sin\omega t.$$

So,

$$i = C\frac{d}{dt}[\sqrt{2}V\sin(\omega t)\,dt] = \sqrt{2}\omega CV\cos(\omega t) = \sqrt{2}\omega CV\sin(\omega t + 90^\circ) = I_m\sin(\omega t + 90^\circ)$$

$$= \sqrt{2}I\sin(\omega t + 90^\circ).$$

The rms value of the current is given by

$$I\angle 90^\circ = (0 + jI) = (\omega C)V\angle 90^\circ = \frac{V}{\frac{1}{\omega C}}\angle 90^\circ, \text{ where } V\angle 0^\circ = (V + j0).$$

The impedance of the circuit is

$$Z\angle -90^\circ = \frac{V\angle 0^\circ}{I\angle 90^\circ} = \frac{V}{I}\angle -90^\circ.$$

Using phasor algebra (division) on the rectangular form, the impedance is given by,

$$Z\angle -90^\circ = \frac{V}{jI} = \frac{1}{j\omega C} = -j\frac{1}{\omega C} = (0 - jX_C) = X_C\angle -90^\circ$$

where the capacitive reactance in Ω, is given by

$$X_C = \frac{1}{\omega C} = \frac{1}{2\pi f C},$$

with angular frequency, ω as given earlier.

Note that the current leads the voltage by $\phi = -90^\circ$, with ϕ as negative according to the convention followed here.

This can be observed both from the phasor diagram (Fig. 5.3(b)), and the voltage and current waveforms (Fig. 5.3(c)). As the circuit has only a capacitive reactance with no resistance,

$$X_C = -\frac{1}{\omega C}$$

(which is taken as negative according to convention), the impedance, Z is only in the y-axis (imaginary).

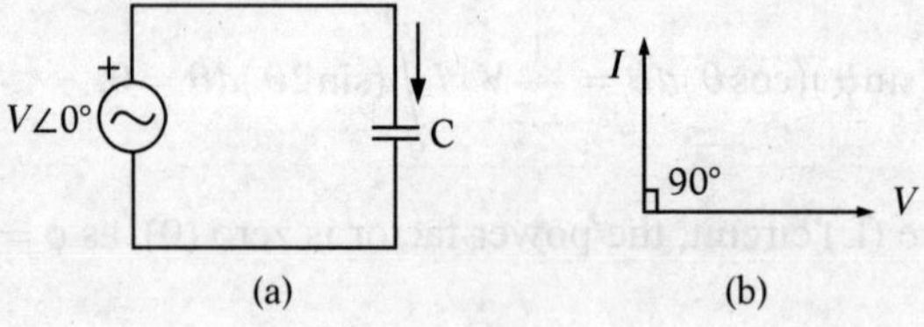

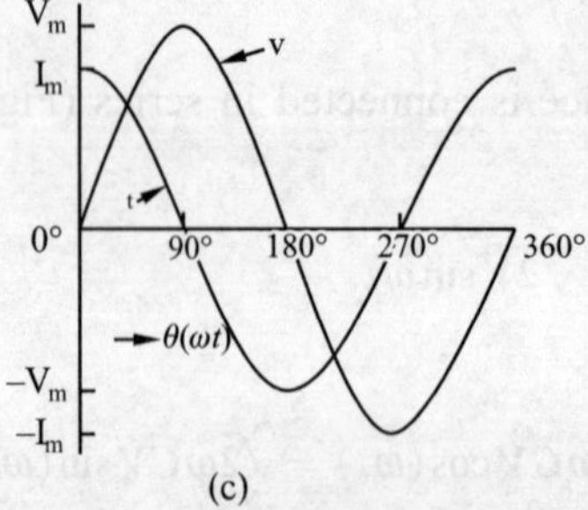

Fig. 5.3 *Circuit with capacitance (C): (a) Circuit diagram; (b) phasor diagram; (c) waveforms (voltage/current)*

Power consumed and power factor

From the waveform of instantaneous power ($W = v.i$) shown in Fig. 5.3(c), for the above circuit, the average power is,

$$W = \frac{1}{2\pi}\int_0^{2\pi}(v\cdot i)\,d\theta = \frac{1}{2\pi}\int_0^{2\pi}[(\sqrt{2}V\sin\theta)\cdot\{\sqrt{2}I\sin(\theta+\frac{\pi}{2})\}]\,d\theta$$

$$= \frac{1}{2\pi}\int_0^{2\pi}V\cdot I.2(\sin\theta)(\cos\theta)d\theta = \frac{1}{2\pi}V\cdot I\int_0^{2\pi}(\sin 2\theta)d\theta = 0.$$

For the purely capacitive (C) circuit, the power factor is zero (0), as $\phi = -90°$, and the average power is zero.

5.2 Series Circuits

INDUCTIVE CIRCUIT (R AND L IN SERIES)

The voltage balance equation for the R-L series circuit (Fig. 5.4(a)) is,

$$v = R\cdot i + L\cdot\frac{di}{dt},$$

R I L + ~
(a)

V A O φ I IjX_L IR C
(b)

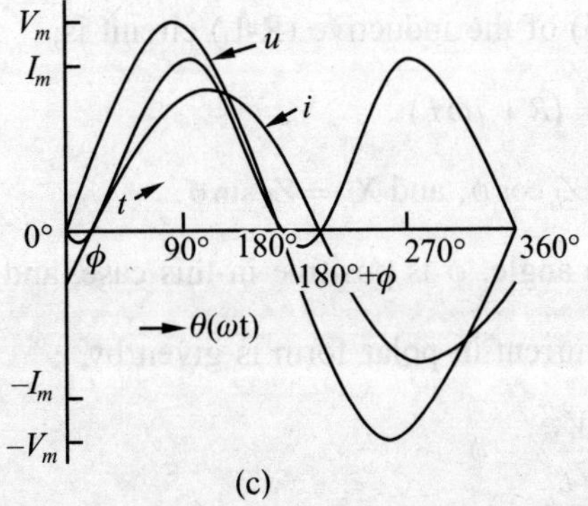

Fig. 5.4 *Circuit with resistance (R) and inductance (L) in series: (a) Circuit diagram; (b) phasor diagram; (c) waveforms (voltage/current)*

where $v = \sqrt{2}V\sin\omega t = V_m\sin\omega t = \sqrt{2}V\sin\theta$, with $\theta = \omega t = 2\pi ft$.

The current, i (in steady state) can be found as,

$$i = \sqrt{2}I\sin(\omega t - \phi) = I_m\sin(\omega t - \phi) = \sqrt{2}I\sin(\theta - \phi).$$

The steps needed to find the rms value of the current, I, are given here.

Substituting the expressions of i and v in the voltage balance equation, this expression is obtained as,

$$\begin{aligned}\sqrt{2}V\sin\theta &= R\cdot\sqrt{2}I\sin(\theta-\phi) + \omega L\cdot\sqrt{2}I\cos(\theta-\phi)\\ &= R\cdot\sqrt{2}I(\sin\theta\cos\phi - \cos\theta\sin\phi) + \omega L\cdot\sqrt{2}I(\cos\theta\cos\phi + \sin\theta\sin\phi)\\ &= R\cdot\sqrt{2}I(\sin\theta\cos\phi - \cos\theta\sin\phi) + X_L\cdot\sqrt{2}I(\cos\theta\cos\phi + \sin\theta\sin\phi),\end{aligned}$$

where $X_L = \omega L$, as given earlier.

Equating the terms of $\sin\theta$ and $\cos\theta$ of the LHS and RHS of the expression, we get,

$$V = R\cdot I\cos\phi + X_L\cdot I\sin\phi,\ \text{and}\ 0 = -R\cdot I\sin\phi + X_L\cdot I\cos\phi.$$

We will first find the magnitude and phase angle of the impedance. The LHS and RHS of the two expressions are divided by I to get

$$\left|\frac{V}{I}\right| = |Z_L| = R\cdot\cos\phi + X_L\cdot\sin\phi\ \text{and}\ 0 = -R\cdot\sin\phi + X_L\cdot\cos\phi.$$

Now, the squares of the two expressions are added and its square root is taken to find the magnitude of the impedance. Its angle is easily derived from the second expression.

$$|Z_L| = \sqrt{[R^2 + (X_L)^2]} = \sqrt{[R^2 + (\omega L)^2]},\ \text{and}\ \phi = \tan^{-1}\left(\frac{X_L}{R}\right) = \tan^{-1}\left(\frac{\omega L}{R}\right).$$

The impedance (Fig. 5.5) of the inductive (R-L) circuit is,

$$Z_L\angle\phi = (R + jX_L) = (R + j\omega L)$$

from which we get, $R = Z_L\cos\phi$, and $X_L = Z_L\sin\phi$.

Note that the impedance angle, ϕ is positive in this case, and ϕ can be written as ϕ_L.

The expression for the current in polar form is given by,

$$I\angle-\phi = \frac{V\angle 0^\circ}{Z_L\angle\phi} = \frac{V}{Z_L}\angle-\phi$$

where $|I| = \left|\frac{V}{Z_L}\right| = \frac{V}{\sqrt{[R^2 + (X_L)^2}} = \frac{V}{\sqrt{[R^2 + (\omega L)^2]}}.$

Using phasor algebra (division) on the rectangular form, the current is given by,

$$I\angle -\phi = \frac{V+j0}{R+jX_L} = \frac{V+j0}{R+j\omega L},$$

which when simplified, will give the same expression for the magnitude and phase angle of the current, I.

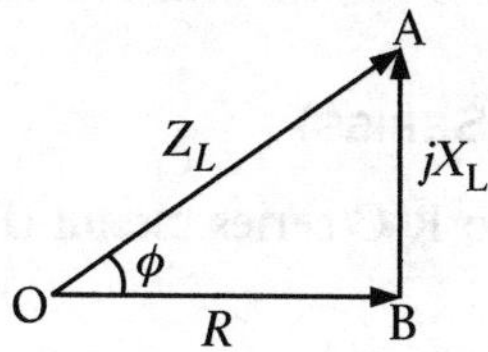

Fig. 5.5

Note that the current lags behind the voltage by ϕ, which is given earlier. This can be observed both from phasor diagram (Fig. 5.4(b)), and the voltage and current waveforms (Fig. 5.4(c)). The inductive reactance is positive. In the phasor diagram, as one moves from the voltage phasor to the current phasor, one has to go in the *clockwise* direction, which means that phase angle, ϕ is taken as positive, though both phasors are assumed to move in the anti-clockwise direction as shown earlier. The complete phasor diagram is shown in Fig. 5.4(b), with the voltage drops across the two components, the input (supply) voltage (OA), and the current (OB). The voltage phasor is taken as reference. It may be observed that

$$V_{OC}(=I\cdot R)+V_{CA}(=I\cdot X_L)=V_{OA}(=I\cdot Z_L),$$

using Kirchhoff's second law relating to the voltage in a closed loop. The phasor diagram can also be drawn with the current phasor as reference, as shown later.

Power consumed and power factor

From the waveform of instantaneous power ($W = v.i$), shown in Fig. 5.4(c), for the above circuit, the average power is,

$$W = \frac{1}{2\pi}\int_0^{2\pi}(v\cdot i)\,d\theta = \frac{1}{2\pi}\int_0^{2\pi}[(\sqrt{2}V\sin\theta)\cdot(\cos\phi)2\pi - \frac{1}{2}[\sin(4\pi-\phi)-\sin\phi]\,d\theta$$

$$= \frac{1}{2\pi}\int_0^{2\pi}V\cdot I[\cos\phi - \cos(2\theta-\phi)]\,d\theta = \frac{1}{2\pi}\left[(V\cdot I)\left\{(\cos\phi)\cdot\theta|_0^{2\pi} - \frac{1}{2}[\sin(2\theta-\phi)|_0^{2\pi}]\right\}\right]$$

$$= \frac{1}{2\pi}\left[(V\cdot I)\left\{(\cos\phi)\cdot 2\pi - \frac{1}{2}[\sin(4\pi-\phi)-\sin\phi\right\}\right] = V\cdot I\cos\phi.$$

Note that the power is only consumed in resistance, R, not in the inductance, L.

So, $W = I^2R$, and the power factor is the ratio of the average or active power to apparent power, which is found as

$$\frac{V \cdot I \cos\phi}{V \cdot I} = \cos\phi = \frac{R}{Z_L} = \frac{R}{\sqrt{[R^2 + (\omega L)^2]}}.$$

The power factor (pf) in this circuit is less than unity (1.0), its range is $0 < \text{pf}\,(\cos\phi) < 1$. The range of the phase angle is $0° < \phi < 90°$, with ϕ being positive as given above.

CAPACITIVE CIRCUIT (R AND C IN SERIES)

The voltage balance equation for the R-C series circuit (Fig. 5.6(a)) is,

$$v = R \cdot i + \frac{1}{C} \cdot \int i \cdot dt = \sqrt{2}V \sin\omega t,$$

The current, I (in steady state) can be derived as

$$i = \sqrt{2}I \sin(\omega t + \phi) = \sqrt{2}I \sin(\theta + \phi)$$

The impedance of the circuit with R and C in series, is

$$Z_C\angle -\phi = (R - jX_C) = \left(R - j\frac{1}{\omega C}\right)$$

where $|Z_C| = \sqrt{[R^2 + (X_C)^2]} = \sqrt{\left[R^2 + \left(\frac{1}{\omega C}\right)^2\right]}$,

and $\phi = \tan^{-1}\left(-\frac{X_C}{R}\right) = \tan^{-1}\left(\frac{1}{\omega CR}\right).$

Note that the impedance angle, ϕ is negative here, and can be written as ϕ_C.

As shown in the series R-L circuit, the impedance can be derived by substituting the expressions for voltage and current in the voltage balance equation (not repeated here).

Voltage, V is taken as reference, with the current, I leading by ϕ. The equations involving voltage and current as phasors, and the impedance as a complex quantity, are given as,

$$V\angle 0° = I\angle\phi \cdot Z_C\angle -\phi \quad \text{or,} \quad Z_C\angle -\phi = \frac{V\angle 0°}{I\angle\phi} = \frac{V}{I}\angle -\phi$$

Using phasor algebra (division) on the rectangular form, the current is given by,

$$I\angle\phi = \frac{V + j0}{R - jX_C} = \frac{V + j0}{R - j\frac{1}{\omega C}},$$

which when simplified, will give the same expression for the magnitude and phase angle of the current, I, the magnitude of which is given by,

$$|I| = \left|\frac{V}{Z}\right| = \frac{V}{\sqrt{[R^2 + (X_C)^2]}} = \frac{V}{\sqrt{\left[R^2 + \left(\frac{1}{\omega C}\right)^2\right]}},$$

its phase angle remaining the same as given earlier. Note that the current leads the voltage by ϕ (the value as given above). This can be observed both from the phasor diagram (Fig. 5.6(b)), and the voltage and current waveforms (Fig. 5.6(c)). The capacitive reactance, X_C is negative. In the phasor diagram, as one moves from the voltage phasor to the current phasor, one has to go in the *anti-clockwise* direction, which means that the phase angle, ϕ is taken as negative. This is in contrast to the case described earlier. The complete phasor diagram is shown in Fig. 5.6(b), with the voltage drops across the two components, the input (supply) voltage and the current. The voltage phasor is taken as reference. The power factor in this circuit is less than unity (1), with ϕ being the same as given earlier. The expression for the average power is $P = V.I.\cos\phi$, which can be found by the method shown earlier. Power is only consumed in the resistance, R, not in the capacitance, C.

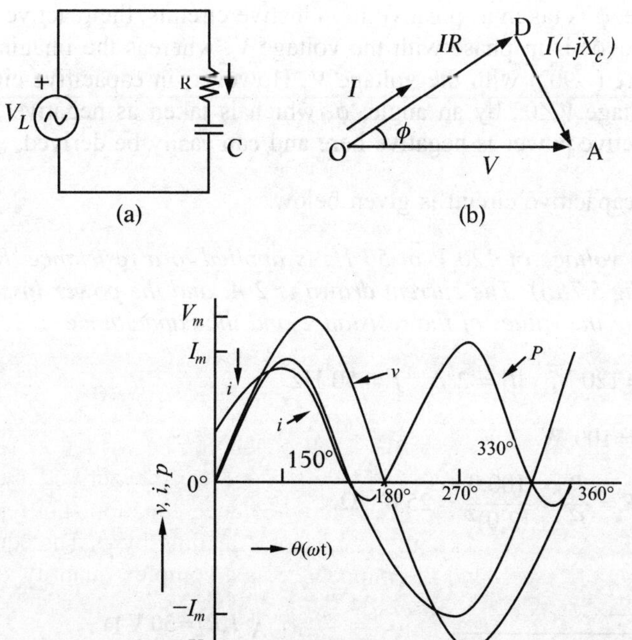

Fig. 5.6 *Circuit with resistance (R) and capacitance (C) in series: (a) Circuit diagram; (b) phasor diagram; (c) waveforms (voltage/current)*

Complex power, apparent power (VI) and reactive power (VAR)

The complex power is the product of the voltage and the complex conjugate of the current, both in phasor form. For the inductive circuit, described earlier, the voltage, ($V\angle 0° = V + j0$) is taken as reference and the current $I\angle - \phi = (I\cos\phi - jI\sin\phi)$ lags behind the voltage by an angle, ϕ.

The complex power is

$$S\angle\phi = V\cdot I^* = V\cdot\angle 0^\circ\cdot I\angle\phi = (V\cdot I)\angle\phi = (V\cdot I\cos\phi + jV\cdot I\sin\phi) = (P + jQ).$$

The magnitude of volt-amperes (S) is the product of the magnitudes of the voltage and the current. Hence,

$$|S| = |(V\cdot I)| = \sqrt{P^2 + Q^2}.$$

It is expressed in VA. The active power (W) is

$$P = \text{Re}(S) = \text{Re}(V\cdot I^*) = V\cdot I\cos\phi, \text{ as derived earlier.}$$

The reactive power (VAR) is given by

$$Q = \text{Im}(S) = \text{Im}(V\cdot I^*) = V\cdot I\sin\phi.$$

As the phase angle, ϕ is taken as positive in inductive circuits, the reactive power is positive. The real part, $(I.\cos\phi)$ is in phase with the voltage V, whereas the imaginary part, $(I.\sin\phi)$ is in the quadrature (-90°) with the voltage V. However, in capacitive circuits, the current $I\angle\phi$ leads the voltage $V\angle 0^\circ$ by an angle, ϕ, which is taken as negative. Hence, it can be stated that the reactive power is negative here and can easily be derived.

An example of a capacitive circuit is given below.

EXAMPLE 5.1 *A voltage of 120 V at 50 Hz is applied to a resistance, R in series with a capacitance, C (Fig.5.7(a)). The current drawn is 2 A, and the power loss in the resistance is 100 W. Calculate the values of the resistance and the capacitance.*

SOLUTION $|V| = 120$ V, $|I| = 2$ A, $f = 50$ Hz.

Power $= P = I^2R = 100$ W.

The resistance $= R = \dfrac{P}{I^2} = \dfrac{100.0}{(2.0)^2} = 25.0\ \ \Omega.$

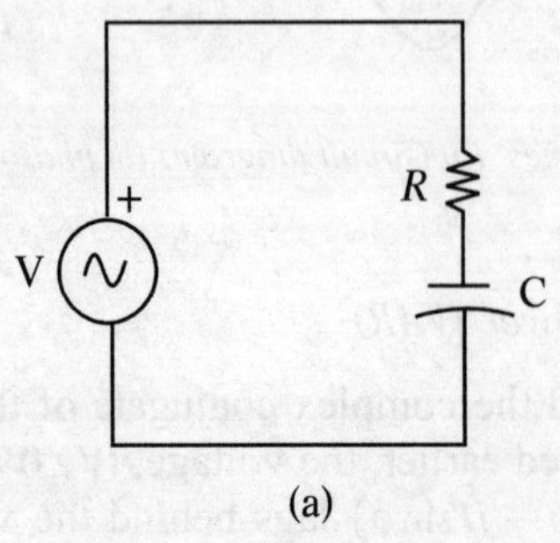

(a)

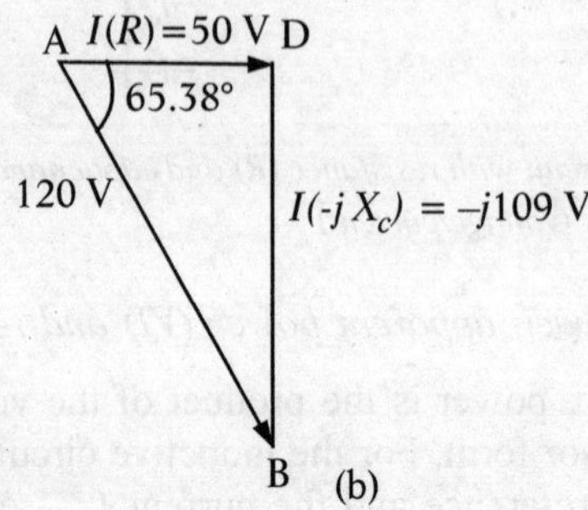

(b)

Fig. 5.7

The magnitude of the impedance $= |Z| = \sqrt{[R^2 + (X_C)^2} = \left|\dfrac{V}{I}\right| = \dfrac{120.0}{2.0} = 60.0 \quad \Omega.$

The capacitive reactance $= X_C = \dfrac{1}{\omega C} = \sqrt{Z^2 - R^2} = \sqrt{(60.0)^2 - (25.0)^2} = 54.54\,\Omega.$

The capacitance $= C = \dfrac{1}{\omega X_C} = \dfrac{1}{2\pi f X_C} = \dfrac{1}{2\pi \times 50.0 \times 54.54} = 58.36\,\mu\text{F}.$

The power factor $= \cos\phi = \dfrac{R}{Z} = \dfrac{25.0}{60.0} = 0.4167.$

The phase angle is $\phi = \cos^{-1}(0.4167) = 65.38°.$

The phasor diagram, with current as reference, is shown in Fig. 5.7(b).

Examples of circuits with lossy inductance coil (r in series with L), and resistance, R, is given later. The R-L-C series circuit fed from a single phase ac supply is presented here.

R-L-C Series Circuit

The voltage balance equation for the R-L-C series circuit (Fig. 5.8(a)) is,

$$v = R\cdot i + L\cdot\frac{di}{dt} + \frac{1}{C}\cdot\int i\cdot dt = \sqrt{2}V\sin\omega t,$$

The current, I (in the steady state) can be derived as

$$i = \sqrt{2}I\sin(\omega t \mp \phi) = \sqrt{2}I\sin(\theta \mp \phi).$$

The impedance of the circuit is

$$Z\angle\pm\phi = [R + j(X_L - X_C)] = \left[R + j\left(\omega L - \frac{1}{\omega C}\right)\right]$$

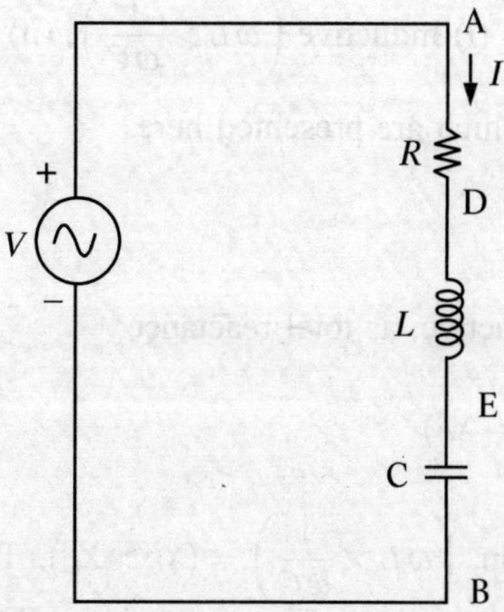

Fig. 5.8(a) *Circuit diagram*

where, $|Z| = \sqrt{[R^2 + (X_L - X_C)^2]} = \sqrt{\left[R^2 + \left(\omega L - \dfrac{1}{\omega C}\right)^2\right]}$,

and

$$\phi = \tan^{-1}\left(\frac{X_L - X_C}{R}\right) = \tan^{-1}\left[\frac{\left(\omega L - \frac{1}{\omega C}\right)}{R}\right].$$

As given earlier for series R-L circuits, impedance can be derived by substituting the expressions for voltage and current in the voltage balance equation.

The voltage, V is taken as reference, with the current, I lagging/leading by $\pm\phi$, as the case may be. The equations involving the voltage and current as phasors, and impedance as a complex quantity are given as,

$$V\angle 0^\circ = I\angle\mp\phi \cdot Z\angle\pm\phi \quad \text{or} \quad Z\angle\pm\phi = \frac{V\angle 0^\circ}{I\angle\mp\phi} = \frac{V}{I}\angle\pm\phi.$$

Using phasor algebra (division) on the rectangular form, the current is given by,

$$I\angle\mp\phi = \frac{V + j0}{R + j(X_L - X_C)} = \frac{V + j0}{R + j\left(\omega L - \frac{1}{\omega C}\right)},$$

which when simplified will give the same expression for the magnitude and phase angle of the current, I, the magnitude of which is given by,

$$|I| = \left|\frac{V}{Z}\right| = \frac{V}{\sqrt{[R^2 + (X_L - X_C)^2]}} = \frac{V}{\sqrt{\left[R^2 + \left(\omega L - \frac{1}{\omega C}\right)^2\right]}},$$

its phase angle remaining the same as given earlier.

There are three types of cases: (i) Inductive $\left(\omega L > \dfrac{1}{\omega C}\right)$, (ii) capacitive $\left(\omega L < \dfrac{1}{\omega C}\right)$, and (iii) resistive $\left(\omega L = \dfrac{1}{\omega C}\right)$, which are presented here.

Inductive

In this case, the circuit is inductive, as total reactance,

$$X = \left(\omega L - \frac{1}{\omega C}\right) = (X_L - X_C)$$

is positive under the condition, $\left(\omega L > \dfrac{1}{\omega C}\right) = (X_L > X_C)$. The current, I lags behind the voltage by an angle, ϕ, which is positive. The power factor (lagging) is less than unity (1) as $0^\circ < \phi < 90^\circ$. The complete phasor diagram, with the voltage drops across the components, the input voltage (OA), and the current (OB), is shown in Fig. 5.8(b). The voltage phasor is taken as reference, in all cases. It may be observed that

$$V_{OC}(=I\cdot R)+V_{CD}(=I\cdot X_L)+V_{DA}(=I\cdot X_C)=V_{OA}(=I\cdot Z),$$

using Kirchhoff's second law relating to the voltage in a closed loop. The phasor diagram can also be drawn with the current phasor as reference, as will be shown in the example given after this section. The expression for the average or active power is $W = V\cdot I\cos\phi$. In all three cases given here, this power is only consumed in the resistance, R, not in inductance/capacitance (L/C).

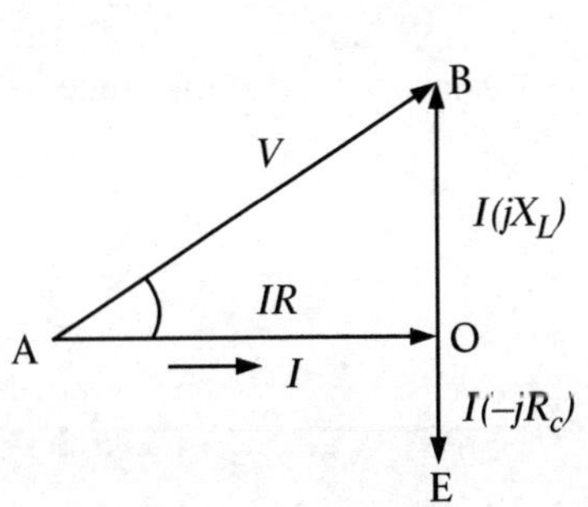

Fig. 5.8(b) *Phasor diagram: Inductive*

Capacitive ($X_L < X_C$)

Fig. 5.8(c) *Phasor diagram: Capacitive*

Capacitive

The circuit is capacitive, as total reactance,

$$X=\left(\omega L-\frac{1}{\omega C}\right)=(X_L-X_C),$$

is negative under the condition, $\left(\omega L<\frac{1}{\omega C}\right)=(X_L<X_C).$

The current, I leads the voltage by an angle, ϕ, which is negative according to the convention given earlier. The complete phasor diagram, with the voltage drops across the components, the input voltage and current, is shown in Fig. 5.8(c). The power factor (leading) is less than unity (1) as the magnitude of the angle, neglecting its sign, is $0° < \phi < 90°$, ϕ being negative. The expression for the average or active power remains the same as given earlier.

Resistive

The third case is *resistive*, as total reactance,

$$X=\left(\omega L-\frac{1}{\omega C}\right)=(X_L-X_C)=0$$

under the condition $\omega L = \dfrac{1}{\omega C}$ or $X_L = X_C$. The impedance is $Z\angle 0° = (R + j0)$.

The current is now at unity power factor ($\phi = 0°$), i.e., the current and the voltage are in phase. The complete phasor diagram, with the voltage drops across the components, the input (supply) voltage and current, is shown in Fig. 5.8(d).

It may be observed that two cases of series (R-L and R-C) circuits can obtained in the following way. The first one (inductive) is by making (i) C very large, i.e., $\dfrac{1}{\omega C} \approx 0$, which means that C is not there. The second one (capacitive) is by making (ii) L vanish (L or $\omega L = 0$).

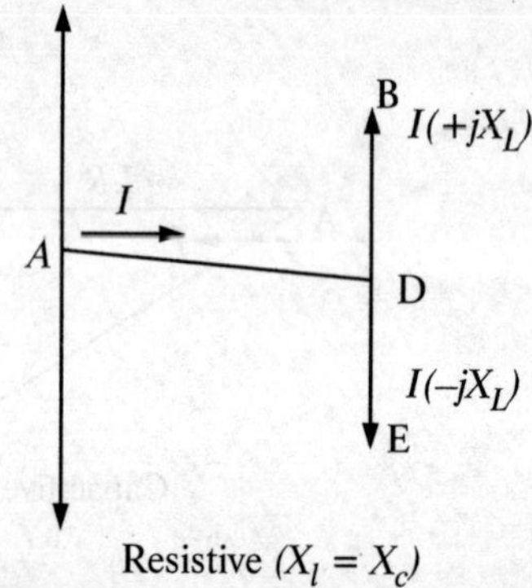

Fig. 5.8(d) *Phasor diagram: Resitive*

EXAMPLE 5.2 A resistor of resistance, R is connected in series with an iron-cored choke coil (in series with L). The circuit (Fig. 5.9(a)) draws a current of 5 A at 240 V, 50 Hz. The voltages across the resistance and the coil are 120 V and 200 V respectively. Calculate (a) the resistance, reactance and impedance of the coil, (b) the power absorbed by the coil, and (c) the power factor (pf) of the input current.

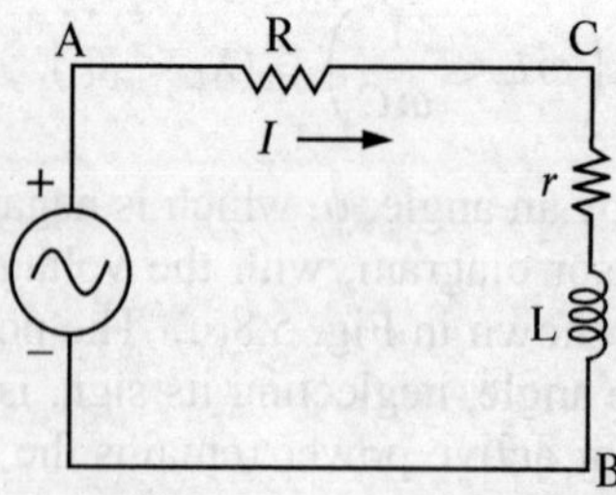

Fig. 5.9(a) *Circuit diagram*

SOLUTION $|V_S| = 240$ V, $|I| = 5$ A, $f = 50$ Hz.

The angular frequency $= \omega = 2\pi f = 2\pi \times 50.0 = 314.16$ rad/s.

The phasor diagram of the current and the voltage drops in the resistor of resistance, R, and the coil (r in series with L), is shown in Fig. 5.9(b). Note that the current, I in the circuit is taken as reference, which is the practice in ac series circuit.

The voltage drop across the resistor, $R = |V_1| = I \cdot R = 120$ V.

The resistance $= R = \dfrac{V_1}{I} = \dfrac{120}{5} = 24.0\,\Omega$.

The voltage drop across the coil $= |V_2| = I \cdot |Z_L| = 200$ V.

The magnitude of its impedance $= |Z_L| = \sqrt{r^2 + (X_L)^2} = \left|\dfrac{V_2}{I}\right| = \dfrac{200}{5} = 40.0\,\Omega$

The supply voltage $= |V_S| = I \cdot |Z| = 240$ V.

The magnitude of the total impedance of the circuit

$$= |Z| = \sqrt{(R+r)^2 + (X_L)^2} = \left|\frac{V_S}{I}\right| = \frac{240}{5} = 48.0\ \ \Omega.$$

In the phasor diagram (Fig. 5.9(b)), the supply voltage, and the voltage drops as given earlier, are

$\text{OB} = V_s = 240$ V, $\text{OC} = V_1 = 120$ V,

$\text{CB} = V_2 = 200$ V, $\text{OA} = I = 5$ A.

The voltage drops across r and X_L are CD and DB respectively, with $\angle$ CDB = 90°.

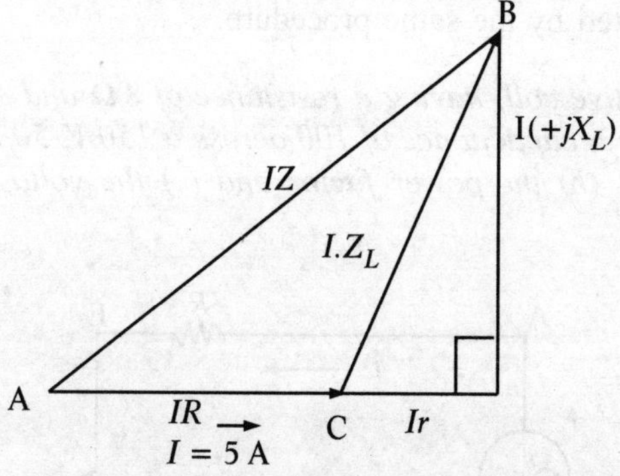

Fig. 5.9(b) *Phasor diagram*

If the angle $\angle$ BOC is taken as ϕ, then, using the formula, we get,

$$\cos\phi = \cos\angle\text{BOC} = \frac{[(\text{OB})^2 + (\text{OC})^2 - (\text{BC})^2]}{2 \cdot \text{OB} \cdot \text{OC}} = \frac{[(240)^2 + (120)^2 - (200)^2]}{2 \times 240 \times 120}$$

$$= \frac{32,000}{57.600} = 0.556.$$

The phase angle of the total circuit $= \phi = \cos^{-1}(0.556) = 56.25°$.

The impedance of the total circuit

$$= Z\angle\phi = [(R+r) + jX_L] = 48.0\angle 56.25° = (26.67 + j39.9)\ \Omega.$$

$$R + r = 26.67\ \Omega.$$

The resistance of the coil $= r = 26.67 - 24.0 = 2.67\ \Omega$.

The reactance of the coil $= X_L = \omega L = 2\pi f L = 39.9\ \ \Omega$.

The inductance of the coil $= L = \dfrac{X_L}{\omega} = \dfrac{X_L}{2\pi f} = \dfrac{39.9}{314.16} = 0.127\ \text{H} = 127\ \text{mH}$.

The power factor (pf) of the input current $= \cos\phi d = \cos 56.25° = 0.556$ (lag).

To check, the impedance of the coil

$$= Z_L\angle\phi_L = (r + jX_L) = (2.67 + j39.9) \approx 40.0\angle 86.172°\ \Omega.$$

The power factor of the coil $= \cos\phi_L = \cos 86.172° = 0.0668 \approx 0.067$ (lag).

The copper loss in the coil $= I^2 r = (5.0)^2 \times 2.67 = 66.75$ W.

Alternatively, the problem can be solved in the following manner. Firstly, the angle, $\phi = \angle$BCD is found, with BD drawn perpendicular to OC, intersecting at D. Then, all other values can also be computed by the same procedure.

EXAMPLE 5.3 *An inductive coil, having a resistance of* 8 Ω *and an inductance of 80 mH, is connected in series with a capacitance of 100 across a 150 V, 50 Hz supply (Fig. 5.10(a)). Calculate (a) the current, (b) the power factor and (c) the voltage drops in the coil and capacitance respectively.*

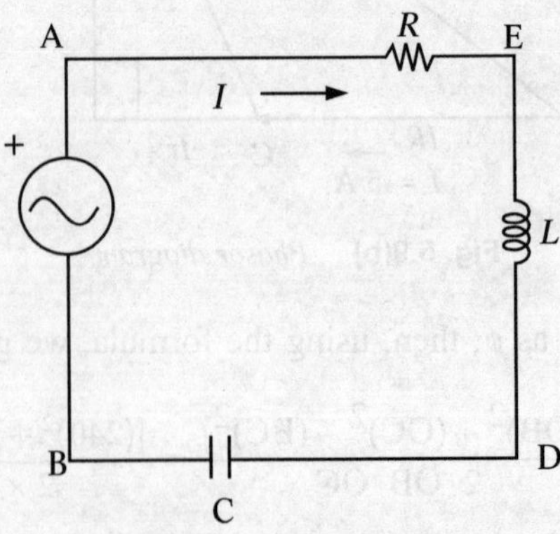

Fig. 5.10(a) *Circuit diagram*

SOLUTION $|V_S| = 150$ V, $f = 50$ Hz.

The angular frequency $= \omega = 2\pi f = 2\pi \times 50.0 = 314.16$ rad/s.

$$R = 8\ \Omega L = 80\text{ mH} = 80.0 \times 10^{-3} = 0.08\text{ H}.$$

$$X_L = \omega L = 2\pi f L = 314.16 \times 0.08 = 25.13\ \Omega.$$

$$C = 100\ \mu\text{F} = 100.0 \times 10^{-6}\text{ F}.$$

$$X_C = \frac{1}{\omega C} = \frac{1}{2\pi f C} = \frac{1}{314.16 \times 100.0 \times 10^{-6}} = 31.83\ \Omega.$$

The impedance of the coil $= Z_L\angle\phi_L = (R + jX_L) = (8.0 + j25.13) = 26.373\angle72.34°\Omega$.

The power factor (pf) of the coil $= \cos\phi_L = \cos 72.34° = 0.303$ (lag).

The total impedance of the circuit

$$= Z\angle-\phi = [R + j(X_L - X_C)]$$

$$= [8.0 + j(25.13 - 31.83)] = (8.0 - j6.7) = 10.435\angle-39.95°\Omega.$$

Note that the circuit is capacitive in nature, the current, I leads the supply voltage, V_S by an angle, ϕ.

The current drawn from the supply

$$= I\angle\phi = \frac{V_S\angle0°}{Z\angle-\phi} = \frac{150.0}{10.435}\angle39.95° = 14.375\angle39.95° = (11.02 + j9.23)\text{ A}.$$

The power factor (pf) $= \cos\phi = \cos 39.95° = 0.767$ (lead).

Please note that the current phasor is taken as reference in the phasor diagram (Fig. 5.10(b)). The voltage drop in the coil is

$$= V_1\angle\phi_L = (I\angle0° \cdot Z_L\angle\phi_L)$$

$$= (14.375 \times 26.373)\angle72.34° = 379.1\angle72.34° = (115.0 + j361.2)\text{ V}.$$

The voltage drop in the capacitance

$$= V_2\angle-90° = (I\angle0° \cdot X_C\angle-90°)$$

$$= (14.375 \times 31.83)\angle-90.0° = 457.8\angle-90.0° = (0.0 - j457.8)\text{ V}.$$

Alternatively, two voltage drops are computed as,

$$V_1\angle\phi_L = (I\angle0° \cdot Z_L\angle\phi_L)$$

$$= 14.375 \times (8.0 + j25.13) = (115.0 + j361.2) = 379.1\angle72.34°\text{ V},$$

$$V_2\angle-90° = (I\angle0° \cdot X_C\angle-90°)$$

$$= 14.375 \times (0.0 - j31.83) = (0.0 - j457.6) = 457.6\angle-90.0° \text{ V}.$$

as the current $= I\angle0° = 14.375\angle0° = (14.375 + j0.0)$ A.

The supply voltage, with the current as reference

$$V_S\angle-\phi = 150.0\angle-39.95° = (115.0 - j96.3) \text{ V}.$$

This voltage is the phasor sum of two voltages, and is found as,

$$V_S\angle-\phi = (V_1\angle\phi_L + V_2\angle-90°) = [(115.0 + j361.2) + (0.0 - j457.8)]$$

$$= (115.0 - j96.6) = 150.2\angle-40.0° \text{ V},$$

which is nearly equal to that obtained earlier.

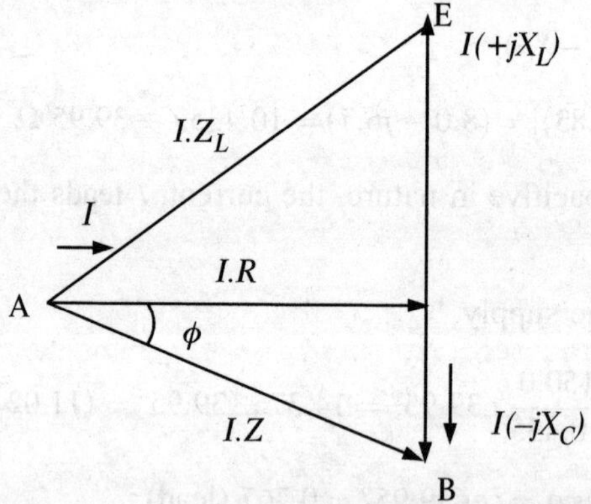

Fig. 5.10(b)

5.3 Resonance in R-L-C Series Circuits

Resonance occurs in R-L-C circuits because energy is stored in the capacitor as well as in the inductor but in different ways—in an electric field when the capacitor is charging and in a magnetic field when current flows through the inductor. Oscillation occurs when energy is transferred from one to the other within the circuit. The resistance in an R-L-C circuit will 'damp' the oscillation, diminishing it with time if there is no driving AC power source in the circuit.

Consider a series R-L-C circuit fed from a variable frequency ac supply (Fig. 5.11(a)). The phasor diagram is shown in Fig. 5.11(b). The total impedance of the circuit, the supply frequency being f in Hz, is given by,

$$Z\angle\phi = [R + j(X_L - X_C)] = \left[R + j\left(\omega L - \frac{1}{\omega C}\right)\right]$$

where angular frequency $= \omega = 2\pi \cdot f$ (rad/s).

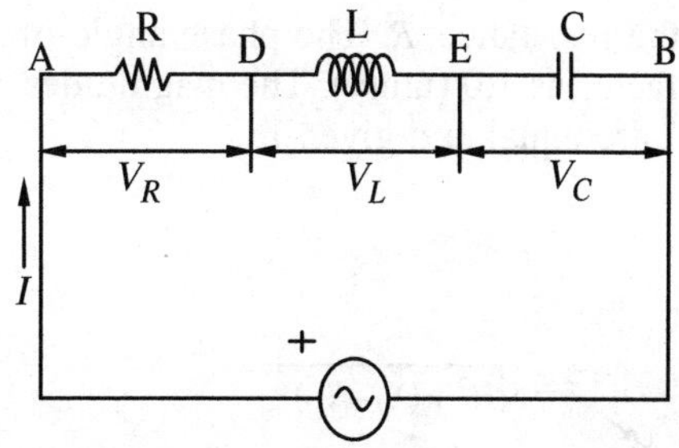

Fig. 5.11(a)

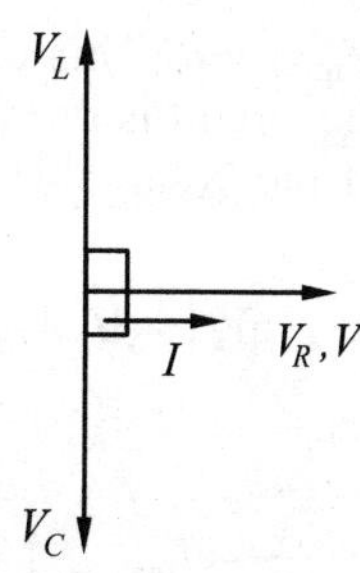

Fig. 5.11(b)

The magnitude of the impedance $= |Z| = \sqrt{(R)^2 + (X_L - X_C)^2} = \sqrt{(R)^2 + \left(\omega L - \dfrac{1}{\omega C}\right)^2}$.

The angle of the impedance $= \phi = \tan^{-1}\left(\dfrac{X_L - X_C}{R}\right) = \tan^{-1}\left(\dfrac{\omega L - \frac{1}{\omega C}}{R}\right)$.

The current, I in the circuit, with the supply voltage being $V\angle 0°$, taken as reference, is,

$$I\angle - \phi = \frac{V\angle 0°}{Z\angle\phi} = \left|\frac{V}{Z}\right|\angle - \phi.$$

The magnitude of the current

$$= |I| = \left|\frac{V}{Z}\right| = \frac{V}{\sqrt{(R)^2 + (X_L - X_C)^2}} = \frac{V}{\sqrt{(R)^2 + (\omega L - \frac{1}{\omega C})^2}},$$

its phase angle, ϕ being the same as given earlier.

The magnitude of the current is maximum at resonant frequency, f_0, the magnitude of the impedance being minimum. Hence, the condition for the circuit to be in resonance is,

$$X_{L0} - X_{C0} = 0, \text{ or } X_{L0} = X_{C0}, \text{ or } \omega_0 L = \frac{1}{\omega_0 C}.$$

That is, the inductive reactance is equal to the capacitive reactance at resonant frequency, f_0.

The angular resonant frequency (rad/s) is found as $\omega_0 = 2\pi f_0 = \dfrac{1}{\sqrt{LC}}$.

The resonant frequency in Hz $= f_0 = \dfrac{\omega_0}{2\pi} = \dfrac{1}{2\pi\sqrt{LC}}$.

The resistance, R, inductance, L and capacitance, C are in Ω, H and F respectively.

It has been shown earlier that the inductive reactance is equal to the capacitive reactance at resonant frequency, f_0. The magnitude of the maximum value of the current as shown in Fig. 5.11(c) under this condition is $|I_{max}| = \left|\frac{V}{R}\right|$, and the minimum value of impedance (Ω) is

$|Z_{min}| = R$. This current is only limited by the value of the resistance, R. The phase angle of the circuit is $\phi = 0°$. The circuit is resistive, with power factor as 1.0 (unity). The magnitudes of the voltage drops in the inductance, L/capacitance, C are equal and given by,

$$|V_L| = |I \cdot X_L| = |I \cdot \omega L| = |V_C| = |I \cdot X_C| = \left| I \cdot \frac{1}{\omega C} \right|.$$

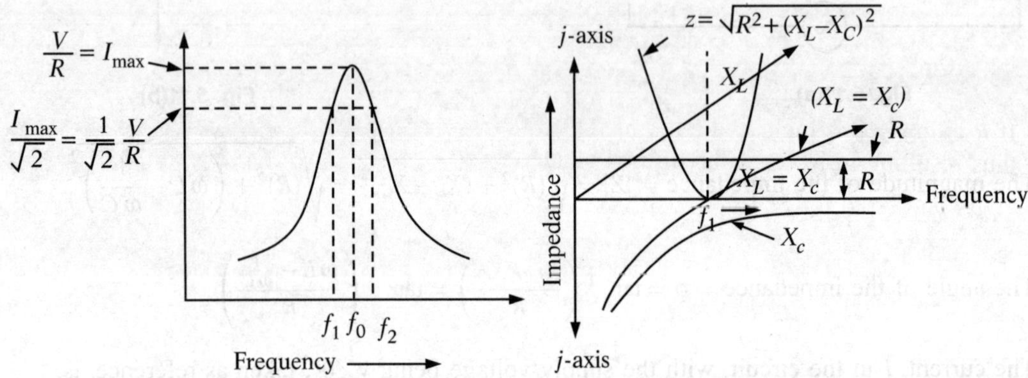

Fig. 5.11(c) **Fig. 5.11(d)**

This is the case when all elements are constant. Figure 5.11(d) shows the curve of resistance, inductance, capacitance and overall impedance at varying frequency and resonant frequency. The impedance of the circuit with the constant values of inductance, L, and capacitance, C, is minimum at resonant frequency, and increases as the frequency is changed (increased/decreased), ($f > f_0$ or $f < f_0$), or $f \neq f_0$. If the supply frequency, f is lower than the resonant one ($f < f_0$), then $X_L < X_C$. As $X_L = \omega \cdot L = 2\pi \cdot fL$, and $X_{L0} = \omega_0 \cdot L = 2\pi \cdot f_0 \cdot L$, with $f < f_0, X_L < X_{L0}$. Similarly, $X_C X_{C0}$, as

$$X_C = \frac{1}{\omega \cdot C} = \frac{1}{2\pi \cdot f \cdot C} \text{ and } X_{C0} = \frac{1}{\omega_0 \cdot C} = \frac{1}{2\pi \cdot f_0 \cdot C}.$$

Hence, $X_L < X_C$ as $X_{L0} = X_{C0}$. If the supply frequency, f is higher than the resonant one ($f > f_0$), then $X_L > X_C$. Similarly, it can be shown that $X_L > X_{L0}$, and $X_C < X_{C0}$, in this condition. Hence, $X_L > X_C$, as $X_{L0} = X_{C0}$. Thus, $|Z| > |Z_0|$, as $|Z_0| = R$, which means that the magnitude of the impedance is minimum at resonant frequency, and increases with change in frequency. In either case, $(X_L - X_C)$ has a certain value or magnitude, irrespective of its sign. The current is maximum at $f = f_0$.

QUALITY FACTOR

The term 'quality factor' or in short, Q-factor, of an inductance coil is defined as the ratio of the voltage drop across its inductive reactance and the applied voltage, at the resonant frequency (f_0), the coil being connected in series with the capacitor across the supply. As the current, I is the same in a series circuit, the voltage drop across its inductive reactance is $V_L = IX_{L0} = I\omega_0 L$, where the resonant angular frequency, $\omega_0 = 2\pi f_0$. The applied voltage is $V = IR$, as $X_{L0} = \omega_0 L = X_{C0} = \frac{1}{\omega_0 C}$, i.e., the voltage drops across the inductance and the capacitance cancel each other. The Q-factor of the coil at resonant frequency is given by

$$Q_0 = \left|\frac{V_L}{V}\right| = \frac{\omega_0 L}{R}.$$

It can be shown that the voltage drop across the inductance is magnified. The magnification ratio is

$$Q_0 = \frac{\omega_0 L}{R} = \frac{2\pi f_0 L}{R} = \frac{1}{R}\sqrt{\frac{L}{C}}.$$

If R is low, Q_0 is high, as higher magnification is desirable. If R is higher, Q_0 is decreased, thus resulting in lower value of the voltage drop.

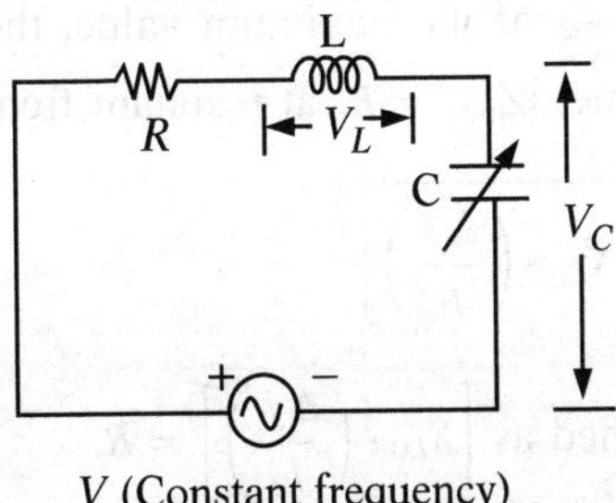

Fig. 5.12

Consider the circuit when one of the elements, say C is variable, while the other two elements (R and L) remain constant (Fig. 5.12). The supply frequency, f is also kept constant. The value of the capacitance needed to produce resonance under this condition, is given by,

$$C = \frac{1}{\omega^2 L} = \frac{1}{(2\pi f)^2 L}$$

This can be derived by equating,

$$X_L = \omega L = 2\pi f L = X_C = \frac{1}{\omega C} = \frac{1}{2\pi f C}.$$

All other conditions are as given earlier.

BANDWIDTH AND HALF POWER POINTS

The magnitude of the maximum value of the current $\left(I_{max} = \frac{V}{R}\right)$ in the circuit (Fig. 5.11(a)) is obtained (Fig. 5.11(c)) at the resonant frequency, f_0, as given earlier. If the magnitude of the current is reduced to $\left(\frac{1}{\sqrt{2}} = 0.707\right)$ of its maximum value, the power consumed in the resistance, R, is half of that with the maximum current, as power is (I^2R). Hence, these points are termed as *half power points*. If two frequencies at two points, are taken as f_1 and f_2, where $f_1 = f_0 - \frac{\Delta f}{2}$, and $f_2 = f_0 + \frac{\Delta f}{2}$, with the difference of the frequencies being

$\Delta f = f_2 - f_1$, the magnitude of the impedances at the two frequencies, f_1 and f_2, is given by,

$$|Z| = \sqrt{R^2 + \left[2\pi\left(f_0 \pm \frac{\Delta f}{2}\right)L - \frac{1}{2\pi\left(f_0 \pm \frac{\Delta f}{2}\right)C}\right]^2}.$$

This term is simplified, using

$$X_{L0} = 2\pi f_0 L = X_{C0} = \frac{1}{2\pi f_0 C}$$

and assuming the ratio of $\frac{\Delta f}{2f_0}$ as small. The reactance at these frequencies is $X = X_{L0} \cdot \frac{\Delta f}{f_0}$. As the magnitude of the current is $\frac{1}{\sqrt{2}}$ of its maximum value, the magnitude of the impedance is $\sqrt{2}$ times of its minimum value, $|Z_{\text{min}}| = R$, at resonant frequency.

$$\text{So, } |Z| = \sqrt{2} \cdot R = \sqrt{R^2 + \left[X_{L0} \cdot \left(\frac{\Delta f}{f_0}\right)\right]^2}.$$

From the above, it can be obtained as $\left[X_{L0} \cdot \left(\frac{\Delta f}{f_0}\right)\right] = R.$

The *bandwidth*, which is the difference between two frequencies known as the upper and lower half power points, is given by

$$\Delta f = f_2 - f_1 = \frac{R \cdot f_0}{2\pi f_0 L} = \frac{R}{2\pi L} = \frac{f_0}{Q_0}.$$

The value is small, being inversely proportional to the quality factor, Q_0, if R is small.

EXAMPLE 5.4 *A constant voltage of frequency 1 MHz, is applied to a lossy coil (r in series with L), which is connected in series with a variable capacitor, C (Fig. 5.13(a)). The current is maximum, when C = 400 pF. It is reduced to $\frac{1}{\sqrt{2}}$ of the above value, when C = 450 pF. Find the values of r and L. Also calculate the quality factor of the coil, and the bandwidth, with C = 400 pF.*

SOLUTION r = resistance of the lossy coil (Ω), other symbols used here are standard ones.

The circuit is given in Fig. 5.13(a).

$F = 1\ \text{MHz} = 1.0 \times 10^6\ \text{Hz}$, $C_0 = 400\ \text{pF} = 400 \times 10^{-12}\ \text{F}$.

Angular frequency $= \omega = 2\pi \cdot f = 2\pi \times 1.0 \times 10^6 = 6.2832 \times 10^6$ rad/s

$$X_{C0} = \frac{1}{\omega C_0} = \frac{1}{2\pi f C_0} = \frac{1}{6.2832 \times 10^6 \times 400.0 \times 10^{-12}} \approx 398.0\ \Omega$$

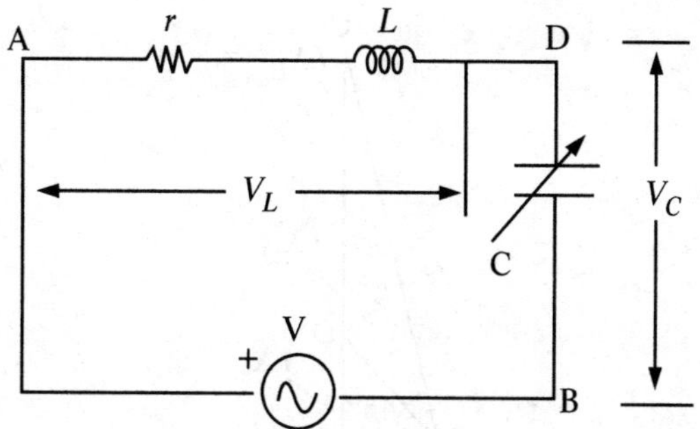

Fig. 5.13(a)

As the current is maximum $\left(I_{\max} = \frac{V}{r}\right)$, with this value of capacitance,

$$X_L = X_{L0} = \omega \cdot L = 2\pi f \cdot L = X_{C0} = 398.0 \quad \Omega$$

or, $$L = \frac{X_L}{\omega} = \frac{X_L}{2\pi f} = \frac{398.0}{6.2832 \times 10^6} = 63.34 \times 10^{-6} \text{ H} = 63.34 \ \mu\text{H}.$$

Alternatively, the formula as derived earlier, is changed to

$$L = \frac{1}{\omega^2 C} = \frac{1}{(2\pi f)^2 C}.$$

Substituting the values, we get,

$$L = \frac{1}{\omega^2 C} = \frac{1}{(6.2832 \times 10^6)^2 \times 400.0 \times 10^{-12}} = 63.33 \ \mu\text{H},$$

which is nearly the same as obtained earlier.

$$C_1 = 450 \text{ pF} = 450 \times 10^{-12} \text{ F}.$$

$$X_{C1} = \frac{1}{\omega C_1} = \frac{1}{2\pi f C_1} = \frac{1}{6.2832 \times 10^6 \times 450.0 \times 10^{-12}} = 353.7 \quad \Omega$$

$$Z_1 \angle \phi = R + j(X_L - X_{C1}) = [r + j(398.0 - 353.7)] = (r + j44.3)$$

or, $$|Z_1| = \sqrt{r^2 + (44.3)^2} \ \Omega.$$

The current $= |I| = \left|\frac{I_m}{\sqrt{2}}\right| = \frac{V}{\sqrt{2} \cdot r} = \left|\frac{V}{Z_1}\right| = \frac{V}{\sqrt{r^2 + (44.3)^2}}.$

From the above, $\sqrt{2} \cdot r = \sqrt{r^2 + (44.3)^2}$

or, $2r^2 = [r^2 + (44.3)^2]$ or, $r = 44.3 \ \Omega$

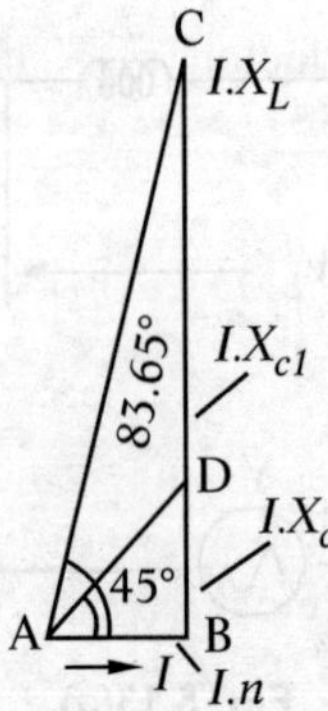

Fig. 5.13(b)

The quality factor of the coil, with

$$C_0 = 400 \text{ pF}, \; Q_0 = \frac{X_L}{r} = \frac{398.0}{44.3} = 8.984 \approx 9.0.$$

$$\text{The bandwidth} = \Delta f = (f_2 - f_1) = \frac{f}{Q_0} = \frac{1.0 \times 10^6}{9.0} = 111.0 \times 10^3 \text{ Hz} = 111.0 \text{ kHz}.$$

The capacitive reactance $= -jX_C = -j353.7 = 353.7\angle -90°\Omega$.

Using the current in the series circuit as reference, with $C_2 = 450\,\mu$F, the voltage drop in the coil

$$= V_L\angle\phi_L = (I\angle 0° \cdot Z_L\angle\phi_L).$$

The voltage drop in the capacitor

$$V_C\angle -90° = (I\angle 0° \cdot X_{C2}\angle -90°) = (353.7 \cdot I)\angle 90° \text{ V}.$$

The applied voltage is the sum of the two voltage drops, which is,

$$V\angle 45° = (V_L\angle\phi_L + V_C\angle -90°) = [I \cdot \{(44.3 + j398.0) - j353.7\}] = [I \cdot (44.3 + j44.3)]$$

$$= (I \cdot 62.65)\angle 45° = [I \cdot (\sqrt{2} \cdot 44.3)]\angle 45° \text{ V}.$$

As stated earlier, the impedance of the circuit, with $C_1 = 400\,\mu$F, at resonant frequency

$$= Z\angle 0° = (r + j0.0) = 44.3\angle 0°\Omega, \text{ as } X_L = X_{C1}.$$

The magnitude of the impedance is minimum and that of current is maximum, which is given by

$$|I_m| = \left|\frac{V}{r}\right| = \left|\frac{V}{44.3}\right|.$$

Also this current is at unity (1) power factor (pf). The impedance of the circuit with $C_2 =$ 450 μF, is inductive, and the current lags behind the voltage by 45° (pf = 0.707 lag).

The magnitude of the current

$$= |I_m| = \left|\frac{V}{Z}\right| = \left|\frac{V}{62.65}\right| = \left|\frac{V}{\sqrt{2}\cdot 44.3}\right| = \left|\frac{I_m}{\sqrt{2}}\right|,$$

The magnitude of impedance is $\sqrt{2}$ times its magnitude, with $C_1 = 400\ \mu$F.

The complete phasor diagram of the circuit, with $C_2 = 450\ \mu$F , is given in Fig. 5.13(b).

It may be noted here that the impedance diagram (complex quantity) is the same as that of the voltage phasor diagram.

Additional Problems

Example 5.5 *A circuit draws a current of 4 A at a power factor of 0.707 lagging when connected to a 115 V, 50 Hz supply. Another circuit takes a current of 10 A at a power factor of 0.8 leading when connected to the same supply, after the first one is removed. If the two circuits are now connected in series across the 230 V at the same frequency (Fig. 5.14(a)), calculate, (a) the current drawn, (b) its power factor, and (c) the power consumed. Also find the voltage drops across the two impedances connected in series, under this condition. Draw the complete phasor diagram.*

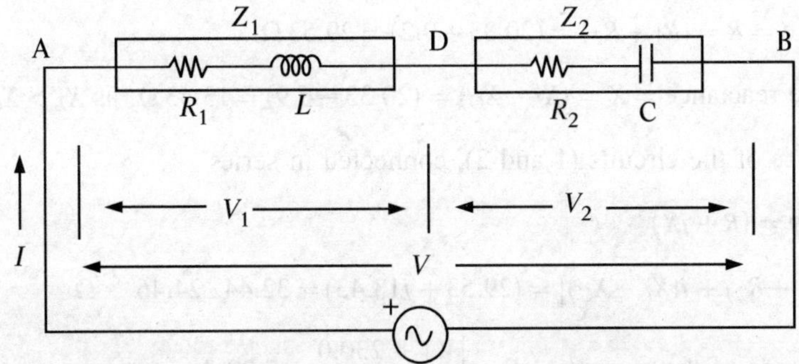

Fig. 5.14(a)

Solution The two circuits, connected in series, and fed from a 230 V supply is shown in Fig. 5.14(a). f = 50 Hz, which is the same in all cases.

First circuit: $|V_1| = 115$ V, $\quad |I_1| = 4$ A.

If the supply voltage is taken as reference (in all cases), then this current is at lagging power factor ($\cos\phi_1 = 0.707$).

$$\phi_1 = \cos^{-1}(0.707) = 45°,$$

which is positive. This circuit is inductive.

Impedance 1

$$= Z_1\angle\phi_1 = (R_1 + jX_L)$$

$$= \frac{V_1\angle 0^\circ}{I_1\angle -\phi_1} = \frac{115.0}{4.0}\angle 45.0^\circ = 28.75\angle 45.0^\circ = (20.33 + j20.33) \quad \Omega.$$

Second circuit: $|V_2| = 115$ V, $|I_2| = 10$ A.

This current is at leading power factor ($\cos\phi_2 = 0.8$). This circuit is capacitive.

$$\phi_2 = \cos^{-1}(0.8) = 36.87^\circ,$$

which is negative.

Impedance 2

$$= Z_2\angle -\phi_2 = (R_2 - jX_C)$$

$$= \frac{V_2\angle 0^\circ}{I_2\angle\phi_2} = \frac{115.0}{10.0}\angle -36.87^\circ = 11.5\angle -36.87^\circ = (9.2 - j6.9)\Omega.$$

The two circuits connected in series: (Fig. 5.14(a)), $|V| = 230$ V.

Total resistance $= R = (R_1 + R_2) = (20.33 + 9.2) = 29.53\ \Omega$.

Total inductive reactance $= X = (X_L - X_C) = (20.33 - 6.9) = 13.43\ \Omega$, as $X_L > X_C$.

Total impedance of the circuits (1 and 2), connected in series

$$= Z\angle\phi = (R + jX)$$

$$= [(R_1 + R_2) + j(X_L - X_C)] = (29.53 + j13.43) = 32.44\angle 24.46^\circ \quad \Omega$$

(a) The magnitude of the current $= |I| = \left|\frac{V}{Z}\right| = \frac{230.0}{32.44} = 7.09$ A.

The current lags behind the voltage, by an angle, $\phi = 24.46^\circ$, same as the angle of total impedance, Z, which is inductive. This will be checked later.

The current $= I\angle -\phi = 7.09\angle 24.46^\circ = (6.45 + j2.94)$ A.

(b) Power factor $= \cos\phi = \cos 24.46^\circ = 0.91$ (lag).

(c) Power consumed $= P = VI\cos\phi = 230.0 \times 7.09 \times 0.91 = 1.484$ kW.

Normally, the current in the series circuit is used as reference, i.e.,

$$I\angle 0^\circ = 7.09\angle 0^\circ = (7.09 + j0.0) \text{ A},$$

using the magnitude of the current obtained earlier.

The voltage drop across impedance 1

$$= V'\angle\phi' = I\angle-\phi\cdot Z_1\angle\phi_1$$

$$= (7.09\times 28.75)\angle(-24.46^\circ + 45.0^\circ) = 203.84\angle 20.54^\circ = (190.9 + j71.5)\text{ V.}$$

The voltage drop across impedance 2

$$= V''\angle\phi'' = I\angle-\phi\cdot Z_2\angle-\phi_2$$

$$= (7.09\times 11.5)\angle-(24.46^\circ + 36.87^\circ) = 81.54\angle-61.33^\circ = (39.1 - j71.5)\text{ V.}$$

It can be shown that the applied voltage is the sum of the two voltage drops, which is

$$V\angle 0^\circ = (V'\angle\phi' + V''\angle\phi''),$$

$$= [(190.9 + j71.5) + (39.1 - j71.5)] = (230.0 + j0.0) = 230.0\angle 0^\circ\text{V},$$

the magnitude being the same as found earlier.

The angle of lag is also checked here. The complete phasor diagram, under this condition, is given in Fig. 5.14(b).

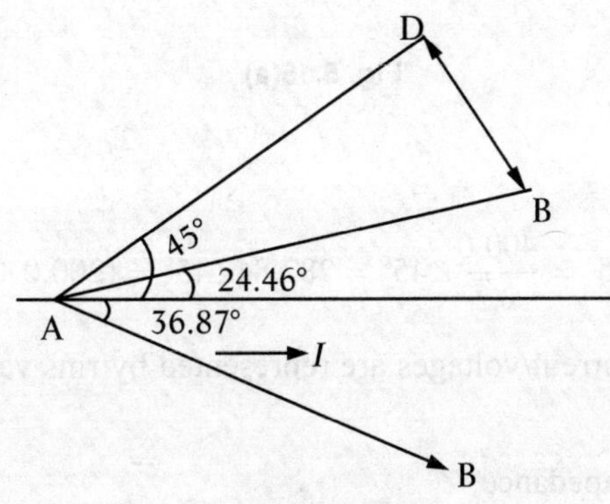

Fig. 5.14(b)

EXAMPLE 5.6 *When a supply voltage (V) of*

$$v = 400\sin\left(\theta + \frac{\pi}{4}\right)$$

is applied to three impedances, Z_1, Z_2 and Z_3 in series, the voltages across Z_1 and Z_2 are

$$v_1 = 100\sin\theta \textit{ and } v_2 = 50\sin\left(\theta - \frac{\pi}{4}\right)$$

respectively (Fig. 5.15(a)). Determine the voltage drop across Z_3. If the supply current (A) is $i = 30\sin\theta$, find the circuit impedance and the three impedances. The angle (rad) is $\theta = \omega t$, with ω being the angular frequency. Calculate the active power, the reactive volt-amperes, and the apparent power in the three impedances, and also the total impedance.

SOLUTION Three impedances, connected in series, are fed from an ac supply (Fig. 5.15(a)).

Taking the current (A), $I = 30\sin\omega t$, as reference, we get,

$$I\angle 0^\circ = (I + j0) = \frac{30.0}{\sqrt{2}}\angle 0^\circ = 21.213\angle 0^\circ = (21.213 + j0.0) \text{ A.}$$

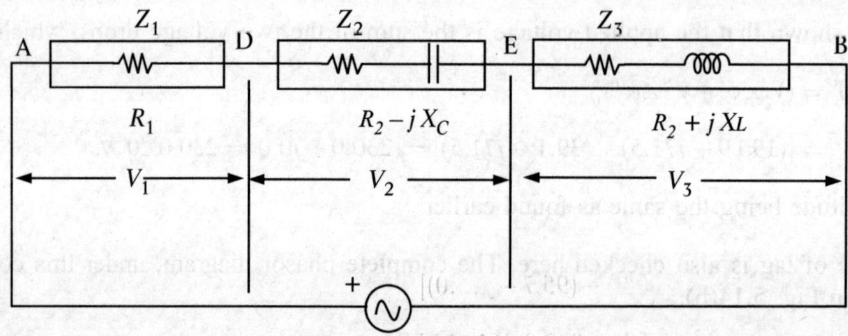

Fig. 5.15(a)

$$\frac{\pi}{4} \text{ rad.} = \frac{180^\circ}{4} = 45^\circ.$$

$$\text{The supply voltage} = V\angle 45^\circ = \frac{400.0}{\sqrt{2}}\angle 45^\circ = 282.84\angle 45^\circ = (200.0 + j200.0) \text{ V.}$$

Normally, the phasors of current/voltages are represented by rms values. The phasor diagram is shown in Fig. 5.15(b).

The voltage drop across impedance, Z_1

$$= V_1\angle 0^\circ = \frac{100.0}{\sqrt{2}}\angle 0^\circ = 70.71\angle 0^\circ = (70.71 + j0.0) \text{ V.}$$

The voltage drop across impedance, Z_2

$$= V_2\angle -45^\circ = \frac{50.0}{\sqrt{2}}\angle -45^\circ = 35.36\angle -45^\circ = (25.0 - j25.0) \text{ V.}$$

$$(V_1\angle 0^\circ + V_2\angle -45^\circ) = [(70.71 + j0.0) + (25.0 - j25.0)] \approx (95.7 - j25.0)$$

$$= 98.9\angle -14.64^\circ \text{ V}$$

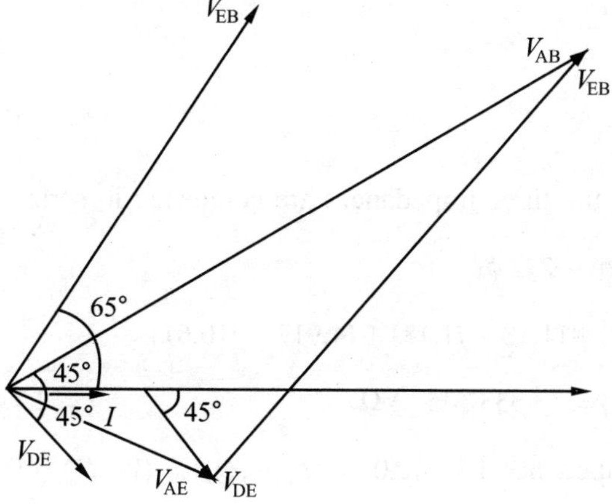

Fig. 5.15(b)

The voltage drop across the impedance, Z_3

$$V_3\angle\phi_3 = [V\angle 45^\circ - (V_1\angle 0^\circ + V_2\angle -45^\circ)]$$

$$= [(200.0 + j200.0) - (95.7 - j25.0)]$$

$$= (104.3 + j225.0) = 248.0\angle 65.13^\circ \text{ V}$$

In time domain, the voltage,

$$v_3 = (248.0 \times \sqrt{2})\sin(\omega t + 65.13^\circ) = 350.72\sin(\omega t + 65.13^\circ) \text{ V.}$$

$$65.13^\circ = 65.13 \times \frac{\pi}{180} = 1.137 \text{ rad.}$$

The impedance of the circuit

$$= Z\angle 45^\circ = (R + jX)$$

$$= \frac{V\angle 45^\circ}{I\angle 0^\circ} = \frac{282.84}{21.213}\angle 45^\circ = 13.333\angle 45^\circ = (9.43 + j9.43) \quad \Omega.$$

The three impedances are also found as,

$$Z_1\angle 0^\circ = (R_1 + j0)$$

$$= \frac{V_1\angle 0^\circ}{I\angle 0^\circ} = \frac{70.71}{21.213}\angle 0^\circ = 3.333\angle 0^\circ = (3.333 + j0.0) \quad \Omega.$$

$$Z_2\angle -45^\circ = \frac{V_2\angle -45^\circ}{I\angle 0^\circ}$$

$$= \frac{35.36}{21.213}\angle -45^\circ = 1.667\angle -45^\circ = (1.18 - j1.18) \quad \Omega.$$

$$Z_3\angle 65.13^\circ = \frac{V_3\angle 65.13^\circ}{I\angle 0^\circ}$$

$$= \frac{248.0}{21.213}\angle 65.13^\circ = 11.69\angle 65.13^\circ = (4.917 + j10.61) \quad \Omega.$$

The circuit impedance, as the three impedances are connected in series,

$$Z\angle\phi = Z_1\angle 0^\circ + Z_2\angle\phi_2 + Z_3\angle\phi_3$$

$$= (3.333 + j0.0) + (1.18 - j1.18) + (4.917 + j10.61)$$

$$= (9.43 + j9.43) = 13.333\angle 45^\circ \quad \Omega.$$

The apparent power in impedance $1 = S_1\angle 0^\circ = (P_1 + jQ_1) = (V_1 \cdot I^*) = (V_1\angle 0^\circ \cdot I\angle 0^\circ)$

$$= (70.71 \times 21.213)\angle(0^\circ + 0^\circ) = 1.5\angle 0^\circ = (1.5 + j0.0) \text{ kVA}.$$

The phase angle between the voltage, V and current, $I_1 = \phi_1 = (\phi_1 - 0^\circ) = 0^\circ$, with V_1 in phase with I.

The power factor $= \cos\phi_1 = \cos 0^\circ = 1.0$ (unity).

The active power $= P_1 = 1.5$ kW.

The reactive volt-amperes $= Q_1 = 0.0$ kVAR.

The apparent power in impedance $2 = S_2\angle\phi_2 = (P_2 + jQ_2) = (V_2 \cdot I^*) = (V_2\angle\phi_2 \cdot I\angle 0^\circ)$

$$= (35.36 \times 21.213)\angle(-45^\circ + 0^\circ) = 0.75\angle -45^\circ = (0.53 - j0.53) \text{ kVA}.$$

The phase angle between the voltage and current $= \phi_2 = (\phi_2 + 0^\circ) = 45^\circ$, with V_2 lagging I, or I leading V_2.

The power factor $= \cos\phi_2 = \cos 45^\circ = 0.707$ (lead).

The active power $=$ 0.53 kW.

The reactive volt-amperes $= -$ 0.53 kVAR.

The apparent power in impedance $3 = \quad S_3\angle\phi_3 = (P_3 + jQ_3) = (V_3 \cdot I^*) = (V_3\angle\phi_3 \cdot I\angle 0^\circ)$.

$$= (248.0 \times 21.213)\angle(65.13^\circ + 0^\circ) = 5.26\angle 65.13^\circ = (2.213 - j4.773) \text{ kVA}.$$

The phase angle between the voltage and current $= \phi_3 = (\phi_3 + 0^\circ) = 65.13^\circ$, with V_3 leading I, or I lagging V_3.

The power factor $= \cos\phi_3 = \cos 65.13^\circ = 0.42$ (lag).

The active power $=$ 2.213 kW.

The reactive volt-amperes = 4.773 kVAR.

The active and reactive power can also be computed as,

$$P_3 = (I)^2 \cdot R_3 = (21.213)^2 \times 4.917 = 2.213 \text{ kW, and}$$

$$Q_3 = (I)^2 \cdot X_3 = (21.213)^2 \times 10.61 = 4.775 \text{ kVAR, respectively,}$$

which are nearly the same as obtained earlier.

The apparent power of the total circuit

$$= S\angle\phi = (P + jQ_2) = (V \cdot I^*) = (V\angle\phi \cdot I\angle 0°)$$

$$= (282.84 \times 21.213)\angle(45° + 0°) = 6.0\angle 45° = (4.24 + j4.24) \text{ kVA.}$$

The phase angle between the voltage and current

$$\phi = (\phi + 0°) = 45°, \text{ with } V \text{ leading } I, \text{ or } I \text{ lagging } V.$$

The power factor $= \cos\phi = \cos 45° = 0.707$ (lag).

The active power = 4.24 kW.

The reactive volt-amperes = 4.24 kVAR.

Alternatively, using the Cartesian form,

The apparent power $= S\angle\phi_5 = (V \cdot I^*) = [(200.0 + j200.0) \times (21.213 + j0.0)]$

$$= (4.24 + 4.24) = 6.0\angle 45° \text{ kVA}$$

The active and reactive power of the total circuit are given as,

$$P = P_1 + P_2 + P_3 = 1.5 + 0.53 + 2.213 = 4.243 \text{ kW, and}$$

$$Q = Q_1 + Q_2 + Q_3 = 0.0 - 0.53 + 4.773 = 4.243 \text{ kVAR, respectively.}$$

The apparent power of the total circuit is given as $S\angle\phi = S_1\angle 0° + S_2\angle\phi_2 + S_3\angle\phi_3$, which can be easily computed as given earlier; the value obtained will also be the same.

Please note the following:

1. The three impedances, 1, 2 and 3 connected in series across the supply, are resistive only, capacitive, and inductive, respectively. The total circuit impedance is inductive. These values have been computed earlier.

2. This circuit being a series one, current is taken as reference.

EXAMPLE 5.7 *A series resonant circuit has a resistance of 15 Ω, and a capacitance of 4 μF (Fig. 5.16). If the bandwidth is 500 rad/s, determine (a) the resonant frequency, f_0 in Hz, (b) the quality factor (Q_0), (c) the inductance (L), and (d) the upper and lower half power frequencies (ω_1 and ω_2).*

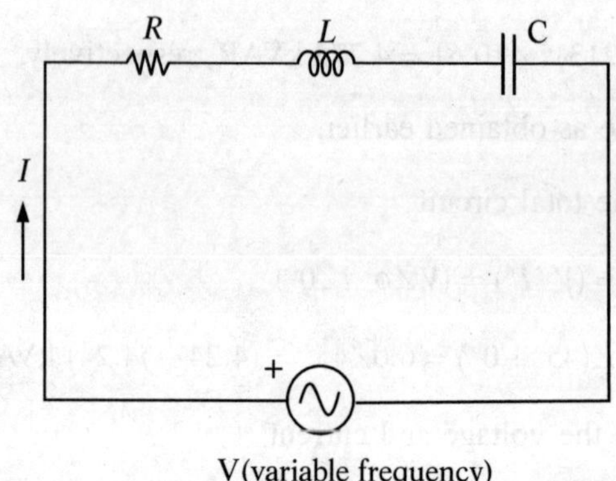

Fig. 5.16

SOLUTION The R-L-C series circuit, fed from a variable frequency supply, is shown in Fig. 5.16.

$$R = 15\,\Omega,\ C = 4\,\mu\text{F} = 4 \times 10^{-6}\ \text{F}.$$

Quality factor (at resonant frequency) $= Q_0 = \dfrac{\omega_0 L}{R}$, or, $\dfrac{Q_0}{\omega_0} = \dfrac{L}{R}$.

Bandwidth for the circuit (Fig. 5.16) $= \omega_b = 2\pi f_b = \dfrac{\omega_0}{Q_0} = \dfrac{R}{L} = 500$ rad/s.

$$f_b = \frac{\omega_b}{2\pi} = \frac{500.0}{2\pi} = 79.58\ \text{Hz}.$$

(c) The inductance $= L = \dfrac{R}{\omega_b} = \dfrac{15.0}{500.0} = 0.03\ \text{H} = 30\ \text{mH}.$

(a) The resonant frequency $= f_0 = \dfrac{1}{2\pi\sqrt{LC}} = \dfrac{1.0}{2 \times \pi \times \sqrt{0.03 \times 4 \times 10^{-6}}}$

$$= 459.4\ \text{Hz}.$$

The resonant angular frequency

$$\omega_0 = \frac{1}{\sqrt{LC}} = (2\pi f_0) = (2.0 \times \pi \times 459.4) = 2,886.5\ \text{rad/s}$$

(b) The quality factor $= Q_0 = \dfrac{\omega_0}{\omega_b} = \dfrac{\omega_0 \cdot L}{R} = \dfrac{2\pi f_0 L}{R} = \dfrac{2,886.5 \times 0.03}{15.0} = \dfrac{86.6}{15.0} = 5.773.$

The two (upper and lower) half power frequencies are first determined.

Half of the bandwidth (angular frequency) $= \dfrac{\omega_b}{2} = \dfrac{500}{2} = 250$ rad/s

If the simplified relations are used,

$$\omega_1 = \omega_0 - \frac{\omega_b}{2} = 2,886.5 - 250.0 = 2,636.5 \text{ rad/s}$$

and $\omega_2 = \omega_0 + \dfrac{\omega_b}{2} = 2,886.5 + 250.0 = 3,136.5$ rad/s.

The lower and upper half power frequencies in Hz, are found as,

$$\text{Lower half power frequency} = f_1 = \frac{\omega_1}{2\pi} = \frac{2,636.5}{2\pi} = 419.6 \text{ Hz.}$$

$$\text{Upper half power frequency} = f_2 = \frac{\omega_2}{2\pi} = \frac{3,136.5}{2\pi} = 499.2 \text{ Hz.}$$

Using the exact formula, the values obtained are given here.

$$\text{The lower half power frequency} = \omega_1 = -\frac{\omega_b}{2} + \sqrt{(\omega_0)^2 + \left(\frac{\omega_b}{2}\right)^2}$$

$$= -250 + \sqrt{(2,886.5)^2 + (250)^2} = -250.0 + \sqrt{[(8.3319 + 0.0625) \times 10^6]}$$

$$= -250.0 + \sqrt{(8.3944 \times 10^6)} = (-250.0 + 2,897.3) = 2,647.3 \text{ rad/s.}$$

$$\text{The upper half power frequency} = \omega_2 = \frac{\omega_b}{2} + \sqrt{(\omega_0)^2 + \left(\frac{\omega_b}{2}\right)^2}$$

$$= 250.0 + \sqrt{(2,886.5)^2 + (250.0)^2} = 250.0 + \sqrt{(8.3944 \times 10^6)} = (250.0 + 2,897.3)$$

$$= 3.147.3 \text{ rad/s.}$$

In both cases, the values obtained are higher than the previous ones.

The errors in the two cases (upper and lower half power frequencies) are given here.

$$\text{For the first (lower) one} = \frac{2,647.3 - 2,636.5}{2,647.3} = \frac{10.8}{2,647.3} = 0.0041 = 0.41\%$$

$$\text{For the second (upper) one} = \frac{3,147.3 - 3,136.5}{3,147.3} = \frac{10.8}{3,147.3} = 0.0034 = 0.34\%$$

The errors in both cases are very small.

Conclusion

In this chapter, we studied the currents in circuits being fed from a single-phase ac supply at line frequency. The circuits had single elements (R/L/C), two elements (R and L or C), and lastly, three elements (R, L and C), all elements connected in series. In all cases, phasor quantities were used, and in almost all examples, phasor diagrams were included. Resonance in series R-L-C circuits, fed from a variable frequency supply, is also presented. Parallel and series–parallel circuits fed from a single-phase ac supply, along with resonance phenomenon in parallel circuits, will be described with examples of each type, in the next chapter.

Multiple Choice Type Questions

Tick the correct option(s), out of the four choices provided in each question.

5.1 A 1 V ac voltage feeds to a load of $(1j)$. Then the power dissipation due to load is

(a) 0.5 W (b) 1 W (c) 2 W (d) None of (a)–(c)

5.2 When current supplied by a voltage source $v = 150\cos(\omega t + 10°)$ in a circuit is $i = 5.0\cos(\omega t + 50°)$, the active and reactive power flow in the circuit is

(a) 187.5 W, 324.8 Var (b) 225.2 W, 400 Var
(c) 321 W, 225 Var (d) None of (a)–(c)

5.3 A series circuit contains a 10 Ω resistor, a 20 mH inductor and a voltage source v. If the current through the circuit is $i = 5.0\cos(500t + 10°)$, then voltage v is

(a) $70.7\cos(500t + 55°)$ (b) $35.0\cos(300t + 50°)$
(c) $25.0\cos(\omega t + 45°)$ (d) None of (a)–(c)

5.4 In the circuit shown in Fig. MCQ5.4, the current through the inductor is [GATE]

(a) 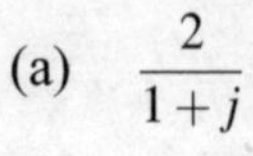 $\frac{2}{1+j}$ (b) $\frac{-1}{1+j}$ (c) $\frac{1}{1+j}$ (d) 0

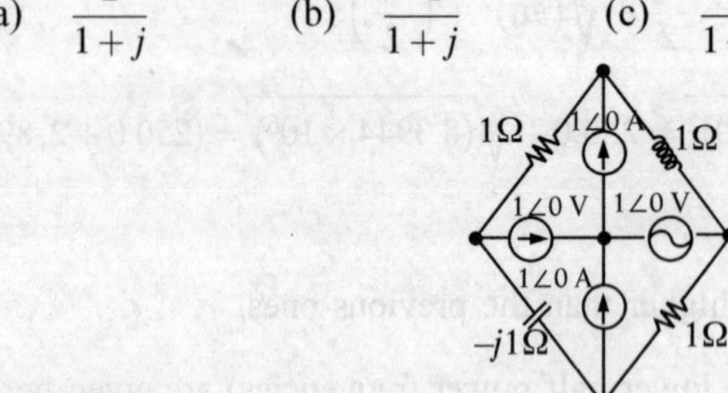

Fig. MCQ5.4

5.5 In the circuit shown in Fig. MCQ5.5, the readings in the three voltmeters are $V_1 = 220$ V, $V_2 = 122$ V, $V_3 = 136$ V. The power factor of the load is [GATE]

(a) 0.45 (b) 0.50 (c) 0.55 (d) 0.60

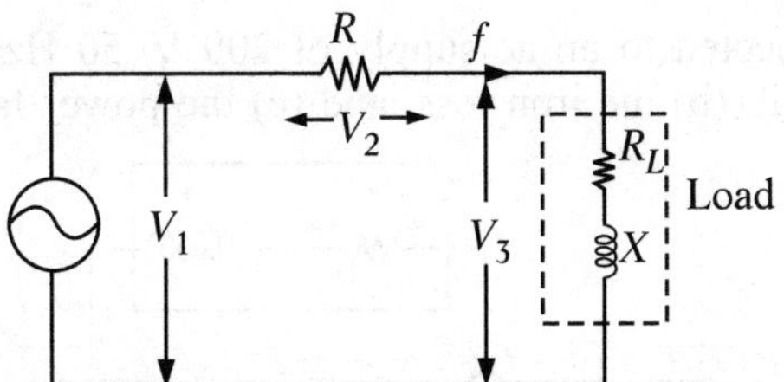

Fig. MCQ5.5

5.6 If $R_L = 5\ \Omega$, the approximate power consumption in the load is [GATE]

(a) 700 W (b) 750 W (c) 800 W (d) 850 W

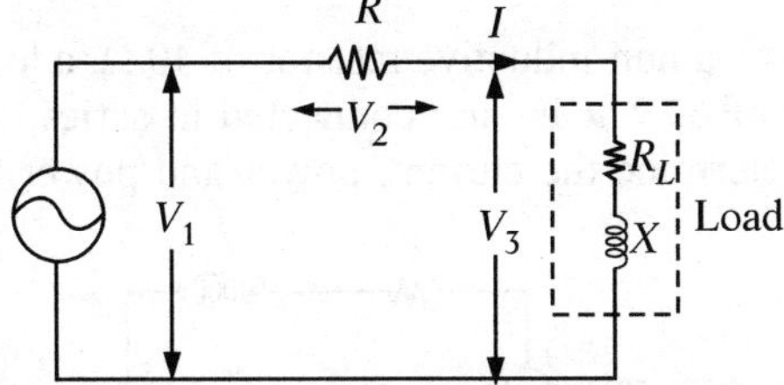

Fig. MCQ5.6

5.7 The rms value of the current $i(t)$ in the circuit shown in Fig. MCQ5.7 is

(a) $\frac{1}{2}$ A (b) $\frac{1}{\sqrt{2}}$ A (c) 1 A (d) $\sqrt{2}$ A [GATE]

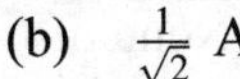

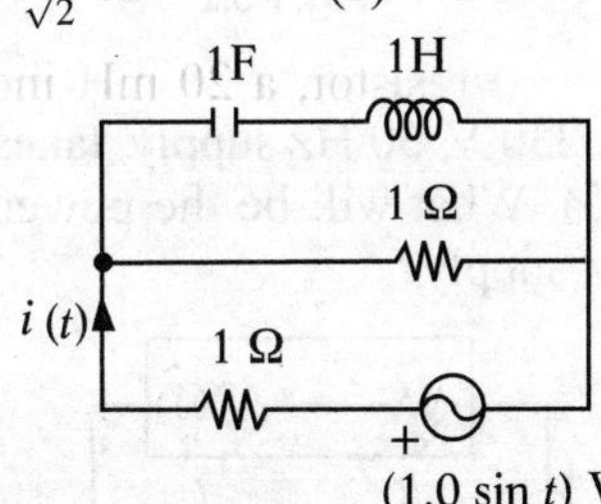

Fig. MCQ5.7

5.8 The voltage applied to a circuit is $100\sqrt{2}\cos(100\pi t)$ volts and the circuit draws a current of $10\sqrt{2}\sin(100\pi t + \pi/4)$ amperes. Taking the voltage as the reference phasor, the phasor representation of the current in amperes is

(a) $10\sqrt{2} < \frac{\pi}{-4}$ (b) $10 < \frac{\pi}{-4}$ (c) $10 < \frac{\pi}{-4}$ (d) $10 < \frac{\pi}{-4}$

Answers

5.1 (a) 5.2 (a) 5.3 (a) 5.4 (c) 5.5 (a)

5.6 (b) 5.7 (b) 5.8 (b)

Problems

5.1 An iron-cored choking coil has the circuit equivalent of a resistor of 4 Ω (which represents the iron loss in the coil), and an unknown inductance, L. It draws a current of 10 A, when connected to an ac supply of 200 V, 50 Hz (Fig. P5.1). Find (a) the inductance of the coil, (b) the iron loss, and (c) the power factor (pf).

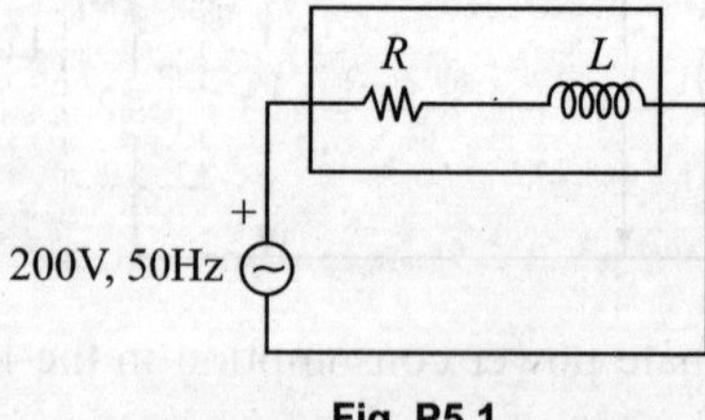

Fig. P5.1

5.2 Three circuit elements—a non-inductive resistor of 40 Ω, a lossless inductance of 0.254 H, and a capacitance of 63.7 μ F—are connected in series, to an ac supply of 230 V, 50 Hz (Fig. P5.2). Determine the current, power and power factor.

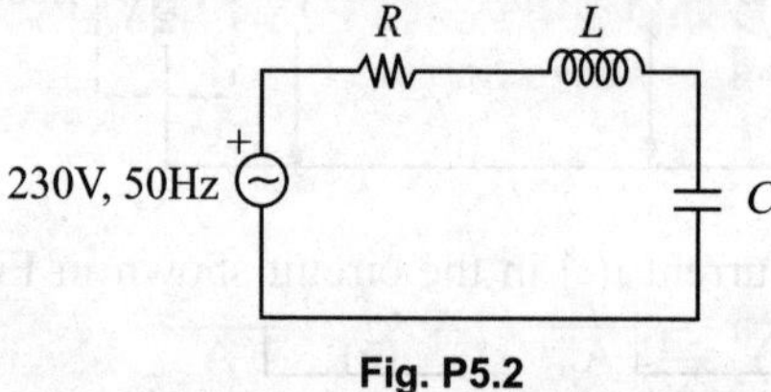

Fig. P5.2

5.3 A choke coil, connected to a 230 V, 50 Hz supply, takes a current of 10 A at 0.8 power factor (pf) lagging (Fig. P5.3). What will be the power taken by the choke, when it is connected to a 200 V, 25 Hz supply?

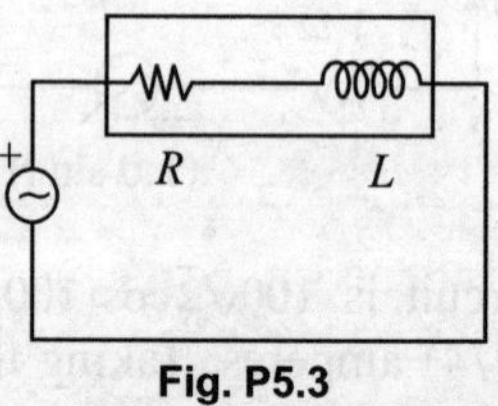

Fig. P5.3

5.4 A non-inductive resistor, R_1 is connected in series with a lossy coil, across a 230 V, 50 Hz supply (Fig. P5.4). The voltages across the resistor and the coil, are 100 V and 180 V respectively. The current is 5 A. Calculate (a) the resistance, R_1, (b) the resistance, inductance and impedance of the coil, (c) the total circuit impedance, (d) the total power factor and (e) the total power loss in the circuit.

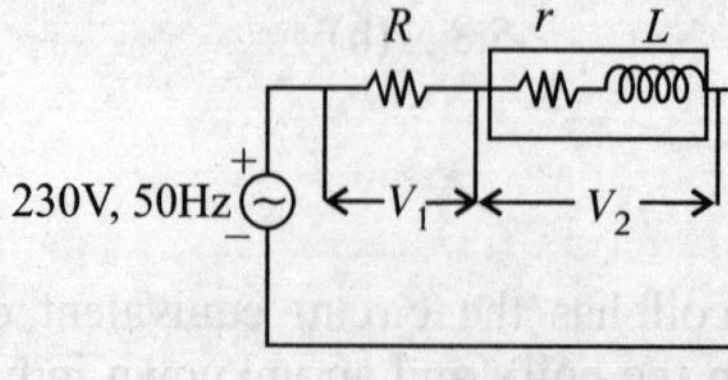

Fig. P5.4

5.5 An iron-cored coil has a resistance (dc) of 7.5 Ω. When connected to a 220 V, 50 Hz supply (Fig. P5.5), it takes a current of 5 A, and dissipates a power of 500 W. Determine the core loss, the inductance of the coil, power factor (pf), and the value of the series resistance, which is equivalent to the effect of the core loss.

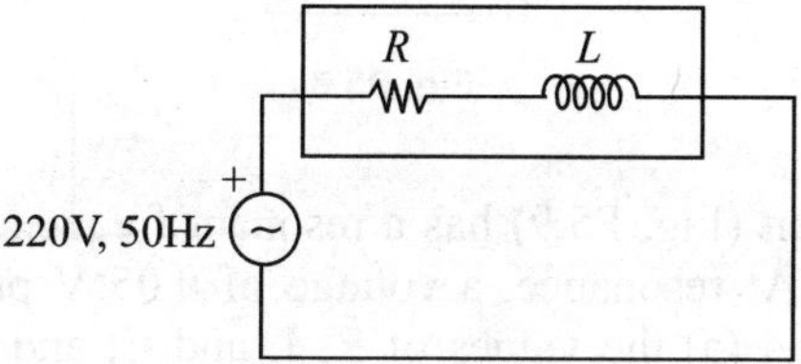

Fig. P5.5

5.6 When a supply voltage (V) of $v = 400\sin\left(\theta + \frac{\pi}{4}\right)$ is applied to three impedances, Z_1, Z_2 and Z_3 in series, the voltages across Z_1 and Z_2 are $v_1 = 100\sin\theta$ and $v_2 = 50\sin\left(\theta - \frac{\pi}{4}\right)$, respectively. Determine the voltage drop across the impedance, Z_3. If the supply current (A) is $i = 30\sin\theta$, find the circuit impedance. The angle (rad) is $\theta = \omega t$, with ω being angular frequency.

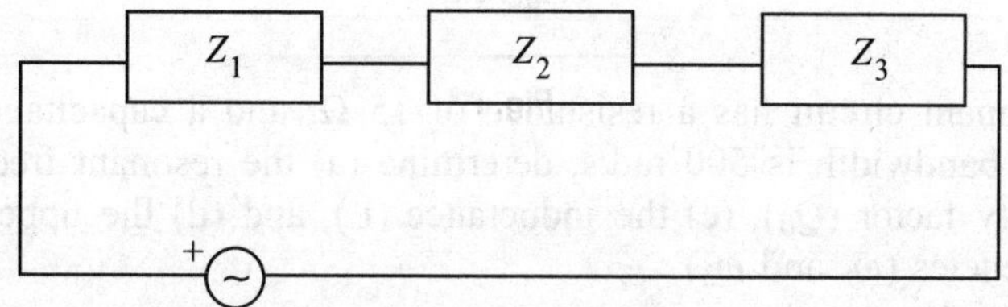

Fig. P5.6

5.7 The voltage across a circuit is given by $(300 + j60)$ V, and the current through it by $(10 - j5)$A. Determine (a) the active power, (b) the reactive volt-amperes, and (c) the apparent power. Also find the circuit impedance.

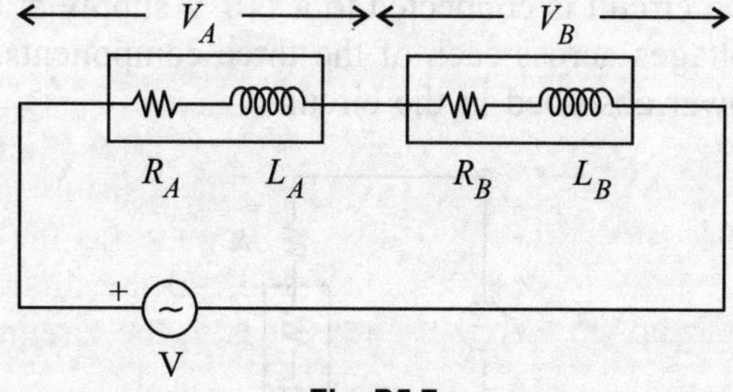

Fig. P5.7

5.8 Two lossy coils, A and B, are connected in series to a 220 V, 50 Hz supply (Fig. P5.8). The resistance of coil A is $R_A = 7\ \Omega$, and the inductance of coil B is $L_B = 10$ mH. If the active and reactive VA, drawn from the supply, are 2,500 W and 1,500 VAR respectively, find the values of the resistance of coil B (R_B), and the inductance of coil A (L_A). Calculate also the voltage drops across the two coils. Draw the phasor diagram.

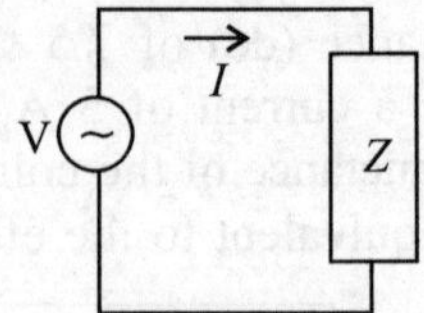

Fig. P5.8

5.9 An R-L-C series circuit (Fig. P5.9) has a resonant frequency of $\omega_0 = 10^6$ rad/s, and a bandwidth of 1 kHz. At resonance, a voltage of 0.05 V produces a current of 5 mA (both rms values). Find (a) the values of R, L and C, and (b) the voltages across the inductance, L and capacitance, C (separately) at resonance.

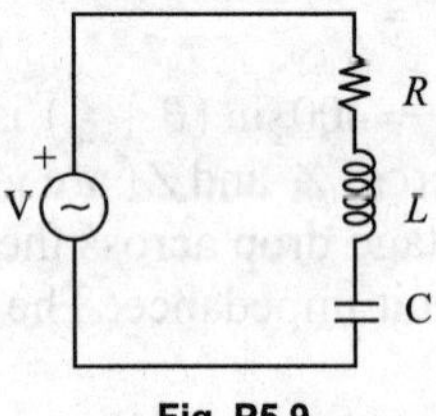

Fig. P5.9

5.10 A series resonant circuit has a resistance of 15 Ω, and a capacitance of 4 μF (Fig. P5.9). If the bandwidth is 500 rad/s, determine (a) the resonant frequency, f_0 in Hz, (b) the quality factor (Q_0), (c) the inductance (L), and (d) the upper and lower half power frequencies (ω_1 and ω_2).

5.11 A coil of resistance, 30 Ω, and inductance, 0.5 H, in series with a condenser, is fed from a constant voltage variable frequency source (Fig. P5.9). The maximum current is 1.25 A at 100 Hz. Find the frequency, when the current is 0.884 A.

5.12 A series circuit consists of a capacitor, 0.5 μF, a coil of resistance, 20 Ω, and inductance, 0.4 H, and a non-inductive resistor of 40 Ω (Fig. P5.12). Find the resonant frequency of the circuit. When the circuit is connected to a 110 V supply at this resonant frequency, calculate (a) the voltages across each of the three components, (b) the current drawn, and (b) the total power absorbed in the circuit.

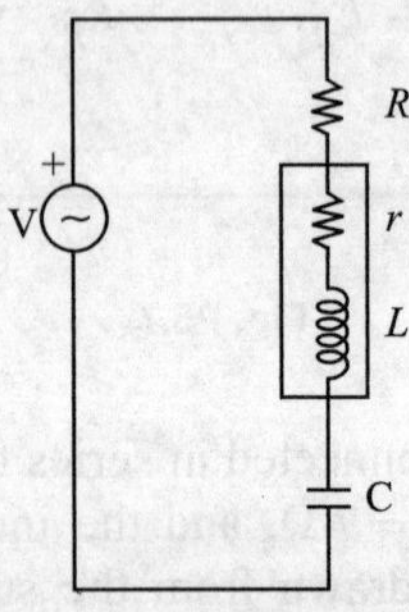

Fig. P5.12

CHAPTER 6

SINGLE PHASE AC PARALLEL AND SERIES–PARALLEL CIRCUITS

In the previous chapter, we studied current flowing through various combinations of elements, all connected in series and fed from a single phase ac supply at line frequency.

In this chapter, the currents and voltage drops in the branches of parallel and series–parallel circuits, all being fed from a single phase ac supply, are determined. The phasor diagrams, involving currents and voltages, are included in the examples, which are given to illustrate the cases. Lastly, the resonance phenomenon in parallel circuits fed from a variable frequency ac supply is presented in detail, with examples.

6.1 Parallel Circuits

Consider a circuit with all three elements, R, L and C connected in parallel (Fig. 6.1), and fed from an ac supply. The current from the supply can be computed by various methods, of which two are described here.

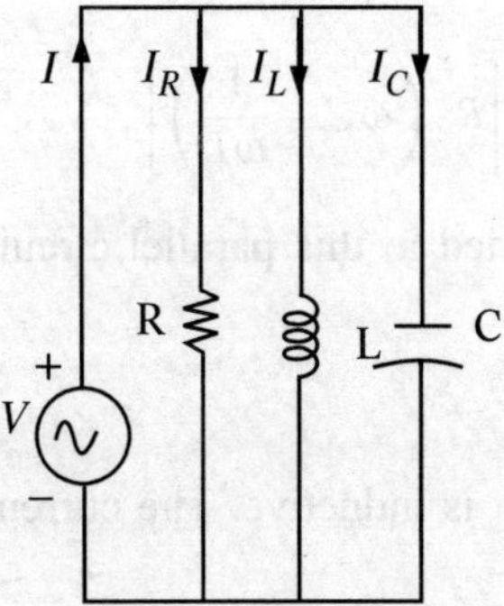

Fig. 6.1 *Circuit diagram*

FIRST METHOD

The current in the three branches are first computed. The total current drawn from the supply is the phasor sum of all three branch currents—this is computed by using Kirchhoff's first law related to the currents at the node. The voltage phasor (V) is taken as reference. The line frequency used is $f = 50$ Hz.

The branch currents are found as,

$$I_R\angle 0^\circ = (I_R + j0) = \frac{V\angle 0^\circ}{R} = \frac{V}{R}\angle 0^\circ,$$

$$I_L\angle -90^\circ = (0 - jI_L) = \frac{V\angle 0^\circ}{X_L\angle 90^\circ} = \frac{V\angle 0^\circ}{\omega L\angle 90^\circ} = \frac{V\angle 0^\circ}{2\pi f L\angle 90^\circ} = \frac{V}{X_L}\angle -90^\circ$$

$$= \frac{(V + j0)}{(0 + jX_L)} = \frac{V}{jX_L} = -j\frac{V}{X_L}, \text{ and}$$

$$I_C\angle 90^\circ = (0 + jI_C) = \frac{V\angle 0^\circ}{X_C\angle -90^\circ} = \frac{V\angle 0^\circ}{\frac{1}{\omega C}\angle 90^\circ} = \frac{V\angle 0^\circ}{\frac{1}{2\pi f C}\angle 90^\circ} = \frac{V}{X_C}\angle 90^\circ$$

$$= \frac{(V + j0)}{(0 - jX_C)} = \frac{V}{-jX_C} = j\frac{V}{X_C}.$$

Of the three branches, the first one consists of resistance only. Here, the current, I_R is in phase with the voltage (V). In the second branch, the current, I_L lags behind the voltage by 90°, as there is only an inductance, while the third branch having only a capacitance, the current, I_C leads the voltage by 90°. All these cases have been presented in Chapter 5 (Section 5.1). The total current is found as,

$$I\angle \pm\phi = [I_R + j(I_C - I_L)] = V\cdot\left[\frac{1}{R} + j\left(\omega C - \frac{1}{\omega L}\right)\right],$$

where, $|I| = \sqrt{(I_R)^2 + (I_C - I_L)^2} = V\cdot\sqrt{\left(\frac{1}{R}\right)^2 + \left(\omega C - \frac{1}{\omega L}\right)^2}$, and

$$\phi = \tan^{-1}\left(\frac{I_C - I_L}{I_R}\right) = \tan^{-1}\left[R\cdot\left(\omega C - \frac{1}{\omega L}\right)\right].$$

Let us now see the power consumed in this parallel circuit in contrast to the series circuits studied in Chapter 5.

Inductive

Consider the case when the circuit is inductive. The current lags behind the voltage by ϕ (ϕ being positive), as

$$I_L > I_C, \quad \text{i.e.,} \quad \frac{1}{\omega L} > \omega C, \quad \text{or} \quad (X_L = \omega L) > \left(X_C = \frac{1}{\omega C}\right).$$

This condition is in contrast to that derived in the case of series circuit (Chapter 5). The power factor is less than unity (1). The complete phasor diagram, with the three branch currents along with total current, and also the voltage, is shown in Fig. 6.2. The voltage phasor is taken as reference in all cases. The currents are OD = I_R, DC = I_L, CB = I_C, and OB = I. Kirchhoff's first law related to the currents at the node is applied as stated above.

The expression for the average power is

$$W = VI\cos\phi = (I_R)^2 R = \frac{V^2}{R}.$$

Power is only consumed in the resistor, R; not in the inductor/capacitor (L and C), in all three cases.

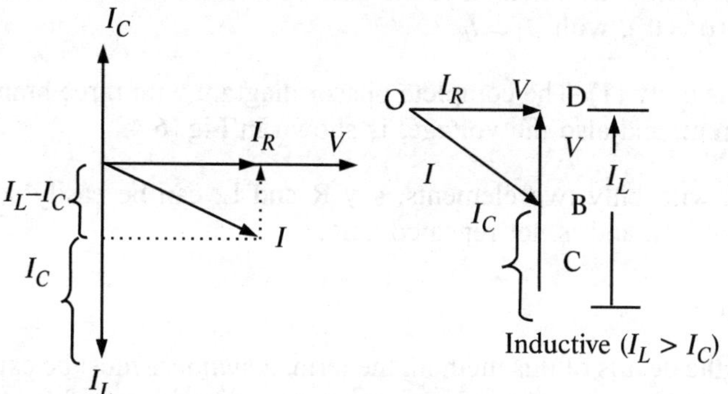

Fig. 6.2 *Phasor diagram of an inductive circuit*

Capacitive

Consider the case when the circuit is capacitive. The current leads the voltage by ϕ (ϕ being negative), as

$$I_C > I_L, \quad \text{i.e.,} \quad \omega C > \frac{1}{\omega L}, \quad \text{or} \quad (X_L = \omega L) < \left(X_C = \frac{1}{\omega C}\right).$$

This condition is in contrast to that derived in the case of series circuit. The power factor is less than unity (1). The complete phasor diagram, with the three branch currents along with total current, and also the voltage is shown in Fig. 6.3.

Fig. 6.3 *Phasor diagram of a capactive circuit* **Fig. 6.4** *Phasor diagram of a resistive circuit*

Resistive

Consider the case when the circuit is resistive.

$$I_L = I_C, \text{ i.e., } \frac{1}{\omega L} = \omega C, \quad \text{or} \quad (X_L = \omega L) = \left(X_C = \frac{1}{\omega C}\right).$$

This is the same condition as obtained in the case of a series circuit. The current is in phase with the voltage, ($\phi = 0^\circ$), with $|I| = I_R$.

The power factor is unity (1). The complete phasor diagram with three branch currents along with the total current, and also the voltage, is shown in Fig. 6.4.

NOTE: A circuit with only two elements, say R and L, can be easily solved, as given in Chapter 5 (Section 5.3), and is not repeated here.

SECOND METHOD

Before going into the details of this method, the term, *admittance* must be explained. When two resistors are connected in series, the equivalent resistance is the sum of the two resistances, the resistance being scalar (positive). If two impedances are connected in series, the equivalent impedance is the sum of the two impedances, all impedances being complex quantities, which is $Z_{eq}\angle\phi = (Z_1\angle\phi_1 + Z_2\angle\phi_2)$. Please note that the two terms, real and imaginary of the two impedances and also the equivalent one, may be positive or negative. This was explained in Chapter 4.

If two resistors are connected in parallel, the inverse of the equivalent resistance is the sum of the inverse of the two resistors

$$\frac{1}{R_{eq}} = \frac{1}{R_1} + \frac{1}{R_2},$$

as given in Chapter 1. The inverse or reciprocal of the impedance is termed 'admittance', which is a complex quantity, given by $Y\angle-\phi = \frac{1}{Z\angle\phi}$, where impedance, Z is also complex. If two impedances are connected in parallel, the equivalent admittance or impedance is given by,

$$Y_{eq}\angle-\phi = \frac{1}{Z_{eq}\angle\phi} = \left(\frac{1}{Z_1\angle\phi_1} + \frac{1}{Z_2\angle\phi_2}\right) = (Y_1\angle-\phi_1 + Y_2\angle-\phi_2).$$

Its admittance (Y) is complex. The real and imaginary parts are called conductance (G) and susceptance (B) respectively. So, $Y = G + jB$. If the impedance is $Z\angle\phi = (R + jX)$ with X being positive, then the admittance is,

$$Y\angle-\phi = (G - jB) = \frac{1}{Z\angle\phi} = \frac{1}{(R+jX)} = \frac{(R-jX)}{(R+jX)\cdot(R+jX)} = \frac{(R-jX)}{(R^2+X^2)},$$

where $G = \dfrac{R}{(R^2+X^2)}$, and $B = \dfrac{X}{(R^2+X^2)}$.

Please note the way in which the result of the division of two complex quantities is obtained.

Both the numerator and the denominator are multiplied by the complex conjugate of the denominator so as to make the denominator a real quantity. It may be stated here that the units of admittance, and also conductance and susceptance are Ω^{-1}.

The magnitude and phase angle of Z and Y are

$$|Z| = \sqrt{R^2 + X^2}, \qquad \phi = \tan^{-1}\left(\frac{X}{R}\right), \text{ and}$$

$$|Y| = \sqrt{G^2 + B^2}, \quad \phi = \tan^{-1}\left(\frac{B}{G}\right) = \tan^{-1}\left(\frac{X}{R}\right)$$

Let us now obtain the current in the circuit (Fig. 6.1). The admittances of the three branches are,

$$Y_1\angle 0^\circ = \frac{1}{Z_1\angle 0^\circ} = \frac{1}{R}, \quad Y_2\angle -90^\circ = \frac{1}{Z_2\angle 90^\circ} = \frac{1}{jX_L} = -j\frac{1}{\omega L}, \text{ and}$$

$$Y_3\angle 90^\circ = \frac{1}{Z_3\angle -90^\circ} = \frac{1}{-jX_C} = j\omega C.$$

The total admittance is obtained using the phasor sum of the three branch admittances as,

$$Y_{eq}\angle \pm\phi = (Y_1\angle 0^\circ + Y_2\angle -90^\circ + Y_3\angle 90^\circ) = \left[\frac{1}{R} + j\left(\omega C - \frac{1}{\omega L}\right)\right] = (G + jB).$$

The magnitude and phase angle of the total or equivalent admittance is given by,

$$|Y| = \sqrt{\left(\frac{1}{R}\right)^2 + \left(\omega C - \frac{1}{\omega L}\right)^2} \quad \text{and} \quad \phi = \tan^{-1}\left[R \cdot \left(\omega C - \frac{1}{\omega L}\right)\right],$$

where $\quad G = \dfrac{1}{R}, \quad$ and $\quad B = \left(\omega C - \dfrac{1}{\omega L}\right).$

The total impedance of the circuit is,

$$Z\angle \mp\phi = (R \mp jX) = \frac{1}{Y\angle \pm\phi} = \frac{1}{(G \pm jB)} = \frac{G}{(G^2 + B^2)} \mp \frac{B}{(G^2 + B^2)}.$$

The total current in the circuit is obtained as,

$$I\angle \pm\phi = \frac{V\angle 0^\circ}{Z\angle \mp\phi} = (V\angle 0^\circ \cdot Y\angle \pm\phi) = (V \cdot Y)\angle \pm\phi,$$

where the magnitude of current is $|I| = |V \cdot Y| = \left|\dfrac{V}{Z}\right|.$

The current is the same as obtained earlier, with the value of Y substituted in the above equation.

Note that the above circuit is an ideal case where all the elements used are assumed to ideal—specially, both L and C are lossless. However, in actual cases, inductive coils are lossy (r in series with L). Hence, even when two parallel circuits are there, both the branches may contain resistance, in addition to inductance/capacitance. This is best illustrated with an example.

EXAMPLE 6.1 *A circuit, having two impedances of $Z_1 = (8 + j15)\ \Omega$ and $Z_2 = (6 + j8)\ \Omega$ in parallel, is connected to a single phase ac supply (Fig. 6.5(a)), and the current drawn is 10 A. Find the currents in each branch, both in magnitude and phase, and also the supply voltage.*

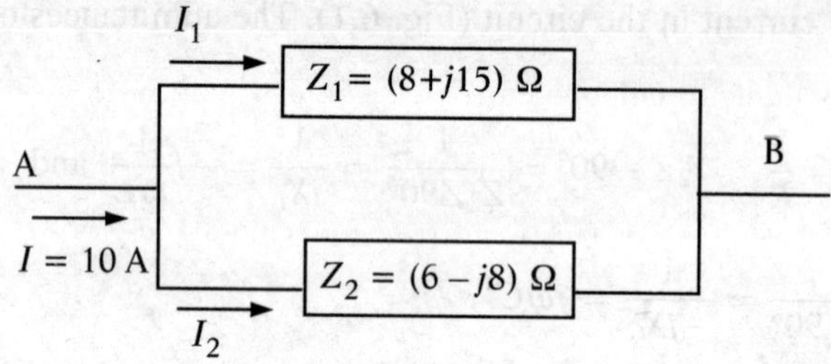

Fig. 6.5(a) *Circuit diagram*

SOLUTION The circuit is given in Fig. 6.5(a).

The impedance of branch 1 $= Z_1\angle\phi_1 = (R_1 + jX_1) = (8 + j15) = 17.0\angle 61.93°\ \Omega$.

This impedance is inductive, i.e., it contains a resistor (R_1) and an inductor (L).

The impedance of branch 2 $= Z_2\angle -\phi_2 = (R_2 - jX_2) = (6 - j8) = 10.0\angle -53.13°\ \Omega$.

This impedance is capacitive, i.e., it contains a resistor (R_2) and a capacitor (C).

Total current $= I\angle 0° = 10.0\angle 0° = (10.0 + j0.0)$ A, which is taken as reference here. The admittances, using impedances in *rectangular* form, are,

$$Y_1\angle -\phi_1 = \frac{1}{Z_1\angle\phi_1} = \frac{1}{(8+j15)} = \frac{(8-j15)}{(8^2+15^2)} = \frac{(8-j15)}{289.0} = (27.68 - j51.9) \times 10^{-3}\ \Omega^{-1}.$$

$$Y_2\angle\phi_2 = \frac{1}{Z_2\angle -\phi_2} = \frac{1}{(6-j8)} = \frac{(6+j8)}{(6^2+8^2)} = \frac{(6+j8)}{100.0} = (60.0 + j80.0) \times 10^{-3}\ \Omega^{-1}.$$

Alternatively, using impedances in *polar* form, the admittances are,

$$Y_1\angle -\phi_1 = \frac{1}{Z_1\angle\phi_1} = \frac{1}{17.0\angle 61.93°} = 0.05882\angle -61.93° = (27.68 - j51.9) \times 10^{-3}\ \Omega^{-1}.$$

$$Y_2\angle\phi_2 = \frac{1}{Z_2\angle -\phi_2} = \frac{1}{10.0\angle -53.13°} = 0.1\angle 53.13° = (60.0 + j80.0) \times 10^{-3}\ \Omega^{-1}.$$

The total admittance is,

$$Y\angle\phi = (Y_1\angle-\phi_1 + Y_2\angle\phi_2)$$

$$= [(27.68 - j51.9) + (60.0 + j80.0)] \times 10^{-3}$$

$$= (87.68 + j28.1) \times 10^{-3} = 0.09207\angle 17.77^\circ\,\Omega^{-1}.$$

The total impedance is,

$$Z\angle-\phi = \frac{1}{Y\angle\phi} = \frac{1}{0.09207\angle 17.77^\circ} = 10.86\angle-17.77^\circ = (10.34 - j3.314)\,\Omega.$$

Alternatively, the equivalent impedance can be found as given here.

$$Z'\angle\phi_5 = (Z_1\angle\phi_1 + Z_2\angle-\phi_2)$$

$$= [(8.0 + j15.0) + (6.0 - j8.0)] = (14.0 + j7.0)$$

$$= 15.65\angle 26.565^\circ\,\Omega.$$

$$Z\angle-\phi = \frac{Z_1\angle\phi_1 \cdot Z_2\angle-\phi_2}{(Z_1\angle\phi_1 + Z_2\angle-\phi_2)} = \frac{Z_1\angle\phi_1 \cdot Z_2\angle-\phi_2}{Z'\angle\phi_5}$$

$$= \left(\frac{17.0 \times 10.0}{15.65}\right)\angle(61.93^\circ - 53.13^\circ - 26.565^\circ) = 10.86\angle-17.765^\circ$$

$$= (10.34 - j3.314)\,\Omega.$$

The supply voltage is,

$$V\angle-\phi = (I\angle 0^\circ \cdot Z\angle-\phi) = (10.0 \times 10.86)\angle-17.77^\circ$$

$$= 108.6\angle-17.77^\circ = (103.4 - 33.14)\text{ V}.$$

Note that the voltage is lagging behind the current by an angle, ϕ, or the current leads the voltage.

The power factor (pf) of the total current $= \cos\phi = \cos 17.77^\circ = 0.952$ (lead) .

The two branch currents are,

$$I_1\angle-\phi_3 = \frac{V\angle-\phi}{Z_1\angle\phi_1} = \frac{108.6}{17.0}\angle-(17.77^\circ + 61.93^\circ)$$

$$= 6.39\angle-79.7^\circ = (1.143 - j6.29)\text{ A}.$$

$$I_2\angle\phi_4 = (I\angle 0^\circ - I_1\angle-\phi_3) = [(10.0 + j0.0) - (1.143 - j6.29)]$$

$$= (8.857 + j6.29) = 10.86\angle 35.38^\circ\text{ A}.$$

Alternatively, this current is obtained as,

$$I_2\angle\phi_4 = \frac{V\angle-\phi}{Z_2\angle-\phi_2} = \frac{108.6}{10.0}\angle(-17.77^\circ + 53.13^\circ)$$

$$= 10.86\angle 35.36^\circ = (8.857 + j6.285)\text{ A},$$

which is nearly equal to the value obtained earlier.

The phasor diagram with the total (input) current as reference is shown in Fig. 6.5(b).

Alternatively, the two branch currents can be found by the procedure given below.

$$I_1\angle-\phi_3 = \frac{I\angle 0^\circ \cdot Z_2\angle-\phi_2}{(Z_1\angle\phi_1 + Z_2\angle-\phi_2)} = \frac{I\angle 0^\circ \cdot Z_2\angle-\phi_2}{Z'\angle\phi_5} = \left(\frac{I\cdot Z_2}{Z'}\right)\angle-(\phi_2+\phi_5)$$

$$= \left(\frac{10.0\times 10.0}{15.65}\right)\angle-(53.13^\circ + 26.565^\circ) = 6.39\angle-79.7^\circ = (1.143 - j6.29)\text{ A}.$$

$$I_2\angle\phi_4 = \frac{I\angle 0^\circ \cdot Z_1\angle\phi_1}{(Z_1\angle\phi_1 + Z_2\angle-\phi_2)} = \frac{I\angle 0^\circ \cdot Z_1\angle\phi_1}{Z'\angle\phi_5} = \left(\frac{I\cdot Z_1}{Z'}\right)\angle(\phi_1-\phi_5)$$

$$= \left(\frac{10.0\times 17.0}{15.65}\right)\angle(61.93^\circ - 26.565^\circ) = 10.86\angle 35.365^\circ = (8.856 + j6.29)\text{ A}.$$

Note that these values are the same as obtained earlier.

$10.86\angle-35.365^\circ = 10.86\angle 35.38^\circ I = 10.0\angle 0^\circ.$

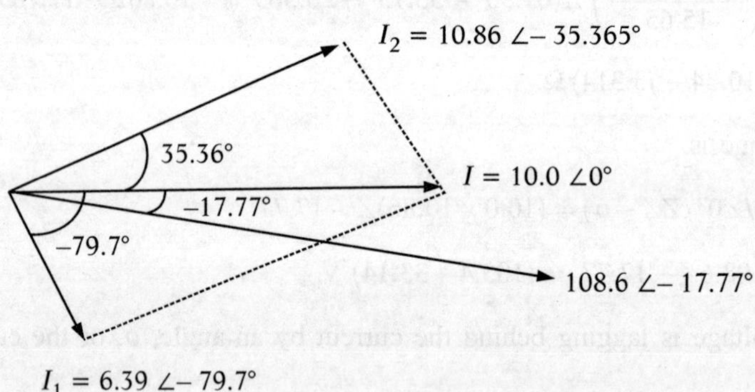

Fig. 6.5(b) *Phasor diagram*

6.2 Necessity of Power Factor (PF) Improvement in AC Circuits

When a consumer takes power supply from companies, different types of loads, such as, ac motors (mainly, two types—three-phase and single-phase induction motors), arc lamps, electric discharge lamps, and induction heating furnace, are used. Some of these run at low power factor, which affects the total system power factor and results in the consumer paying a higher tariff for the power consumed.

For a single phase system, as given in Chapter 5, the power drawn is given by $P = VI\cos\phi$, where $\cos\phi$ is the power factor, its value being less than 1, nearly, say 0.8. The symbols used are the same. For the same power consumed by the load, the current is higher, because of low power factor, which is given by

$$I = \frac{P}{V\cos\phi}.$$

For higher current, larger cross-sectional area of the conductor is required and more energy is lost in the distribution system resulting in higher cost. Because of the costs of larger equipment and wasted energy, electrical companies will usually charge a higher cost to industrial or commercial customers where there is a low power factor.

Hence, the total power factor is to be increased. This can be done by connecting a capacitor in parallel across the supply (Fig. 6.6(a)), which takes current at a leading power factor. The total power factor can be increased to unity (1.0), as shown in the phasor diagram (Fig. 6.6(b)). Sometimes, the power factor is increased only close to 1.0, say 0.95, to reduce the cost of the capacitor required, as the current in the capacitor is now decreased (Fig. 6.6(c)).

The power factor can also be expressed as

$$\cos\phi = \frac{\text{kW}}{\text{kVA}},$$

where $kVA = VI.10^{-3}$. The reactive $VA(kVAR)$ is given by $kVA\sin\phi$. The VA supplied by the capacitor (kVAR) in the above two cases is shown in Fig. 6.6(c), which shows that the VA rating in kVAR of the capacitor used is lower when the total power factor is raised to 0.95, not 1.0. This will be illustrated in the examples given later.

EXAMPLE 6.2 *The power consumed in the inductive load (Fig. 6.6(a)) is 2.5 kW at 0.71 lagging power factor (pf). The input voltage is 230 V, 50 Hz. Find the value of the capacitor C which needs to be connected in parallel with the above load so that the resultant power factor of the input current is 0.866 lagging. If the power factor of the input current is now changed to 0.95 lagging, find the new value of the capacitor, C_1.*

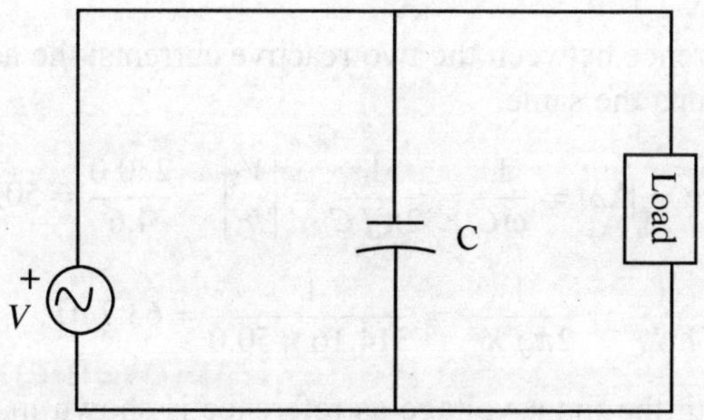

Fig. 6.6(a)

SOLUTION: The circuit is shown in Fig. 6.6(a).

$f = 50$ Hz, $|V| = 230$ V.

$\omega = 2\pi.f = 2\pi \times 50.0 = 314.16$ rad/s, $P = 2.5$ kW = 2,500 W.

The power factor (pf) of the inductive branch (1) = $\cos\phi_L = 0.71$ (lag).

The phase angle of the above current = $\phi_L = \cos^{-1}(0.71) \approx 45°$.

$$P = V.I_L.\cos\phi_L = 2,500 \text{ W}.$$

The magnitude of current in the inductive load

$$= |I_L| = \frac{P}{V\cdot\cos\phi_L} = \frac{2,500}{230.0\times 0.71} = 15.3 \text{ A}.$$

The current $= I_L\angle-\phi = 15.3\angle-45° = (10.87 - j10.87)$ A

The power consumed in the circuit remains the same as the capacitor, C (branch 2) connected in parallel with the inductive load which does not consume any power. However, the reactive power in the circuit changes. The active component of the total current remains the same as computed earlier.

$$\text{Power} = P = V.I.\cos\phi = 2,500 \text{ W}.$$

The power factor (pf) of total current = $\cos\phi = 0.866$ (lag).

The phase angle of total current = $\phi = \cos^{-1}(0.866) = 30°$.

The magnitude of total current $= |I| = \dfrac{P}{V\cdot\cos\phi} = \dfrac{2,500}{230.0\times 0.866} = 12.55$ A

The current $= I\angle-\phi = 12.55\angle-30° = (10.87 - j6.275)$ A.

The current in the capacitance, C (branch 2) $= I_C\angle-90° = I_2\angle-90° = (I\angle\phi - I_1\angle\phi_L)$

$$= [(10.87 - j6.275) - (10.87 - j10.87)] \approx j4.6 = 4.6\angle 90° \text{ A}.$$

This current is the difference between the two reactive currents; the active currents are equal, as the total power remains the same.

The capacitive reactance $= |X_C| = \dfrac{1}{\omega C} = \dfrac{1}{2\pi f C} = \left|\dfrac{V}{I_C}\right| = \dfrac{230.0}{4.6} = 50.0\,\Omega$.

The capacitance $= C = \dfrac{1}{\omega X_C} = \dfrac{1}{2\pi f X_C} = \dfrac{1}{314.16\times 50.0} = 63.7\,\mu$F.

The phasor diagram with the input voltage as reference is shown in Fig. 6.6(b).

In the second case, the power factor (pf) of total current $\cos\phi_1 = 0.95$ (lag).

The phase angle of total current = $\phi_1 = \cos^{-1}(0.95) = 18.195°$.

The magnitude of total current $= |I_3| = \dfrac{P}{V.\cos\phi_1} = \dfrac{2,500}{230.0\times 0.95} = 11.44$ A.

The current $= I_3\angle-\phi_1 = 11.44\angle-18.195° = (10.87 - j3.57)$ A

The current in the capacitance, C_1 (branch 2)

$$= I_{Cl}\angle 90° = I_4\angle 90° = (I\angle-\phi - I_1\angle-\phi_L)$$

$$= [(10.87 - j3.57) - (10.87 - j10.87)] \approx j7.3 = 7.3 < 90° \text{ A}$$

The capacitive reactance $= |X_{Cl}| = \dfrac{1}{\omega C_1} = \dfrac{1}{2\pi f C_1} = \left|\dfrac{V}{I_{Cl}}\right| = \dfrac{230.0}{7.3} = 31.51\,\Omega$

The capacitance $= C_1 = \dfrac{1}{\omega X_{Cl}} = \dfrac{1}{2\pi f X_{Cl}} = \dfrac{1}{314.16 \times 31.51} = 101.0\,\mu\text{F}$

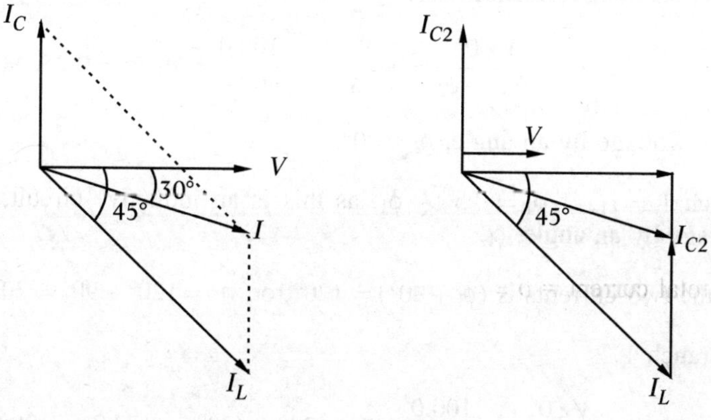

Fig. 6.6(b) Fig. 6.6(c)

EXAMPLE 6.3: *An inductive load (R in series with L) is connected in parallel with a capacitance C of 12.5 μF (Fig. 6.7(a)). The input voltage to the circuit is 100 V at 31.8 Hz. The phase angle between the two branch currents, ($I_1 = I_L$), and ($I_2 = I_C$), is 120°, and the current in the first branch is $I_1 = I_L$= 0.5 A. Find the total current, and also the values of r and L.*

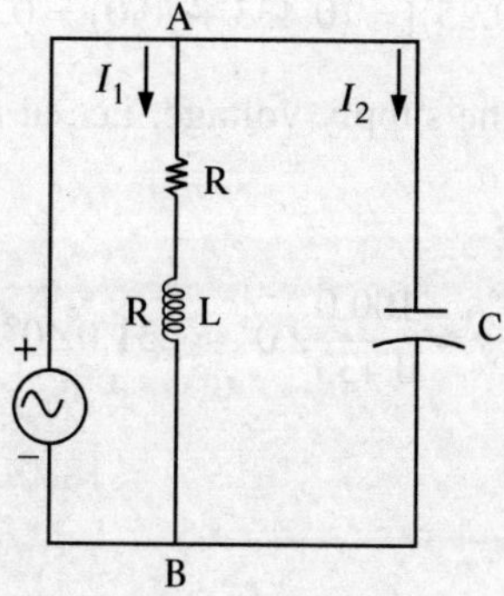

Fig. 6.7(a) *Circuit diagram*

SOLUTION The circuit is given in Fig. 6.7(a).

$f = 31.8$ Hz $|V| = 100$ V, $|I_1| = 0.5$ A.

$\omega = 2\pi . f = 2\pi \times 31.8 \approx 200$ rad/s.

C = 12.5 μF = 12.5×10^{-6} F.

Reactance of capacitor $= X_C = \dfrac{1}{\omega C} = \dfrac{1}{2\pi f C} = \dfrac{1}{200.0 \times 12.5 \times 10^{-6}} = 400.0\ \Omega$.

This is also the magnitude of the impedance in branch 2.

The supply voltage is taken as reference here.

The current in branch 2 $= I_2\angle\phi_2 = \dfrac{V\angle 0^\circ}{Z_2\angle-\phi_2} = \dfrac{V}{-jX_C} = j\dfrac{100.0}{400.0} = j0.25 = 0.25\angle 90^\circ$ A.

This current leads the voltage by an angle, $\phi_2 = 90^\circ$.

The current in branch 1 $= I_1\angle-\phi_1 = 0.5\angle-\phi_1$, as this is an inductive circuit, where the current lags the voltage by an angle, ϕ_1.

The angle between the two currents is $(\phi_1 + 90^\circ) = 120^\circ$, or, $\phi_1 = 120^\circ - 90^\circ = 30^\circ$.

The impedance of branch 1

$$= Z_1\angle\phi_1 = (R + jX_L) = \frac{V\angle 0^\circ}{I_1\angle-\phi_1} = \frac{100.0}{0.5}\angle 30^\circ = 200.0\angle 30^\circ = (173.2 + j100.0)\ \Omega$$

Resistance in branch 1 $= R = 173.2\ \Omega$.

Inductance in branch 1 $= L = \dfrac{X_L}{\omega} = \dfrac{X_L}{2\pi f} = \dfrac{100.0}{200.0} = 0.5$ H = 500 mH.

The current drawn from the supply $= I\angle\phi = (I_1\angle-\phi_1 + I_2\angle\phi_2) = (0.5\angle-30^\circ + 0.25\angle 90^\circ)$

$$= [(0.433 - j0.25) + (0.0 + j0.25)] = (0.433 + j0.0) = 0.433\angle 0^\circ \text{ A.}$$

The total current is in phase with the supply voltage, i.e., at unity power factor ($\cos 0^\circ = 1.0$).

The total or equivalent impedance

$$= Z\angle 0^\circ = (R' + j0.0) = \frac{V\angle 0^\circ}{I\angle 0^\circ} = \frac{100.0}{0.433}\angle 0^\circ \approx 231.0\angle 0^\circ = (231.0 + j0.0)\ \Omega.$$

This impedance is resistive only.

The total admittance is,

$$Y\angle 0^\circ = \frac{1}{Z\angle 0^\circ} = (Y_1\angle -\phi_1 + Y_2\angle \phi_2) = \left(\frac{1}{Z_1\angle \phi_1} + \frac{1}{Z_2\angle -\phi_2}\right)$$

$$= \left(\frac{1}{200.0\angle 30^\circ} + \frac{1}{400.0\angle -90^\circ}\right) = [(5.0\angle -30^\circ + 2.5\angle 90^\circ) \times 10^{-3}]$$

$$= [\{(4.33 - j2.5) + (0.0 + j0.25)\} \times 10^{-3}] = [(4.33 + j0.0) \times 10^{-3}]$$

$$= (4.33 \times 10^{-3})\angle 0^\circ \Omega^{-1}.$$

The total impedance is,

$$Z\angle 0^\circ = \frac{1}{Y\angle 0^\circ} = \frac{1.0}{4.33 \times 10^{-3}}\angle 0^\circ \approx 231.0\angle 0^\circ = (231.0 + j0.0)\,\Omega$$

which is the same as obtained earlier.

Alternatively, the equivalent impedance can be found as,

$$Z'\angle \phi_3 = (Z_1\angle \phi_1 + Z_2\angle -\phi_2) = [(173.2 + j100.0) + (0.0 - j400.0)]$$

$$= (173.2.0 - j300.0) = 346.4\angle -60.0^\circ\ \Omega.$$

$$Z\angle -\phi = \frac{Z_1\angle \phi_1 \cdot Z_2\angle -\phi_2}{(Z_1\angle \phi_1 + Z_2\angle -\phi_2)} = \frac{Z_1\angle \phi_1 \cdot Z_2\angle -\phi_2}{Z'\angle \phi_3}$$

$$= \left(\frac{200.0 \times 400.0}{346.4}\right)\angle(30.0^\circ - 90.0^\circ + 60.0^\circ)$$

$$\approx 231.0\angle 0^\circ = (231.0 + j0.0)\Omega.$$

The phasor diagram is drawn in Fig. 6.7(b).

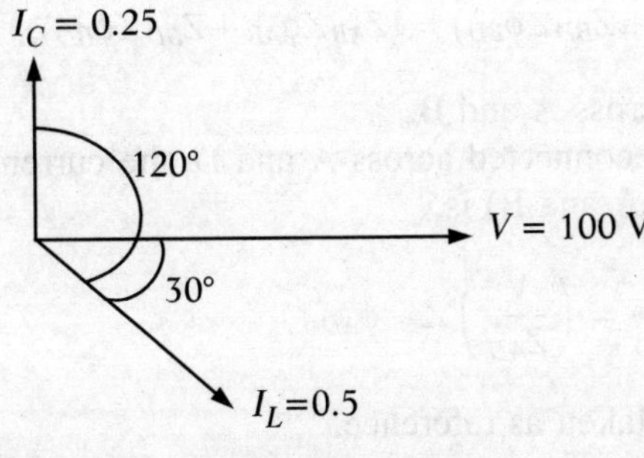

Fig. 6.7(b)

6.3 Series-Parallel Circuits

Consider a circuit with a branch having impedance Z_1 in series with two parallel branches having impedances, Z_2 and Z_3 as shown in Fig. 6.8, connected to a single phase ac supply.

The impedances of the branches (in Ω) are given here.

$$Z_1\angle\phi_1 = (R_1 + jX_1),\quad Z_2\angle\phi_2 = (R_2 + jX_2), \text{ and } Z_3\angle\phi_3 = (R_3 + jX_3).$$

Note that the signs of the three reactances can be positive or negative, as the case may be, i.e., inductive or capacitive. The angles also may change signs accordingly.

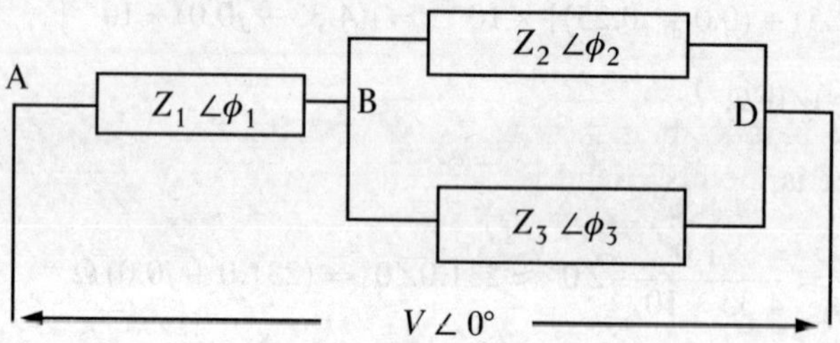

Fig. 6.8

As the branches 2 and 3 are connected in parallel across B and D, the admittance is,

$$Y_{BD}\angle-\phi_{BD} = \frac{1}{Z_{BD}\angle\phi_{BD}} = (Y_2\angle-\phi_2 + Y_3\angle-\phi_2) = \left(\frac{1}{Z_2\angle\phi_2} + \frac{1}{Z_3\angle\phi_3}\right).$$

If $Z'\angle\phi = (Z_1\angle\phi_1 + Z_2\angle\phi_2)$, then impedance is found as,

$$Z_{BD}\angle\phi_{BD} = \frac{1}{Y_{BD}\angle-\phi_{BD}} = \frac{Z_2\angle\phi_2\cdot Z_3\angle\phi_3}{(Z_2\angle\phi_2 + Z_3\angle\phi_3)}$$

$$= \frac{Z_2\angle\phi_2\cdot Z_3\angle\phi_3}{Z'\angle\phi} = \left(\frac{Z_2\cdot Z_3}{Z'}\right)\angle(\phi_2+\phi_3-\phi).$$

The total impedance of the circuit (across A and D)

$$= Z_{AD}\angle\phi_{AD} = (Z_1\angle\phi_1 + Z_{BD}\angle\phi_{BD}) = (Z_{AB}\angle\phi_{AB} + Z_{BD}\angle\phi_{BD}),$$

as branch 1 is connected across A and B.

As the supply voltage is connected across A and D, the current drawn, which is also the current in branch 1 (across A and B) is,

$$I_{AB}\angle-\phi_{AD} = \frac{V\angle 0°}{Z_{AD}\angle\phi_{AD}} = \left(\frac{V}{Z_{AD}}\right)\angle-\phi_{AD},$$

with the supply voltage, V taken as reference.

The current in branch 2 (across B and D)

$$= \quad I_2\angle\phi_5 = \frac{I_{AB}\angle-\phi_{AD}\cdot Z_3\angle\phi_3}{(Z_1\angle\phi_1 + Z_2\angle\phi_2)} = \frac{I_{AB}\angle-\phi_{AD}\cdot Z_3\angle\phi_3}{Z'\angle\phi} = \left(\frac{I_{AB}\cdot Z_3}{Z'}\right)\angle(-\phi_{AD}+\phi_3-\phi).$$

The current in branch 3 (across the same points, B and D) $= I_3\angle\phi_6 = (I_{AB}\angle-\phi_{AB} - I_2\angle\phi_5)$, as the total current is the sum of two currents in branches 2 and 3, connected across B and D.

Alternatively, the current in branch 3 (across B and D) can be obtained as,

$$= \quad I_3\angle\phi_6 = \frac{I_{AB}\angle-\phi_{AD}\cdot Z_2\angle\phi_2}{(Z_1\angle\phi_1+Z_2\angle\phi_2)} = \frac{I_{AB}\angle-\phi_{AD}\cdot Z_2\angle\phi_2}{Z'\angle\phi}.$$

$$= \left(\frac{I_{AB}\cdot Z_2}{Z'}\right)\angle(-\phi_{AD}+\phi_2-\phi).$$

The currents in the parallel branches (across B and D) can be found by first finding the voltage drop across the points B and D, as,

$$V_{BD}\angle\phi_7 = (I_{AB}\angle-\phi_{AD}\cdot Z_{BD}\angle\phi_{BD}) = (I_{AB}\cdot Z_{BD})\angle(-\phi_{AD}+\phi_{BD}).$$

Then the branch currents is found as (we will find only in one branch, the rest is left as an exercise to the reader),

The current in branch 2 = $\quad I_2\angle\phi_5 = \dfrac{V_{BD}\angle\phi_7}{Z_2\angle\phi_2} = \left(\dfrac{V_{BD}}{Z_2}\right)\angle(\phi_7-\phi_2).$

Thus, the currents along with the voltage drops in all the branches can be calculated. The phasor diagram is best illustrated with the following examples, where the complete phasor diagram will be drawn in each case.

EXAMPLE 6.4 *Find the input voltage at 50 Hz to be applied to the circuit shown in Fig. 6.9, such that the current in the capacitor is 8 A?*

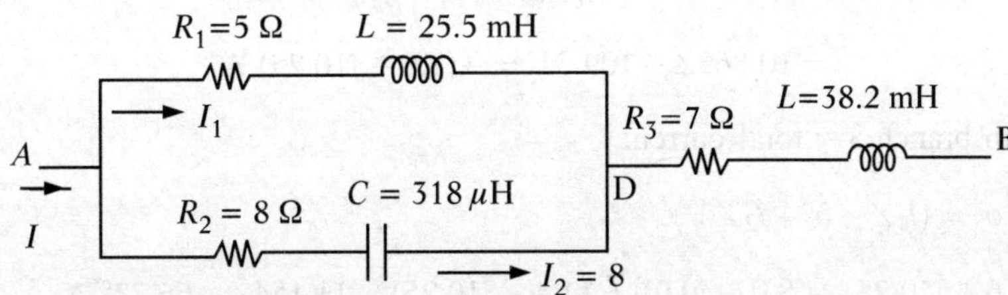

Fig. 6.9 *Circuit diagram*

SOLUTION The circuit diagram is shown in Fig. 6.9.

$$f = 50 \text{ Hz}, \ \omega = 2\pi.f = 2\pi\times 50.0 = 314.16 \text{ rad/s}.$$

$$L_1 = 0.0255 \text{ H} = 25.5 \text{ mH}, \ X_1 = \omega.L_1 = 2\pi f.L_1 = 314.16\times 0.0255 = 8.0\ \Omega.$$

$$C = 318.0\,\mu\text{F} = 318.0\times 10^{-6} \text{ F}.$$

$$X_2 = \frac{1}{\omega C} = \frac{1}{2\pi f C} = \frac{1}{314.16\times 318.0\times 10^{-6}} = 10.0 \quad \Omega.$$

$$L_3 = 0.0382 \text{ H} = 38.2 \text{ mH}, \ X_3 = \omega.L_3 = 2\pi f.L_3 = 314.16\times 0.0382 = 12.0\,\Omega.$$

The current in branch 2 $= I_2\angle 0° = 8.0\angle 0° = (8.0 + j0.0)$ A.

This current is taken as reference.

The impedance of the three branches are as follows:

$$Z_1\angle\phi_1 = (R_1 + jX_1) = (5 + j8) = 9.434\angle 58.0°\ \Omega.$$

$$Z_2\angle -\phi_2 = (R_2 - jX_2) = (8 - j10) = 12.81\angle -51.34°\ \Omega.$$

$$Z_3\angle\phi_3 = (R_3 + jX_3) = (7 + j12) = 13.89\angle 59.74°\ \Omega.$$

Readers are requested to draw the phasor diagram as given before.

The voltage drop in branches 1 and 2 (across A and D)

$$= V_{AD}\angle -\phi_2 = (I\angle 0° \cdot Z_2\angle -\phi_2)$$

$$= (8.0 \times 12.81)\angle -51.34° = 102.5\angle -51.34° = (64.0 - j80.0)\ \text{V}$$

Alternatively, it can be found as,

$$V_{AD}\angle -\phi_2 = (I\angle 0° \cdot Z_2\angle -\phi_2)$$

$$= [(8.0 + j0.0) \times (8.0 - j10.0)] = (64.0 - j80.0) = 102.5\angle -51.34°\ \text{V}.$$

The current in branch 1 $= I_1\angle -\phi_4 = \dfrac{V_{AD}\angle -\phi_2}{Z_1\angle\phi_1} = \dfrac{102.5}{9.434}\angle -(51.34° + 58.0°)$

$$= 10.865\angle -109.34° = -(3.6 + j10.25)\ \text{A}$$

The current in branch 3 = total current

$$= I\angle -\phi_5 = (I_1\angle -\phi_4 + I_2\angle 0°).$$

$$= [-(3.6 + j10.25) + (8.0 + j0.0)] = (4.4 - j10.25) = 11.154\angle -66.77°\ \text{A}.$$

The voltage drop in branch 3 (across D and B)

$$= V_{DB}\angle -\phi_6 = (I\angle -\phi_5 \cdot Z_3\angle\phi_3)$$

$$= (11.154 \times 13.89)\angle(-66.77° + 59.74°) = 154.93\angle -7.03° = (153.77 - j18.96)\ \text{V}.$$

Total voltage drop across A and B

$$= V_{AB}\angle -\phi_7 = (V_{AD}\angle -\phi_2 + V_{DB}\angle -\phi_6)$$

$$= [(64.0 - j80.0) + (153.77 - j18.96)] = (217.77 - j98.96) = 239.2\angle -24.44°\ \text{V}.$$

EXAMPLE 6.5 *A resistor of 50 Ω in parallel with an inductor of 0.3 H is connected in series with a capacitor, C (Fig. 6.10). A voltage of 220 V, 50 Hz is applied to the circuit.*

Find (a) the value of C so that the power factor is unity, (b) the total current and (c) the current in the inductor.

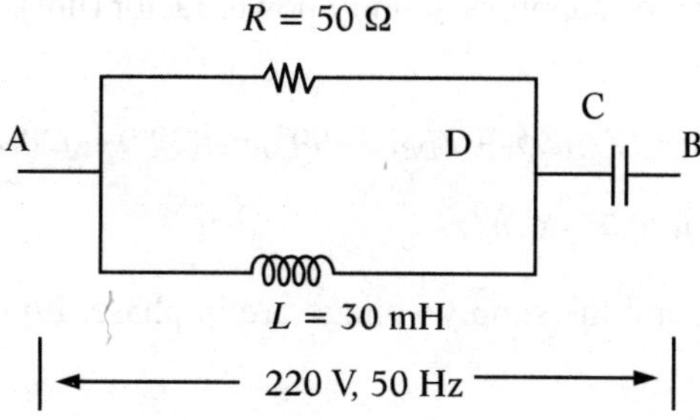

Fig. 6.10 *Circuit diagram*

SOLUTION The circuit diagram is given in Fig. 6.10.

$$f = 50 \text{ Hz}, \qquad \omega = 2\pi.f = 2\pi \times 50.0 = 314.16 \text{ rad/s}.$$

$$R = 50\,\Omega, \qquad |V| = 220 \text{ V}.$$

$$L_1 = 0.3 \text{ H} = 300 \text{ mH}, X_L = \omega.L = 2\pi f.L = 314.16 \times 0.3 = 94.25\,\Omega$$

The admittance between the points, A and D

$$= Y_{AD}\angle -\phi_1 = \left(\frac{1}{R} + \frac{1}{jX_L}\right) = \left(\frac{1}{50.0} + \frac{1}{j94.25}\right) = [(20.0 - j10.61) \times 10^{-3}]$$

$$= 0.02264\angle -27.95°\Omega^{-1}.$$

The impedance

$$= Z_{AD}\angle\phi_1 = \frac{1}{Y_{AD}\angle -\phi_1} = \frac{1}{0.02264\angle -27.95°} = 44.17\angle 27.95° = (39.0 + j20.7) \quad \Omega.$$

Alternatively, the equivalent impedance can be found as given below.

$$Z_1\angle 0° = (R + j0.0) = (50.0 + j0.0) = 50.0\angle 0° \quad \Omega.$$

$$Z_2\angle 90° = (0.0 + jX_L) = (0.0 + j94.25) = 94.25\angle 90°\ \Omega.$$

$$Z'\angle\phi' = (Z_1\angle 0° + Z_2\angle 90°) = (50.0 + j94.25) = 106.7\angle 62.05°\ \Omega.$$

$$Z_{AD}\angle\phi_1 = \frac{Z_1\angle 0° \cdot Z_2\angle 90°}{(Z_1\angle 0° + Z_2\angle 90°)} = \frac{Z_1\angle 0° \cdot Z_2\angle 90°}{Z'\angle\phi'} = \left(\frac{50.0 \times 94.25}{106.7}\right)\angle(90° - 62.05°)$$

$$= 44.17\angle 27.95° = (39.0 + j20.7)\,\Omega.$$

The impedance between the points D and B

$$= Z_{DB}\angle -90^\circ = X_C\angle -90^\circ = -jX_C = -j\frac{1}{\omega C} = -j\frac{1}{2\pi f C} \quad \Omega.$$

As total current drawn from the supply is at unity power factor (upf), total impedance between the points, A and B

$$= Z_{AB}\angle 0^\circ = (R' + j0) = (Z_{AD}\angle\phi_1 + Z_{DB}\angle -90^\circ) = [(39.0 + j20.7) - jX_C]$$

$$= [39.0 + j(20.7 - X_C)]\ \Omega.$$

Note that the total current and the supply voltage are in phase. Equating the imaginary part, we get,

$$X_C = \frac{1}{\omega C} - 20.7\ \Omega$$

$$\text{The capacitance} = C = \frac{1}{\omega X_C} = \frac{1}{314.16 \times 20.7} = 153.8 \quad \mu\text{F}.$$

Now, total impedance $= Z_{AB}\angle 0^\circ = (R' + j0) = (39.0 + j0.0) = 39.0\angle 0^\circ \quad \Omega.$

Total current drawn from the supply

$$= I\angle 0^\circ = \frac{V\angle 0^\circ}{Z_{AB}\angle 0^\circ} = \frac{220.0}{39.0}\angle 0^\circ = 5.64\angle 0^\circ = (5.64 + j0.0)\ \text{A}.$$

The voltage drop between the points, A and D $= V_{AD}\angle\phi_1 = (I_{AD}\angle 0^\circ \cdot Z_{AD}\angle\phi_1)$

$$= (5.64 \times 44.17)\angle 27.95^\circ = 249.1\angle 27.95^\circ = (220.0 + j116.75)\ \text{V}.$$

The current in the resistor, R

$$= I_R\angle\phi_1 = \frac{V_{AD}\angle\phi_1}{R} = \frac{249.1}{50.0}\angle 27.95^\circ = 4.982\angle 27.95^\circ = (4.4 + j2.335)\ \text{A}.$$

The current in the inductor, L

$$= \quad I_L\angle -\phi_2 = \frac{V_{AD}\angle\phi_1}{X_L\angle 90^\circ} = \frac{249.1}{94.25}\angle(27.95^\circ - 90^\circ)$$

$$= 2.643\angle -62.05^\circ = (1.24 - j2.335)\ \text{A}.$$

If two currents are added, the total current can be easily obtained.

Alternatively, the branch currents (only one example is included) can be found by the procedure given below.

$$I_R\angle\phi_1 = I_1\angle\phi_1 = \frac{I\angle 0^\circ \cdot Z_2\angle 90^\circ}{(Z_1\angle 0^\circ + Z_2\angle 90^\circ)} = \frac{I\angle 0^\circ \cdot Z_2\angle 90^\circ}{Z'\angle\phi'} = \left(\frac{I \cdot Z_2}{Z'}\right)\angle(90^\circ - \phi')$$

$$= \left(\frac{5.64 \times 94.25}{106.7}\right) \angle(90^\circ - 62.05^\circ) = 4.982\angle 27.95^\circ = (4.4 + j2.335)\ \text{A},$$

which is the same as found earlier.

The voltage drop between the points, D and B $= V_{DB}\angle -90^\circ = (I_{AD}\angle 0^\circ \cdot X_C\angle -90^\circ)$

$$= (5.64 \times 20.7)\angle -90^\circ = 116.75\angle -90^\circ = (0.0 - j116.75)\ \text{V}.$$

It can be easily shown that the supply voltage is the sum of the two voltage drops, as given earlier.

EXAMPLE 6.6 *In a circuit (Fig. 6.11), the wattmeter reads 960 W and the ammeter reads 6 A. Calculate the values of V_S, V_C, I_C, I_L and X_C.*

SOLUTION In this circuit (Fig. 6.11), the power is consumed in two resistors, R_1 and R_2 only; no power is consumed in inductance L, and capacitance C. These two components affect only the reactive power.

$$R_1 = 10\ \Omega, R_2 = 6\ \Omega$$

$$X_L = 8\ \Omega, |I| = 6\ \text{A}, P = 960\ \text{W}.$$

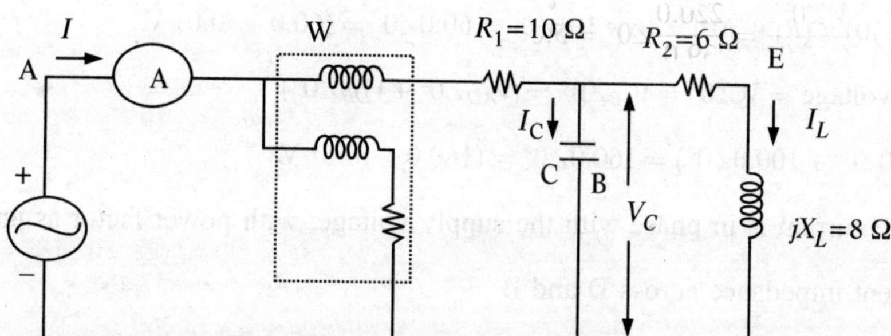

Fig. 6.11 *Circuit diagram*

Total power $= P = [(I^2 \cdot R_1) + \{(I_L)^2 \cdot R_2\}] = [(6^2 \times 10) + \{(I_L)^2 \times 6\}]$

$$= [360 + \{(I_L)^2 \times 6\}] = 960\ \text{W}$$

or, $(I_L)^2 = \dfrac{960 - 360}{6} = 100$ or, $|I_L| = \sqrt{100} = 10$ A.

The impedance of the inductive branch 2 $= \quad Z_L\angle\phi_L = (R_2 + jX_L) = (6 + j8) = 10.0\angle 53.13^\circ\ \Omega.$

The magnitude of the voltage in the inductive branch

$$= |V_{DB}| = |V_C| = |I_L \cdot Z_L| = 10 \times 10 = 100.0\ \text{V}.$$

If the above voltage is taken as reference, i.e., $V_{DB} = 100.0\angle 0^\circ$, the current in that branch

$= I_L\angle-\phi_L = 10.0\angle-53.13^\circ = (6.0 - j8.0)$ A.

The current in capacitance 1

$$= I_C\angle 90^\circ = jI_C = \frac{V_{DB}\angle 0^\circ}{X_C\angle -90^\circ} = \frac{V_{DB}}{-jX_C} = j\frac{V_{DB}}{X_C} = \frac{V_{DB}}{X_C}\angle 90^\circ \text{ A.}$$

Total current $= I\angle\phi = (I_L\angle-\phi_L + I_C\angle 90^\circ) = [(6 - j8) + jI_C] = [6 + j(I_C - 8)]$ A.

The magnitude of the total current $= |I| = \sqrt{6^2 + (I_C - 8)^2} = 6$ A

or, $36 + (I_C - 8)^2 = 6^2 = 36$ or, $(I_C - 8) = 0$ or, $|I_C| = 8.0$ A.

The capacitive reactance $= X_C = \left|\frac{V_{DB}}{I_C}\right| = \frac{100.0}{8.0} = 12.5\ \Omega.$

Total current $= I\angle\phi = (6 + j0) = 6.0\angle 0^\circ$ A, substituting the value of I_C in the earlier expression.

So, total current is in phase with the voltage, V_{DB}.

The voltage drop across A and D $= V_{AD}\angle 0^\circ = (I\angle 0^\circ \cdot Z_{AD}\angle 0^\circ)$

$= [(I + j0) \times (R_1 + j0)] = (6 \times 10)\angle 0^\circ = 60.0\angle 0^\circ = (60.0 + j0.0)$ V.

The supply voltage $= V_S\angle 0^\circ = V_{AB}\angle 0^\circ = (V_{AD}\angle 0^\circ + V_{DB}\angle 0^\circ)$

$= (60.0\angle 0^\circ + 100.0\angle 0^\circ) = 160.0\angle 0^\circ = (160.0 + j0.0)$ V.

Note that the current is in phase with the supply voltage, with power factor as unity (1).

The equivalent impedance across D and B

$$= Z_{DB}\angle 0^\circ = \frac{V_{DB}\angle 0^\circ}{I\angle 0^\circ} = \frac{100.0}{6.0}\angle 0^\circ = 16.67\angle 0^\circ = (16.67 + j0.0)\Omega.$$

Alternatively, the equivalent impedance can be found using two branch impedances connected in parallel across D and B, as given earlier.

The impedance of the circuit (across A and B) $= Z_{AB}\angle 0^\circ = (Z_{AD}\angle 0^\circ + Z_{DB}\angle 0^\circ)$

$= [(10.0 + j0.0) + (16.67 + j0.0)] = (26.67 + j0.0) = 26.67\angle 0^\circ.$

Each of the above impedances can be easily obtained from the voltage drops and the current. The total power consumed can also be checked using supply voltage and current.

6.4 Resonance in Parallel Circuits

As may have been inferred from the previous chapter, a circuit is in resonance if the voltage is in phase with the current and the impedence is purely resistive. In this chapter on parallel circuits, we will consider two cases—the first one in which all three elements, R, L and C, are connected in parallel, across an ac supply and the second one in which a lossy L and C are connected in parallel to the R.

FIRST CASE

Consider a circuit where three *ideal* elements, R, L and C, are connected in parallel across an ac supply of variable frequency, with voltage, V taken as reference (Fig. 6.12(a)). This means that resistance, R is non-inductive, and both inductance, L and capacitance, C are lossless. The total current drawn from the supply, is,

$$I\angle-\phi = [I_R - j(I_L - I_C)] = \left[\frac{V}{R}\angle 0^\circ + \left(\frac{V}{X_L}\angle -90^\circ + \frac{V}{X_L}\angle 90^\circ\right)\right]$$

$$= \left[\frac{V}{R} - j\left(\frac{V}{X_L} - \frac{V}{X_C}\right)\right].$$

As discussed in Chapter 5, Section 5.3, the condition for resonance in an R-L-C series circuit fed from the same variable frequency ac supply, is given by

$$X_L = \omega L = X_C = \frac{1}{\omega C},$$

when the resonant frequency is

$$f_0 = \frac{\omega_0}{2\pi} = \frac{1}{2\pi\sqrt{LC}}.$$

The same condition, if used in this case, gives

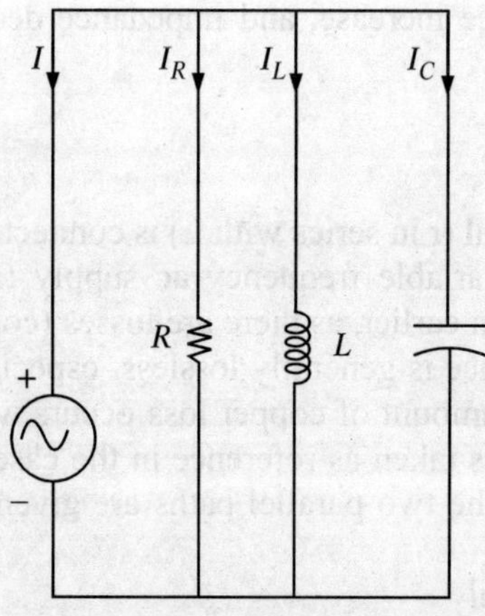

Fig 6.12(a)

$$|I_L| = \left|\frac{V}{X_L}\right| = |I_C| = \left|\frac{V}{X_C}\right|,$$

When substituted in the earlier equation, the total current is

$$I\angle 0^\circ = I_R\angle 0^\circ = \frac{V}{R}\angle 0^\circ.$$

It may be noted that the total current here is minimum, in contrast to the earlier one in series circuits, where the current is maximum. The current is in phase with the supply voltage, with the power factor as unity (1.0). The magnitudes of the current in the inductor, L and the capacitor, C are equal, and substituting the resonant frequency in the expression, we get,

$$|I_C| = |I_L| = \frac{V}{2\pi f_0 L} = \frac{V}{L}\cdot\sqrt{LC} = V\cdot\sqrt{\frac{C}{L}}.$$

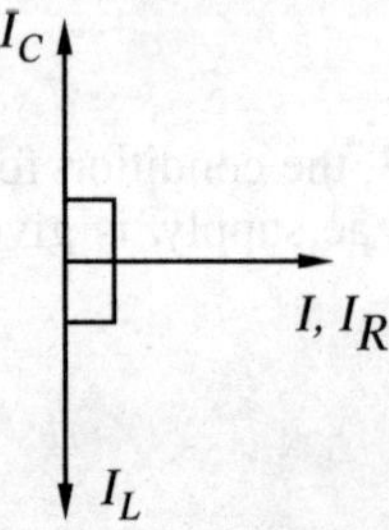

Fig 6.12(b)

The magnitude of admittance is also minimum, with the supply voltage (V) remaining the same. The impedance of the circuit is maximum, while it was minimum in the other case. The phasor diagram is shown in Fig. 6.12(b). If the frequency is changed, both the magnitude of current and circuit admittance increase, and impedance decreases, as described in detail in Chapter 5, Section. 5.3.

SECOND CASE

In this case, a lossy inductive coil (r in series with L) is connected in parallel with capacitance, C, being fed from the same variable frequency ac supply (Fig. 6.13(a)). This is real, in contrast to the ideal circuit given earlier, as there are losses (copper and iron) in an iron-cored inductive coil, but the capacitance is generally lossless, especially at lower frequency. If it is an air-cored coil, only a small amount of copper loss occurs with no iron loss, as there is no iron here. The voltage applied is taken as reference in the case of simple parallel circuits, as shown earlier. The currents in the two parallel paths are given below.

The current in the inductive coil

$$= I_L\angle -\phi_L = (I_L\cos\phi_L - jI_L\sin\phi_L) = \frac{V\angle 0^\circ}{Z_L\angle\phi_L}$$

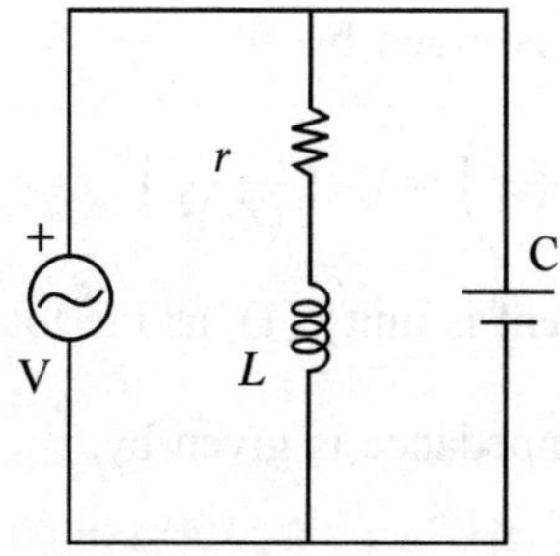

Fig. 6.13(a)

$$= \frac{V\angle 0^\circ}{(r + jX_L)} = \frac{V\angle 0^\circ}{(r + j\cdot 2\pi f L)} = \left[\frac{V\cdot r}{[r^2 + (X_L)^2]} - j\cdot\frac{V\cdot X_L}{[r^2 + (X_L)^2]}\right].$$

The current in the capacitance

$$= I_C\angle 90^\circ = (0 + jI_C) = \frac{V\angle 0^\circ}{X_C\angle -90^\circ} = \frac{(V + j0)}{-jX_C} = j\cdot\frac{V}{X_C}$$

The total current drawn is the sum of the two currents, and is given by,

$$= I\angle\phi_1 = (I_L\angle -\phi_L + I_C\angle 90^\circ) = [(I_L\cos\phi_L - jI_L\sin\phi_L) + jI_C]$$

$$= [I_L\cos\phi_L + j(I_C - jI_L\cos\phi_L)] = V\cdot\left[\frac{r}{[r^2 + (X_L)^2]} - j\cdot\left\{\frac{X_L}{[r^2 + (X_L)^2]} - \frac{1}{X_C}\right\}\right]$$

At resonant frequency, the total current is at unity power factor, as given earlier, with $\phi_1 = 0^\circ$.

So, $I\angle 0^\circ = (I + j0)$, from which the condition derived is given by,

$$I_L\sin\phi_L = I_C, \text{ where } |I_L| = \left|\frac{V}{Z_L}\right| = \frac{V}{\sqrt{r^2 + (X_L)^2}}.$$

After simplification and using $\phi_L = \sin^{-1}\left(\frac{X_L}{Z_L}\right)$, we get,

$$\left(\frac{X_L}{[r^2 + (X_L)^2]} - \frac{1}{X_C}\right) = 0 \quad \text{or,} \quad X_C = \frac{1}{\omega C} = \frac{[r^2 + (X_L)^2]}{X_L} = \frac{(Z_L)^2}{\omega L}$$

The value of C is found as,

$$C = \frac{L}{[r^2 + (2\pi f L)^2]}.$$

In this case, the reactive part of the current ($I_L\sin\phi_L$) in the inductive coil, which is negative, is equal to the reactive part of the current (I_C) in the capacitance, which is positive. Here, it is assumed that the capacitance, C is variable, with the supply being constant.

The magnitude of total current is given by,

$$|I| = I_L \cos\phi_L = V \cdot \left(\frac{r}{[r^2 + (X_L)^2]}\right) = V \cdot \left(\frac{r}{(Z_L)^2}\right).$$

The power factor of the total current is unity (1), as the total current and supply voltage are in phase.

The magnitude of the circuit impedance is given by,

$$|Z| = \left(\frac{[(Z_L)^2]}{r}\right) = \left(\frac{[r^2 + (X_L)^2]}{r}\right) = \left[r + \left(\frac{(X_L)^2}{r}\right)\right].$$

This impedance is maximum as total current is minimum.

The admittance of the circuit is proportional to the current as the applied voltage remains the same as C. All values, as derived for the current, are also applicable for admittance. The phasor diagram is shown in Fig. 6.13(b).

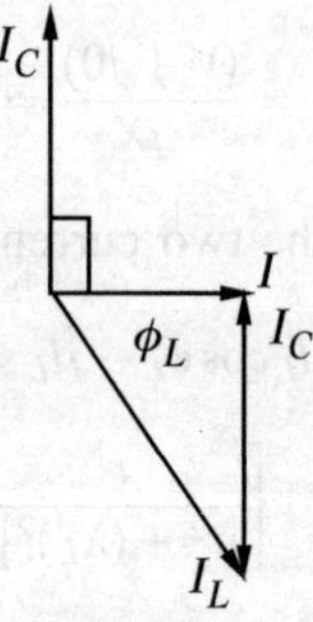

Fig. 6.13(b)

The resonant frequency for constant values of r, L and C, is given by,

$$f_0 = \frac{\omega_0}{2\pi} = \frac{1}{2\pi} \cdot \left(\frac{1}{\sqrt{LC}} - \frac{r^2}{L^2}\right).$$

All values as found earlier are same under this condition.

Other cases of parallel and series–parallel circuits, fed from an ac supply, are not considered here, as they can be solved by the methods given in Section 6.2 and 6.3.

The resistor, R is included in the circuits, to control or limit, 'dampen' the current at resonant frequency.

6.5 Quality Factor

For the circuit with a lossy inductance coil in parallel with a lossless capacitor (EXAMPLE 6.8, Fig. 6.15(a)), the magnitude of the current fed from a variable frequency ac supply is minimum at resonant frequency. The quality factor, termed in short as Q-factor, in this case, is defined as the ratio of the current in the inductance coil and the total current at the resonant frequency (f_0). The Q-factor at that frequency is given by,

$$Q_0 = \left|\frac{I_{L0}}{I_0}\right| = \left|\frac{1}{Z_{L0}} \times \frac{(Z_{L0})^2}{r}\right| = \left|\frac{Z_{L0}}{r}\right| = \frac{\sqrt{[r^2 + (X_{L0})^2]}}{r} = \sqrt{1 + \left(\frac{\omega_o L}{r}\right)^2}.$$

If the resistance of the coil is assumed to be low, then $\frac{X_{L0}}{r} >> 1$.

Using this condition,

$$Q_0 = \frac{\omega_0 L}{r} = \frac{X_{L0}}{r},$$

which is the same as found earlier in Chapter 5, where the Q-factor of the coil, in series with a capacitor, has been calculated.

The magnitude of the coil impedance is nearly equal to the inductive reactance, i.e., $|Z_{L0}| \approx X_{L0}$. It may be noted that the Q-factor is also the current magnification factor. If the value of the resistance, r is increased, Q-factor decreases, resulting in lower magnification.

EXAMPLE 6.7 *A coil of resistance, 20 Ω, and inductance, 0.6 H, is connected in series with a condenser. The combination draws maximum current, when a fixed amplitude sinusoidal voltage of 50 Hz is applied. A second condenser is now connected in parallel with this circuit (Fig. 6.14(a)). What should be the capacitance for the combined circuit to act as a non-inductive resistance at 100 Hz? Determine the current drawn by the combined circuit, if the applied voltage is 220 V.*

SOLUTION $V = 220.0\angle 0° = (220.0 + j0.0)$ V, $f_1 = 50$ Hz.

$$R = 20\,\Omega, \quad L = 0.6 \text{ H} = 600 \text{ mH}.$$

$$\omega_1 = (2\pi f_1) = (2 \times \pi \times 50.0) = 314.16 \text{ rad/s}.$$

The maximum current in the circuit (Fig. 6.14(a), with the switch, S open) is drawn at resonance, when,

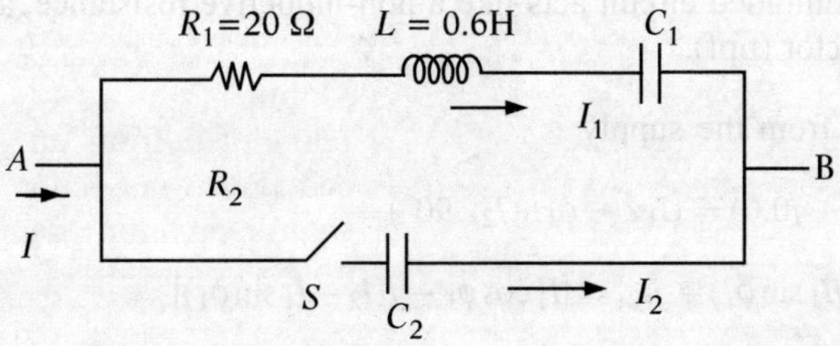

Fig. 6.14(a)

$$X_{L1} = \omega_1 L = X_{C1} = \frac{1}{\omega_1 C_1} = 314.16 \times 0.6 = 188.5 \ \Omega$$

or, $C_1 = \dfrac{1}{\omega_1 X_{C1}} = \dfrac{1}{(\omega_1)^2 \cdot L} = \dfrac{1}{(314.16)^2 \times 0.6} = 16.9\,\mu\text{F}.$

The impedance at this frequency = $Z'\angle 0^\circ = R + j(X_{L1} - X_{C1}) = R + j0.$

The maximum current

$$= I\angle 0^\circ = \frac{V\angle 0^\circ}{Z'\angle 0^\circ} = \frac{V}{R}\angle 0^\circ = \frac{220.0}{22.0}\angle 0^\circ = 11.0\angle 0^\circ = (11.0 + j0.0)\ \text{A}.$$

The phasor diagram for this (first case), being simple, is not included.

$$f_2 = 100\ \text{Hz},\ \omega_2 = (2\pi f_2) = (2 \times \pi \times 100.0) = 628.3\ \text{rad/s}.$$

The impedance of the first branch with the switch, S closed

$$= Z_1\angle\phi_1 = R + j\left(\omega_2 L - \frac{1}{\omega_2 C_1}\right) = 30.0 + j\left((628.3 \times 0.6) - \frac{1}{628.3 \times 16.9 \times 10^{-6}}\right)$$

$$= 30.0 + j(377.0 - 94.2) = (30.0 + j282.8) = 284.4\angle 83.945^\circ\ \Omega.$$

Note that the impedance is inductive.

The current in the first branch

$$= I_1\angle -\phi_1 = (I_1\cos\phi_1 - jI_1\sin\phi_1) = \frac{V\angle 0^\circ}{Z_1\angle\phi_1}$$

$$= \frac{220.0}{284.4}\angle -83.945^\circ = 0.7736\angle -83.945^\circ = (0.0816 - j0.7693)\ \text{A}.$$

The current in the second branch, having a capacitor, C_2 only

$$= I_2\angle 90^\circ = jI_2 = \frac{V\angle 0^\circ}{X_{C2}\angle -90^\circ} = \frac{V}{X_{C2}}\angle 90^\circ = j\frac{V}{X_{C2}}\ \text{A}.$$

Also note that the combined circuit acts like a non-inductive resistance, and the total current is at unity power factor (upf).

Total current drawn from the supply

$$= \quad I\angle 0^\circ = (I + j0.0) = (I_1\angle -\phi_1 + I_2\angle 90^\circ)$$

$$= [(I_1\cos\phi_1 - jI_1\sin\phi_1) + jI_2] = [I_1\cos\phi_1 + j(I_2 - I_1\sin\phi_1)]$$

$$= [(0.0816 - j0.7693) + jI_2] = [0.0816 + j(I_2 - 0.7693)].$$

Equating the imaginary terms in the above expression, we get,

$$(I_2 - 0.7693) = 0.0 \quad \text{or,} \quad |I_2| = \frac{V}{X_{C2}} = 0.7693\ \text{A}$$

or, $$X_{C2} = \frac{1}{\omega_2 C_2} = \frac{V}{I_2} = \frac{220.0}{0.7693} = 286.0\ \Omega.$$

The capacitance is obtained as $C_2 = \dfrac{1}{\omega_2 X_{C2}} = \dfrac{1.0}{628.3 \times 286.0} = 5.565\ \mu\text{F}.$

Now, total current $= I' \angle 0° = (0.0816 + j0.0) = 0.0816\angle 0°$ A.

The equivalent impedance of the circuit

$$= Z' \angle 0° = \frac{V\angle 0°}{I'\angle 0°} = \frac{220.0}{0.0816}\angle 0° = 2{,}696.0\angle 0° = (2.696 + j0.0)\ \text{k}\Omega.$$

The phasor diagram for this (second case) is shown in Fig. 6.14(b).

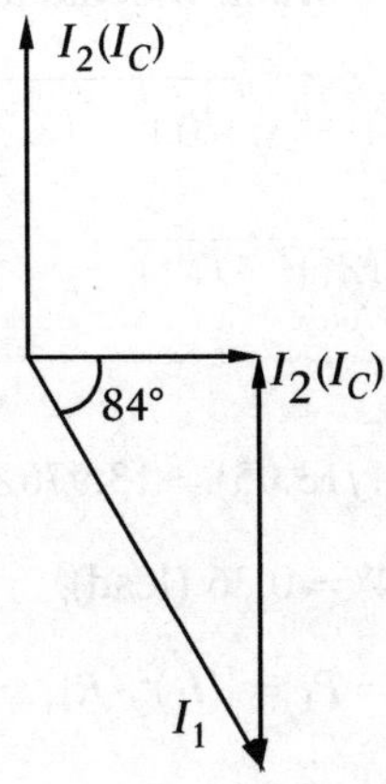

Fig. 6.14(b)

Additional Problems

Example 6.8 *Two impedances, Z_1 and Z_2, are connected in parallel (Fig. 6.15(a)). The first branch takes a leading current of 16 A, and has a resistance of 5 Ω, while the second branch takes a lagging current at a power factor (pf) of 0.8. The total power supplied is 5 kW, the applied voltage being (100 + j200) V. Determine the complex expressions for the branch and total currents, and for the circuit constants.*

Solution The circuit is shown in Fig. 6.15(a).

$$V\angle\phi_1 = (100.0 + j200.0) = 223.61\angle 63.435°\ \text{V}.$$

$$R_1 = 5\ \Omega, \quad |I_1| = 16.0\ \text{A}.$$

$$\cos\phi_2 = 0.8\ \text{(lag)}, \ \phi_2 = \cos^{-1}(0.8) = 36.87°.$$

Total power $= P = 5$ kW.

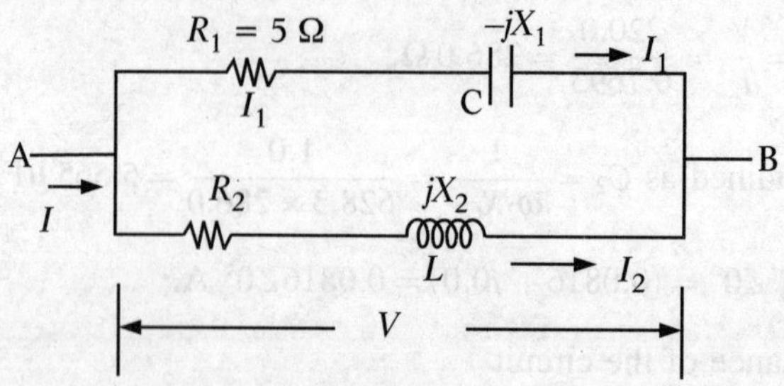

Fig. 6.15(a)

Branch I

Since the first branch takes a leading current, it contains a capacitive reactance, X_C.

The magnitude of the impedance $= |Z_1| = \sqrt{(R_1)^2 + (X_C)^2} = \left|\dfrac{V}{I_1}\right| = \dfrac{223.61}{16.0} = 13.976\ \Omega.$

The capacitive reactance $= \quad X_C = \sqrt{(Z_1)^2 - (R_1)^2} = \sqrt{(13.976)^2 - (5.0)^2} = 13.05\ \Omega.$

The impedance of branch I

$$= Z_1\angle -\phi_1 = (R_1 - jX_C) = (5.0 - j13.05) = 13.976\angle -69.04^\circ\ \Omega.$$

Power factor (pf) $= \cos\phi_1 = \cos\ 69.04^\circ = 0.36$ (lead).

Power absorbed by the resistance, $R_1 = P_1 = [(I_1)^2 \cdot R_1] = [(16.0)^2 \times 5.0] = 1.28$ kW.

The current $= I_1\angle\phi_4 = \dfrac{V\angle\phi_3}{Z_1\angle\phi_1} = \dfrac{223.61}{13.976}\angle(63.435^\circ + 69.04^\circ) = 16.0\angle 132.47^\circ$

$$= (-10.8 + j11.8)\ \text{A}$$

Alternatively, $I_1\angle\phi_4 = \dfrac{(100.0 + j200.0)}{(5.0 - j13.05)} = \dfrac{(100.0 + j200.0)\cdot(5.0 + j13.05)}{195.33}$

$$= \frac{[\{(0.5 - 2.61) + j(1.305 + 1.0)\} \times 10^3]}{195.33} = \frac{(-2,110.0 + j2,305.0)}{195.33}$$

$$= (-10.8 + j11.8) = 16.0\angle 132.47^\circ\ \text{A}.$$

Branch II

Power absorbed in the second circuit $= P_2 = (VI_2\cos\phi_2) = (P - P_1) = (5.0 - 1.28)$

$$= 3.72\ \text{kW} = 3,720\ \text{W}.$$

The magnitude of the current $= |I_2| = \dfrac{P_2}{V\cdot\cos\phi_2} = \dfrac{3,720}{223.6 \times 0.8} = 20.8$ A.

The current $= \quad I_2\angle-\phi_2 = 20.8\angle-36.87° = (16.64 - j12.48)$ A, which lags behind the voltage, taken as reference here, by ϕ_2.

The impedance $= Z_2\angle\phi_2 = (R_2 + jX_L) = \dfrac{V\angle 0°}{I_2\angle-\phi_2} = \dfrac{223.61}{20.8}\angle 36.87° = 10.75\angle 36.87°$

$= (8.6 + j6.45) \quad \Omega.$

The current $= I_2\angle\phi_5 = \dfrac{V\angle\phi_3}{Z_2\angle\phi_2} = \dfrac{223.61}{10.75}\angle(63.435° - 36.87°) = 20.8\angle 26.565°$

$= (18.6 + j9.3)$ A.

Total circuit

The current $= I\angle\phi_6 = (I_1\angle\phi_4 + I_2\angle\phi_5) = [(-10.8 + j11.8) + (18.6 + j9.3)]$

$= (7.8 + j21.1) = 22.5\angle 69.71°$ A.

The phasor diagram is shown in Fig. 6.15(b).

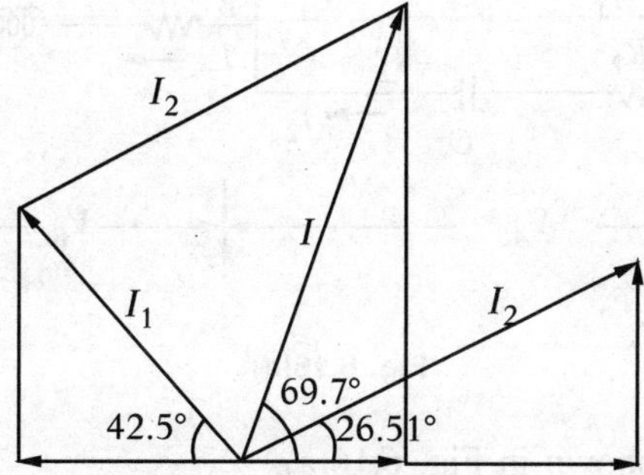

Fig. 6.15(b)

The equivalent impedance and power supplied are also computed here.

The equivalent impedance $= Z_{eq}\angle-\phi_7 = \dfrac{V\angle\phi_3}{I\angle\phi_6} = \dfrac{223.61}{22.5}\angle(63.435° - 69.71°)$

$= 9.94\angle-6.275° = (9.88 - j1.086) \ \Omega.$

Power factor (pf) $= \cos\,\phi_7 = \cos\,6.275° = 0.994$ (lead).

Total power $= P = (VI\cos\,\phi_7) = (223.61 \times 22.5 \times 0.994) = 5.0$ kW,

which is the same as given in the problem.

The equivalent impedance is also found as,

$(Z_1\angle-\phi_1 + Z_2\angle\phi_2) = [(5.0 - j13.05) + (8.6 + j6.45)] = (13.6 - j6.6)$

$$= 15.117\angle -25.89^\circ \quad \Omega$$

$$\text{The equivalent impedance} = Z_{eq}\angle\phi_7 = (R + jX) = \frac{Z_1\angle -\phi_1 \cdot Z_2\angle\phi_2}{(Z_1\angle -\phi_1 + Z_2\angle\phi_2)}$$

$$= \left(\frac{13.976 \times 10.75}{15.117}\right)\angle(-69.04^\circ + 36.87^\circ + 25.89^\circ) = 9.94\angle -6.28^\circ$$

$$= (9.88 - j1.087)\,\Omega,$$

which is the same as obtained earlier.

EXAMPLE 6.9 *In a series–parallel circuit, shown in Fig. 6.16(a), the parallel branches, A and B, are in series with the branch, C. Three impedances in Ω are $Z_A = (4 + j3)$, $Z_B = (5 - j6)$, and $Z_C = (2 + j8)$. If the current in the branch C is 25 A, calculate the branch currents and voltages, and also the total voltage. Hence, compute the complex power (the active and reactive VA) for each branch, and the complete circuit. Draw the phasor diagram.*

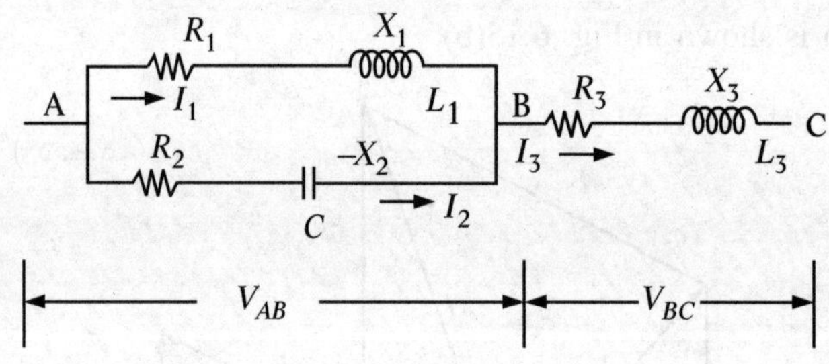

Fig. 6.16(a)

SOLUTION The circuit is shown in Fig. 6.16(a).

As all reactance values of the branch impedances are given, the frequency data is not used here. The branches are now numbered as 1, 2 and 3, instead of A, B and C respectively.

The branch impedances are,

$$Z_1\angle\phi_1 = (R_1 + jX_1) = (4 + j3) = 5.0\angle 36.87^\circ\ \Omega.$$

$$Z_2\angle -\phi_2 = (R_2 - jX_2) = (5 - j6) = 7.81\angle -50.194^\circ \quad \Omega.$$

$$Z_3\angle\phi_3 = (R_3 + jX_3) = ((2 + j8) = 8.246\angle 75.964^\circ \quad \Omega.$$

The current in branch 3 $= I_3\angle 0^\circ = 25.0\angle 0^\circ = (25.0 + j0.0)$A.

$$(Z_1\angle\phi_1 + Z_2\angle -\phi_2) = [(4 + j3) + (5 - j6)] = (9 - j3) = 9.487\angle -18.435^\circ \quad \Omega.$$

The branches 1 and 2 are connected in parallel (Fig. 6.16(a)).

The equivalent impedance $= Z_{eq}\angle\phi_4 = \dfrac{Z_1\angle\phi_2 \cdot Z_2\angle-\phi_2}{(Z_1\angle\phi_1 + Z_2\angle-\phi_2)}$

$$= \frac{5.0\times 7.81}{9.487}\angle(36.87^\circ - 50.194^\circ + 18.435^\circ) = 4.116\angle 5.111^\circ = (4.1 + j0.367)\ \Omega.$$

Total impedance $= Z'\angle\phi_5 = (Z_{eq}\angle\phi_4 + Z_3\angle\phi_3) = [(4.1 + j0.367) + (2.0 + j8.0)]$

$$= (6.1 + j8.367) = 10.355\angle 53.906^\circ\ \Omega.$$

The applied voltage $= V\angle\phi_5 = (I_3\angle 0^\circ \cdot Z'\angle\phi_5) = (25.0\angle 0^\circ \times 10.355\angle 53.906^\circ)$

$$= 258.9\angle 53.906^\circ = (152.5 + j209.2)\ \text{V}.$$

The voltage drop across branch 3 $= V_3\angle\phi_3 = (I_3\angle 0^\circ \cdot Z_3\angle\phi_3)$

$$= (25.0\angle 0^\circ \times 8.246\angle 75.964^\circ) = 206.2\angle 75.964^\circ = (50.0 + j200.0)\ \text{V}.$$

The current in branch 1 $= I_1\angle\phi_6 = I_3\angle 0^\circ \cdot \dfrac{Z_2\angle-\phi_2}{(Z_1\angle\phi_1 + Z_2\angle-\phi_2)}$

$$= \frac{25.0\angle 0^\circ \times 7.81\angle-50.194^\circ}{9.487\angle-18.435^\circ} = \frac{25.0\times 7.81}{9.487}\angle(0^\circ - 50.194^\circ + 18.435^\circ)$$

$$= 20.58\angle-31.76^\circ = (17.5 - j10.83)\ \text{A}.$$

Alternatively, $I_1\angle\phi_6 = \dfrac{25.0\times(5-j6)}{(9-j3)} = \dfrac{25.0\times[(5-j6)\cdot(9+j3)]}{90.0}$

$$= \frac{25.0\times[(45+18)+j(15-54)]}{90.0} = \frac{25.0\times(63.0-j39.0)}{90.0} = (17.5 - j10.83)$$

$= 20.58\angle-31.76^\circ$ A, which is the same as obtained earlier.

The current in branch 2 $= I_2\angle\phi_7 = (I_3\angle 0^\circ - I_1\angle\phi_6)$

$$= [(25.0 + j0.0) - (17.5 - j10.83)] = (7.5 + j10.83) = 13.17\angle 55.3^\circ\ \text{A}.$$

The voltage drop across branch 1 = the voltage drop across branch 2

$$= V_1\angle\phi_4 = V_2\angle\phi_4 = (I_1\angle\phi_6 \cdot Z_1\angle\phi_1) = (20.58\angle-31.76^\circ \times 5.0\angle 36.87^\circ) = 102.9\angle 5.11^\circ$$

$$= (102.5 + j9.2)\ \text{V}.$$

The above voltages can also be obtained by the two formulas given below.

$$V_1\angle\phi_4 = V_2\angle\phi_4 = (V\angle\phi_5 - V_3\angle\phi_3),$$

or, $\quad V_1\angle\phi_4 = V_2\angle\phi_4 = (I_3\angle 0^\circ \cdot Z_{eq}\angle\phi_4) = (I_2\angle\phi_7 \cdot Z_2\angle-\phi_2).$

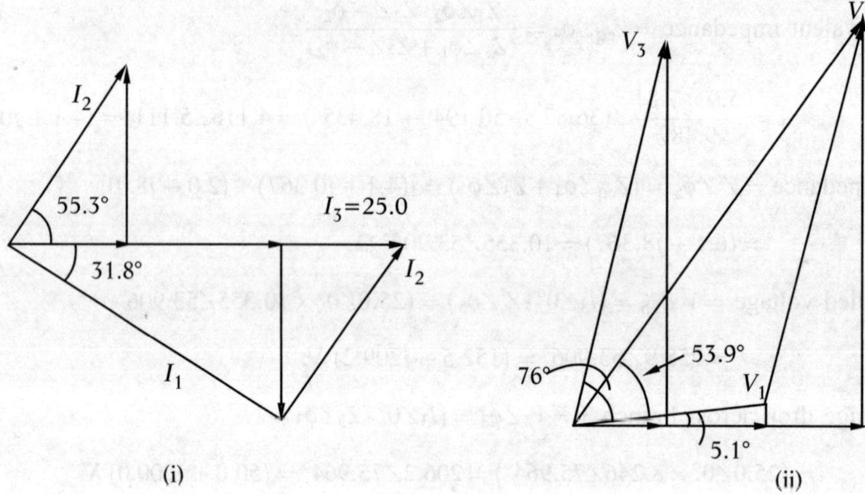

Fig. 6.16(b)

The phasor diagram for (i) branch currents, and (ii) branch voltages, including total ones is shown in Fig. 6.16(b), with the current in branch 3 (C) as reference. Students can draw the total phasor diagram, after combining both of them. The applied voltage may also be taken as reference.

The complex power (the active (W) and reactive VA), for each branch, and the complete circuit, are computed now.

The complex power for branch 1(A) $= S_1\angle\phi_1 = (P_1 + jQ_1) = [V_1\angle\phi_4 \cdot (I_1\angle\phi_6)^*]$

$$= (102.9\angle 5.11^\circ \times 20.58\angle 31.76^\circ) = 2,117.7\angle 36.87^\circ = (1,694.0 + j1,271.0) \text{ VA}.$$

The complex power for branch 2 (B) $= S_2\angle -\phi_2 = (P_2 - jQ_2) = [V_2\angle\phi_4 \cdot (I_2\angle\phi_7)^*]$

$$= (102.9\angle 5.11^\circ \times 13.17\angle -55.3^\circ) = 1,355.2\angle -50.19^\circ = (867.6 - j1,041.1) \text{ VA}.$$

The complex power for branch 3 (C) $= S_3\angle\phi_3 = (P_3 + jQ_3) = [V_3\angle\phi_3 \cdot (I_3\angle 0^\circ)^*]$

$$= (206.2\angle 75.964^\circ \times 25.0\angle 0^\circ) = 5,155.0\angle 75.964^\circ = (1,250.0 + j5,000.0) \text{ VA}.$$

The complex power for the complete circuit $= S\angle\phi_5 = (P + jQ) = [V\angle\phi_5 \cdot (I_3\angle 0^\circ)^*]$

$$= (258.9\angle 53.906^\circ \times 25.0\angle 0^\circ) = 6,472.5\angle 53.906^\circ = (3,813.0 + j5,230.1) \text{ VA}.$$

Notes

1. The reactive VAR is capacitive (–ve) for branch C, as the branch is capacitive in nature, while for the other two branches (A and B), the reactive VAR is inductive (+ve), as the branches are inductive. The reactive VAR for the complete circuit is inductive as the circuit is inductive in nature.

2. The complex power (the active power, *W* and reactive *VA*), for the complete circuit is the sum of the complex power in three branches (A, B and C), which can be easily shown.

3. The complex power (the active power, *W* and reactive *VA*), in any of the branches can be obtained by the procedure given here. The values for only one branch (A) are calculated.

The complex power for branch 1 $= S_1\angle\phi_1 = (P_1 + jQ_1) = [(I_1)^2 \cdot Z_1(\phi_1)]$

$$= [(I_1)^2 \cdot (R_1 + jX_1)] = [(20.58)^2 \cdot (4 + j3)] = (1{,}694.0 + j1{,}271.0)$$

$$= 2{,}117.8\angle 36.87^\circ \text{ VA},$$

which is the same as obtained earlier.

Conclusion

In this chapter, we studied the parallel and series–parallel circuits, all being fed from a single-phase ac supply. The phasor diagrams with branch currents and voltage drops were discussed. The problem of resonance in parallel circuits fed from a variable frequency supply was also studied. In the next chapter, we will study three-phase ac circuits.

Multiple Choice Type Questions

Tick the correct option(s), out of the four choices provided in each question.

6.1 Power factor of a purely resistive circuit is

(a) 1 (b) 0 (c) 0.1 (d) None of (a)–(c)

6.2 The reading of the voltmeter as shown in Fig. MCQ6.2 is

(a) 3 V (b) 1 V (c) 2 V (d) None of (a)–(c)

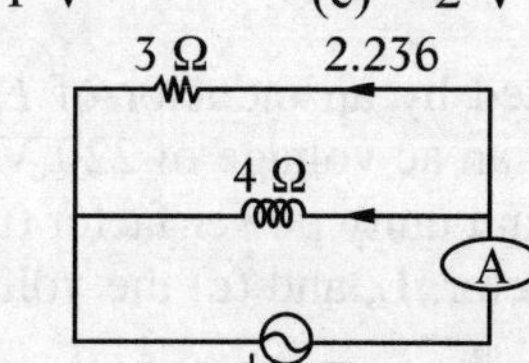

Fig. MCQ6.2

6.3 The reading of the ammeter as shown in Fig. MCQ6.3 is

(a) 3 A (b) 1 A (c) 2 A (d) None of (a)–(c)

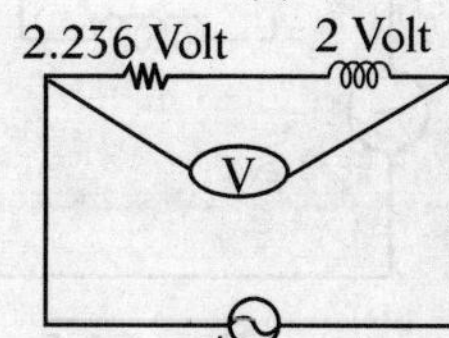

Fig. MCQ6.3

6.4 The RL circuit of Fig. MCQ6.4 is fed from a constant magnitude, variable frequency sinusoidal voltage source V at 100 Hz; the R and L elements each have a voltage drop, u_{RMS}. If the frequency of the source is changed to 50 Hz, the new voltage drop across R is:

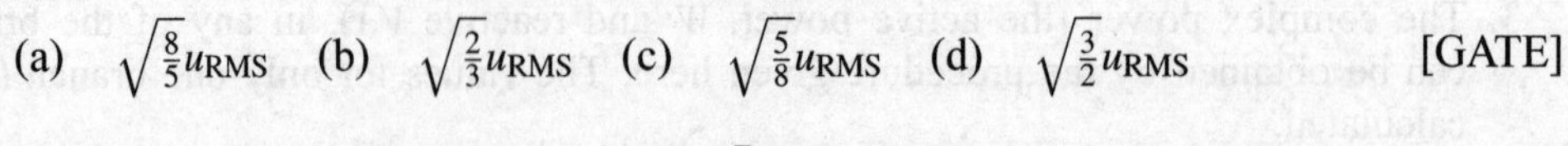

(a) $\sqrt{\frac{8}{5}}u_{RMS}$ (b) $\sqrt{\frac{2}{3}}u_{RMS}$ (c) $\sqrt{\frac{5}{8}}u_{RMS}$ (d) $\sqrt{\frac{3}{2}}u_{RMS}$ [GATE]

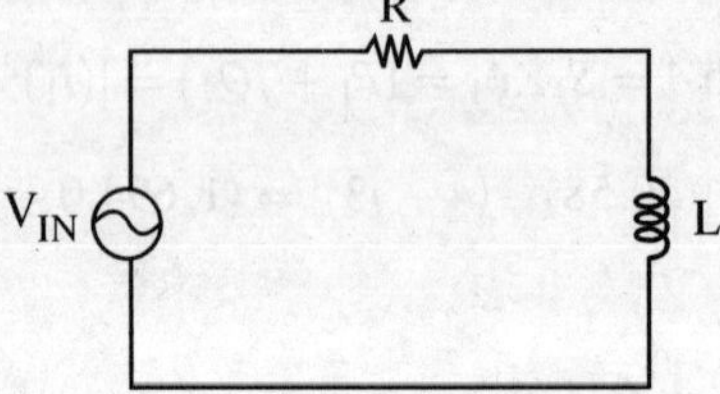

Fig. MCQ6.4

Answers

6.1 (a) 6.2 (a) 6.3 (a) 6.4 (c)

Problems

6.1 Three circuit elements—a non-inductive resistance of 100 Ω, a lossless inductance of 0.2 H, and a capacitance of 50 μF—are connected in parallel to an ac supply of 220 V, 50 Hz (Fig. P6.1). Compute the current drawn from the supply, and its power factor.

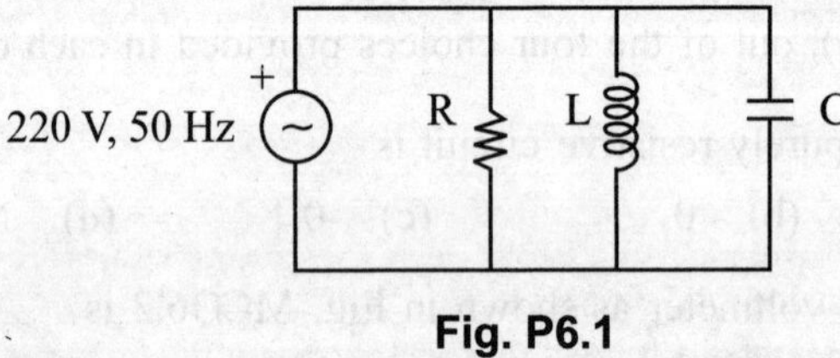

Fig. P6.1

6.2 A resistor of $R = 100\,\Omega$, shunted by an inductor of $L = 0.4$ H, is connected in series with a variable capacitor, C, to an ac voltage of 220 V, 50 Hz (Fig. P6.2). Find (a) the value of the capacitor, C to obtain unity power factor (upf) in the circuit, (b) the current in the resistor, R, and the inductor, L, and (c) the voltage across the capacitor, C.

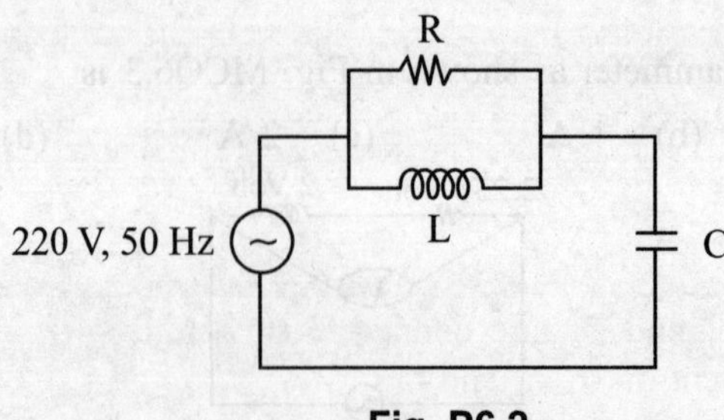

Fig. P6.2

6.3 A circuit diagram is shown in Fig. P6.3. The voltage drop across the inductance, is $10\angle 0°$ V. The supply frequency is 50 Hz.

(a) Obtain the current and voltage phasors in all the elements, and the current source, I_s.

(b) Compute the complex power of the current source, I_s,

(c) Determine the power absorbed by the resistors,

and also show that their sum is equal to the power supplied by the current source.

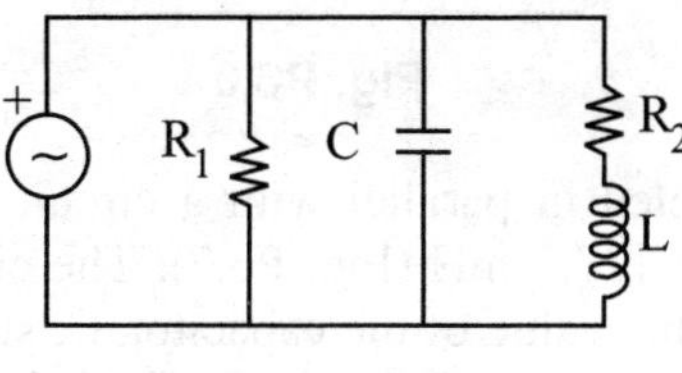

Fig. P6.3

6.4 A resistor of 60 Ω and a capacitor, C, are connected in parallel across a 220 V, 50 Hz supply (Fig. P6.4). The current drawn from the supply is 6 A. Calculate the capacitance of the capacitor. Find the currents in the resistor and the capacitor.

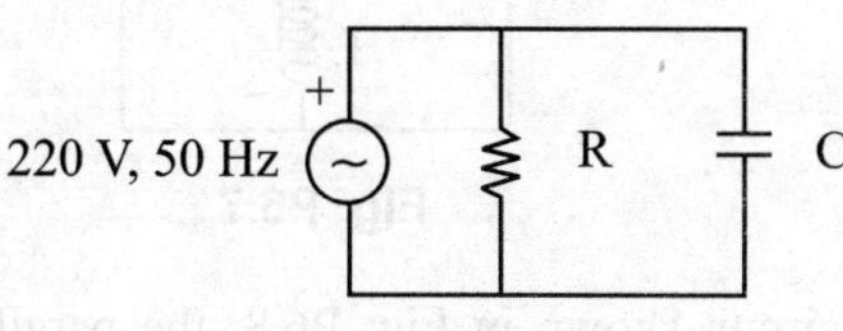

Fig. P6.4

6.5 A coil of resistance, 50 Ω and inductance, 0.318 H, is connected in parallel with a circuit having a resistor of 75 Ω in series with a capacitor of 159 μF (Fig. P6.5). The total circuit is connected to a 230 V, 50 Hz supply. Determine,

(a) the current in each branch, (b) the current drawn from the supply,
(c) its phase angle, and (d) the impedance of the total circuit.

Find also the resistance and reactance of an equivalent series circuit which will take the same current at the same power factor (pf), as the parallel circuit.

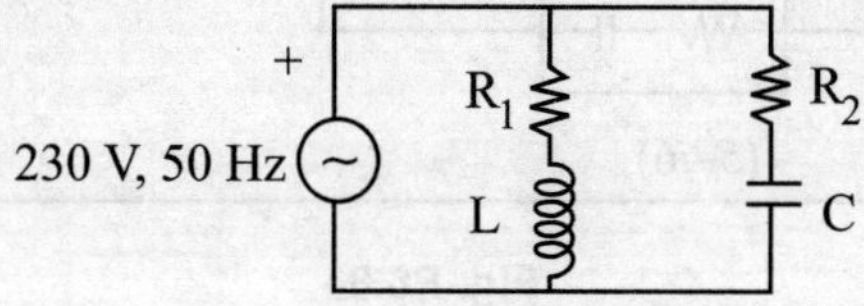

Fig. P6.5

6.6 Two impedances, Z_1 and Z_2, are connected in parallel (Fig. P6.6). The first branch takes a leading current of 16 A, and has a resistance of 5 Ω, while the second branch takes a lagging current at a power factor (pf) of 0.8. The total power supplied is 5 kW, the applied voltage being $(100 + j200)$ V. Determine the complex expressions for the branch and total currents, and for the circuit constants.

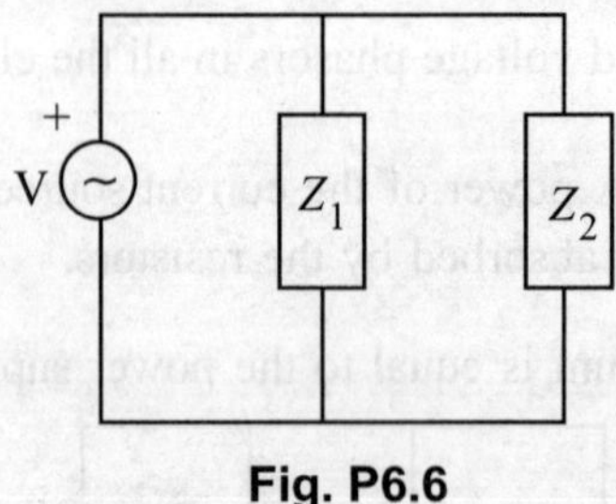

Fig. P6.6

6.7 A capacitor, C, is connected in parallel with a circuit, having a resistor of 30 Ω in series with an inductor of 127.3 mH (Fig. P6.7). The circuit is connected to a 220 V, 50 Hz supply. Determine the value of the capacitor, C, such that the total current drawn from the supply is at unity power factor (upf). Find the current in two branches, and total current.

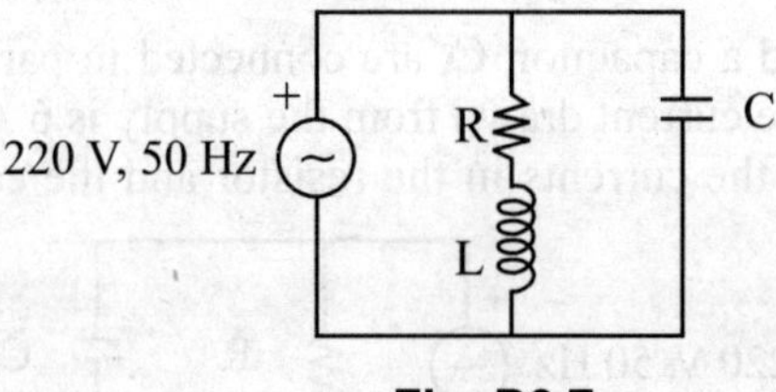

Fig. P6.7

6.8 In the series–parallel circuit shown in Fig. P6.8, the parallel branches, A and B, are in series with the branch, C. Three impedances in Ω, are, $Z_A = (4 + j3), Z_B = (5 - j6)$, and $Z_C = (2 + j8)$. If the current in branch C is 25 A, calculate the branch currents and voltages, and also the total voltage. Hence, compute the complex power (the active *W*, and reactive *VA*) for each branch, and the complete circuit. Draw the phasor diagram.

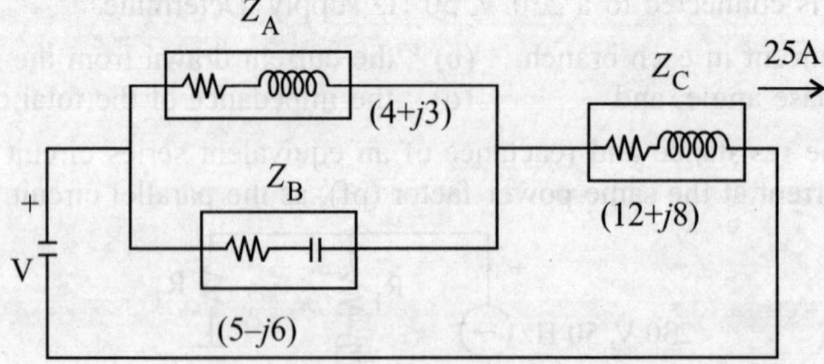

Fig. P6.8

6.9 Find all branch currents in the circuit shown in Fig. P6.9, when it is supplied from an ac voltage of 100 V. Compute the total power, and also the power absorbed in the branches.

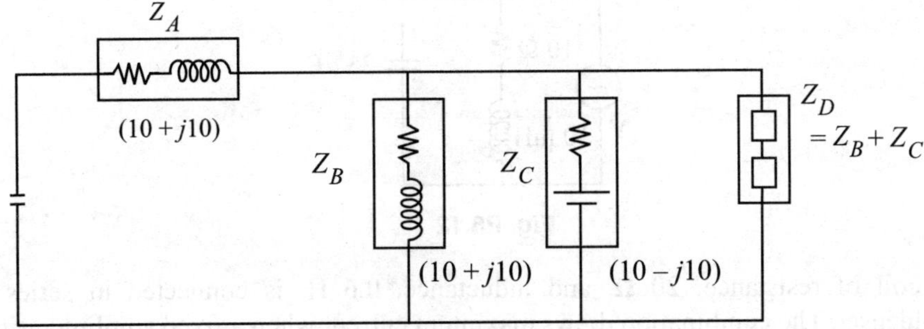

Fig. P6.9

6.10 Three impedances in Ω, $(6 + j5)$, $(8 - j6)$, and $(8 + j10)$, are connected in parallel (Fig. P6.10). Calculate the current in each branch, when the total current is 20 A.

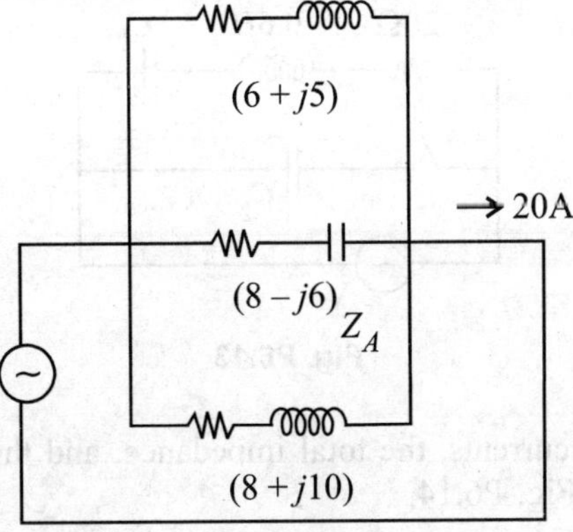

Fig. P6.10

6.11 The circuit, shown in Fig. 6.11, is connected to 220 V, 50 Hz supply. Find the current in the inductance, L_2, and the total current drawn from the supply.

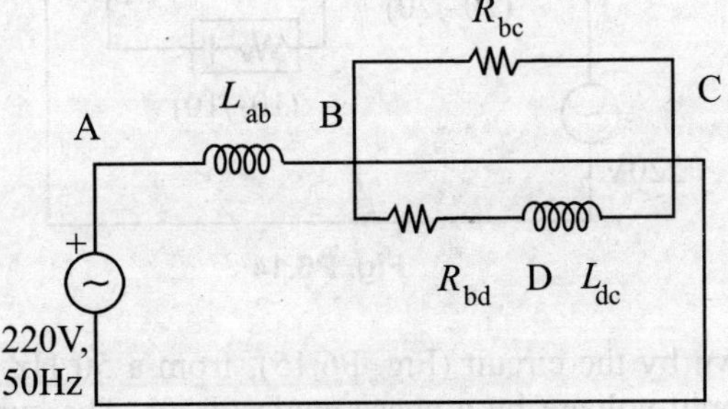

Fig. P6.11

6.12 A coil of resistance, 10 Ω, and inductance, 10 mH, is connected in parallel with a 25 μF capacitor (Fig. P6.12). Calculate the frequency at resonance, and the effective impedance of the circuit.

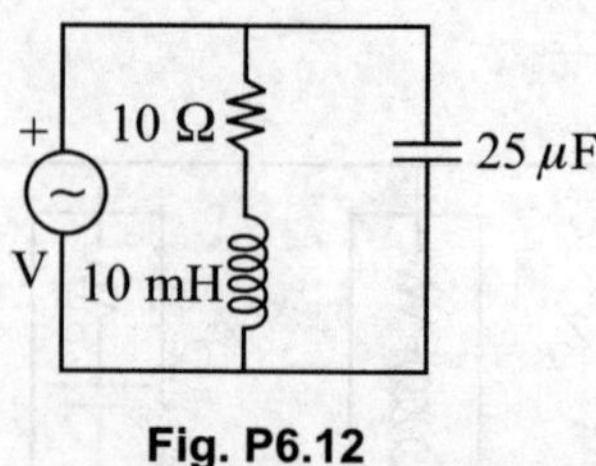

Fig. P6.12

6.13 A coil of resistance, 20 Ω, and inductance, 0.6 H, is connected in series with a condenser. The combination draws maximum current when a fixed amplitude sinusoidal voltage of 50 Hz is applied. A second condenser is now connected in parallel with this circuit (Fig. P6.13). What should be the capacitance for the combined circuit to act as a non-inductive resistance at 100 Hz? Determine the current drawn by the combined circuit, if the applied voltage is 220 V.

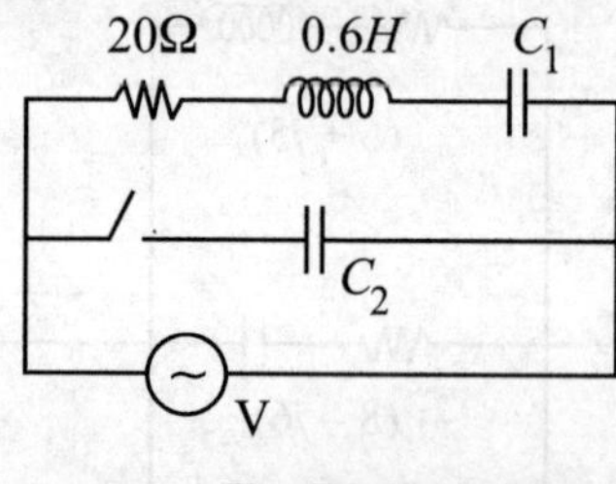

Fig. P6.13

6.14 Determine all branch currents, the total impedance, and the total power consumed, in the circuit, shown in Fig. P6.14.

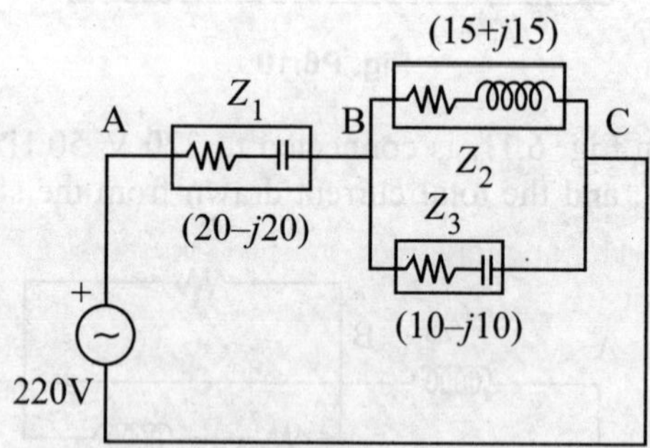

Fig. P6.14

6.15 Total current drawn by the circuit (Fig. P6.15), from a 50 Hz sinusoidal supply is 12 A leading the supply voltage by a phase angle of 30°. The current in the first branch is 8 A. Calculate,

(a) the supply voltage, (b) the unknown branch current,

(c) the values of the resistance, R and the capacitance, C, and

(d) the resistance and reactance of the equivalent series circuit.

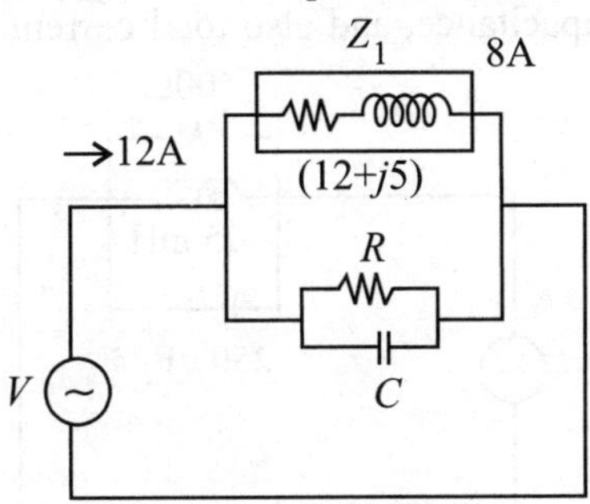

Fig. P6.15

6.16 A resistance of 5 Ω is in parallel with an inductance, the reactance of which is $X_L = 8$ Ω. The resistance of the inductance is neglected. This circuit is connected in series with another consisting of a resistance of 2 Ω in series with a capacitive reactance of 4 Ω. The combination (Fig. P6.16) is connected to a 110 V supply. Find (a) the current taken from the supply, and its power factor, and (b) the potential difference (pd) across each circuit.

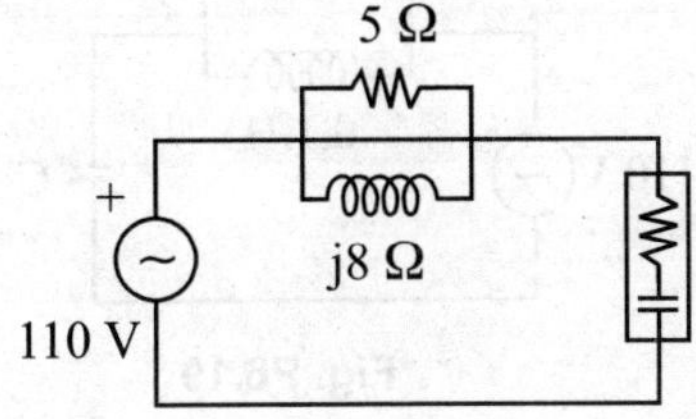

Fig. P6.16

6.17 Three circuit elements—a non-inductive resistance of 500 Ω, a lossless inductance of 25 mH, and a capacitance of 250 μ F—are connected in parallel to a current source of 1 A (Fig. P6.17). The angular frequency is 400 rad/s. Compute the currents drawn by the circuit elements.

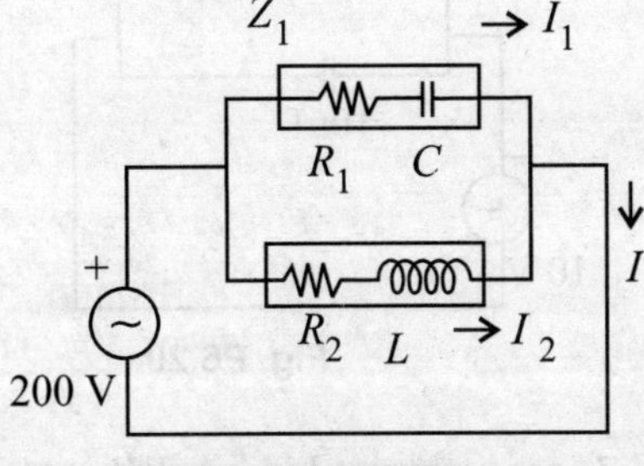

Fig. P6.17

6.18 A resistance of 25 Ω is connected in parallel with a lossless inductance of 0.1 H. A capacitance, C is to be connected in series with the above circuit (Fig. P6.18), such that the current drawn from a supply of 110 V, 50 Hz, is at unity power factor (upf). Find the value of the capacitance, and also total current.

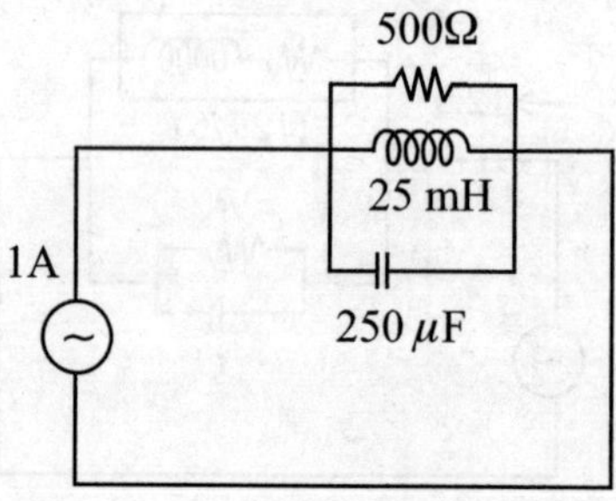

Fig. P6.18

6.19 A non-inductive resistance of 1 Ω is in series with a lossless inductance of 10 μH. This circuit is connected in parallel with a capacitance of 10 nF. The combination (Fig. P6.19) is connected to a 10 V supply. Calculate the total current (a) at the resonant frequency, and (b) at 90% of the resonant frequency.

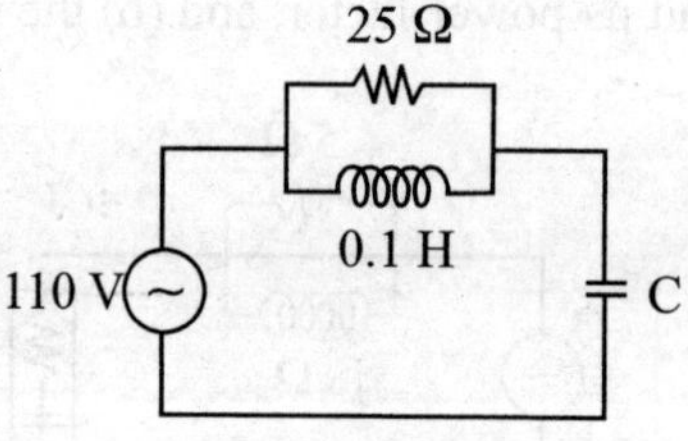

Fig. P6.19

6.20 If the reading of the voltmeter connected across the resistor (R_2) of 4 Ω in branch 2 of the circuit (Fig. P6.20) is 60 V, find the reading of the ammeter connected to measure the current drawn from the ac supply. Determine (a) the active power, (b) the reactive volt-amperes, and (c) the apparent power.

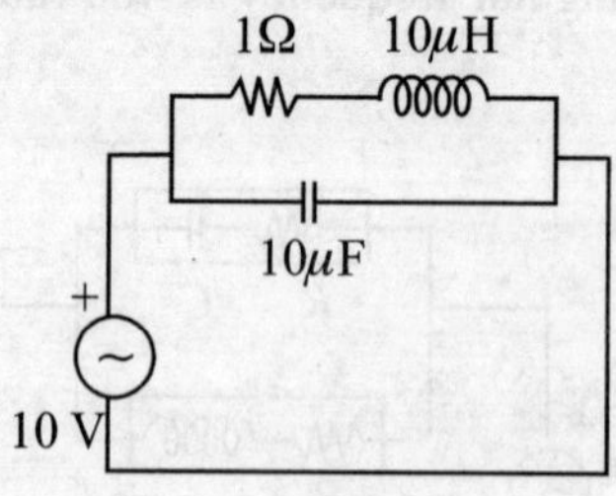

Fig. P6.20

6.21 Two impedances, Z_A and Z_B are connected in parallel, which are given in Ω, as $(6 + j12)$ and $(15 - j20)$ respectively. A capacitance, C is to be connected in series with the above circuit (Fig. P6.21), such that the current drawn from the supply of 220 V, 50 Hz, is at unity power factor (upf). Find the value of the capacitance, and also total current.

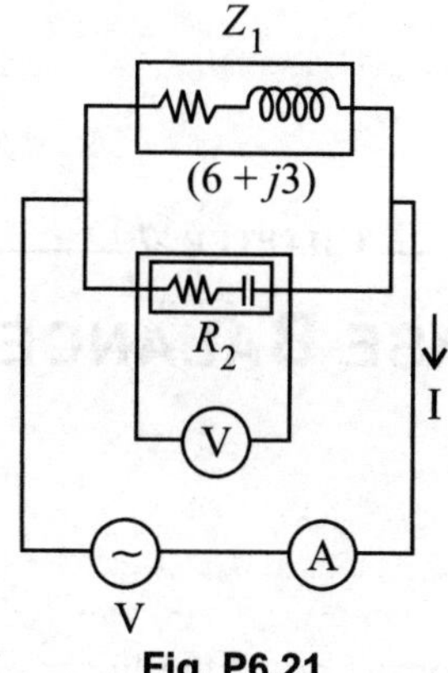

Fig. P6.21

6.22 The impedance, Z_1 is connected in series with two impedances, Z_2 and Z_3 connected in parallel (Fig. P6.22), which are given in Ω, as $(6+j8), (12-j18)$ and $(10+j15)$ respectively. If the voltage drop across Z_1 is 100 V, calculate the value of the applied voltage to the circuit. Draw the phasor diagram.

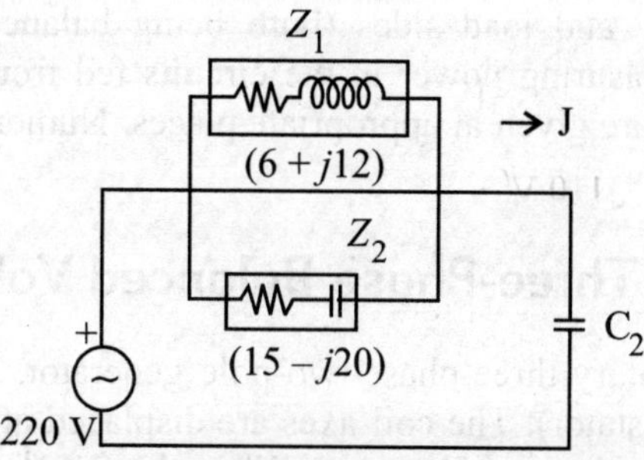

Fig. P6.22

6.23 A circuit having a resistor ($R_1 = 6\ \Omega$) and inductor $L = 25.5$ mH, is connected in parallel with a circuit having a resistor ($R_2 = 4\ \Omega$) in series with a capacitor, C, (Fig. P6.23). The circuit is connected to a 220 V, 50 Hz supply. Determine the value of the capacitor, C, such that the total current drawn from the supply is at unity power factor (upf). Find also (a) the current in each branch, and the (b) total current drawn from the supply.

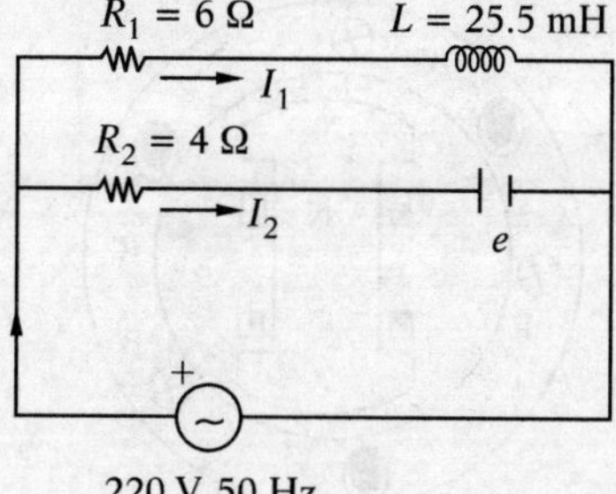

Fig. P6.23

CHAPTER 7

THREE-PHASE BALANCED SUPPLY

In Chapter 6, we studied circuits consisting of the linear elements—resistors, inductors and capacitors—fed from a single phase ac supply. In this chapter, which may be termed as an extension of the previous one, balanced circuits, fed from a three-phase ac supply, along with the measurement of power, will be described. Firstly, we will see how three-phase balanced voltages are generated. Then, the two types of connections (star and delta), normally used for the above supply, followed by line and phase quantities (voltages and currents) for the connections, in both supply and load sides (both being balanced), are described. Lastly, the different methods of measuring power in the circuits fed from a three-phase supply are explained. Phasor diagrams are given at appropriate places. Numerical examples are included to illustrate the material.

7.1 Generation of Three-Phase Balanced Voltages

Figure 7.1 shows an elementary three-phase two-pole generator. Three coils RR′, YY′, BB′ are placed on the armature (stator). The coil axes are displaced at an angle of 120° in space from each other. The field is placed at the rotor. When the field is excited by dc and rotated, voltage will be generated in each phase according to Faraday's law. The flux linking any phase will vary sinusoidally with time and the sinusoidal voltage will also be in three phases, if the field structure is so designed that the flux is distributed sinusoidally over the poles. Coils, RR′, YY′, BB′, are identical, each having equal number of turns and wound with wire of the same type of material, and same cross-section. The induced emfs in the coils have the same frequency, but are displaced by 120°.

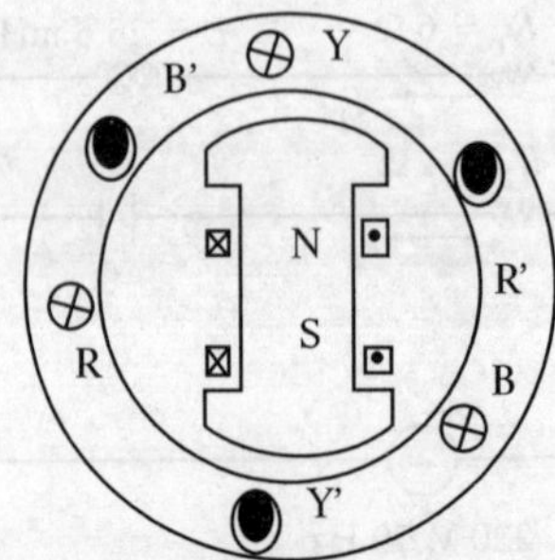

Fig. 7.1

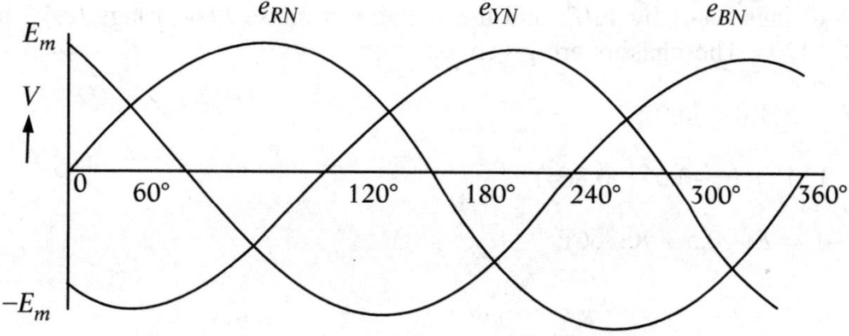

Fig. 7.2

The instantaneous values of emf (Fig. 7.2) induced in coil RR′, YY′, BB′ are:

$e_{RN} = E_m \sin \omega t,\ e_{YN} = E_m \sin(\omega t - 120)$, and

$e_{BN} = E_m \sin(\omega t - 240°) = E_m \sin(\omega t + 120°)$,

7.2 Three-phase Voltages for Star Connection

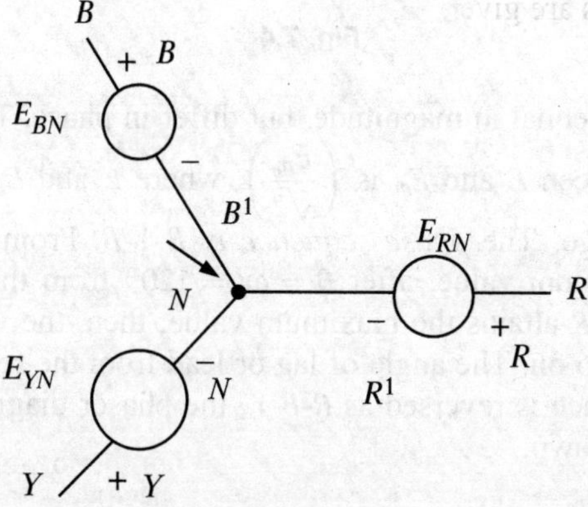

Fig. 7.3

The connection diagram of the three-phase two-pole generator in Fig. 7.1 which is star (Y) connected is shown in Fig. 7.3. The terminals of the three windings for the three phases are $R(+)$ and $R'(-)$, $Y(+)$ and $Y'(-)$ and $B(+)$ and $B'(-)$. Taking the winding of one phase, say phase R as an example, R with sign (+) is taken as the start, and R' with sign (–) is taken as the finish. The same is the case with the two other phases. For making the star (Y) connection, R', Y' and B' are connected together, and the point is taken as neutral, N. The phasor representation of the voltages is shown in Fig. 7.4. The three phase voltages are

$$e_{RN} = E_m \sin\theta;\ e_{YN} = E_m \sin(\theta - 120°);\ e_{BN} = E_m \sin(\theta - 240°) = E_m \sin(\theta + 120°).$$

It may be noted that if the voltage in phase $R(e_{RN})$ is taken as reference, then the voltage in phase $Y(e_{YN})$ lags (e_{RN}) by 120°, and the voltage in phase $B(e_{BN})$ lags (e_{YN}) by 120°, or leads (e_{RN}) by 120°. The phasors are given as:

$$E_{RN}\angle 0° = E(1.0 + j0.0);$$

$$E_{YN}\angle -120° = E(-0.5 - j0.866);$$

$$E_{BN}\angle 120° = E(-0.5 + j0.866);$$

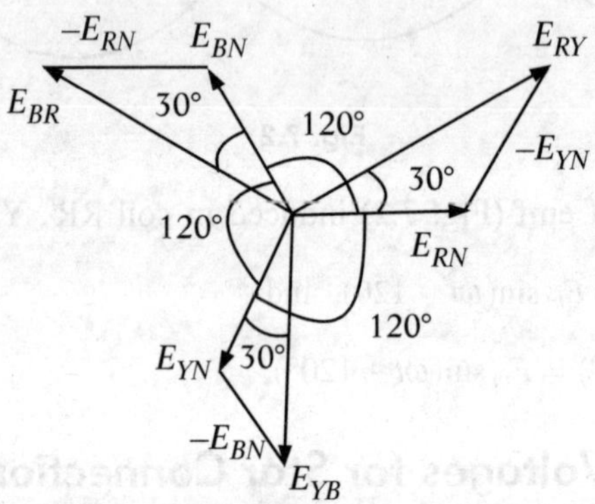

Fig. 7.4

The phase voltages are all equal in magnitude, but differ in phase. This is also shown in Fig. 7.4. The relationship between E and E_m is $\left(\dfrac{E_m}{\sqrt{2}}\right)$, where E and E_m are rms and maximum values of the phase voltage. The *phase sequence is R-Y-B*. From Fig. 7.2, the voltage in phase Y attains the maximum value, after $\theta = \omega t = 120°$ from the time or angle, $\theta = 0°$, after the voltage in phase R attains the maximum value; then, the voltage in phase B attains the maximum value, and so on. The angle of lag or lead from the reference phase, R is stated earlier. If the phase sequence is reversed as *R-B-Y*, the phasor diagram is shown in Fig. 7.5. The waveforms are not shown.

RELATION BETWEEN PHASE AND LINE VOLTAGES FOR STAR CONNECTION

Three line voltages (Fig.7.5) are obtained by the following procedure. The line voltages (potential difference between two lines of different phase) are,

$$E_{RY} = (E_{RN} - E_{YN}) = (E\angle 0° - E\angle -120°) = E[(1.0 + j0.0) - (-0.5 - j0.866)]$$
$$= E(1.5 + j0.866) = \sqrt{3}E\angle 30°.$$

The magnitude of the line voltage, E_{RY} is $\sqrt{3}$ times the magnitude of the phase voltage (potential difference between a line and a neutral), E_{RN} and E_{RY} leads E_{RN} by 30°.

Let us use the phasor diagram (Fig. 7.5) to find the line voltage, E_{RY}. The phase voltages, E_{RN} and E_{YN} are drawn first, and then, E_{RY} is found by using the relation, $E_{RY} = E_{RN} - E_{YN}$. The triangle so formed is an isosceles one, and the magnitude of the line voltage is E_{RY}, which is the same expression as obtained earlier. The line voltage, $E_{RY} = 2.E_{RN}.\cos 30° = \sqrt{3}.E_{RN}$, which is the same expression as obtained earlier. Line voltage, E_{RY} leads the phase voltage, E_{RN} by 30°.

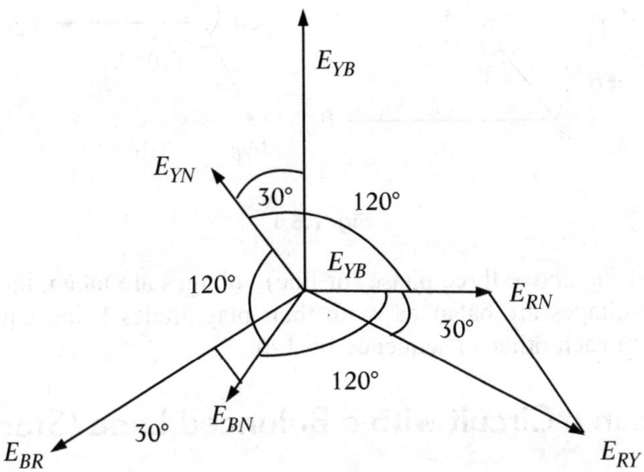

Fig. 7.5

The two line voltages, E_{YB} and E_{BR} can be obtained similarly.

$$E_{YB} = (E_{YN} - E_{BN}) = (E\angle -120° - E\angle +120°)$$

$$= E[(-0.5 - j0.866) - (-0.5 + j0.866)] = E(-j1.732) = \sqrt{3}E\angle -90°.$$

$$E_{BR} = (E_{BN} - E_{RN}) = (E\angle +120° - E\angle 0°) = E[(-0.5 + j0.866) - (1.0 + j0.0)]$$

$$= E(-1.5 + j0.866) = \sqrt{3}E\angle +150°.$$

Hence, the three line voltages are balanced, with their magnitudes being equal, and the phase angle being displaced from each other in sequence by 120°. Moreover, the line voltage, say, E_{RY}, leads the corresponding phase voltage, E_{RN} by 30°.

7.3 Relation between Phase and Line Voltages for Delta Connection

The connection diagram of a delta (Δ) connected three-phase system is shown in Fig. 7.6(a), along with the phasor representation of the voltages (Fig. 7.6(b)). For making a delta (Δ) connection, the start of one winding is connected to the finish of the next one in sequence. For instance, starting from phase R, R' is to connected to Y, and then Y' to B, and so on (Fig. 7.6(a)). The line and phase voltages are the same in this case, and given as,

$$E_{RY} = E\angle 0; \quad E_{YB} = E\angle -120°; E_{BR} = E\angle +120°.$$

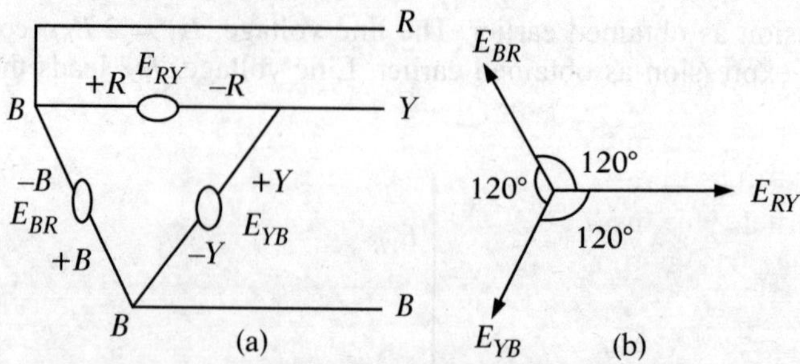

Fig. 7.6

If the phasor sum of the above three phase (or line) voltages are taken, the result is zero (0). The phase or line voltages are balanced, with their magnitudes being equal, and the phase being displaced from each other in sequence by 120°.

7.4 Currents in a Circuit with a Balanced Load (Star Connected)

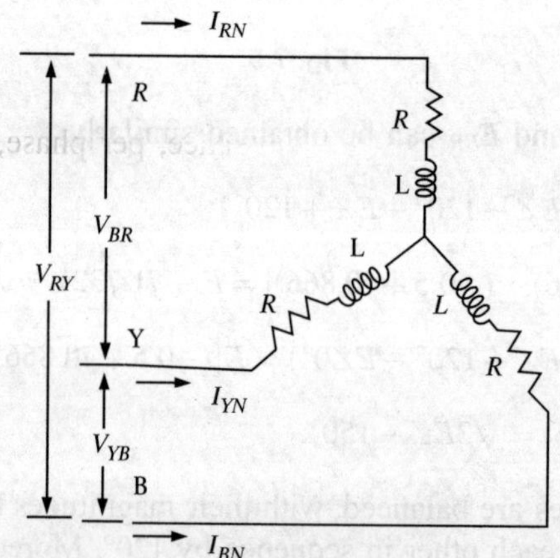

Fig. 7.7(a)

A three-phase star (Y) connected balanced load (Fig. 7.7(a)) is fed from a balanced three-phase supply, which may be a three-wired one. A balanced load means that the magnitude of the impedance per phase is the same, i.e.,

$$|Z_p| = |Z_{RN'}| = |Z_{YN'}| = |Z_{BN'}|,$$

and their angle is also the same, $\phi_p = \phi_{RN'} = \phi_{BN'}$. In other words, if the impedance per phase is given as,

$$Z_p \angle \phi_p = (R_p + jX_p),$$

then $R_p = R_{RN'} = R_{YN'} = R_{BN'}$, and $X_p = X_{RN'} = X_{YN'} = X_{BN'}$. The magnitude and phase angle of the impedance per phase are:

$$Z_p = \sqrt{(R_p)^2 + (X_p)^2}, \text{ and } \phi_p = \tan^{-1}\left(\frac{X_p}{R_p}\right),$$

For a balanced load, the magnitude of the phase voltages, $|V_p| = |V_{RN'}| = |V_{YN'}| = |V_{BN'}|$ are the same as those of the source voltages per phase, $|V_{RN}| = |V_{YN}| = |V_{BN}|$, if it is connected in star, as given earlier. This means that the star point, N' is the same as the star point, N on the source side. The phase currents are obtained as,

$$I_{RN'}\angle{-\phi_p} = \frac{V_{RN}\angle 0^\circ}{Z'_{RN}\angle\phi_p} = \frac{V_{RN}}{Z_{RN'}}\angle-\phi_p.$$

$$I_{YN'}\angle{-(120^\circ+\phi_p)} = \frac{V_{YN}\angle-120^\circ}{Z'_{YN}\angle\phi_p} = \frac{V_{YN}}{Z_{YN'}}\angle-(120^\circ+\phi_p).$$

$$I_{BN'}\angle(120^\circ-\phi_p) = \frac{V_{BN}\angle+120^\circ}{Z'_{BN}\angle\phi_p} = \frac{V_{BN}}{Z_{BN'}}\angle-(120^\circ-\phi_p)$$

The complete phasor diagram is shown in Fig. 7.7(b). In this case, the phase voltage V_{RN}, is taken as reference. The phase currents are equal in magnitude, i.e.,

$$(|I_p| = |I_{RN'}| = |I_{YN'}| = |I_{BN'}|),$$

as the magnitudes of the voltage and load impedance, per phase, are the same, with their phase angles displaced from each other in sequence by 120°. The magnitude of the phase currents is expressed as $\left(|I_p| = \left|\frac{V_p}{Z_p}\right|\right)$. The phase currents are also equal to the line currents $(|I_L| = |I_R| = |I_Y| = |I_B|)$ in this case.

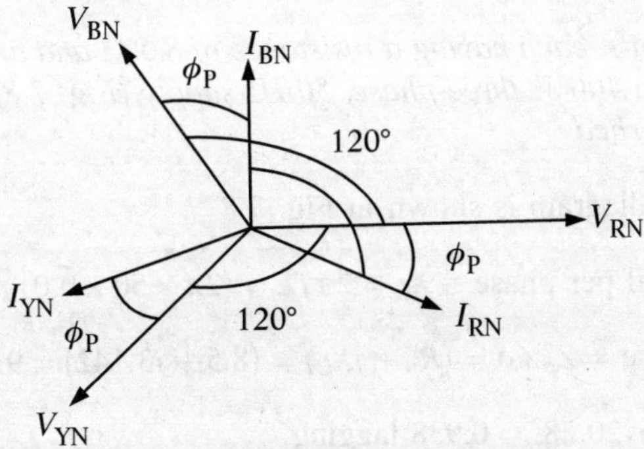

Fig. 7.7(b)

7.5 Total Power Consumed in a Circuit with a Star Connected Balanced Load

The power consumed in a circuit fed from a single phase supply was presented in Chapters 5 and 6. Using the same expression for the star connected balanced circuit, fed from a three-phase supply shown in Fig. 7.7(a)–(b), the power consumed per phase is given by

$$W_p = V_p.I_p \cos\phi_p = V_p.I_p \cos(C_p, I_p).$$

It has been shown earlier that the magnitude of the phase voltage is given by $|V_p| = \frac{|V_L|}{\sqrt{3}}$, where the magnitude of the line voltage is $|V_L|$. The magnitudes of the phase and line currents are the same, i.e., $|I_p| = |I_L|$ Substituting the two expressions, the total power consumed is obtained as,

$$W = 3.\left(\frac{V_L}{\sqrt{3}}\right) I_L \cos\phi_p = \sqrt{3} \times (V_L \cdot I_L) \cos\phi_p.$$

It may be noted that the phase angle, ϕ_p is the angle between the phase voltage, V_p and the phase current, I_p.

EXAMPLE 7.1 *The emf generated in each phase of an alternator (star connected) is 6,000 V. Determine (i) the line voltage and (ii) the total power output when the current in each phase is 400 A and the power factor is 0.85 lagging.*

SOLUTION The alternator is star connected. Hence, $\sqrt{3}E_{\text{ph}} = V_L; I_{\text{ph}} = I_L$.

Here, $E_{\text{ph}} = 6{,}000$ V; $I_{\text{ph}} = I_L = 400$ A; power factor $= \cos\phi = 0.85$ lagging.

(i) Line voltage, $V_L = \sqrt{3}E_{\text{ph}} = (\sqrt{3} \times 6{,}000) \approx 10.392$ kV.

(ii) Total power, $P = (\sqrt{3}V_L I_L \cos\phi) = \sqrt{3} \times (10.392 \times 400.0 \times 0.85) = 6.12$ MW

or, $P = 6.12$ MW.

EXAMPLE 7.2 *Three coils, each having a resistance of 8.5 Ω and an inductance of 0.01 H, are connected in star to a 400 V, three-phase, 50 Hz supply (Fig. 7.8). Find the line current and the total power absorbed.*

SOLUTION The circuit diagram is shown in Fig. 7.8.

The reactance of each coil per phase $= X_L = 2\pi f L = (2\pi \times 50 \times 0.01) = 3.142\ \Omega$.

Load impedance per phase $= Z_p\angle\phi = (R_p + jX_L) = (8.5 + j3.142) = 9.0621\angle 20.28°\ \Omega$.

Power factor $= \cos\phi = \cos 20.28° = 0.938$ lagging.

Phase voltage $= V_{\text{ph}} = \frac{v_L}{\sqrt{3}} = \frac{400.0}{\sqrt{3}} = 231.0$ A.

Therefore, phase current $= I_{\text{ph}} = \frac{v_{ph}}{z_{\text{ph}}} = \frac{231.0}{9.0621} = 25.49$ A.

Current $= I_L = I_{ph} = 25.49$ A.

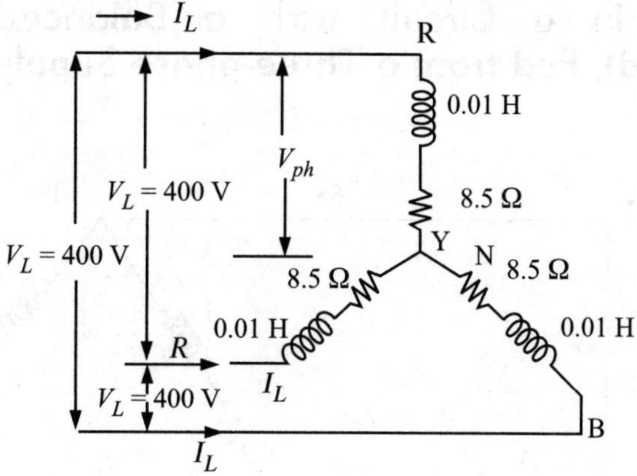

Fig. 7.8

Power absorbed $= P = \sqrt{3}(V_L I_L \cos\phi) = (\sqrt{3} \times 400 \times 25.49 \times 0.938) = 16.56$ kW.

Power absorbed can be computed as $P = 3(I_{\text{ph}})^2 R_p = 3 \times (25.49)^2 \times 8.5 = 16.56$ kW,

which is nearly the same as found earlier.

EXAMPLE 7.3 *Determine the active and reactive components of currents in each phase of a star connected 33 KV, 3-phase generator supplying 1500 kW at a lagging power factor of 0.85. Find the new output if the current is maintained at the same value but the power factor is raised to 0.9 lagging.*

SOLUTION Line voltage, $V_L = 33$ kV $= 33000$ volts; power supplied by the generator is 1500 kW; power factor, $\cos\phi = 0.85$.

Since, power supplied, $P = (\sqrt{3} V_L I_L \cos\phi)$.

$$\text{Line current, } I_L = \frac{P}{\sqrt{3} V_L \cos\phi} = \frac{(1500 \times 10^3)}{(\sqrt{3} \times 33000) \times 0.85} = \frac{(1500 \times 10^3)}{48582.6} = 30.875 \text{ A.}$$

Phase current, $I_{\text{ph}} = I_L = 30.875$ A.

Active component of current, $I_{\text{active}} = (I_{\text{ph}} \cos\phi) = (30.875 \times 0.85) = 26.24$ A.

Similarly, reactive component of current, $I_{\text{reactive}} = (I_{\text{ph}} \times \sin\phi) = (30.875 \times \sqrt{1 - (0.85)^2})$.

$$I_{\text{reactive}} = 16.26 \text{ A.}$$

Power output when power factor is raised to 0.95 lagging will be as

$$P = (\sqrt{3} V_L I_L \cos\phi) = \sqrt{3} \times (33000 \times 30.875 \times 0.95) = 1676.45 \text{ kW.}$$

7.6 Currents in a Circuit with a Balanced Load (Delta Connected), Fed from a Three-phase Supply

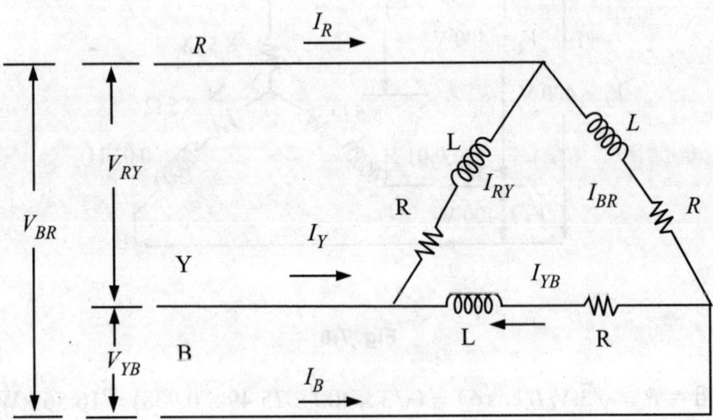

Fig. 7.9(a)

A three-phase delta (Δ) connected balanced load (Fig. 7.9(a)) is fed from a balanced three-phase supply. As mentioned before, a balanced load means that the magnitude of the impedance per phase, is the same, i.e.,

$$|Z_{VB}| = |Z_{RY}| = |Z_{YB}| = |Z_{BR}|,$$

and their angles are also the same, as $\phi_p = \phi_{RY} = \phi_{YB} = \phi_{BR}$. In other words, if the impedance per phase is given as $Z_p\angle\phi_p = (R_p + jX_p)$, then $R_p = R_{RY} = R_{YB} = R_{BR}$, and also $X_p = X_{RY} = X_{YB} = X_{BR}$. The magnitude and phase angle of the impedance per phase are:

$$Z_p = \sqrt{(R_p)^2 + (X_p)^2} \quad \text{and} \quad \phi_p = \tan^{-1}\left(\frac{X_p}{R_p}\right).$$

In this case, the magnitudes of the phase voltages $|V_p|$ are the same as those of the line voltages $|V_L| = |V_{RY}| = |V_{YB}| = |V_{BR}|$. The phase currents (Fig. 7.9(b)) are obtained as,

$$I_{RY}\angle(-\phi_p) = \left(\frac{V_{RY}\angle 0^\circ}{Z_{RY}\angle\phi_p}\right) = \frac{V_{RY}}{Z_{RY}}(\angle - \phi_p).$$

$$I_{YB}\angle\{-(120^\circ + \phi_p)\} = \left(\frac{V_{YB}\angle - 120^\circ}{Z_{YB}\angle\phi_p}\right) = \frac{V_{YB}}{Z_{YB}}\angle\{-(120^\circ + \phi_p)\}.$$

$$I_{BR}\angle\{+(120^\circ + \phi_p)\} = \left(\frac{V_{BR}\angle + 120^\circ}{Z_{BR}\angle\phi_p}\right) = \frac{V_{BR}}{Z_{BR}}\angle\{-(120^\circ + \phi_p)\}.$$

In this case, the phase voltage, V_{RY} is taken as reference. The phase currents are equal in magnitude, i.e.,

$$(|I_P| = |I_{RY}| = |I_{YB}| = |I_{BR}|),$$

as the magnitudes of the voltage and the load impedance, per phase, are the same, with their phase angles displaced from each other in sequence by 120°. The magnitude of the phase currents is expressed as

$$|I_P| = \left|\frac{V_p}{Z_p}\right|.$$

The line currents (Fig. 7.9(b)) are found as,

$$I_R\angle(-\theta_R) = (I_{RY} - I_{BR}) = [I_p\angle(-\theta_p) - I_p\angle(120° - \phi_p)]$$

or, $$I_R\angle(-\theta_R) = [I_p(\cos\phi_p - j\sin\phi_p)] - I_p[\cos(120° - \phi_p) + j\sin(120° - \phi_p)]$$

$$= I_p[\{\cos\phi_p + \cos(60° + \phi_p)\} - j\{\sin\phi_p + \sin(60° + \phi_p)\}]$$

$$= (I_p.2\cos 30°).[\cos(30° + \phi_p) - j\sin(30° + \phi_p)]$$

$$= (\sqrt{3}.I_p)\angle\{-(30° + \phi_p)\}.$$

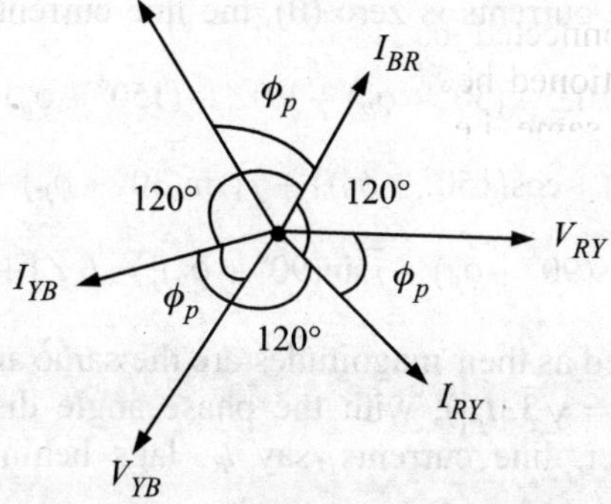

Fig. 7.9(b)

Therefore, $I_R\angle(-\theta_R) = (\sqrt{3}.I_p)\angle\{-(30° + \phi_p)\} = I_L < \{-(30° + \phi_p)\}$.

Let us look at an alternative procedure for finding the line current, I_R—using the phasor diagram as given in Section 7.2 for finding the line voltages from the phase voltages. Let us also recapitulate the formula for conversion of phasor or complex quantity from polar form to rectangular (Cartesian) form given in Chapter 4,

$$c\angle\theta = (a + jb) = ce^{j\theta}.$$

The expression for line current can then be given by,

$$I_R\angle(-\theta_R) = I_p[1\angle(-\phi_p) - 1 < 120° - \phi_p)] = I_p.[e^{-j\phi_p} - e^{-j(120° - \phi_p)}]$$

$$= I_p e^{-j\phi_p}.[(1.0 + j0.0) - e^{-j120°}] = I_p e^{-j\phi_p}[1.5 - j0.866] = \sqrt{3}.I_p e^{-j(30° + \phi_p)}$$

$$= \sqrt{3}.I_p\angle-(30^\circ+\phi_p).$$

This is the same expression as found earlier. Now, to obtain the expression for line current, I_Y and I_B,

$$I_Y\angle(-\theta_Y) = (I_{YB} - I_{RY}) = [I_p\angle\{-(120^\circ+\phi_p)\} - I_p\angle(-\phi_p)]$$

or, $$I_Y\angle(-\theta_Y) = I_p[-\{\cos(60^\circ-\phi_p)+\cos\phi_p\} + j\{\sin(60^\circ-\phi_p)-\sin\phi_p\}]$$

$$= -(I_p.2\cos 30^\circ).[\cos(30^\circ-\phi_p)+j\sin(30^\circ-\phi_p)]$$

$$= (\sqrt{3}.I_p)\angle\{(-180^\circ+30^\circ-\phi_p)\} = (\sqrt{3}.I_p)\angle\{-(150^\circ+\phi_p)\}.$$

Therefore, $I_Y\angle(-\theta_Y) = (\sqrt{3}.I_p)\angle\{-(150^\circ+\phi_p)\} = I_L\angle\{-150^\circ+\phi_p)\}$

Similarly, the expression for I_B is obtained as,

$$I_B = (\sqrt{3}.I_p)\angle\{+(90^\circ-\phi_p)\}.$$

Therefore, $I_B\angle(-\phi_B) = (\sqrt{3}.I_p)\angle\{+(90^\circ-\phi_p)\} = I_L\angle\{+(90^\circ-\phi_p)\}.$

As the sum of the three line currents is zero (0), the line current, I_B can also be found as,

$$I_B = -(I_R+I_Y) = -I_L[1.0\angle-(30^\circ+\phi_p) + 1.0\angle-(150^\circ+\phi_p)]$$

$$= I_L[-\{\cos(30^\circ+\phi_p)+\cos(150^\circ+\phi_p)\} + j\{\sin(30^\circ+\phi_p)+\sin(150^\circ+\phi_p)\}]$$

$$= I_L.(2\cos 60^\circ).[-\cos(90^\circ+\phi_p)+j\sin(90^\circ+\phi_p)] = I_L\angle\{+(90^\circ-\phi_p)\}.$$

The line currents are balanced as their magnitudes are the same and $\sqrt{3}$ times the magnitudes of the phase currents, ($|I_L| = \sqrt{3}.|I_p|$), with the phase angle displaced from each other in sequence by 120°. Moreover, line currents, say I_R, lags behind the corresponding phase current, I_{RY} by 30°.

If the phase current, I_{RY} is taken as reference, the phase currents are

$$I_{RY}(\angle 0^\circ) = I_p(1.0+j0.0); \qquad I_{YB}(\angle-120^\circ) = I_p(-0.5-j0.866);$$

$$I_{BR}(\angle+120^\circ) = I_p(-0.5+j0.866)$$

The line currents are obtained as

$$I_R = \{I_{RY}(<0^\circ) - I_{BR}(\angle+120^\circ)\} = I_p\{(1.0+j0.0)-(-0.5+j0.866)\}$$

$$= I_p(1.5-j0.866) = \sqrt{3}I_p\{\angle(-30^\circ)\} = I_L\{\angle(-30^\circ)\}.$$

$$I_Y = \{I_{YB}(\angle-120^\circ) - I_{RY}(\angle 0^\circ)\} = I_p\{(-0.5-j0.866)-(1.0+j0.0)\}$$

$$= I_p(1.5-j0.866) = \sqrt{3}I_p\{\angle(-150^\circ)\} = I_L\{\angle(-150^\circ)\}.$$

$$I_B = \{I_{BR}(\angle+120°) - I_{YB}(\angle-120°)\} = I_p\{(-0.5+j0.866)-(-0.5-j0.866)\}$$

$$= I_p(0.0+j1.732) = \sqrt{3}I_p\{\angle(+90°)\} = I_L\{\angle(+90°)\}.$$

7.7 Total Power Consumed in a Delta Connected Circuit with a Balanced Load

In Section 7.5, the equation for the power consumed in a star connected balanced circuit fed from a three-phase supply, was presented. The power consumed per phase for a delta connected balanced circuit, is given by

$$W_P = V_P.I_P \cos\phi_P = V_P.I_P \cos(V_P.I_P).$$

It has been shown earlier that the magnitude of the phase voltages are the same, i.e., $|V_p| = |V_L|$. The magnitude of the phase current is $\frac{1}{\sqrt{3}}$ times the magnitude of the line current, i.e., $|I_p| = \left(\frac{|I_L|}{\sqrt{3}}\right)$. Substituting the two expressions, the total power consumed is obtained as

$$W = 3.V_L\left(\frac{I_L}{\sqrt{3}}\right)\cos\phi_P = \sqrt{3}\times(V_L.I_L)\cos\phi_P.$$

It may be observed that the phase angle, ϕ_p is the angle between the phase voltage V_p and the phase current, I_p. Moreover, it may be noted that the expression for the total power in a three-phase balanced circuit is the same, whatever be the type of connection—*star or delta.*

EXAMPLE 7.4 *A balanced three-phase delta connected load has an impedance of* $(10+j15)$ Ω *per phase (Fig. 7.10). If a 440 V, three-phase supply is connected to this load, find (i) phase current (ii) line current and (iii) the power supplied to the load.*

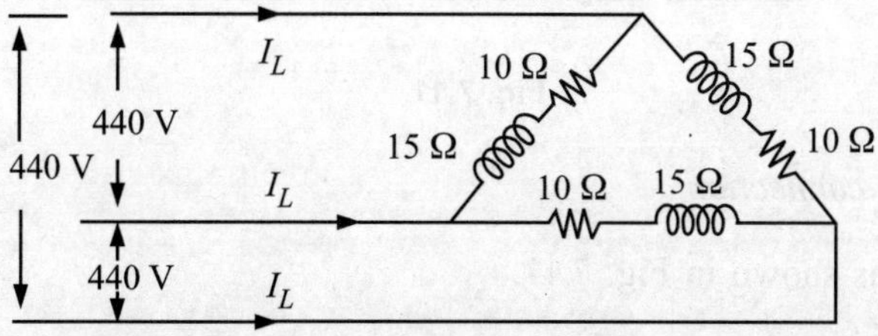

Fig. 7.10

SOLUTION The circuit diagram is shown in Fig. 7.10.

Load impedance per phase $= Z_p\angle\phi = (R_p + jX_L) = (10.0 + j15.0) = 18.027\angle 56.309°\ \Omega$.

(i) Phase current $= |I_{ph}| = \left|\frac{V_{ph}}{Z_{ph}}\right| = \frac{440.0}{18.027} = 24.407$ A.

(ii) Line current, $I_L = (\sqrt{3}\times I_{ph}) = (\sqrt{3}\times 24.407) = 42.272.$

(iii) Power factor of the load = $\cos\phi = \cos 56.309° = 0.554$ lagging.

Therefore, power supplied to the load,

$$P = \sqrt{3}(V_L I_L \cos\phi) = \sqrt{3} \times (440 \times 42.272 \times 0.554) \text{ or, } P = 17.846 \text{ kW.}$$

Power supplied can be computed as

$$P = 3(I_{ph})^2 R_p = 3 \times (24.407)^2 \times 10.0 = 17.87 \text{ kW,}$$

which is nearly the same as found earlier.

EXAMPLE 7.5 *A balanced three-phase load consists of three coils each of resistances 4 Ω and an inductive reactance of 5 Ω. Determine the current and power absorbed when the coils are (i) star connected and (ii) delta connected, across a 400 V, three-phase supply.*

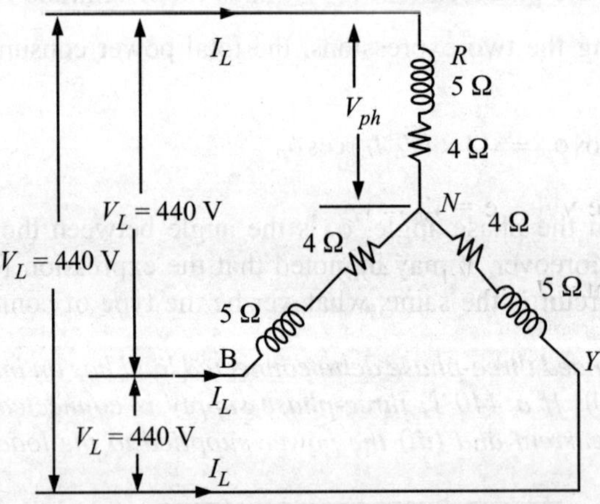

Fig. 7.11

SOLUTION (i) *Star connection*

The circuit diagram is shown in Fig. 7.11.

Load impedance per phase = $Z_p \angle\phi = (R_p + jX_L) = (4.0 + j5.0) = 6.4\angle 51.34°$ Ω.

Power factor = $\cos\phi = \cos 51.34° = 0.625$ lagging.

Phase voltage = $V_{\text{ph}} = \dfrac{V_L}{\sqrt{3}} = \dfrac{400.0}{\sqrt{3}} = 231.0$ A.

Therefore, phase current = $I_{\text{ph}} = \dfrac{V_{\text{ph}}}{Z_{\text{ph}}} = \dfrac{231.0}{6.4} = 36.09$ A.

Line current $= I_L = I_{\text{ph}} = 36.09$ A.

Power absorbed $= P = \sqrt{3}(V_L I_L \cos\phi) = (\sqrt{3} \times 400.0 \times 36.09 \times 0.625) = 15.62$ kW.

(ii) *Delta connection*

The three equal load impedances are now reconnected in delta, as shown in Fig. 7.12.

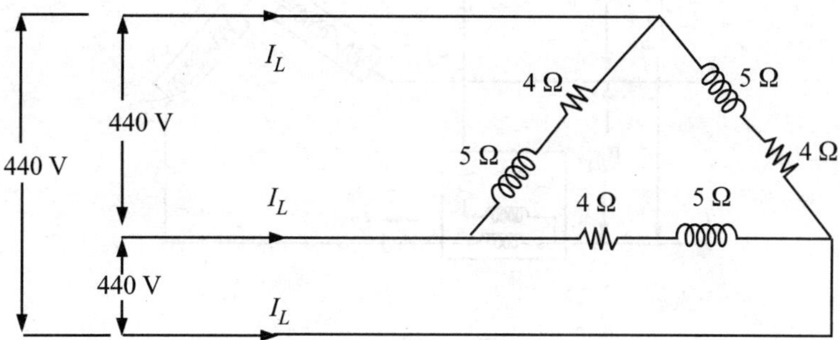

Fig. 7.12

The magnitude and angle of the impedances remain the same as earlier.

Phase voltage = Line voltage $= V_{ph} = V_L = 400.0$ V.

Therefore, phase current $= I_{ph} = \dfrac{V_{ph}}{Z_{ph}} = \dfrac{400.0}{6.4} = 62.5$ A.

Line current $= I_L = \sqrt{3} I_{ph} = \sqrt{3} \times 62.5 = 108.25$ A.

Power absorbed $= P = \sqrt{3}(V_L I_L \cos\phi) = (\sqrt{3} \times 400.0 \times 108.25 \times 0.625) = 46.87$ kW.

This is nearly three times the power absorbed when three equal load impedances are connected in star as computed earlier. The usefulness of delta connected loads is thus understood.

7.8 Measurement of Power in a Three-phase Circuit

7.8.1 Two Wattmeter Method of Power Measurement

The connection diagram for the measurement of power in a three-phase circuit using two wattmeters is given in Fig. 7.13. We can use this method irrespective of the circuit connection, which may be star or delta. The circuit may be taken as an unbalanced one, with the balanced type being only a special case. Please note the connection of the two wattmeters. The current coils of wattmeters 1 and 2 are in series with the two phases, *R* and *B* with the pressure or voltage coils being connected across *R* and *Y*, and *B* and *Y* respectively. *Y* is the third phase, in which no current coil is connected.

If a star connected circuit is taken as an example, the total instantaneous power consumed in the circuit is given as,

$$W = (i_{RN'}) \times (V_{RN'}) + (i_{YN'}) \times (V_{YN'}) + (i_{BN'}) \times (V_{BN'})$$

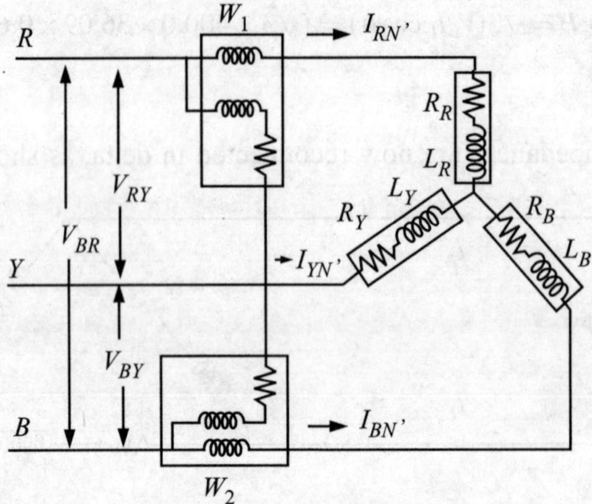

Fig. 7.13

Each of the terms in the above expression is the instantaneous power consumed for the phases. From the connection diagram, the current in, and the voltage across the respective (current and pressure or voltage) coils in wattmeter W_1 are $i_{RN'}$ and $V_{RY} = (V_{RN'} - V_{YN'})$. On the other hand, the current in and the voltage across the respective (current and pressure or voltage) coils in wattmeter W_2 are $i_{BN'}$ and $V_{BY} = (V_{BN'} - V_{YN'})$. Hence, the instantaneous power measured by wattmeter, W_1 is,

$$W_1 = (i_{RN'}) \times (V_{RY}) = (i_{RN'}) \times (V_{RN'} - V_{YN'})$$

Similarly, the instantaneous power measured by wattmeter, W_2 is,

$$W_2 = (i_{BN'}) \times (V_{BY}) = (i_{BN'}) \times (V_{BN'} - V_{YN'})$$

The sum of the two readings as given above is,

$$(W_1 + W_2) = \{(i_{RN'}) \times (V_{RN'} - V_{YN'}) + (i_{BN'}) \times (V_{BN'} - V_{YN'})\}$$

$$= \{(i_{RN'}) \times (V_{RN'} - V_{YN'}) + (i_{BN'}) \times (V_{BN'} - V_{YN'})\}$$

$$= \{(i_{RN'})(V_{RN'}) - (i_{RN'})(V_{YN'}) + (i_{BN'})(V_{BN'}) - (i_{BN'})(V_{YN'})\}.$$

Or, $(W_1 + W_2) = \{(i_{RN'})(V_{RN'}) + (i_{BN'})(V_{BN'})\} - \{V_{YN'} \times (i_{RN'} + i_{BN'})\}$.

Since, $(i_{RN'} + i_{YN'} + i_{BN'}) = 0, (i_{YN'}) = \{-(i_{RN'} + i_{BN'})\}$.

Substituting the above expression for $i_{YN'}$ in the expression for sum of instantaneous power,

$$(W_1 + W_2) = \{(i_{RN'}(V_{RN'}) + (i_{BN'})(V_{BN'})\} + \{(i_{YN'}) \times (V_{YN'})\}.$$

If this expression is compared with the earlier expression for the total instantaneous power consumed in the circuit, they are found to be the same. Hence, it can be concluded that the sum of the two wattmeter readings is the total power consumed in the three phase circuit (assumed here as a star connected one). Though we have used a star connected circuit, the same may also be easily proved for a delta connected circuit. As no other condition is imposed, the circuit can be taken as an unbalanced one, the balanced type being only a special case, as stated earlier.

PHASOR DIAGRAM FOR A THREE-PHASE BALANCED STAR CONNECTED CIRCUIT

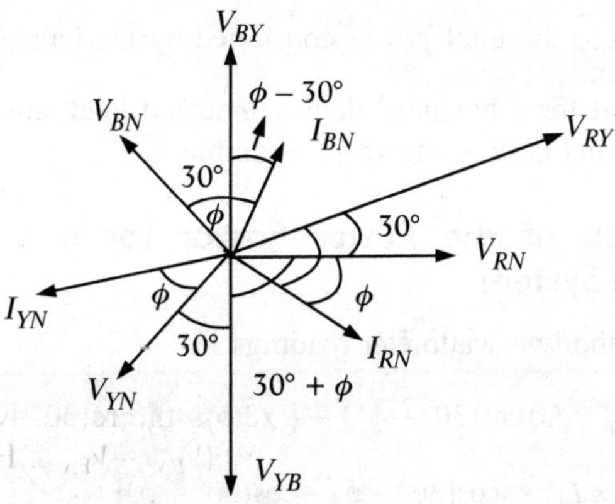

Fig. 7.14

The phasor diagram using the two wattmeter method, for a three-phase balanced star connected circuit is shown in Fig. 7.14. As discussed in Section 7.5, the phase currents lag behind the respective phase voltages by $\phi = \phi_p$, the angle of the load impedance per phase. The phase angle, ϕ is taken as positive for the inductive load. Moreover, the neutral point on the load N′ is the same as the neutral point on the source (N), if it is assumed to be connected in star. The voltage at that point is zero (0).

The readings of the two wattmeters are found as,

$$W_1 = \{(V_{RY}) \times (I_{RN})\}\cos\{(V_{RY}),(I_{RN})\} = \{(V_{RY}) \times (I_{RN})\}\cos(30° + \phi)$$

$$= \sqrt{3}(V_p \times I_p)\cos(30° + \phi).$$

$$W_2 = \{(V_{BY}) \times (I_{BN})\}\cos\{(V_{BY}),(I_{BN})\} = \{(V_{BY}) \times (I_{BN})\}\cos(30° - \phi)$$

$$= \sqrt{3}(V_p \times I_p)\cos(30° - \phi)$$

The line voltage, V_{RY} leads the respective phase voltage, V_{RN} by 30°, and the phase voltage, V_{RN} leads the phase current, I_{RN} by ϕ. Hence, the phase difference between V_{RY} and I_{RN} is $(30° + \phi)$. Similarly, the phase difference between V_{BY} and I_{BN} in the second case can be found and also checked from the phasor diagram.

The sum of the two wattmeter readings is,

$$(W_1 + W_2) = \{\sqrt{3}(V_p \times I_p)\cos(30^\circ + \phi)\} + \{\sqrt{3}(V_p \times I_p)\cos(30^\circ - \phi)\}$$

$$= \sqrt{3}(V_p \times I_p) \times [\cos(30^\circ + \phi) + \cos(30^\circ - \phi)]$$

$$= \sqrt{3}(V_p \times I_p) \times (2 \times \cos 30^\circ \cos\phi) = \sqrt{3}(V_p \times I_p) \times (2 \times \frac{\sqrt{3}}{2} \cos\phi).$$

Or, $(W_1 + W_2) = 3(V_p \times I_p)\cos\phi = \sqrt{3}(V_L \times I_L)\cos\phi).$

So, $(W_1 + W_2)$ is equal to the total power consumed by the balanced load.

This method is also valid for a balanced delta connected load, and can be easily obtained. The phasor diagram for this case is shown in Example 7.6.

7.8.2 Determination of the Power Factor for a Balanced Load in a Three-Phase System

The difference between the two wattmeter readings is,

$$(W_2 - W_1) = \{\sqrt{3}(V_p \times I_p)\cos(30^\circ - \phi)\} - \{\sqrt{3}(V_p \times I_p \cos(30^\circ + \phi)\}$$

$$= \sqrt{3}(V_p \times I_p) \times [\cos(30^\circ - \phi) - \cos(30^\circ + \phi)]$$

$$= \sqrt{3}(V_p \times I_p) \times (2 \times \sin 30^\circ \sin\phi) = \sqrt{3}(V_p \times I_p) \times (2 \times \tfrac{1}{2} \times \sin\phi).$$

Or, $(W_2 - W_1) = \sqrt{3}(V_p \times I_p)\sin\phi.$

If the two sides are multiplied by $\sqrt{3}$, we get

$$\sqrt{3}(W_2 - W_1) = \sqrt{3}\{\sqrt{3}(V_p \times I_p)\sin\phi = 3(V_P \times V_P)\sin\phi\,\sqrt{3}(V_L \times I_L)\sin\phi$$

From the sum and difference of the two wattmeter readings, we get,

$$\left(\frac{W_2 - W_1}{W_2 + W_1}\right) = \frac{\sqrt{3}(V_p \times I_p)\sin\phi}{3(V_p \times I_p)\cos\phi} = \frac{1}{\sqrt{3}}\tan\phi$$

Or,

$$\phi = \tan^{-1}\left[\sqrt{3} \times \left(\frac{W_2 - W_1}{W_2 + W_1}\right)\right].$$

The power factor, $\cos\phi$ of the balanced load,

$$\cos\phi = \frac{1}{\sqrt{\{1 + (\tan\phi)^2\}}} = \frac{1}{\sqrt{1 + \left\{\sqrt{3} \times \left(\frac{W_2 - W_1}{W_2 + W_1}\right)\right\}^2}}$$

$$= \frac{1}{\sqrt{1+3\left\{\left(\frac{W_2-W_1}{W_2+W_1}\right)\right\}^2}} = \frac{1}{\sqrt{1+3\left\{\left(\frac{1-y}{1+y}\right)\right\}^2}}$$

where, $y = \dfrac{W_1}{W_2}$

Similarly, the expression for $\sin\phi$ can be obtained.

Comments on Two Wattmeter Readings

When the balanced load is only resistive ($\phi = 0°$), i.e., power factor ($\cos\phi = 1.0$), the readings of the two wattmeters

$$\{(W_1 = W_2) \propto (\cos 30°) = 0.866(+\text{ve})\},$$

are *equal and positive.*

Before taking the case of a purely reactive (inductive/capacitive) load, let us consider first a lagging power factor (for example, $\cos\phi = 0.5$), $\phi = 60°$. Under this condition,

$$W_1 = 0.0, \text{ as } W_1 \propto \cos(30° + 60°) = \cos 90° = 0.0, \text{ and}$$

$$W_2 \propto \cos(30° - 60°) = \cos(-30°) = \cos 30° = 0.866, \text{ (positive)}.$$

Also the magnitudes of the phase or line voltage and also phase current are assumed to be constant, which means that the magnitude of the load impedance (inductive) is constant, but the phase angle, ϕ varies as stated.

As the lagging power factor decreases from 1.0 to 0.5, with ϕ increasing from 0° to 60°, the reading of the first wattmeter W_1 decreases from a certain positive value to zero (0). However, the reading of the second wattmeter W_2 increases from a certain positive value to the positive maximum, as the lagging power factor is decreased from 1.0 to 0.866 ($= \cos 30°$), with ϕ increasing from 0 degree to (+30) degree. As the lagging power factor decreases from 0.866 to 0.5, with ϕ increasing from (+30°) to (+60°), the reading of the second wattmeter W_2 decreases from the positive maximum to a certain positive value. It may be noted that in all these cases, $W_2 > W_1$, with both the readings being positive.

If the lagging power factor is 0.0 ($\phi = +90°$), the circuit being purely inductive, the readings of the two wattmeters,

$$\{(W_1 = -W_2) \propto \cos(30° + 90°) = \cos(120°) = -0.5\}, \text{ (negative)},$$

are equal and opposite, i.e., W_1 is negative and W_2 is positive. The total power consumed is zero, being the sum of the two wattmeter readings, as the circuit is purely inductive. This means that as the lagging power decreases from 0.5 to 0.0, with ϕ increasing from (+60°) to (+90 °), the reading of wattmeter W_1 decreases from zero (0) to a certain negative value, while the reading of wattmeter W_2 decreases from a certain positive value to a lower positive one. It may be noted that $|W_2| > |W_1|$, which means that the total power consumed, i.e., $(W_2 + W_1)$ is positive, with only W_1 being negative.

It may be noted that if the power factor is leading, ($\phi = -$ve), the circuit being capacitive, the readings of the two wattmeters change with the readings interchanging, i.e., W_1 taking the value of W_2, and vice versa.

7.8.3 Power Measurement Using One Wattmeter Only for a Balanced Load

The circuit diagram for measuring power for a balanced three-phase load using one wattmeter is shown in Fig. 7.15. The only assumption made is that either the neutral point on the load or the source side is available. The wattmeter measures the power consumed for one phase only, and the reading is $W = (V_p \times I_p)\cos\phi$. The total power is three times the above reading, as the circuit is balanced. Hence, the load must be star connected with the load neutral point being available. The load may also be a delta connected balanced one, if the neutral pinpoint on the source side is available.

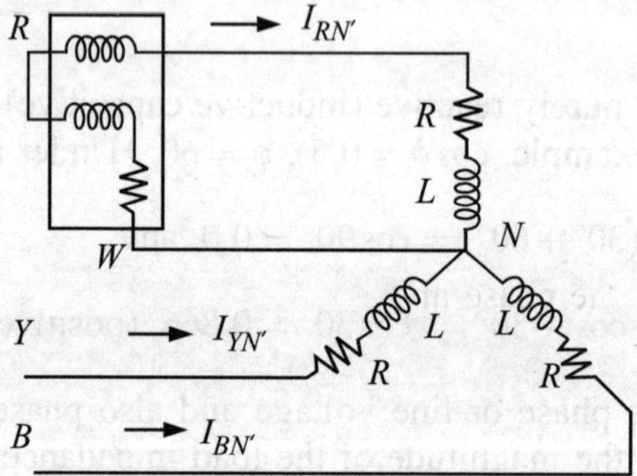

Fig. 7.15

EXAMPLE 7.6 Calculate the readings of the two wattmeters (W_1 and W_2) connected to measure the total power for a balanced star connected load shown in Fig. 7.15, fed from a three-phase, 400 V balanced supply with the phase sequence as R-Y-B. The load impedance per phase is $(20 + j15)\ \Omega$. Also find the line and phase currents, power factor, total power, total reactive VA and total VA.

SOLUTION The circuit diagram is shown in Fig. 7.15.

$$|V_L| = 400\text{ V};\ Z_{RN} = Z_p = (20 + j15) = 25.0\angle +36.87^\circ\ \Omega$$

As the star connected load is balanced, the magnitude of the phase voltage is,

$$|V_p| = \frac{|V_L|}{\sqrt{3}} = \frac{400}{\sqrt{3}} = 231\text{V}$$

Taking the phase voltage, V_{RN} as reference, the phase voltages are,

$$V_{RN} = (231\angle 0^\circ);\ V_{YN} = (231\angle -120^\circ);\ V_{BN} = (231\angle +120^\circ).$$

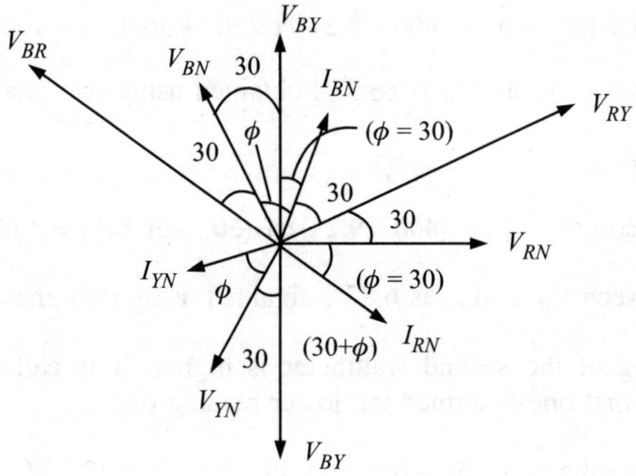

Fig. 7.16

The phasor diagram is shown in Fig. 7.16. It has been shown that the line voltage, V_{RY} leads the corresponding phase voltage, V_{RN} by $30°$. So, the line voltages are,

$$V_{RY} = (400\angle+30°); V_{YB} = (400\angle-90°); V_{BR} = (400\angle+150°).$$

For a star connected load, the phase and the line currents are the same.

The current in R phase is,

$$I_{RN} = I_R = \left(\frac{V_{RN}}{Z_p}\right) = \left(\frac{231.0\angle+0°}{25.0\angle+36.87°}\right) = (9.24\angle-36.87°) = (7.39 - j5.54) \text{ A}.$$

Two other phase and line currents are,

$$I_{YN} = I_Y = \left(\frac{V_{YN}}{Z_p}\right) = \left(\frac{231.0\angle-120°}{25.0\angle+36.87°}\right) = (9.24\angle-156.87°) = (-8.5 - j3.63) \text{ A}.$$

$$I_{BN} = I_B = \left(\frac{V_{BN}}{Z_p}\right) = \left(\frac{231.0\angle+120°}{25.0\angle+36.87°}\right) = (9.24\angle+83.13°) = (1.105 + j9.174) \text{ A}.$$

The power factor of the load is $\cos\phi = \cos 36.87° = 0.8$ lagging with $\phi = +36.87°$, as the load is inductive.

Total VA $= 3.(V_p \times I_p) = 3.(231.0 \times 9.24) = 6.403$ kVA.

This can be taken as $\sqrt{3}(V_L \times I_L) = \sqrt{3}(400 \times 9.24) = 6.402$ kVA.

Total power $= 3(V_P \times I_p)\cos\phi \; \{3(231.0 \times 9.24)\cos 36.87°\} = 5.12$ kW.

Total reactive VA $= 3(V_p \times I_p)\sin\phi = \{3(231.0 \times 9.24) \times \sin 36.87°\} = 3.84$ kVAR.

The readings of the two wattmeters are

$$W_1 = (V_{RY} \times I_R)\cos(30^\circ + \phi) = (400 \times 9.24)\cos(30^\circ + 36.87^\circ) = 1.45 \text{ kW}.$$

The phase angle between V_{RY} and I_R is 66.87°, obtained using two phasors, where

$$V_{BY} = (400\angle + 90^\circ).$$

$$W_2 = (V_{BY} \times I_B)\cos(30^\circ - \phi) = (400 \times 9.24)\cos(30^\circ - 36.87^\circ) = 3.67 \text{ kW}.$$

The phase angle between V_{BY} and I_R is 6.87°, obtained using two phasors.

Note that the reading of the second wattmeter is higher. It is called the higher reading wattmeter, while the first one is termed the lower reading one.

The sum of the two readings is $(W_1 + W_2) = (1.45 + 3.67) = 5.12$ kW, which is the same as the total power computed earlier.

As the power factor is greater than 0.5, the readings of the two wattmeters are positive.

EXAMPLE 7.7 *Two wattmeters are used to measure the power taken by a three-phase induction motor on no-load. The wattmeter readings in W are + 450 and −75. Calculate the power factor of the motor at no-load.*

SOLUTION Reading of the higher reading wattmeter, $W_2 = 450$ W.

Reading of the lower reading wattmeter, $W_1 = -75$ W.

If ϕ is the phase difference in between two wattmeter W_1 and W_2, then

$$\tan\phi = \left[\sqrt{3}\left(\frac{W_2 - W_1}{W_2 + W_1}\right)\right] = \left[\sqrt{3}\left\{\frac{450 - (-75)}{450 + (-75)}\right\}\right] = \left(\sqrt{3} \times \frac{525}{375}\right) = 2.425$$

or, $\phi = \tan^{-1}(2.425) = 67.59^\circ$.

Therefore, no-load power factor = $\cos\theta = \cos 67.59^\circ = 0.381$ lagging, as the current drawn is inductive. Note that the power factor is lower than 0.5. Therefore, the reading of the second wattmeter is negative, while that of the first one is positive.

EXAMPLE 7.8 *A balanced delta-connected load takes a line current of 15 A at a power factor of 0.8 leading from a 440 V, three-phase, 50 Hz supply (Fig. 7.17). Find the resistance and capacitance of the load per phase. Also calculate the reading of two wattmeters connected in the circuit to measure total power.*

SOLUTION The complete circuit diagram is shown in Fig. 7.17.

Phase voltage = $V_p = V_L = 440$ V; line current = $I_L = 15$ A.

Power factor of the load, $\cos\phi = 0.8$ leading.

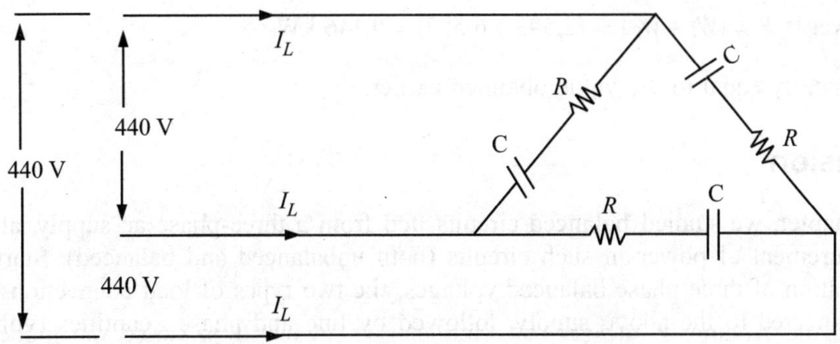

Fig. 7.17

Phase angle $= \phi = \cos^{-1}(0.8) = 36.87°$.

$$\text{Phase current} = I_p = \frac{I_L}{\sqrt{3}} = \frac{15.0}{\sqrt{3}} = 8.66 \text{ A.}$$

$$\text{Impedance per phase} = |Z_{ph}| = \left|\frac{V_{ph}}{I_{ph}}\right| = \frac{440.0}{8.66} = 50.81\,\Omega.$$

The load impedance per phase is expressed as

$$= Z_p\angle - \phi = (R_p - jX_{cp}) = 50.81\angle - 36.87° = (40.65 - j30.49)\,\Omega.$$

Therefore, resistance per phase $= R_{ph} = 40.65\,\Omega$, and capacitance per phase

$$C_{ph} = \frac{1}{(2\pi f.X_{cp})} = \frac{1}{(2\pi \times 50 \times 30.49)} = 104.4\,\mu\text{F}.$$

Let W_1 and W_2 be the two wattmeter readings, W_1 being the reading of the higher reading wattmeter and W_2 being the reading of the lower reading wattmeter. Since the load power factor is greater than 0.5, W_2 (reading of the lower reading wattmeter) is also positive. Also, note that the load is capacitive in this case.

Total power, $P = \sqrt{3}V_L I_L \cos\phi = \sqrt{3} \times (440.0 \times 15.0 \times 0.8) = 9.145$ kW.

The same comment applies to the reading of the two wattmeters connected to measure power.

So, $W_2 = V_L I_L \cos(30° - \phi) = (440.0 \times 15.0) \times \cos(30° + 36.87°) = 6.6 \times \cos 66.87°$

$$= 6.6 \times 0.393 = 2.593 \text{ kW.}$$

$$W_1 = V_L I_L \cos(30° + \phi) = (440.0 \times 15.0) \times \cos(30° - 36.87°) = 6.6 \times \cos 6.87°$$

$$= 6.6 \times 0.993 = 6.553 \text{ kW.}$$

Total power $= P = (W_1 + W_2) = (2.593 + 6.553) = 9.146$ kW,

which is nearly equal to the value obtained earlier.

Conclusion

In this chapter, we studied balanced circuits, fed from a three-phase ac supply, along with the measurement of power in such circuits (both unbalanced and balanced). Starting with the generation of three-phase balanced voltages, the two types of load connections (star and delta) connected to the above supply, followed by line and phase quantities (voltages and currents) for the connections, in both supply and load sides (both being balanced), were discussed. The loads per phase may be inductive, capacitive or resistive. Phasor diagrams were included at appropriate places in all cases. Numerical examples were given to illustrate the material. With this chapter, we complete the study of circuits (dc and single/three-phase ac types). In the next three chapters, we will study the field part (both electro-magnetic and electro-static).

Multiple Choice Type Questions

Tick the correct option(s), out of the four choices provided in each question.

7.1 The current through the neutral wire during the balanced loading condition is

(a) zero (b) non-zero

7.2 The current through the neutral wire during the unbalanced loading condition is

(a) non-zero (b) zero

7.3 Three-phase power can be measured by

(a) two wattmeter method (b) three wattmeter method
(c) non of (a)–(b)

7.4 The readings of the two watt meters used for measuring the balanced load in the three-phase system are W_1 and W_2. The phase angle of the load is

(a) $\tan^{-1}\left[\sqrt{3}\times\left(\dfrac{W_2 - W_1}{W_2 + W_1}\right)\right]$ (b) $\tan^{-1}\left[\sqrt{3}\times\left(\dfrac{W_2 + W_1}{W_2 - W_1}\right)\right]$
(c) none of (a)–(b)

7.5 The readings of the two watt meters used for measuring the balanced load in the three-phase system are W_1 and W_2. If they are equal and positive, then the load are

(a) resistive (b) inductive (c) capacitive

7.6 Two watt meters, which are connected to measure the total power on a three-phase system supplying a balanced load, read 10.5 kW and 2.5 kW, respectively. The total power and power factor , respectively, are

(a) 13.0 kW, 0.334 (b) 13.0 kW, 0.684
(c) 8.0 kW, 0.52 (d) 8.0 kW, 0.344

Answers

7.1. (a) 7.2. (a) 7.3. (c) 7.4. (a) 7.5. (a) 7.6. (b)

Problems

7.1 The emf generated in each phase of an alternator (star connected) is 5716 V (Fig. P.7.5). Determine (i) the line voltage and (ii) the total power output, when the current in each phase is 450 A and the power factor is 0.85 lagging.

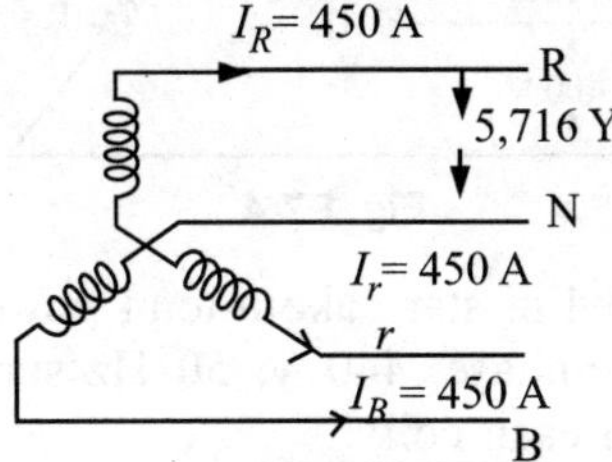

Fig. P7.1

7.2 A 440 V, three-phase, 50 Hz supply is fed to three coils, star connected, each having a resistance of 25 Ω and an inductive reactance of 20 Ω as shown in Fig. P7.2. Calculate (i) the line current, (ii) the power factor, and (iii) the power supplied.

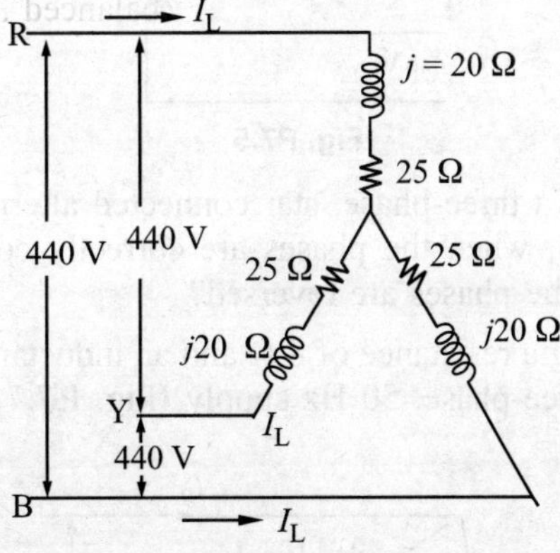

Fig. P7.2

7.3 Determine the active and reactive components of currents in each phase of a star connected 12 kV, three-phase generator supplying 5.5 MW at a lagging power factor of 0.85 (Fig. P.7.3). Find the new output if the current is maintained at the same value, but the power factor is raised to 0.9 lagging.

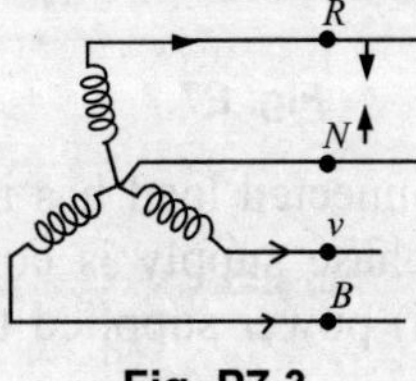

Fig. P7.3

7.4 Three coils, each having a resistance of 8 Ω and an inductance of 0.01 H, are connected in star to a 400 V, three-phase, 50 Hz supply (Fig. P.7.4). Find the line current and the total power absorbed.

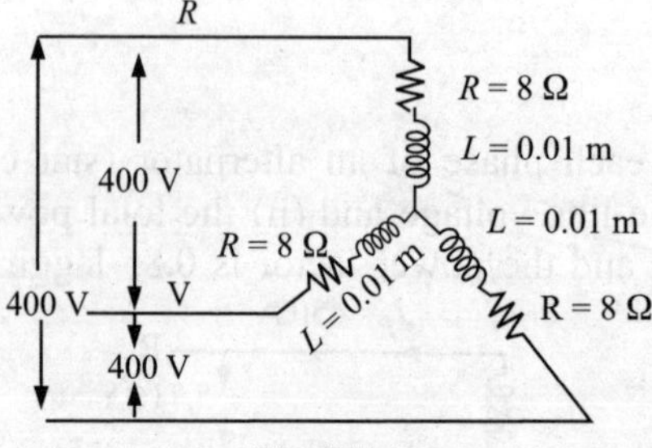

Fig. P7.4

7.5 Three similar coils connected in star, take a total power of 2.0 kW at a power factor of 0.5 lagging from a three-phase, 440 V, 50 Hz supply (Fig. P.7.5). Calculate the resistance and inductance of each coil.

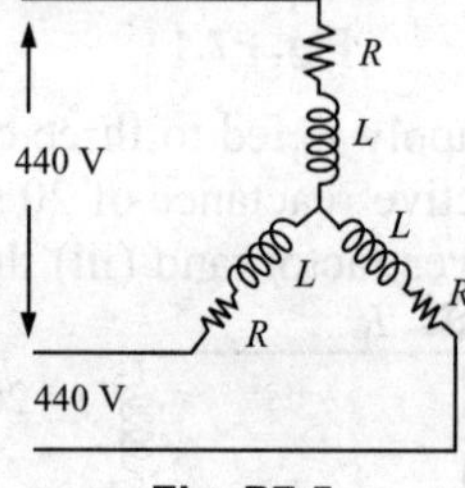

Fig. P7.5

7.6 If the phase voltage of a three-phase, star connected alternator be 254.0 V, what will be (i) the line voltages, when the phases are correctly connected, and (ii) when the connections of one of the phases are reversed?

7.7 Three coils, each having a resistance of 8 Ω and an inductance of 0.01 H are connected in delta to a 400 V, three-phase, 50 Hz supply (Fig. P.7.7). Find the line current and total power absorbed.

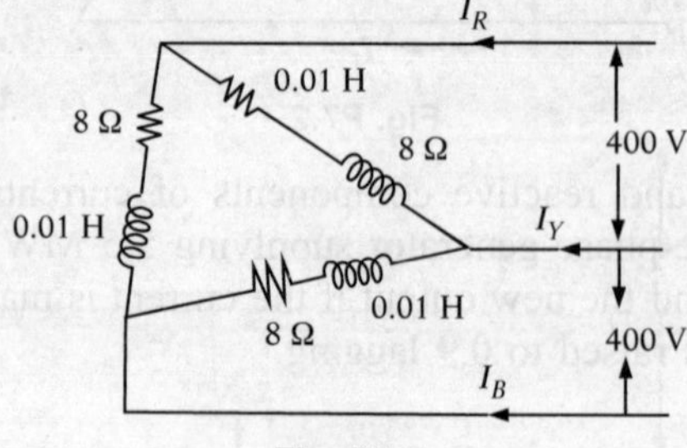

Fig. P7.7

7.8 A balanced three-phase delta connected load has impedance of $(20 + j35)\Omega$ per phase (Fig. P.7.8). If a 440 V, three-phase supply is connected to this load, find (i) phase current, (ii) line current, and (iii) power supplied to the load.

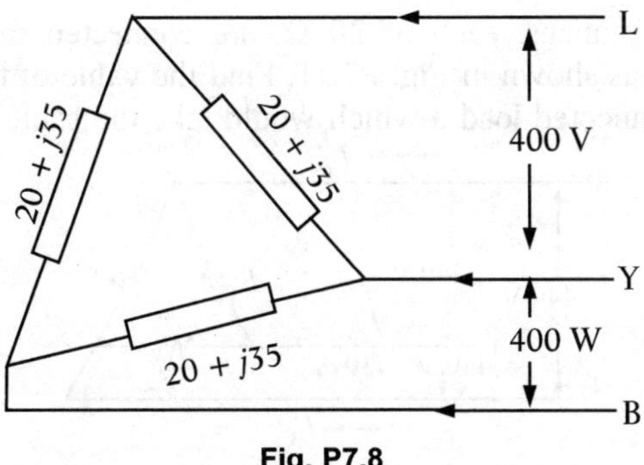

Fig. P7.8

7.9 A balanced three-phase load consists of three coils each of resistances 8 Ω and inductive reactance of 10 Ω (Fig. P.7.9). Determine the current and power absorbed, when the coils are (i) star connected and (ii) delta connected, across a 440 V, three-phase supply.

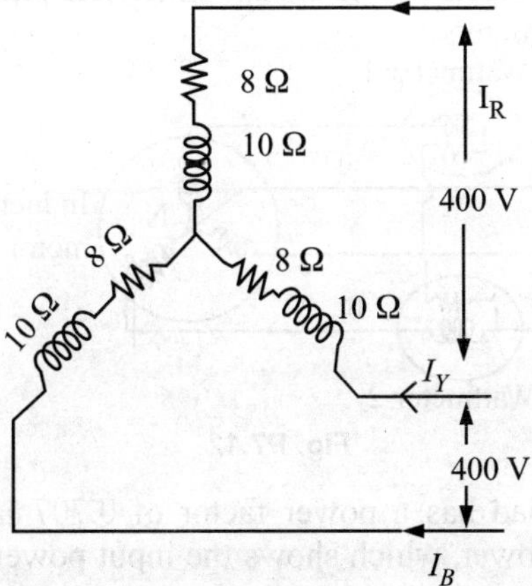

Fig. P7.9

7.10 A balanced delta connected load takes a line current of 15 A at a power factor of 0.8 leading, from a 440 V, three-phase, 50 Hz supply (Fig. P.7.10). Calculate the resistance and capacitance of the load per phase.

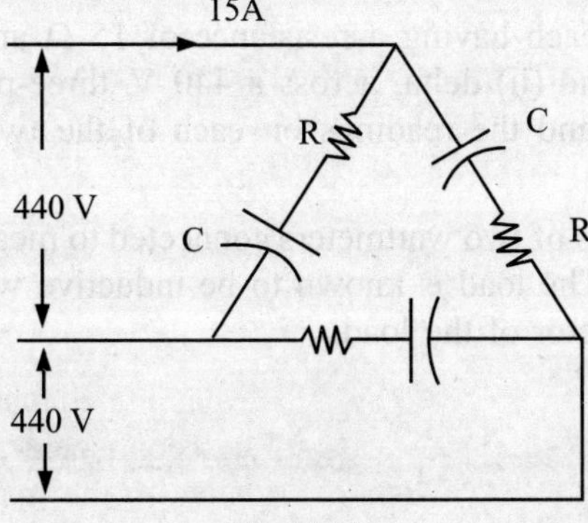

Fig. P7.10

7.11 Three identical resistances, each of 20 Ω, are connected in delta across a 440 V, three-phase supply, as shown in Fig. P7.11. Find the value of the resistance in each leg of balanced star-connected load a which would take the same line current.

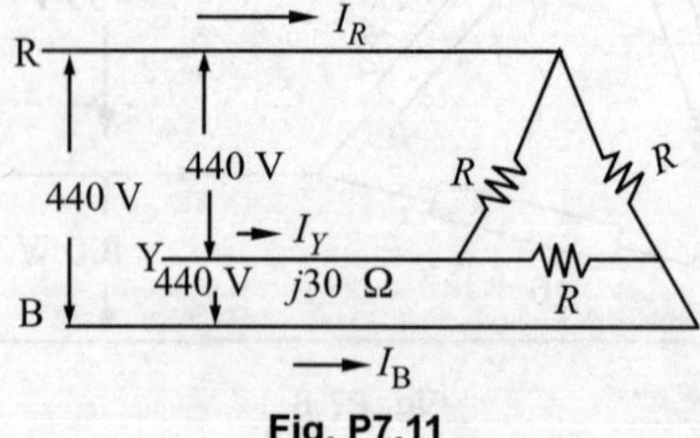

Fig. P7.11

7.12 Two wattmeters are used to measure the power taken by a three-phase induction motor on no-load (Fig. P.7.12). The wattmeter readings in W are +450 and –75. Calculate (i) the power factor of the current at no-load, and (ii) the phase difference of voltage and current in the two wattmeters.

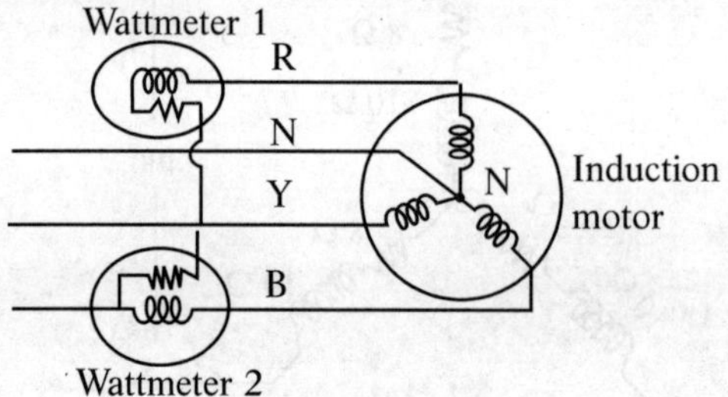

Fig. P7.12

7.13 A three-phase motor load has a power factor of 0.707 lagging. Two wattmeters are connected to measure power, which shows the input power as 25 kW. Find the reading of each wattmeter.

7.14 The power input to a 2.5 kV, 50 Hz, three-phase motor running on full load at an efficiency of 95% is measured by two wattmeters, which indicate 350 kW and 200 kW respectively. Find (i) the input power, (ii) the power factor, and (iii) the line current and output.

7.15 Three identical coils, each having a resistance of 15 Ω and a reactance of 5 Ω, are connected in (i) star and (ii) delta, across a 440 V, three-phase supply. Find, in each case, the line current and the readings on each of the two wattmeters connected to measure the power.

7.16 The ratio of the readings of two wattmeters connected to measure power in a three-phase balanced load, is 3:1. The load is known to be inductive with a lagging power factor. Calculate the power factor of the load.

CHAPTER 8

ELECTRO-MAGNETISM AND MAGNETIC CIRCUITS

INTRODUCTION

In the previous chapters, we studied various aspects of electric circuits. In the next three chapters, the field part—magnetic and electric—is taken up. In this chapter, we will discuss electro-magnetism, i.e., the magnetic field produced by the current in a conductor(s)/coil. Ampere's law and Biot–Savart law are used to find the magnetic field intensity, magnetic flux, and also magnetic flux density. Magnetic circuits are discussed. The magnetic field intensity and also flux density are calculated, when different types of cores are excited by the current in the coil(s) around the core. The core may be of different shapes and sizes, made with mainly magnetic material. It may be filled with air, when the magnetic cores are not there, or removed. A coil consists of a certain number of turns. Numerical examples are included to illustrate the text.

8.1 Magnetism

Magnetism is a force of attraction or replusion that acts at a distance. A magnet is an object that exhibits a strong magnetic field (Fig. 8.1) and will attract materials like iron to it. Magnets have two poles, called the North (N) and South (S) poles. Two magnets will be attracted by their opposite poles, and each will repel the like pole of the other magnet.

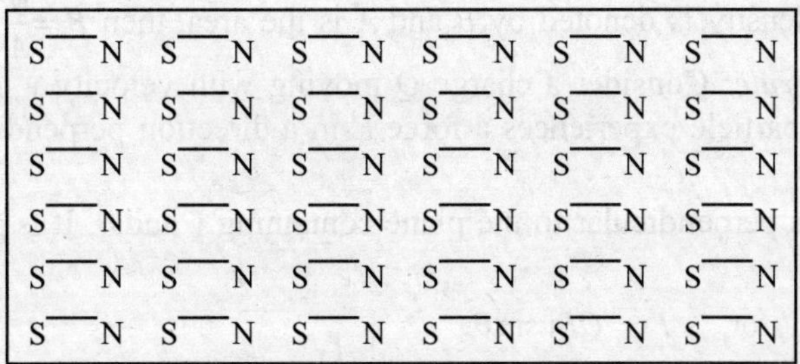

Fig. 8.1 *Magnetic field*

Lodestone and magnetite are two naturally occurring magnets. Magnetite, also known as a natural magnet, is a grey coloured substance composed of iron and oxygen (Fe_3O_4). It was first discovered in Magnesia in Asia Minor. It is also found in Canada, Finland, Norway and Urals. Lodestone, another magnet found in nature is irregular in shape and has weak attraction. Magnetic properties can be developed artificially in some metals or alloys. Artificial magnets can be prepared by two methods. The first method is by rubbing a piece of iron or steel with lodestone. The other method is by passing direct current through a coil wound over a piece of iron or steel.

MOLECULAR THEORY OF MAGNETISM

According to the molecular theory of magnetism, a molecule of magnetic substance is itself a complete magnet. In an unmagnetised specimen, the molecules are arranged either haphazardly, or in closed groups or chains causing no resultant magnetic effect.

The opposite poles neutralise each other throughout the piece. When a magnet is brought near the specimen, the molecules are swung round by the external magnetic force. Due to the process of magnetisation, these molecular magnets arrange themselves in definite lines. When all the little magnets are in line, the magnetisation is complete and the substance is said to be magnetically saturated.

Oersted was the first to establish the connection between electricity and magnetism. He discovered that the space around a current-carrying conductor had a magnetic field, i.e., a current-carrying conductor exerts a force on magnetic poles.

Before we discuss the important laws that govern electro-magnetism, let us define some concepts that are central to the idea of magnetism.

Magnetic field strength (H): The space around the current-carrying conductor is defined as the site of a magnetic field. The magnetic field strength is defined as the force experienced by a unit pole placed in a region under the influence of the magnetic field.

Magnetic flux (ϕ): Magnetic flux is an imaginary line which has no physical significance. It shows the direction flow of the magnetic field.

The magnetic flux θ through a surface S is defined as $\phi = \int_{\vec{B}} dS$.

The flux density is defined as the flux passing per unit area through a plane at right angles to the flux. If flux density is denoted by B and A is the area, then $B = \dfrac{\phi}{A}$ wb/m^2.

Fleming's left-hand rule: Consider a charge Q moving with velocity v in a region where B exists. The charged particle experiences a force F in a direction perpendicular to both v and B.

The direction of F is perpendicular to the plane containing $\vec{v}$ and $\vec{B}$. It is given by Fleming's left-hand rule,

$$F = Q\left(\vec{v} \times \vec{B}\right) \quad \text{or,} \quad F = QBv\sin\theta.$$

The fore finger, the central finger and the thumb of the left hand are stretched in such a way that they are mutually at right angles to each other (Fig. 8.2). If the fore finger points in the direction of the magnetic field vector B and the central finger points in the direction of the current or velocity vector v, then the thumb gives the direction of the deflecting force F on the charge.

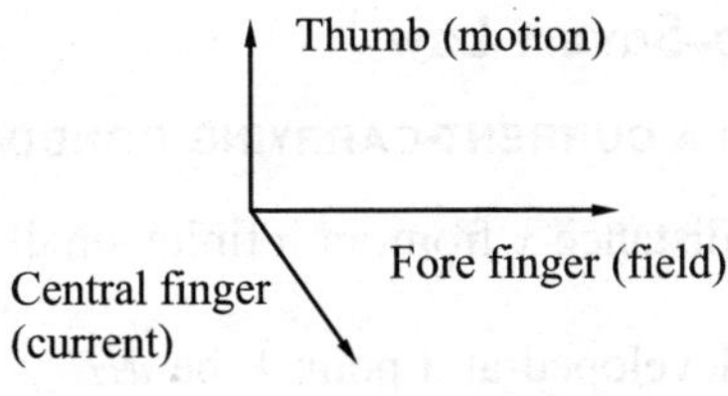

Fig. 8.2

The direction of the force on a negative charge is opposite to that on a positive charge.

8.2 Biot-Savart Law

Consider a conductor XY (Fig. 8.3) carrying a current i, and element AB of length $\vec{dl}$. O is the mid-point of AB. P is a point at a distance $\vec{r}$ from O. θ is the angle between $\vec{dl}$ and $\vec{r}$. The magnetic induction $\vec{dB}$ at point P due to the current element dl and r

$$\vec{dB} = \left(\frac{\mu_0}{4\pi}\right)\left\{\frac{i(dI \times \vec{r})}{r^2}\right\} = \left(\frac{\mu_0}{4\pi}\right)\left(\frac{idI\sin\theta}{r^2}\right).$$

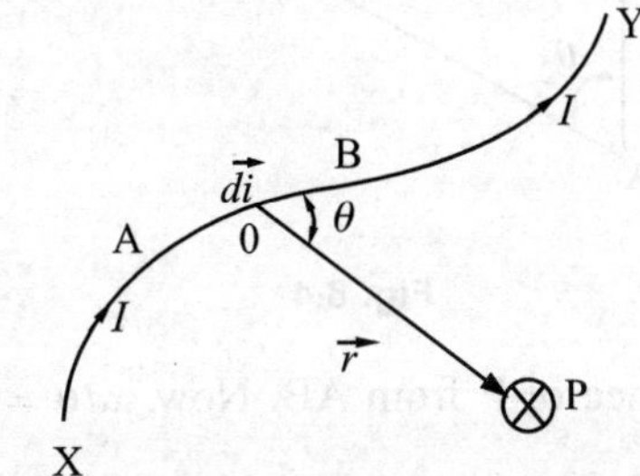

Fig. 8.3

The direction of dB is that of the vector $(dl \times r)$.

In vacuum or free space, B is related to H (magnetic field intensity), by the formula, $B = \mu_0 H$, where μ_0 is a constant, called permeability of free space.

AMPERE'S CIRCUITAL LAW

The law states that the magnetic force induced by a magnetic field around a closed curve is equivalent to the current in the enclosed curve. In other words, the line integral $\oint B.dl$ for a closed curve is equal to μ_0 times the net current i through the area bounded by the curve. That is, $\oint B.dl = \mu_0 I$, where, μ_0 is the permeability constant.

8.2.1 Application of Bio–Savart Law

MAGNETISING FORCE DUE TO A CURRENT-CARRYING CONDUCTOR

Let a point P be situated at a distance x from an infinite small length of the wire AB which is carrying I amperes.

Let the magnetising force developed at a point P be dH.

$$dH = \left(\frac{Idl \sin\theta}{4\pi x^2}\right). \qquad \text{(i)}$$

The force is directed into the plane of the paper as shown by a cross at point P (Fig. 8.4).

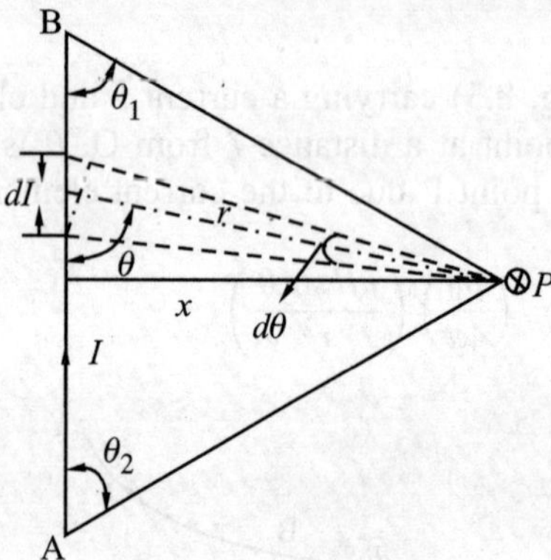

Fig. 8.4

Let r be the perpendicular distance of P from AB. Now, $xd\theta = dl \sin\theta$.

$$x\sin\theta = r \quad \text{or} \quad x = \frac{r}{\sin\theta}.$$

$$dH = \left(\frac{Ixd\theta}{4\pi x^2}\right) = \left(\frac{Id\theta}{4\pi x}\right) = \left(\frac{Id\theta}{4\pi r/\sin\theta}\right)$$

or, $$dH = \left(\frac{I\sin\theta d\theta}{4\pi r}\right) \qquad \text{(ii)}$$

Therefore, the total magnetising force due to a current-carrying conductor AB can be found by the integrating expression (ii) between limit θ_1 to θ_2.

$$\int dH = \int_{\theta_1}^{\theta_2}\left(\frac{I\sin\theta d\theta}{4\pi r}\right) = \left(\frac{I}{4\pi r}\right)\int_{\theta_1}^{\theta_2} \sin\theta d\theta$$

or $$H = \left(\frac{1}{4\pi r}\right).(\cos\theta_1 - \cos\theta_2) \text{ AT/m, or A/m.}$$

If the length of the conductor is long, then $\theta_2 = \pi$ or 180° and $\theta_1 = 0$.

In that case,

$$H = \left(\frac{I}{4\pi r}\right).(\cos 0° - \cos 180°) = \left(\frac{I.(1+1)}{4\pi r}\right) = \left(\frac{I}{2\pi r}\right) \text{ AT/m.}$$

MAGNETISING FORCE ON THE AXIS OF A SQUARE COIL

Let us take a square coil (Fig. 8.5) consisting of four conductors each of length $2l$ and each carrying I ampere.

It is to be noted that the magnetising force at the axial point P due to opposite sides AB and CD are H_{AB} and H_{CD} directed at right angles to the plane at P and AB, and the plane at P and CD respectively—the forces have the same magnitude. We can arrive at a similar conclusion for the sides AD and BC.

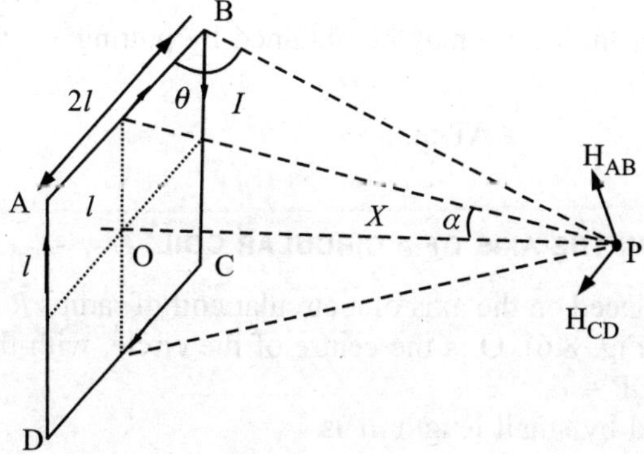

Fig. 8.5

$$H_{AB} = \left(\frac{I}{4\pi r}\right)\{\cos\theta - \cos(180-\theta)\} = \left(\frac{I}{4\pi r}\right)\{\cos\theta - (-\cos\theta)\} = \left(\frac{I}{4\pi r}\right)(2\cos\theta)$$

$$= \left(\frac{I\cos\theta}{2\pi r}\right).$$

$$r = \sqrt{(l^2 + x^2)}.$$

Then ultimately,

$$H_{AB} = \left\{\frac{I\cos\theta}{2\pi\sqrt{(l^2+x^2)}}\right\}.$$

The axial component of H_{AB} is

$$H'_{AB} = H_{AB}(\sin\propto) = \left\{\frac{I\cos\theta}{2\pi\sqrt{(l^2+x^2)}}\right\}(\sin\propto).$$

It may be noted here that all the four sides of the rectangular coil will contribute an equal amount to the resultant magnetic force H at point P.

$$H = (4 \times H'_{AB}) = \left[4 \times \left\{\frac{I\cos\theta}{2\pi\sqrt{(l^2+x^2)}}\right\}(\sin\alpha)\right].$$

But $\cos\theta = \dfrac{l}{\sqrt{(2l)^2+(x)^2}} = \dfrac{l}{\sqrt{4l^2+x^2}}$ and $\sin\alpha = \dfrac{1}{\sqrt{l^2+x^2}}$.

$$\text{Then, } H = \left[4 \times \left\{\frac{\left(l \times \frac{1}{\sqrt{4l^2+x^2}}\right)}{2\pi\sqrt{l^2+x^2}}\right\}\left(\frac{l}{\sqrt{l^2+x^2}}\right)\right] = \left[\frac{2Il^2}{\pi\sqrt{4l^2+x^2}(l^2+x^2)}\right] \text{AT/m.}$$

The magnetising force at the centre may be obtained by putting $x = 0$.

$$\text{Hence, } H_0 = \frac{2Il^2}{\pi(2l)(l)} = \frac{2Il^2}{2\pi l^2} = \frac{I}{\pi} \text{ AT/m.}$$

MAGNETISING FORCE ON THE AXIS OF A CIRCULAR COIL

Consider a unit pole P placed on the axis of a circular coil of radius R and of N turns carrying a current of I amperes (Fig. 8.6). O is the centre of the circle, with the point, P placed on a straight line, such that OP = a.

The force experienced by small length dl is

$$df = \frac{NIdl}{4\pi r^2} \text{ Newtons.}$$

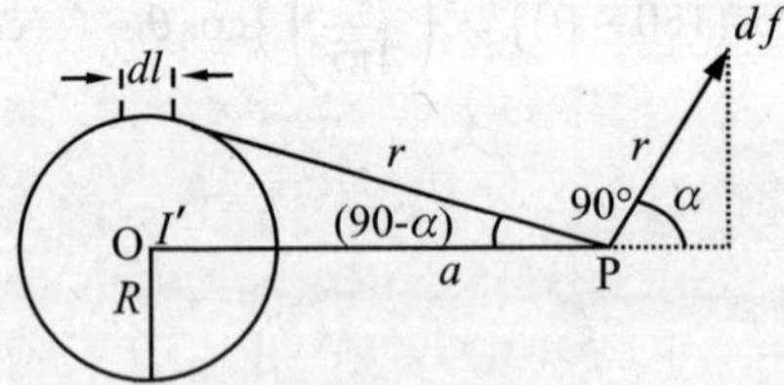

Fig. 8.6

Again, $R = \sqrt{R^2+a^2}$.

$$\text{Therefore, } df = \frac{NIdl}{4\pi(R^2+a^2)}.$$

The force may be resolved into the horizontal component dH along the axis and the vertical component perpendicular to it. If we consider the effect of force for the whole conductor, the vertical component will be cancelled by an equal and opposite component from the other side. Thus only the horizontal component will exist.

Therefore, $dH = \left\{\dfrac{NIdl}{4\pi(R^2+a^2)}\right\}(\cos\propto)$, where, $\cos\propto = \dfrac{R}{\sqrt{R^2+a^2}}$.

Thus, $dH = \left\{\dfrac{NIdl}{4\pi(R^2+a^2)}\right\}\left(\dfrac{R}{\sqrt{R^2+a^2}}\right) = \dfrac{NRIdl}{4\pi(R^2+a^2)^{3/2}}$

$$\int dH = \int_0^{2\pi R}\left\{\left(\frac{NRI}{4\pi(R^2+a^2)^{3/2}}\right)dl\right\} = \left(\frac{NRI}{4\pi(R^2+a^2)^{3/2}}\right)\int_0^{2\pi R} dl \text{ AMP/m.}$$

Or, $H = \left\{\left(\dfrac{NRI}{4\pi(R^2+a^2)^{3/2}}\right)\times(2\pi R)\right\} = \dfrac{NIR^2}{2(R^2+a^2)^{3/2}}$.

Hence, field strength at the centre of the coil, i.e., when $a = 0$.

$$H_0 = \frac{NI}{2R}.$$

FIELD INTENSITY ON THE AXIS OF A SOLENOID

A solenoid is a long, thin wire wrapped around a metallic core which produces a uniform magnetic field when an electric current is passed through it. Let l be the length of N turns of a solenoid (Fig. 8.7) and dx be a small length of solenoid. Thus, the number of turns in dx length of solenoid would be $(N/l)dx$.

Let P be a point on the axis of the solenoid. Hence the force experienced due to an elemental length dx of coil at point P by a unit pole is

$$dH = \frac{\{(N/l)dx\}IR^2}{2(R^2+x^2)^{3/2}} \text{ Amp/m.}$$

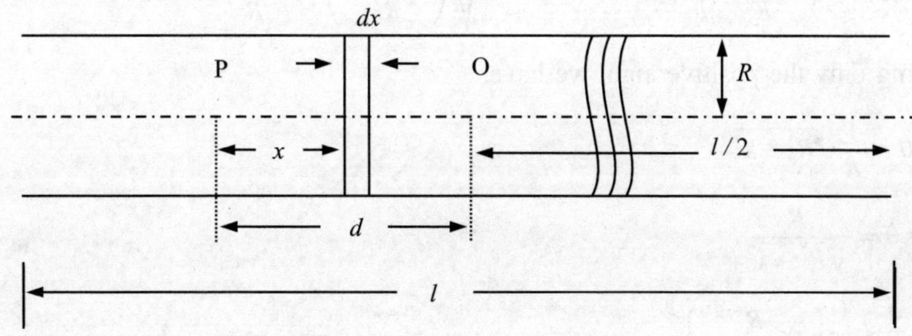

Fig. 8.7

Thus, the field intensity for the whole solenoid is

$$\int dH = \int_{-(\frac{l}{2}-d)}^{(\frac{l}{2}+d)}\left[\frac{\{(N/l)dx\}IR^2}{2(R^2+x^2)^{3/2}}\right].$$

or $$H = \left\{\left(\frac{IR^2}{2}\right) \times \left(\frac{N}{l}\right)\right\} \int_{-(-\frac{l}{2}-d)}^{(\frac{l}{2}+d)} \left[\frac{dx}{(R^2+x^2)^{3/2}}\right] = \left(\frac{NIR^2}{2l}\right) \int_{-(-\frac{l}{2}-d)}^{(\frac{l}{2}+d)} \left[\frac{dx}{(R^2+x^2)^{3/2}}\right].$$

Now, $$\int \frac{dx}{(R^2+x^2)^{3/2}}. \qquad \text{(i)}$$

Let $x = R\tan\theta$, then $\dfrac{dx}{d\theta} = R(\sec\theta)^2$. So, $dx = R(\sec\theta)^2 d\theta$.

Therefore, we have as follows:

$$\int \frac{dx}{(R^2+x^2)^{3/2}} = \int \frac{R(\sec\theta)^2 d\theta}{\{R^2(R\tan\theta)^2\}^{3/2}} = \int \frac{R(\sec\theta)^2 d\theta}{[R^2\{1+(\tan\theta)^2\}]^{3/2}}$$

$$= \frac{1}{R^2}\int \frac{(\sec\theta)^2 d\theta}{\{1+(\tan\theta)^2\}^{3/2}} = \frac{1}{R^2}\int \frac{(\sec\theta)^2 d\theta}{\{(\sec\theta)^2\}^{3/2}}.$$

After simplifying the integral, we have

$$\int \frac{dx}{(R^2+x^2)^{3/2}} = \frac{1}{R^2}\int \frac{d\theta}{\sec\theta} = \frac{1}{R^2}\int (\cos\theta)d\theta = \frac{1}{R^2}(\sin\theta) = \frac{1}{R^2}\left(\frac{x}{\sqrt{R^2+x^2}}\right)$$

[since, $x = R\tan\theta$, $\tan\theta = \dfrac{x}{R}$]

Or, $$(\tan\theta)^2 = \left(\frac{x}{R}\right)^2 \quad \text{or} \quad \{1+(\tan\theta)^2\} = \{1+\left(\frac{x}{R}\right)^2\} = \left(\frac{R^2+x^2}{R^2}\right).$$

Or, $$(\sec\theta)^2 = \left(\frac{R^2+x^2}{R^2}\right) \quad \text{or} \quad \sec\theta = \pm\sqrt{\left(\frac{R^2+x^2}{R^2}\right)} = \pm\frac{1}{R}\sqrt{R^2+x^2}.$$

Considering only the positive sign, we have,

$$\sec\theta = \frac{1}{R}\sqrt{R^2+x^2}.$$

So, $$\cos\theta = \frac{R}{\sqrt{R^2+x^2}}.$$

$$(\cos\theta)^2 = \left\{\frac{R}{\sqrt{R^2+x^2}}\right\}^2.$$

In other words, $$\{1-(\sin\theta)^2\} = \left(\frac{R^2}{R^2+x^2}\right)$$

or, $$(\sin\theta)^2 = \left\{1-\left(\frac{R^2}{R^2+x^2}\right)\right\} = \left(\frac{x^2}{R^2+x^2}\right).$$

or, $\sin\theta = \pm\sqrt{\left(\dfrac{x^2}{R^2+x^2}\right)} = \pm\dfrac{x}{\sqrt{R^2+x^2}}.$

Taking only the positive sign, we ultimately get

$$\sin\theta = \left(\frac{x}{\sqrt{R^2+x^2}}\right).$$

Therefore, $H = \left(\dfrac{NIR^2}{2l}\right)\displaystyle\int_{-(\frac{l}{2}-d)}^{(\frac{l}{2}+d)}\left[\frac{dx}{(R^2+x^2)^{3/2}}\right] = \left(\dfrac{NIR^2}{2l}\right)\times\left[\dfrac{1}{R^2}\left(\dfrac{x}{\sqrt{R^2+x^2}}\right)\right]$

with the lower and upper limit of x as $\left[-\left(\dfrac{l}{2}-d\right)\right]$ and $\left[\left(\dfrac{l}{2}+d\right)\right]$ respectively.

After applying the lower and upper limits of x in the above equation we have,

$$H = \left(\frac{NIR^2}{2l}\right)\left[\frac{1}{R^2}\left(\frac{(\frac{l}{2}+d)}{\sqrt{R^2+(\frac{l}{2}+d)^2}}\right)\right] - \left[\frac{1}{R^2}\left(\frac{-(\frac{l}{2}-d)}{\sqrt{R^2+(-\frac{l}{2}+d)^2}}\right)\right]$$

$$= \left(\frac{NIR^2}{2l}\right)\left[\frac{1}{R^2}\left(\frac{(l+2d)/2}{\sqrt{R^2+\left\{\frac{(l+2d)}{2}\right\}^2}}\right)\right] - \left[\frac{1}{R^2}\left(\frac{-(l-2d)/2}{\sqrt{R^2+\left\{\frac{(-l+2d)}{2}\right\}^2}}\right)\right]$$

$$= \left(\frac{NIR^2}{2l}\right)\left[\frac{1}{R^2}\left(\frac{(l+2d)/2}{\sqrt{\frac{4R^2+(l+2d)^2}{4}}}\right)\right] - \left[\frac{1}{R^2}\left(\frac{-(l-2d)/2}{\sqrt{\frac{4R^2+(l-2d)^2}{4}}}\right)\right]$$

$$= \left[\frac{1}{R^2}\left(\frac{(l+2d)}{\sqrt{4R^2+(l+2d)^2}}\right)\right] + \left[\frac{1}{R^2}\left(\frac{(l-2d)}{\sqrt{4R^2+(l-2d)^2}}\right)\right]$$

$$H = \left(\frac{NIR^2}{2l}\right)\left[\frac{1}{R^2}\left\{\left(\frac{(l+2d)}{\sqrt{4R^2+(l+2d)^2}}\right) + \left(\frac{(l-2d)}{\sqrt{4R^2+(l-2d)^2}}\right)\right\}\right]$$

$$= \left(\frac{NI}{2l}\right) = \left\{\left(\frac{(l+2d)}{\sqrt{4R^2+(l+2d)^2}}\right) + \left(\frac{(l-2d)}{\sqrt{4R^2+(l-2d)^2}}\right)\right\} \quad \text{(ii)}$$

From Fig. 8.8, we can write

$$\cos\alpha_1 = \frac{(\frac{l}{2}+d)}{\sqrt{R^2+(\frac{l}{2}-d)^2}} \quad \text{(iii)}$$

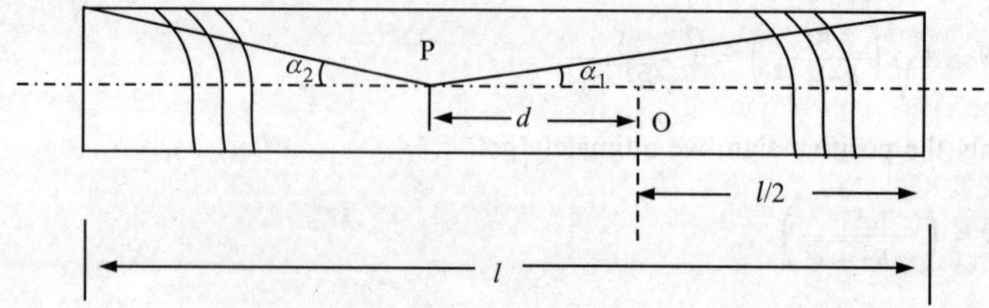

Fig 8.8

and

$$\cos\alpha_2 = \frac{\left(\frac{l}{2}-d\right)}{\sqrt{R^2+\left(\frac{l}{2}-d\right)^2}} \qquad \text{(iv)}$$

Now from Eqs (iii) and (iv), Eq. (ii) may be rewritten as

$$H = \left(\frac{NI}{2l}\right)\{\cos\alpha_1 + \cos\alpha_2\}$$

where α_1 and α_2 are the angles subtended by the end turns at point P.

It may be noted that when the point P is outside the solenoid, α is more than 90 degree and the value of $\cos\alpha_2$ is negative. Thus the value of H for this position is small. On the other hand, α_2 is equal to 90 degree or $\cos\alpha_2 = 0$. If the point P is at the end of the solenoid, magnetic field intensity, $H = \frac{NI}{2l}\cos\alpha_1$.

8.3 Magnetic Circuits

A magnetic circuit is made up of one or more closed loop paths containing a magnetic flux. To begin our discussion on magnetic circuits, let us consider a simple problem where a coil with N number of turns (Fig. 8.9(a)) is wound on a toroidal (circular-ring shaped) core. As the magnetic core is circular in shape, the length of the magnetic path is found to be $l = \pi d$, where d is the mean diameter. There is no air gap, nor any leakage. The current in the exciting coil is I. If the magnetic field intensity, H is assumed to be uniform over the length of the magnetic path, then, using Ampere's circuital law, the excitation, $NI = \oint Hidl = H_t l_i$, which gives the magnetic field intensity,

$$H_i = \frac{NL}{l_i}.$$

The unit of magnetic field intensity H is AT/m or A/m, the number turns, N being a dimensionless quantity. As the coil is excited, magnetic flux, ϕ will be formed inside the magnetic core, whose cross-section is assumed to be rectangular. The magnetic flux density, B is taken as a constant over the cross-sectional area, a, and its value is given by $B = \frac{\phi}{a}$. The units of ϕ and B are Wb and Wb/m^2, respectively.

The concept of a magnetic circuit takes advantage of the similarity of the equations of the magnetic field to that of an electrical circuit. Using this concept, the magnetic fields of complex devices such as transformers can be easily solved using the methods and techniques developed for electrical circuits. Hence, we can draw a circuit equivalent to our coil wound around a toroidal core (Fig. 8.9(b)) with NI as the voltage source and $\Re$ the reluctance (will be explained in detail in the next section).

We will now discuss the $B-H$ characteristics of the magnetic material used in the toroidal core.

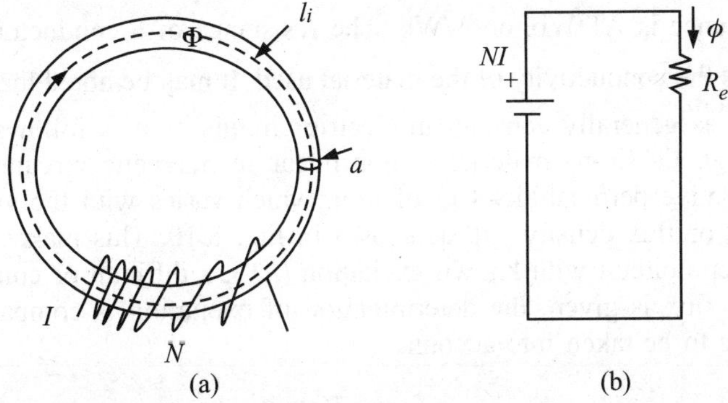

Fig. 8.9

B-H CHARACTERISTICS

As given earlier, the ratio of magnetic flux density to field intensity, is termed as permeability, $\frac{B}{H} = \mu = \mu_0 \mu_r$. The unit of permeability (μ and μ_0) is given by H/m, where H is the unit of inductance (L)–explained in Chapter 9. Of three terms, (μ, μ_0 and μ_r), the first two are permeability of the magnetic material, and free space or air, respectively, whose units are given earlier. The relative permeability, $\mu_r = \frac{\mu}{\mu_0}$, is a dimensionless quantity. Its value is 1.0, for free space or air increases considerably for a magnetic material (fir example iron). It may be noted here that, as the exciting current in the coil increases, the magnetic field intensity, H inside the core increases linearly with the exciting current, I, if the number of turns remains the same. The magnetic flux, ϕ, increases, but not linearly. The magnetic flux density, B also increases, and as given earlier, it is taken as constant over the cross-sectional area. The B–H characteristic of the magnetic material used in the core is non-linear (Fig. 8.10). Hence, the permeability of the magnetic material used varies and is non-linear, as the magnetic field intensity and flux density changes with change in excitation. The slope of the curve changes from point to point.

If we consider the magnetic circuit to be equivalent to an electrical circuit, the term, excitation can also called be magneto-motive force (mmf), as the voltage applied to the electrical circuit consisting of resistances, is called the emf, which causes the currents to flow in the resistances. The currents are found as shown earlier in Chapters 1 and 2, when a dc voltage is applied. Excitation or mmf, i.e., ampere turns (NI), is produced when a current is passed through a coil, wound on a core. As current flows in an electric circuit, when voltage (emf) is applied, in this case, flux is formed in the core, when a current is passed through the coil, or excitation is applied. The resistance of the electric circuit is the ratio of applied voltage (V) to current (I), i.e., $R = \frac{V}{I}$, following Ohm's law. Similarly, the ratio of ampere turns in the coil to the flux developed in the core is termed as reluctance in this case, which is given by,

$$= \frac{NI}{\phi} = \frac{H_i l_i}{Ba} = \frac{l_i}{\mu a} = \frac{l_i}{\mu_0 \mu_r a}.$$

The unit of reluctance is AT/Wb, or A/Wb. The resistance of a conductor is expressed as $R = \frac{\rho l}{a}$, where ρ is the conductivity of the material used. It may be noted that while resistance is generally taken as generally constant in electric circuits as it is follows Ohm's law, the reluctance for magnetic (iron) material is non-linear in magnetic circuits, as it inversely proportional to relative permeability (μ) of iron, which varies with the value of magnetic field intensity (H) or flux density (B), as shown in Fig. 8.10. This makes the computation of flux in a magnetic circuit with known excitation (AT), a difficult or complex task. If the value of magnetic flux is given, the determination of excitation is comparatively easy, but the non-linearity is to be taken into account

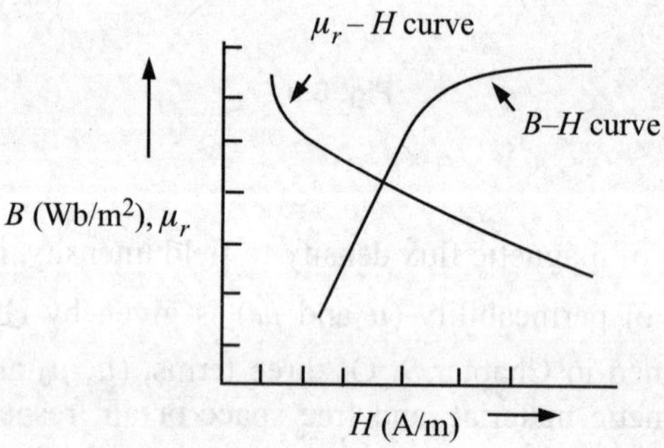

Fig 8.10

Analogy between Electric and Magnetic Circuits

The elements in an electric circuit and there quivalents in a magnetic circuit is shown in Table 8.1.

Table 8.1 *Electric and Magnetic circuits*

Sl. No.	Type	Electric circuit	Magnetic circuit
1	Input	Applied voltage (V)	Excitation (NI)
2	Output	Current (I)	Flux (ϕ)
3	Ratio (output/input)	Resistance (R)	Reluctance ($\Re$)
4	Symbol	Conductance ($1/R$)	Permeance ($1/\Re$)
5	Other symbols	Resistivity (ρ) Conductivity ($1/\rho$)	Inverse of permeability ($1/\mu$) Permeability (μ)

All expressions are given earlier. The length (l) of the conductor is taken as equal to the length (l_i) of the magnetic path in the iron core.

8.3.1 Magnetic Circuits (Series and Series–Parallel Types)

We will consider two types of magnetic circuits in detail – the series circuit and the series – parallel circuit. Let us take the series circuit first. In Fig. 8.9(a), a magnetic material with no air gap was used as the core. Consider now a core with a saw-cut or air gap made in it of an infinitesimally small length (l_a) as shown in Fig. 8.11(a). The magnetic circuit now becomes a series one (Fig. 8.11(b)). Note that the length of the iron part is assumed to remain nearly the same, as the length of the iron part is very large compared to the length of the air gap.

Before proceeding further, *leakage and fringing*, are described briefly. These points have to be considered when flux is formed in a coil wound around a core, with or without an air gap (Fig. 8.11(c)). It may be noted that the coil must be excited by the current passing through it.

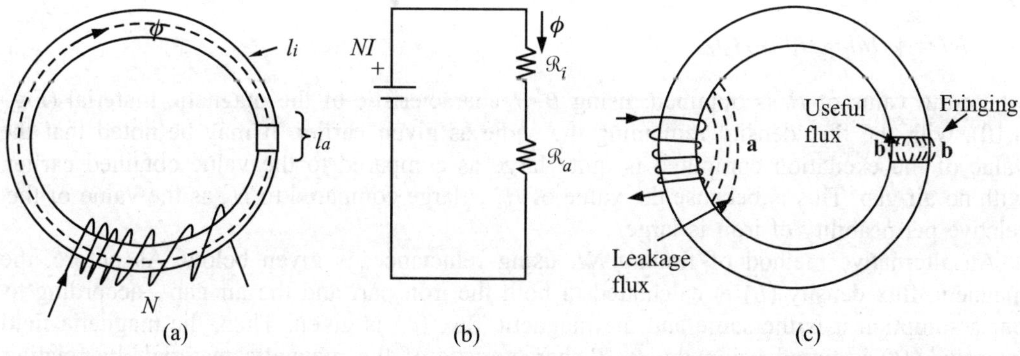

Fig 8.11

Leakage

The *leakage flux* can be taken as the flux that does not follow the desired path in a magnetic circuit. In most practical magnetic circuits, a large part of the flux path is through a magnetic material, as its reluctance is low; the remainder part of the flux path is through air, having high reluctance. The flux in the magnetic material and also air gap, if it is there, is known as useful flux, because itis needed for producing the induced emf in a generator, or torque in a motor, as given in Chapters 9 and 12. The total flux (ϕ_t) produced by the coil wound on a core does not flow through the air gap, as some of it leaks through air (path a) surrounding the iron, as shown in Fig. 8.11(c). These flux lines are termed as leakage flux. The total flux is the sum of useful flux and leakage flux. The leakage coefficient is defined as the total flux (ϕ_i) divided by useful flux (ϕ_g); its value for electricalmachines is between 1.15–1.25. This is not desirable, as it increases both the weight and cost of the machine. If the leakage flux is assumed to be small, the leakage coefficient is 1.0.

FRINGING

When crossing an air gap, magnetic lines of force tend to bulge out, as lines of force at **bb,** in Fig. 8.11(c). This is because lines of force repel each other as it passes through non-magnetic material, such as air. This phenomenon is called fringing. The bulging or fringing results in increasing the effective area of the air gap, and thus decreases the flux density in the air gap. If the air gap is longer, the fringing effect is larger, and vice versa. Hence, if the air gap is shorter, with large cross-sectional area, the fringing may be neglected.

If leakage and fringing is neglected, the magnetic flux and flux density (*B*) is the same in the iron part and the air gap, as the cross-sectional area remains the same in both cases. The magnetic field intensity (*H*) in the air gap is found as,

$$H_a = \frac{B}{\mu_0},$$

the relative permeability of air being 1. Using Ampere's circuital law, the excitation is given by,

$$NI = \oint H dl = H_i l_i + H_a l_a,$$

where the value of H_i is obtained, using *B–H* characteristic of the magnetic material (Fig. 8.10), with the flux density remaining the same as given earlier. It may be noted that the value of the excitation computed is quite large as compared to the value obtained earlier, with no air gap. This is because the value of H_a is large compared to H_i, as the value of the relative permeability of iron is large.

An alternative method of finding *NI*, using reluctance, is given below. As before, the magnetic flux density (*B*) is calculated in both the iron part and the air gap—according to our assumption it is the same and the magnetic flux (ϕ) is given. Then, the magnetic field intensity (H_i) is found using the *B–H* characteristic of the magnetic material. Its relative permeability is determined using the formula, $\mu_r = \dfrac{B_i}{\mu_0 H_i}$. Now, the reluctances of the iron part and the air gap are calculated as

$$\Re_i = \frac{l_i}{\mu_0 \mu_r a}, \quad \text{and} \quad \Re_a = \frac{l_a}{\mu_0 a}.$$

As they are in series as shown in Fig. 8.11(b), total reluctance is the sum of the two reluctances, given by,

$$\Re = \Re_i + \Re_a.$$

Now, the excitation (AT) is found as, $NI = \Re \times \phi$. A numerical example is included at the end of this section.

The second case is that of a series–parallel circuit. A core with a central limb and two outer limbs is shown in Fig. 8.12(a). Magnetic material is used in the core. The width of the central limb is twice that of the outer limbs. The depth of the limbs is the same. The magnetic path is also shown. The cross-sectional area of the outer limbs is half that of the central limb. As the two outer limbs have the same length, the flux in these limbs is also half that of the central limb, as the reluctances of two outer limbs are the same (will be shown later). The magnetic flux density (B) is the same in both the central and two outer limbs, as both the flux and cross-sectional area of the outer limbs are half of those in the central limb. The magnetic field intensity (H) is obtained using the B–H characteristic (Fig. 8.10). The magnetic path is shown in Fig. 8.12(a). Using Ampere's circuital law, the excitation is given by $\oint(Hdl = H.(l_c) + l_o)$, as H is the same in both central and outer limbs. Only the mmf in the central limb is added to the mmf in any of the two outer limbs. It can be observed that mmfs in the two outer limbs—left and right—are the same.

An alternative procedure for finding exitation, using reluctance, is given below. As given earlier, the magnetic flux density (B) is calculated in both the central and two outer limbs, which are the same as the magnetic flux (ϕ) in the central limb is given. Then, the magnetic field intensity (H) is found using the B–H characteristic of the magnetic material. The relative permeability is determined using the formula, $\mu_r = \dfrac{B}{\mu_0 H}$, which is given earlier. The magnetic circuit is shown in Fig. 8.12(b). Now, the reluctances of the central limb and any of two outer limbs are calculated as,

$$\Re_c = \frac{l_c}{\mu_o \mu_r a}, \quad \text{and} \quad \Re_0 - \frac{l_o}{\mu_o \mu \mu_r \frac{a}{2}} - \frac{2l_o}{\mu_o \mu_r a}.$$

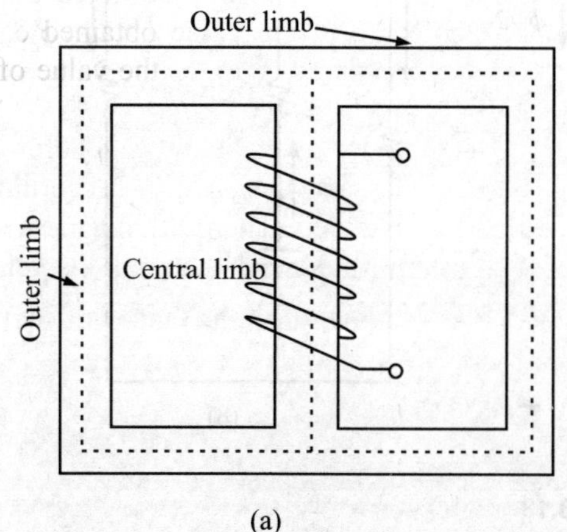

(a)

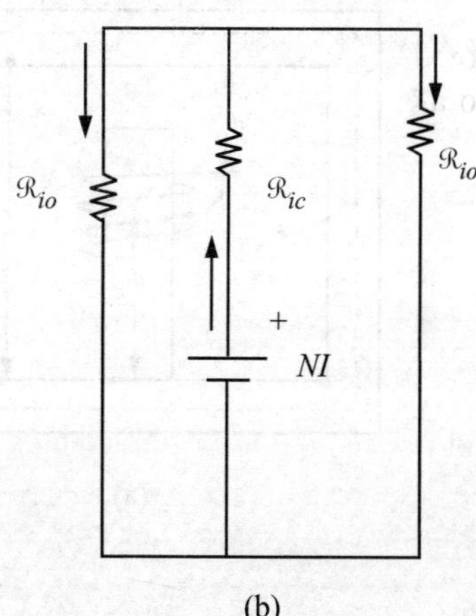

(b)

Fig. 8.12

The reluctances of the two outer limbs are the same, and as given earlier, the flux in any of the outer limbs is half of that in the central limb. The excitation (AT) is found as,

$$= \Re_c \times \phi + \Re_0 \times \frac{\phi}{e} = \frac{l_c + l_o}{\mu_o \mu_{ra}} \times \phi_c/2\,.$$

Please note that the loop is taken to include the central and one of the outer limbs. The total reluctance $= \Re = \dfrac{(l_c + l_o)}{\mu_o \mu_{ra}}$.

We can obtain another series–parallel circuit by introducing a saw cut or air gap of length (l_a) in the central limb, without affecting the two outer limbs (Fig. 8.13(a)). As given earlier, the length of the magnetic path in the central limb remains the same as $l_{ic} >> l_a$. The magnetic circuit which is nearly the same as given earlier, except for the addition of reluctance of air gap, is in series with the reluctance of the central limb, as shown in Fig. 8.13(b). As leakage, and fringing is neglected, the flux and flux density in the air gap are the same as those in the central limb. The magnetic field intensity in the air gap is found as, $H_a = \dfrac{B}{\mu_0}$. The other conditions remain the same, i.e., the magnetic flux density and field intensity are same in central limb, and the two outer limbs. Using ampere's circuital law, the excitation is obtained as

$$I = \oint H dl = H_i(l_{ic} + l_{io}) + H_a l_a.$$

Note that the excitation required to obtain the same flux in the central limb is much higher than that in the previous case. The other method, using reluctance, is not given here, but left to the reader as an exercise.

Energy stored in a magnetic field and also the lifting power in an electro-magnet is explained in Section 9.3 of Chapter 9.

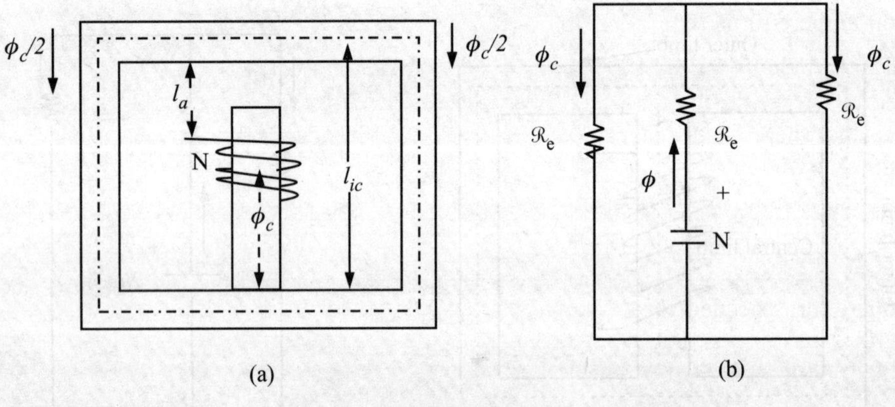

Fig 8.13

ADDITIONAL EXAMPLES

EXAMPLE 8.7 *An iron ring has a cross-sectional area of 5 cm² (Fig. 8.11(a)). An air gap 0.5 mm wide has been introduced by a cut across its section. The effective iron path is 80 cm. The ring is wound with a coil of 400 turns through which a current of 3 A is passed. If the total flux in the air gap is 0.4 mWb, find the relative permeability of iron. Neglect fringing and leakage effects.*

SOLUTION The iron ring with an air gap is shown in Fig. 8.11(a).

The flux is, $\phi = 0.4$ mWb $= 0.4 \times 10^{-3}$ Wb.

Cross-sectional area, $a = 5$ cm$^2 = 5 \times 10^{-4}$ m^2.

Magnetic flux density, $B = \frac{\phi}{a} = \frac{0.4 \times 10^{-3}}{5.0 \times 10^{-4}} = 0.8$ Wb/m^2,

as there is no fringing in the air gap, which means that the cross-sectional area remains the same. This value remains the same in the iron core and the also air gap.

Length of the magnetic path, $l_i = 80$ cm $= 0.8$ m.

Length of the magnetic path, $l_a = 0.5$ mm $= 0.5 \times 10^{-3}$ m.

Permeability of air $= \mu_0 = 4\pi \times 10^{-7}$ H/m.

Relative permeability of air $= \mu_{ra} = 1$.

No. of turns, $N = 400$.

Current in the coil $= I = 3$ A.

Total ampere turns required to produce the given flux is

$$= NI = (H_i \cdot l_i + H_a \cdot l_a) = 400 \times 3 = 1{,}200 \text{ A},$$

Note that there is no leakage, which means that the total ampere turns produce flux in both the iron and air gap.

Magnetic field strength in air, $H_a = \frac{B}{\mu_0 \mu_{ra}} = \frac{0.8}{4\pi \times 10^{-7} \times 1.0} = 636.6 \times 10^3$ A/m.

Ampere turns needed for air gap $= H_a \cdot l_a = 636.6 \times 10^3 \times 0.5 \times 10^3 = 318.3$ A.

Ampere turns needed for iron part

$$= H_i \cdot l_i = N \cdot I - H_a \cdot l_a = 1{,}200 - 381.3 = 818.7 \text{ A}.$$

Magnetic field strength in the iron ring, $H_i = \frac{B}{\mu_0 \mu_{ri}} = \frac{H_a}{\mu_{ri}} = \frac{818.7}{0.8} = 1{,}023.4$ A/m.

Relative permeability of iron $= \quad \mu_{ri} = \frac{H_a}{H_i} = \frac{636.2 \times 10^3}{1{,}023.4} = 622.$

EXAMPLE 8.8 *A cast steel ring has a mean diameter of 20 cm, and a cross-sectional area of 8 cm^2(Fig. 8.11(a)). The coil is wound with 300 turns. Determine the current required to produce a flux of 0.6 mWb (a) with no air gap, and (b) with an air gap of 1 mm. Neglect fringing and leakage effects. The B–H curve of the magnetic material used is given in the table.*

Flux density (Wb/m²)	0.4	0.6	0.8	1.0
Field intensity (A/m)	300	400	600	850

SOLUTION The cast steel ring with an air gap is shown in Fig. 8.11(a).

No. of turns, $N = 300$

The flux is $\phi = 0.6 \text{ mWb} = 0.6 \times 10^{-3}$ Wb.

Cross-sectional area, $a = 8 \text{ cm}^2 = 8.0 \times 10^{-4} \text{ m}^2$.

Magnetic flux density, $B_i = \dfrac{\phi}{a} = \dfrac{0.6 \times 10^{-3}}{8.0 \times 10^{-4}} = 0.75 \text{ Wb/m}^2$

Mean diameter of the cast steel ring $= d = 20$ cm $= 0.2$ m.

Mean circumference of the cast steel ring = length of the magnetic path

$= l_i = \pi . d = \pi \times 0.2 = 0.6283$ m.

The *B–H* curve for the magnetic material is plotted in Fig. 8.14. From this graph, the field intensity, H_i for $B = 0.75 \text{ Wb/m}^2$ is found as 550.0 A/m.

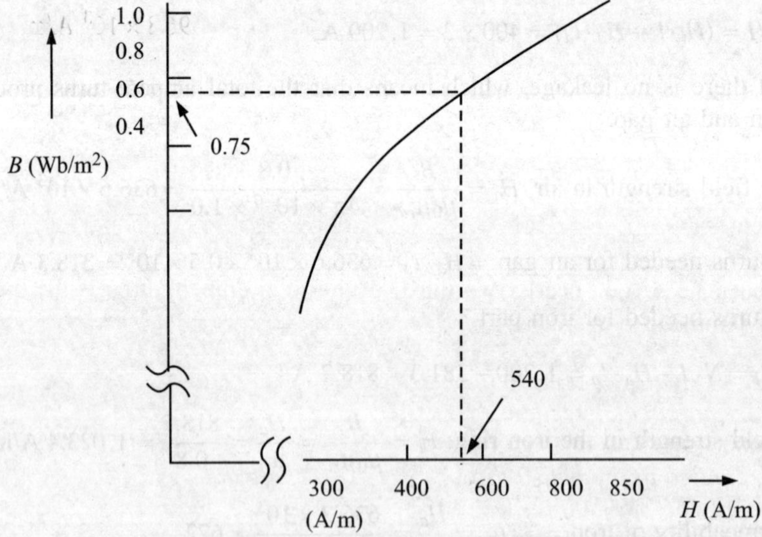

Fig 8.14

We can also find the field intensity using a part of the *B–H* curve. As the value of *B* lies between two points of the curve, 0.6 and 0.8 Wb/m², a straight line can be drawn through the above points, whose slope is

$$\frac{600 - 400}{0.8 - 0.6} = \frac{200.0}{0.2} = 1.000.$$

The value of the field intensity in the magnetic material at $B = 0.75$ Wb/m^2, is obtained as

$$H_i = 400.0 + [1,000.0 \times (0.75 - 0.6)] = 550.0 \text{ A/m}$$

(a) Ampere turns required (A) to produce the given flux is,

$$= NI = H_i \cdot l_i = 550.0 \times 0.6283 = 345.6.$$

Exciting current in the winding $= I = \dfrac{345.6}{300.0} - 1.152 \approx 1.15$ A.

Note that leakage is neglected, and there is no air gap, in this case.

Relative permeability of the magnetic material $= \mu_{ri} = \dfrac{B_i}{\mu_0 H_i} = \dfrac{0.75}{4\pi \times 10^{-7} \times 550.0} = 1,085.$

(b) Length of the air gap, $l_a = 1$ mm $= 1.0 \times 10^{-3}$ m.

Relative permeability of air $= \mu_{ra} = 1$.

As there is no fringing in the air gap, the cross-sectional area remains the same in both the core and the air gap. The values of flux and flux density remain the same.

Magnetic field strength in air, $H_a = \dfrac{B_a}{\mu_0 \mu_{ra}} = \dfrac{0.75}{4\pi \times 10^{-7} \times 1.0} = 596.8 \times 10^3$ A/m.

Total ampere- turns required to produce the given flux is

$$= NI = (H_i \cdot l_i + H_a \cdot l_a) = 345.6 + (596.8 \times 10^3 \times 1.0 \times 10^{-3}) \approx 942.0.$$

Exciting current $= I = \dfrac{942.0}{300} = 3.14$ A.

Please note that 63.5% (i.e., nearly two-third) of the total ampere turns is the mmf drop in the air gap of 1 mm.

EXAMPLE 8.9 *A ring, with a mean diameter of 21 cm, and cross-sectional area of 10 cm^2, is made of two semi-circular sections of cast iron and cast steel respectively, with each joint having reluctance equal to an air gap of 0.2 mm, as shown in Fig. 8.15. Determine the ampere turns required to produce a flux of 0.8 mWb. The relative permeabilities of cast iron and cast steel are 166 and 800 respectively. Neglect fringing and leakage effects.*

SOLUTION The details of the ring with two different magnetic materials—cast iron and cast steel and also two air gapsis shown in Fig. 8.15.

Magnetic flux, $\phi = 0.8$ mWb $= 0.8 \times 10^{-3}$ Wb.

Cross-sectional area, $a = 10$ cm$^2 = 10 \times 10^{-4}$ m^2.

Magnetic flux density, $B = \dfrac{\phi}{a} = \dfrac{0.8 \times 10^{-3}}{10.0 \times 10^{-4}} = 0.8$ Wb/m^2,

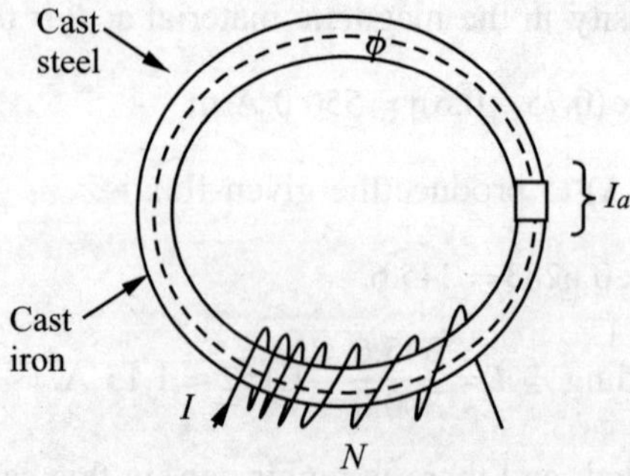

Fig 8.15

As there is no fringing in the air gap, that the cross-sectional area remains both the same. This value remains the same in materials and also the two air gaps.

Mean diameter of the ring = 21 cm = 0.21 m

Relative permeability of cast iron $= \mu_{r1} = 166.$
Relative permeability of cast steel $= \mu_{r2} = 800.$

Relative permeability of air $= \mu_{r3} = 1.$

Magnetic field strength in cast iron, $H_1 = \dfrac{B}{\mu_0 \mu_{r1}} = \dfrac{0.8}{4\pi \times 10^{-7} \times 166} = 3,835.1$ A/m.

Magnetic field strength in cast steel, $H_1 = \dfrac{B}{\mu_0 \mu_{r1}} = \dfrac{0.8}{4\pi \times 10^{-7} \times 800} = 795.8$ A/m.

Magnetic field strength in air, $H_3 = H_a = \dfrac{B}{\mu_0 \mu_{r3}} = \dfrac{0.8}{4\pi \times 10^{-7} \times 1.0} = 636.62 \times 10^3$ A/m.

Total circumference of the ring at the mean length of magnetic path $l = \pi \times 0.21 \approx 0.66$ m.

Length of magnetic path in cast iron, $l_1 = \dfrac{l'}{2} = \dfrac{0.66}{2} = 0.33$ m.

Length of magnetic path in cast steel, $l_2 = \dfrac{l'}{2} = \dfrac{0.66}{2} = 0.33$ m,

which is the same as that of cast iron. Length of magnetic path in two air gaps together,

$$l_3 = l_a = 2 \times 0.2 \times 10^{-3} = 0.4 \times 10^{-3} \text{ m} = 0.4 \text{ mm}.$$

Total ampere turns required to produce the given flux, is,

$$= NI = \sum_{i=1}^{3} (H_i \cdot l_i) = (H_1 \cdot l_1 + H_2 \cdot l_2 + H_3 \cdot l_3)$$

$$= \left[(3,835.1 \times 0.33) + (795.8 \times 0.33) + (0.63662 \times 10^6 \times 0.4 \times 10^{-3})\right]$$

$$= (1,265.6 + 262.6 + 254.6) = 1,782.8 \approx 1,783 \text{ A},$$

as there in no leakage, the total ampere turns produce flux in three parts of the magnetic circuit.

We can also find ampere turns by computing the reluctances of the three parts, i.e., cast iron, cast steel and air gap. The magnetic circuit for this case is shown in Fig. 8.13(b). The unit for reluctance is A/Wb, or H^{-1}.
Reluctance of the cast iron part

$$= \mathfrak{R}_1 = \frac{l_1}{\mu_0 \mu_{r1} a} = \frac{0.33}{4\pi \times 10^{-7} \times 166 \times 10 \times 10^{-4}} = 1.582 \times 10^6 \text{ A/Wb.}$$

Reluctance of the cast steel part

$$= \mathfrak{R}_2 = \frac{l_2}{\mu_0 \mu_{r2} a} = \frac{0.33}{4\pi \times 10^{-7} \times 800 \times 10 \times 10^{-4}} = 0.3283 \times 10^6 \text{ A/Wb}$$

Reluctance of the air gap

$$= Re_3 = Re_a = \frac{l_3}{\mu_0 \mu_{r3} a} = \frac{0.4 \times 10^{-3}}{4\pi \times 10^{-7} \times 1 \times 10 \times 10^{-4}} = 0.3183 \times 10^6 \text{ A/Wb.}$$

These reluctances are connected in series (Fig. 8.11(b)).

$$\text{Total reluctance} = \mathfrak{R} = \mathfrak{R}_1 + \mathfrak{R}_2 + \mathfrak{R}_3 = (1.582 + 0.3283 + 0.3183) \times 10^6$$

$$= 2.2286 \times 10^6 \text{ A/Wb.}$$

$$\text{Ampere turns required} = \phi \cdot \mathfrak{R} = 0.8 \times 10^{-3} \times 2.2286 \times 10^6 \approx 1,783 \text{ A}$$

This value is same as obtained earlier.

EXAMPLE 8.10 *There is a central limb and two outer limbs in the iron core, as shown in Fig. 8.16(a). The winding with 500 turns is placed on the central limb. The mean lengths of the magnetic path in the central limb and any of the two outer limbs are 15 cm and 40 cm respectively. A saw cut or air gap of 1 mm is made on the central limb. The cross-sectional area of the central limb and outer limbs are 10 cm², and 5 cm² respectively. Determine the current required to produce a flux of 1.0 mWb in the central limb. The relative permeability of iron is assumed to be 1,200. Neglect fringing and leakage effects.*

SOLUTION The iron core with a central limb and two outer limbs, and also an air gap, in the central one, is shown in Fig. 8.16.

No. of turns in the winding placed on the central limb $= N = 500$.

The mean lengths of the magnetic path in the central limb $= l_{ic} = 15$ cm $= 0.15$ m.

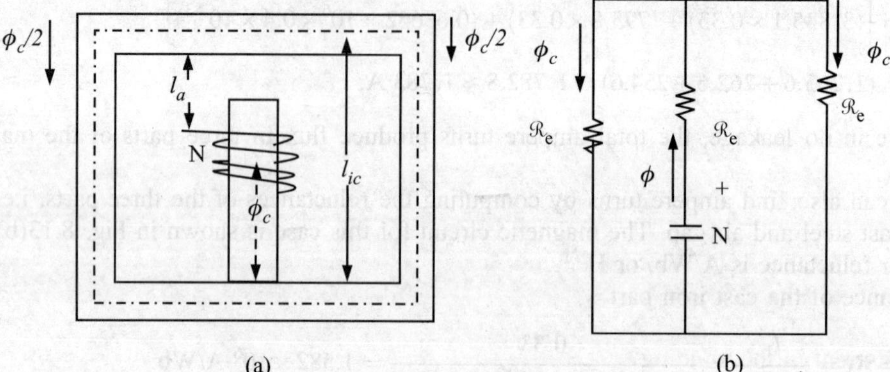

Fig 8.16

The mean lengths of the magnetic path in any of the outer limbs $= l_o = 40$ cm $= 0.4$ m.

The length of the air gap in the central limb $= l_{ac} = 1$ mm $= 0.001$ m.

Note that the mean length of the magnetic path in the central limb remains nearly the same.

The cross-sectional area of the central limb $= a_c = 10\ \text{cm}^2 = 10.0 \times 10^{-4}\ \text{m}^2$.

The cross-sectional area of any of outer limbs $= a_0 = 5\ \text{cm}^2 = 5.0 \times 10^{-4}\ \text{m}^2$.

The flux in the central limb $= \phi_c = 1$ mWb $= 1 \times 10^{-3}$ Wb.

The magnetic flux density in the central limb $= B_{ic} = \dfrac{\phi_c}{a_c} = \dfrac{1 \times 10^{-3}}{10 \times 10^{-4}} = 1.0\ \text{Wb/m}^2$.

The magnetic field intensity in the central limb

$$= H_{ic} = \frac{B_c}{\mu_0 \mu_r} = \frac{1.0}{4\pi \times 10^{-7} \times 1,200} = 663.0\ \text{AT/m or A/m.}$$

As fringing is neglected, the flux in the air gap, and also the flux density remain the same as those in the central limb, with the cross-sectional area being the same.

The magnetic field intensity in the air gap

$$H_{ac} = \frac{B_c}{\mu_0} = \frac{1.0}{4\pi \times 10^{-7}} = 795.8 \times 10^2\ \text{A/m.}$$

As the length of the magnetic path and cross-sectional area ofthe two outer limbs are the same, the reluctances of two paths in the outer limbs are the same, as will be found later. The flux in the outer limbs is half the flux in the central limb. The cross-sectional area of the outer limbs is half of the cross-sectional area of the central limb. This means that both the magnetic flux density and the field intensity are the same as those in the central limb.

As leakage is also neglected, the ampere turns required is

$$NI = \oint H.dl = H_{ic}.l_{ic} + H_{ac}.l_{ac} + H_{io}.l_{io}$$

$$= (663.0 \times 0.15) + (795.8 \times 10^3 \times 0.001) + (663.0 \times 0.4) = 99.5 + 795.8 + 265.2$$

$$= 1.160.5 \text{ AT or A.}$$

Please note that only the central limb including the air gap and one of the outer limbs are used.

The current in the winding $= I = \dfrac{1.1605}{500} \approx 2.32$ A

We can also find the current using the reluctances in the three parts. The magnetic circuit for this case is shown in Fig. 8.16(b).

Reluctance of the central limb (iron part)

$$= \Re_{lc} = \frac{l_{ic}}{\mu_0 \mu_{r1} a_c} = \frac{0.15}{4\pi \times 10^{-7} \times 1,200 \times 10 \times 10^{-4}} = 99.5 \times 10^3 \text{ A/Wb}$$

Reluctance of the central limb (air gap)

$$= \Re_{ac} = \frac{l_1}{\mu_0 a_{ac}} = \frac{0.001}{4\pi \times 10^{-7} \times 10 \times 10^{-4}} = 795.8 \times 10^3 \text{ A/Wb.}$$

Reluctance of the one of the outer limb (iron part)

$$= \Re_{io} = \frac{l_{io}}{\mu_0 \mu_{ri} a_o} = \frac{0.4}{4\pi \times 10^{-7} \times 1,200 \times 5 \times 10^{-4}} = 530.5 \times 10^3 \text{ A/Wb.}$$

The unit of reluctance is also H^{-1}.

Flux in the central limb $= \phi_c = 1 \times 10^{-3}$ Wb

Flux in one of outer limbs $= \dfrac{\phi_0 = \phi_c}{2} = \dfrac{1}{2} \times 10^{-2} = 0.5 \times 10^{-2}$ Wb

Ampere turns (mmf) required $= NI = [\Re_{ic} + \Re_{ac}].\phi_c + \Re_{i0}.\phi_0$

$$= (99.5 + 795.8) \times 10^3 \times 1 \times 10^{-3} + 530.5 \times 10^3 \times 0.5 \times 10^{-3}$$

$$= 895.3 + 265.3 \approx 1,160.5 \text{ AT or A}$$

The mmf required in the coil is the drop in the reluctance in the central limb (both iron part and air gap) and the reluctance in one of the outer limbs. Please note that the flux in the outer limbs is half of that in the central limb. Moreover, the mmf in the two outer limbs is the same, as if the reluctances are in parallel. This problem can also be simplified using the procedure followed in dc circuits.

The current in the winding = $I = \dfrac{1,160.5}{500} \approx 2.32$ A which is the same as obtained earlier.

EXAMPLE 8.11 *An iron ring of 120 cm mean circumference is made from a round core of cross-section 16 cm²(Fig. 8.11(a)). A saw cut of 1 mm width has been made on it. The coil is wound with 400 turns. Determine the current required to produce a flux of 1.6 mWb. The relative permeability of iron is taken as 600. Neglect fringing and leakage effects.*

SOLUTION The iron ring with an air gap is shown in Fig. 8.11(a).

The flux is, $\phi = 1.6$ mWb $= 1.6 \times 10^{-3}$ Wb. Cross-sectional area, a $= 16$ cm$^2 = 16 \times 10^{-4}$ m^2.

$$\text{Magnetic flux density, } B = \frac{\phi}{a} = \frac{1.6 \times 10^{-3}}{16.0 \times 10^{-4}} = 1.0 \text{ Wb/m}^2.$$

As there is no fringing in the air gap, the cross-sectional area remains the same. This value remains the same in the iron core and also the air gap.

Mean circumference of the iron ring = length of the magnetic path (iron), $l_i = 120$ cm $= 1.2$ m.

Length of the magnetic path (air), $l_a = 1$ mm $= 1.0 \times 10^{-3}$ m.

Relative permeability of iron $= \mu_{ri} = 600$. Relative permeability of air $= \mu_{ra} = 1$.

No. of turns, $N = 400$.

$$\text{Magnetic field strength in the iron ring, } H_i = \frac{B}{\mu_0 \mu_{ri}} = \frac{1.0}{4\pi \times 10^{-7} \times 600} = 1,326.3 \text{ A/m}.$$

$$\text{Magnetic field strength in air, } H_a = \frac{B}{\mu_0 \mu_{ra}} = \frac{1.0}{4\pi \times 10^{-7} \times 1.0} = 795.8 \times 10^3 \text{ A/m}.$$

Total ampere turns required to produce the given flux is,

$$NI = (H_i \cdot l_i + H_a \cdot l_a) = (1,326.3 \times 1.2) + (795.8 \times 10^3 \times 1.0 \times 10^{-3})$$

$$= (1,591.6 + 795.8) = 2,387.4 \text{ A}.$$

As there in no leakage, the total ampere turns produce flux in both the iron and air gap.

$$\text{Exciting current} = I = \frac{2,387.4}{400} \approx 5.97 \text{ A}$$

This problem can be solved by computing the total reluctances of both magnetic path and the air gap, as as done in the previous problem.

Conclusion

In this chapter, we were introduced to two concepts—electromagnetism and magnetic circuits. Ampere's law and Biot-Savart law were used to find the magnetic field intensity, magnetic flux, and also magnetic flux density for the magnetic field produced by the current in the conductor(s)/coil. We then studied magnetic circuits. We examined different types of cores of different shapes and sizes, made mainly of magnetic material with or without air gaps. The magnetic field intensity and also flux density were calculated for all these cases. In the next chapter, we will study Faraday's laws of electro-magnetic induction, where emf is induced in the coils due to change in flux caused by various means, and the self- and mutual inductance between coils.

Multiple Choice Type Questions

Tick the correct option(s), out of the four choices provided in each question.

8.1 Magnetic flux passes more readily through

(a) air (b) wood (c) vacuum (d) iron

8.2 The absolute permeability of a magnetic material having a flux density of 1.0 Wb/m^2 is 10^{-3} H/m. The magnetic field intensity in AT/m is

(a) 1,000 (b) 10^{-3} (c) $4\pi \times 10^{-3}$ (d) $4\pi \times 10^{3}$

8.3 A horizontal overhead power line carries a current of 100 A. Find the magnitude of the magnetic field density in μWb/m^2 due to the above current, 2.0 m below the wire.

(a) 5.0 (b) 10.0 (c) 20.0 (d) 0.2

8.4 Two long straight conductors carry a current of 500 A in the same direction. They are placed 2.5 m apart. Determine the magnitude of the force between the conductors in N per m.

(a) 2.0 (b) 0.04 (c) 0.02 (d) 0.5

8.5 The mmf in a magnetic circuit, corresponds to .... in an electric circuit.

(a) voltage drop (b) potential difference
(c) electric field intensity (d) emf

8.6 The magnitude of AT required for an air gap is much greater than that required for the iron part of a magnetic circuit, because.

(a) air is non-magnetic (b) air is a conductor of magnetic flux
(c) air has the lowest relative permeability (d) none of (a) – (c)

8.7 The magnetic field intensity of 600 AT/m will produce a flux density in mT (T = Wb/m^2) in air.

(a) 10^{-3} (b) 0.75 (c) 0.01 (d) 0.5

8.8 A flux density of 1.2 Wb/m^2 is to be produced in an air gap of 2 mm. Calculate the mmf required (AT).

(a) 3,820 (b) 478 (c) 1,910 (d) 955

8.9 A toroidal coil has a magnetic length path of 30 cm and a field intensity of 600 A/m. The current in the coil is 0.25 A. Determine the number of turns of the coil.

(a) 720 (b) 1,440 (c) 2,880 (d) 360

8.10 An iron magnetic circuit has a uniform cross-sectional area of 40 cm^2, and a length of 50 cm. A coil of 150 turns is wound uniformly over the core. When the current in the coil is 2 A, the total flux is 3.2 mWb. Find the relative permeability of iron.

(a) 530 (b) 265 (c) 1,060 (d) 2,120

Answers

8.1. (d)	8.2. (a)	8.3. (b)	8.4. (c)	8.5. (d)
8.6. (a),(c)	8.7. (b)	8.8. (c)	8.9. (a)	8.10. (c)

Problems

8.1 A horizontal power line carries a current of 100 A from east to west. Find the magnetic flux density at a point 2 m below the line (Fig. P8.1).

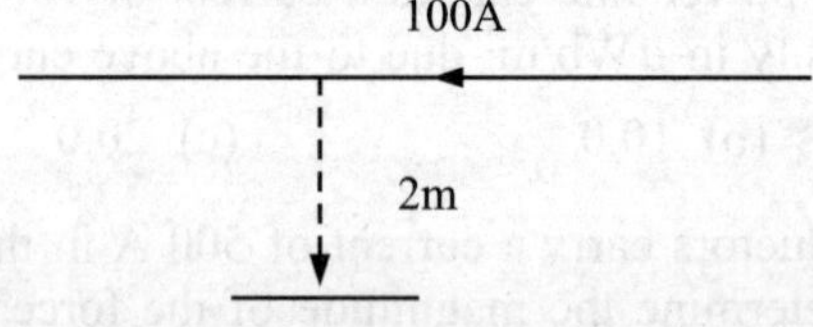

Fig. P8.1

8.2 Calculate the force developed per metre length between two current-carrying conductors, placed 20 cm apart (Fig. P8.2), and carrying currents of 800 A and 1,200 A, respectively.

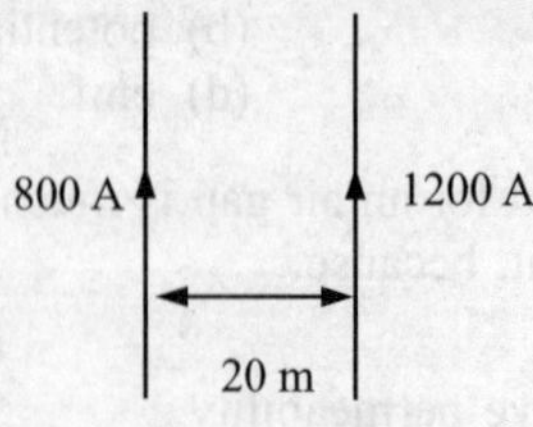

Fig. P8.2

8.3 A solenoid of 300 turns is wound on a continuous ring of iron, the mean diameter of which is 8 cm (Fig. P8.3). The relative permeability of the magnetic material is 1,000. What current is required in the coil such that the flux density in the iron is 1.0 Wb/m^2?

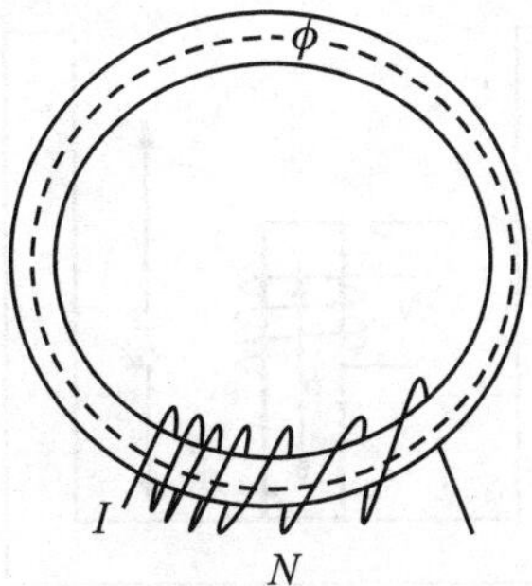

Fig. P8.3

8.4 A straight conductor is carrying 1,500 A, and a point, P is chosen such that the angles formed at the two ends of the conductor are $\phi_1 = 45°$ and $\phi_2 = 60°$. Determine the magnetic field intensity at P, if its perpendicular distance from the conductor is 30 cm.

8.5 A magnetic core, in the form of a closed ring, has mean length of 30 cm, and cross-sectional area of 10 cm^2 (Fig. P8.5). If the flux in the core is 0.8 mWb, find the current in the exciting coil, with the number of turns as 1,000. The relative permeability of the core material is assumed to be 800.

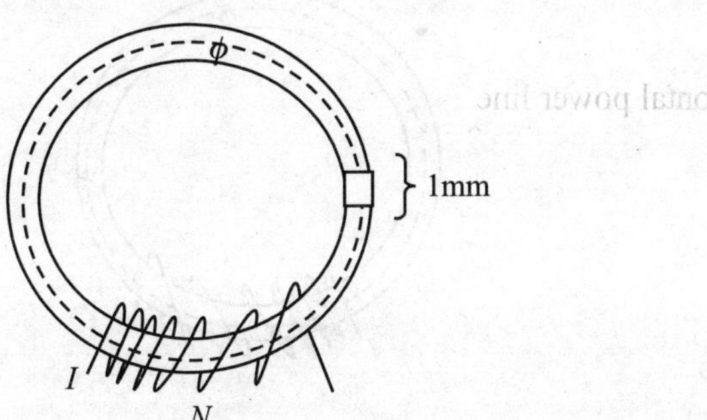

Fig. P8.5

A saw cut of 1 mm wide has been made on it. Determine the new exciting current required to produce the same flux. Neglect fringing and leakage effects.

8.6 An E-shaped core, with a central limb and two outer limbs, is shown in Fig. P8.6. The widths of the central limb and two outer limbs are 4 cm and 2 cm respectively. The thickness is 2 cm—same for all three limbs.

Mean length from A to B through the central limb = 20 cm

Mean length from A to B through any of the outer limbs = 40 cm

There is an air gap of 1 mm only in the central limb. It is wound with 500 turns. Using the following data, find the exciting current required to produce a flux of 0.8 mWb in the central limb.

B (Wb/m^2)	0.9	1.1	1.3
H (A/m)	270	450	600

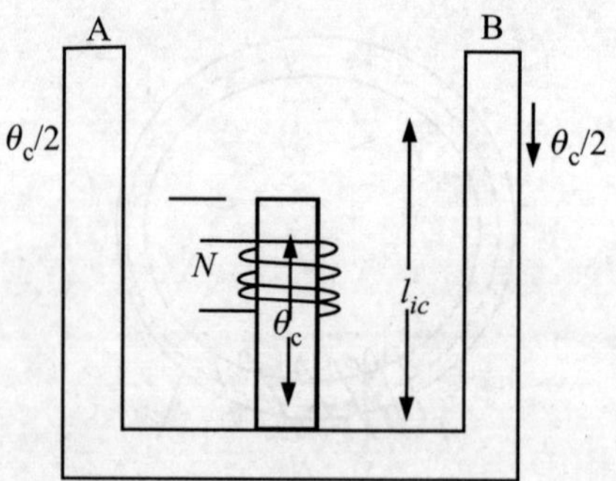

Fig. P8.6

8.7 An iron ring has a square cross-sectional area of 16 cm^2, and a mean circumference of 80 cm. The ring has a winding of 200 turns (Fig. P8.7). Using the following data, calculate the exciting current for a flux of 2 mWb. Also, find the relative permeability. Determine the self-inductance of the coil.

B (Wb/m^2)	0.6	0.8	1.0	1.2	1.4
H (A/m)	75	125	250	500	900

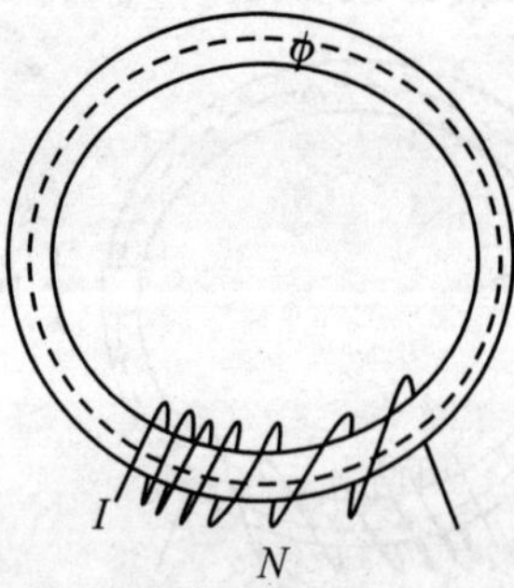

Fig. P8.7

8.8 A circular segment of radius 10 cm subtends an arc of 60° at its centre (Fig. P8.8). A current of 15 A is flowing through it. Find the magnitude and direction of the magnetic field intensity produced at the centre.

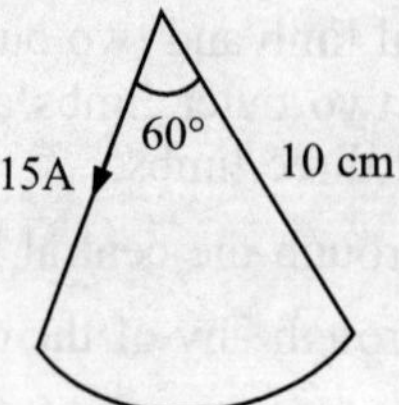

Fig. P8.8

8.9 A current of 5 A is flowing in a wire having the form of an equilateral triangle with sides of 1.5 m (Fig. P8.9). Find the magnetic field intensity produced at the centroid (centre) of the triangle.

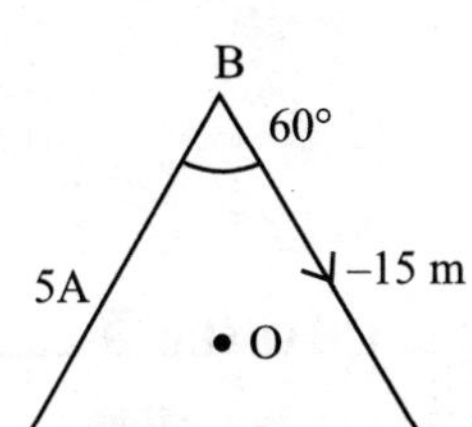

Fig. P8.9

8.10 A rectangular loop, ABCD carrying a current of 15 A in clockwise direction is placed with its longer side parallel to a straight conductor 2 cm apart, carrying a current of 10 A (Fig. P8.10). The sides of the rectangular loop are 12 cm and 8 cm. What is the net force on the loop? Find the magnitude and direction of the net force, if the direction of the current in the loop is reversed.

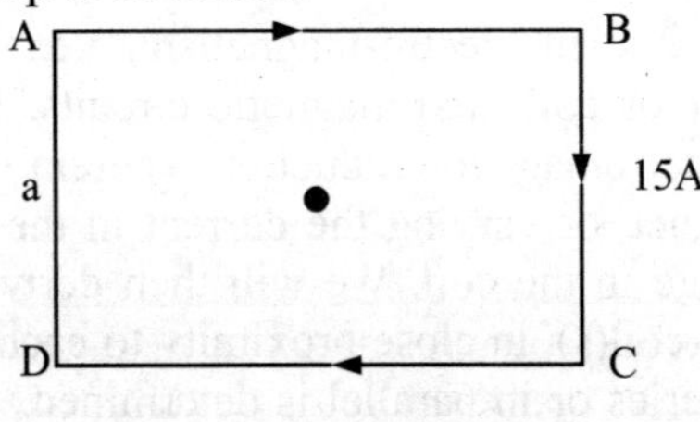

Fig. P8.10

8.11 An iron core in the form of a ring has a closed magnetic path of length 30 cm. The first half of this path has a cross-sectional area of 8 cm^2, while the second half has a cross-sectional area of 4 cm^2 (Fig. P8.11). The core has a winding of 200 turns. Using the following data, calculate the exciting current to produce a flux density of 0.7 Wb/m^2 in the second half. Also, find the relative permeability in both cases.

B (Wb/m^2)	0.2	0.4	0.6	0.8	1.0
H (A/m)	175	300	400	600	850

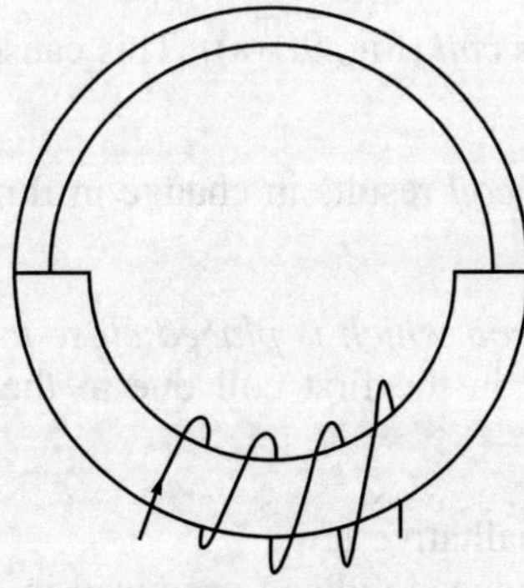

Fig. P8.11

CHAPTER 9

INDUCTANCE

Introduction

In the previous chapter, we studied electro-magnetism, i.e., the magnetic field produced by the current in a conductor or coil, and magnetic circuits. In this chapter, we will first examine Faraday's laws of electromagnetic induction wherein it is shown that the change of flux, either by moving a magnet, or varying the current in the coil, or another coil in close proximity, can induce a voltage in the coil. We will then derive the expressions for the self and mutual inductances of the coil(s), in close proximity to each other. Equivalent inductance of inductances connected in series or in parallel is determined. We will also study the stored energy in the inductance.

9.1 Faraday's Laws of Electro-magnetic Induction

The laws of electro-magnetic induction relate to the voltage produced in a coil when the magnetic field around it changes.

The *first law* states that the voltage is induced in a coil due to the change in the magnetic field around it. Three cases are considered here.

(a) *A magnet is moved near the coil* (Fig. 9.1(a)). This causes change in flux, thus resulting in induced emf in it.

(b) A *change of current in the coil* results in change in flux (Fig. 9.1(b)), again producing an induced emf in it.

(c) If *the current in a second coil which is placed close to the first coil, is changed (Fig. 9.1(c))*, voltage is induced in the first coil due to the change of flux around it (first coil).

The first law is termed as the qualitative law.
The *second law* states that the induced voltage produced in a coil is proportional to the rate of change in the flux. This law is quantitative in nature. Only the expression of its magnitude is given here. Mathematically, the law can be written as,

$$e = N\frac{d\phi}{dt},$$

where,

N = number of turns,

$\frac{d\phi}{dt}$ = rate of change in the flux, and

E = voltage induced in the coil.

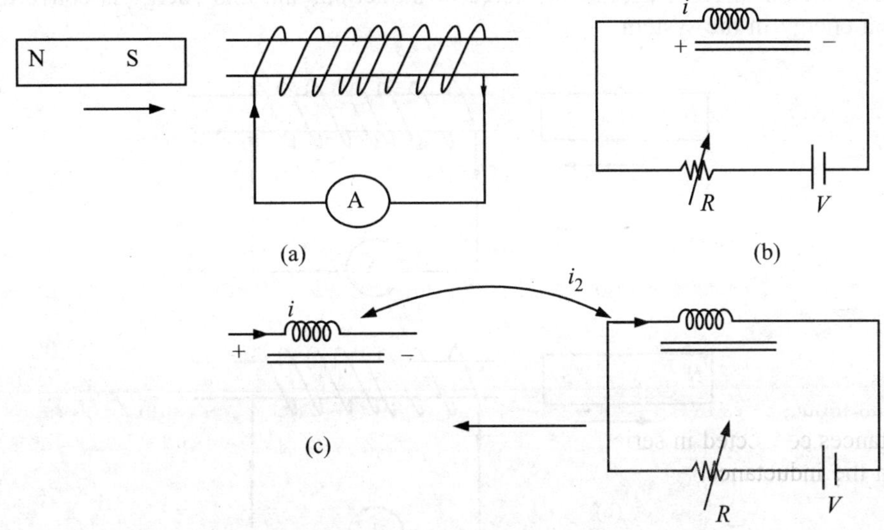

Fig. 9.1

Lenz's law is used to determine the sign of the induced voltage (e) in the coil.

STATEMENT: Lenz's law states that the direction of the induced emf in a coil is such that the current produced in it, when it is short circuited, sets up a magnetic field opposing the motion, or change producing it.

A short circuited coil with an ammeter is shown in Fig. 9.2(a)–(b). When a bar magnet (N–S) is moved towards the coil (Fig. 9.2(a)), an emf is induced in it by the increase in flux due to the movement of the magnet. This causes electric current to flow through the coil in the clockwise direction as shown. Due to the flow of the above current in the coil, magnetic field is produced in the core (iron or air, as the case may be), over which the coil is wound. The direction of the current is such that a north pole is formed in the core, near the north pole of the bar magnet, thus producing a repelling force or action. Similarly, a south pole is formed at the other end. The direction of flux in the coil can be obtained using the right-hand cork screw rule (or the right-hand grip rule where if we curl the fingers of our right hand around a wire with the thumb in the direction of (conventional) current flow, then our fingers point in the same direction as the circular magnetic field around the conductor). Mechanical energy is expended in moving the bar magnet against the force of repulsion, and this energy is converted into electrical energy in this system.

The same result is observed when the bar magnet is moved away from the coil (Fig. 9.2(b)). Then, an emf is induced in the coil by the decrease in flux due to the movement of the magnet. This causes electric current to flow through the coil in the anti-clockwise direction as shown. Due to the flow of the above current in the coil, magnetic field is produced in the core. The direction of the current is such that a south pole is formed in the core near the north pole of the bar magnet, thus producing an attracting force or action. Similarly, a north pole is formed at the other end. As given earlier, the direction of flux in the coil can be obtained by using the right-hand cork screw rule. Mechanical energy is expended in moving the bar magnet against the force of attraction, and this energy is converted into electrical energy in the system.

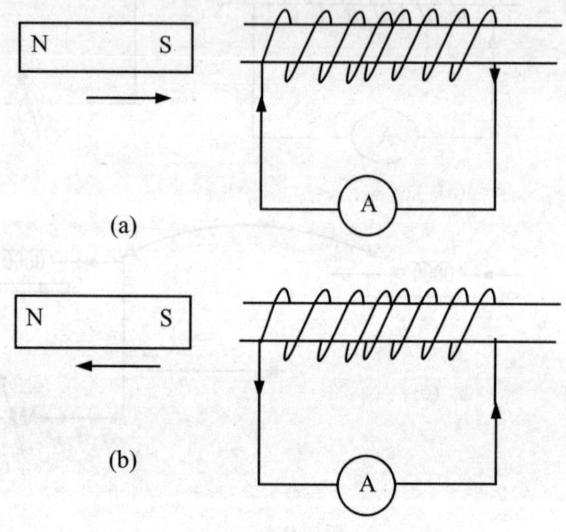

Fig. 9.2

INDUCED VOLTAGE IN A CONDUCTOR, WHEN IT IS MOVED AT CONSTANT VELOCITY IN A UNIFORM MAGNETIC FIELD

If a conductor is moved at a constant velocity in a direction perpendicular to its own length, and at an angle θ to the direction of the magnetic field, assumed to be uniform (Fig. 9.3(a)), by Faraday's law, a voltage is induced in it. The symbols used are:

e = induced voltage in the conductor (V)
B = flux density of the magnetic field (Wb/m^2)
l = length of the conductor (m)
v = constant velocity (m/s)
θ = angle between the direction of the velocity of the conductor, and the uniform magnetic field

The velocity, v is resolved into two components, of which one is along the direction of the magnetic field, and the other is perpendicular to its direction. The values are $v\cos\theta$ and $v\sin\theta$ respectively. The distance, dx, travelled by the conductor perpendicular to the direction of the field in time, dt, is $dx = v\sin\theta \cdot dt$. The area covered is given by,

$$dA = l.dx = l.v\sin\theta \cdot dt.$$

The induced voltage in the conductor (as explained later in this chapter) is proportional to the rate of change of flux linkage (flux linkage (ψ) is defined as the product of the number of turns N and the flux ϕ—in this case, $\psi = \phi$ as $N = 1$, the wire being only a conductor). It is given by,

$$e = \frac{d\phi}{dt} = B.\frac{dA}{dt} = Bl.\frac{dx}{dt} = Blv.\sin\theta.$$

This is a general formula. Two cases are considered here. The first one is when the conductor is moved in a direction perpendicular to the direction of the field at constant velocity (Fig. 9.3(b)). The angle, $\theta = 90°$. Substituting it in the equation derived earlier, the induced emf is given by $e = Blv$, as $\sin 90° = 1$. Here, maximum value of flux is cut by the moving conductor, and the value of the induced emf is maximum. It may be noted that the flux density, B, the velocity of the conductor, v and its length, l, are all at right angles to each other, in three directions, x, y, and z.

The second case we will consider is when the conductor is moved parallel to the direction of the field at constant velocity. The angle is $\theta = 0°$. Substituting this value in the equation derived earlier, the induced emf is $e = 0$, as $\sin 0° = 0$. In this case, no flux is cut by the coil, and emf induced is 0 (zero). Here, the flux density, B, and the velocity of the conductor, v, are in the same direction, x.

The equation derived here forms the basis of the generating emf, mainly in dc and also ac (synchronous) generators.

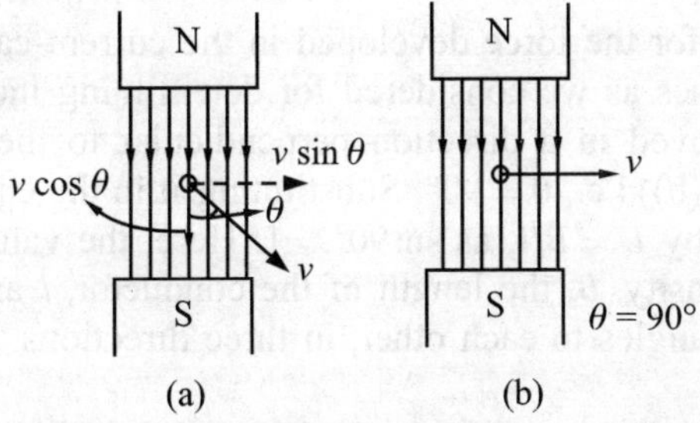

Fig. 9.3

FORCE DEVELOPED IN A CURRENT-CARRYING CONDUCTOR, PLACED IN A UNIFORM MAGNETIC FIELD

If a current-carrying conductor is placed in a uniform magnetic field at an angle θ to the direction of the magnetic field, the force is developed in the same direction as shown in Fig. 9.4(a). The new symbols used here are:

F = force produced in the conductor (N)
I = current in the conductor (A)

Now, if the conductor is moved with a constant velocity, v in the same direction as given earlier, the induced voltage in the moving conductor is given by $e = Blv \cdot \sin\theta$.

The increase in the energy of the moving conductor, in time, dt

$$= \text{Incremental work done} = dW = I.edt$$

or, $$\frac{dW}{dt} = I \cdot e = IBl \cdot v \sin\theta.$$

The instantaneous power, P(W) in the conductor, at time, t, is,

$$P = \frac{dW}{dt} = I \cdot e = IBl \cdot v \sin\theta,$$

from which the increase of energy is $dW = P.dt = e.I.dt = BIl.v\sin\theta.dt$.

The unit of energy or work done is Joules (J) or $W.s$ (when it is electrical work), or $N.m$ (when it is mechanical work).

From the mechanical side, if F is the force (N) produced in the current-carrying conductor, the increase in energy or work done is given by $dW = F \cdot dx' = F \cdot v \cdot dt$, when x' is the distance travelled in the direction with a constant velocity, v (Fig. 9.4(b)). Therefore,

$$dx' = \frac{dx}{\sin\theta} = v.dt.$$

The force produced in the current-carrying conductor is given by,

$$F = \frac{dW}{dx'} = \frac{\frac{dW}{dt}}{\frac{dx'}{dt}} = \frac{BIl.v\sin\theta}{v} = BIl.\sin\theta.$$

This is a general formula for the force developed in the current-carrying conductor. We can consider the same two cases as we considered for determining induced voltage. In the first case, the conductor is moved in a direction perpendicular to the direction of the field at constant velocity (Fig. 9.4(b)) i.e., $\theta = 90°$. Substituting it in the equation derived earlier, the force developed is given by $F = BIl$, as $\sin 90° = 1$. Here, the value of the force developed is maximum. The flux density, B, the length of the conductor, l and the conductor-carrying current, I, are all at right angles to each other, in three directions.

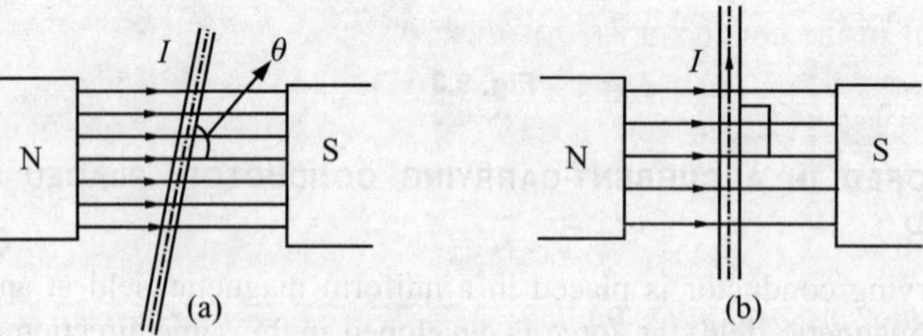

Fig. 9.4

The second case we can consider is when the conductor is moved parallel to the direction of the field at constant velocity. The angle is $\theta = 0°$. Substituting this in the equation derived earlier, the force developed is $F = 0$ as $\sin 0° = 0$. Here, the flux density, B, and the movement of the conductor-carrying current, I, are in the same direction, x.

This equation forms the basis of developing torque action, i.e., motoring, mainly in dc, and also in ac (induction and synchronous) motors.

EXAMPLE 9.1 *A 40 cm long wire (conductor) moves at right angles to its length at 30 m/s in a uniform magnetic field, where the flux density is 0.8 Wb/m^2 (Fig. 9.5). Calculate the induced emf in the conductor when the direction of motion is*

(a) perpendicular to the field, and
(b) inclined at 30° to the direction of the field.

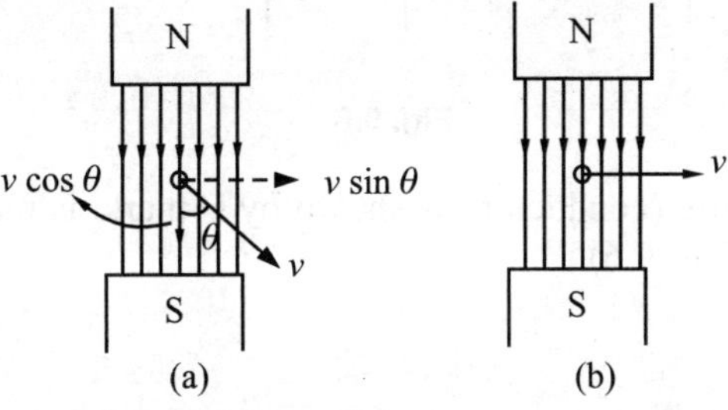

Fig. 9.5

SOLUTION A uniform magnetic field with the conductor at right angle to the field (denoted by • mark), is shown in Fig. 9.5. The conductor moves perpendicular to the direction to the field ($\theta = 90°$) (Fig. 9.5(a)).

Length of the conductor = $l = 40$ cm = 0.4 m.

Velocity = $v = 30$ m/s.

$B = 0.8$ Wb/m^2.

(a) Induced emf in the conductor = $e_1 = Blv\sin\theta$
$= 0.8 \times 0.4 \times 30.0 \times \sin 90° = 9.6$ V.

(b) Induced emf in the conductor when $\theta = 30°$(Fig. 9.5(b)) is $e_2 = Blv\ \sin\theta = 0.8 \times 0.4 \times 30.0 \times \sin 30° = 4.8$ V.

EXAMPLE 9.2 *In a uniform magnetic field of 1 Wb/m^2, a direct current of 50 A is passed through a straight wire (conductor) of 1.2 m, placed perpendicular to the field (Fig. 9.6). Calculate*

(a) *The magnitude of the mechanical force developed in the wire.*
(b) *The prime mover power (i.e., mechanical power) in W, to displace the conductor against the force at a uniform velocity of 6 m/s.*
(c) *The emf generated in the current-carrying wire.*

How do you know that the electrical power produced is the same as the mechanical power developed?

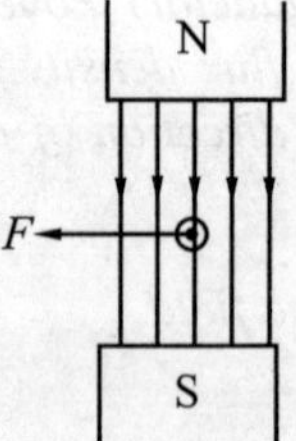

Fig. 9.6

SOLUTION A straight wire (conductor) as shown by • mark, is placed perpendicular to the uniform magnetic field (Fig. 9.6).

$\theta = 90°$.

Before solving this example, let us study Fig. 9.6 given here, which is nearly similar to Fig. 9.3. This view is from the top of Fig. 9.4(b). Here, the magnetic field, its flux density being B, is shown from the top to the bottom on the y-axis. In the straight wire (conductor) of length, l, the current, I, shown by (*) mark, flows from the bottom to the top on the z-axis, with $\theta = 90°$. The force, F on the current-carrying conductor is obtained, using Fleming's left hand rule. Its direction is from the right to the left on the x-axis. If the conductor is placed at an angle, θ, with the magnetic field, the magnitude of the force, F is reduced, as given earlier, its direction remaining the same. Length of the conductor $= l = 1.2$ m.

$B = 1.0$ Wb/m^2.

Current (dc) through the conductor $= I = 50$ A.

(a) Mechanical force developed in the conductor
$= F = BIl \sin\theta = 1.0 \times 50.0 \times 1.2 \times \sin 90°$ N.

(b) Velocity of the conductor $= v = 6$ m/s.
Prime mover, i.e., mechanical power $= F.v = 60 \times 6 = 360$ W (J/s or N.m/s).

(c) Generated emf $= E = Blv \sin\theta = 1.0 \times 1.2 \times 6.0 \times \sin 90° = 7.2$ V,
as the voltage is constant and dc.
The electrical power $= E.I = 7.2 \times 50.0 = 360$ W (= mechanical power).

9.2 Inductance

Let us now study what is meant by inductance—self and mutual. Self inductance is defined as the induction of a voltage in a current-carrying wire when the current in the wire itself is changing. The magnetic field created by a changing current in the circuit itself induces a voltage in the same circuit. If there are two current-carrying wires, with one of the wires placed at a certain distance (not far) from another, the steady current in one wire will produce a magnetic field in the other, though no voltage is induced. However, if the current in one wire changes, the change in magnetic field in the other will induce a voltage—this phenomena is called mutual inductance

Consider a coil is wound on a former (core). The former can be circular, square, or rectangular, as the case may be. There is magnetic material or core, which is placed inside the former. This is called an iron cored coil. If the core is removed, it becomes an air cored coil. If a current (i) is passed through the coil, magnetic flux is produced in the magnetic material, i.e., core. It may be stated that the magnetic flux in the iron core is higher than that in the air core. This has been discussed earlier.

The induced emf in the coil is proportional to the rate of change of flux produced, given by

$$e = N\frac{d\phi}{dt},$$

where,

N = number turns of the coil

ϕ = flux produced in the core (Wb)

e = induced emf in the coil (V)

As before, let us define flux linkage (ψ) as the product of the number of turns and the flux. Its unit is same (Wb) as that of flux, as N is a dimensionless quantity. Hence, the formula for induced emf, can be written as

$$e = \frac{d\psi}{dt} = N\frac{d\phi}{dt}, \text{ where } \psi = N\phi.$$

The term, inductance (L) is defined as the incremental change of the flux linkage divided by the incremental change of the current in the coil.

$$L = \frac{d\psi}{di} = N\frac{d\phi}{di}.$$

Hence, $d\psi = N \cdot d\phi = L \cdot di$,

where, i = current in the coil (A).

The unit of inductance is Henry (H), which is equal to Wb/A. The induced emf can also be written as,

$$e = \frac{d\psi}{dt} = L\frac{di}{dt}.$$

Please note that $L \propto \frac{d\phi}{di}$, as the $\phi - i$ characteristic is non-linear in the case of the material (iron) used in the core—this has been discussed earlier in detail. Only in the linear part of the above characteristic, can L be taken as $L = \frac{N\phi}{I}$. This formula can also be used for air cored coils, as the relative permeability of air is $\mu_r = 1$.

Self inductance

To find the expression for self inductance (Fig. 9.7), the following symbols are used.

N_1

Fig. 9.7 *Self inductance of a coil*

I	=	current in the coil (A)
NI	=	ampere turns of the coil (A)
H	=	field intensity in the magnetic material, i.e., core (A/m))
l	=	length of the magnetic path, i.e., former (m)
B	=	flux density (Wb/m^2)
a	=	cross-sectional area of the core, i.e., former (m^2)
μ	=	$\mu_0\mu_r$ = total permeability of the magnetic material, i.e., core (H/m)
μ_0	=	$4\pi \times 10^{-7}$ = permeability of the non-magnetic material, which may be air (H/m)
μ_r	=	relative permeability of the magnetic material, i.e., core

We start with the relation presented in Chapter 8 on magnetic circuits for ampere turns (MMF) of the coil mmf is the product of the magnetic field intensity core, and length of the magnetic path. It is assumed that the total mmf is dropped only in the length of the core, with a small drop in the lines of force through the air, as the area is very large there.

Hence, $NI = Hl$, or $H = \dfrac{NI}{l}$. The relative permeability (μ_r) of the magnetic material used in the core, $\mu_r = \dfrac{\mu}{\mu_0}$, which is constant. Note that it is a dimensionless quantity, just like the number of turns (N), given earlier. The flux density in the core, is given by $B = \dfrac{\phi}{a}$. The total permeability of the magnetic material is the flux density produced in the core divided by the intensity of the magnetic field, written as $\mu = \mu_0\,\mu_r = \dfrac{B}{H}$, or $B = \mu H$. Substituting the relations obtained earlier, we get,

$$\frac{\phi}{a} = \mu \cdot \frac{NI}{l}, \quad \text{or,} \quad \frac{\phi}{I} = N \cdot \frac{\mu a}{l} = N \cdot \frac{\mu_0 \mu_r a}{l}.$$

The self inductance, L of the coil is found as

$$L = \frac{\psi}{I} = N.\frac{\phi}{I} = N^2.\frac{\mu a}{l} = N^2.\frac{\mu_0 \mu_r a}{l}.$$

If the magnetic material, i.e., core is removed, it becomes an air cored coil, with $\mu_r = 1$.

When we derive the expression for the mutual inductance between two coils in close proximity to each other, the number of turns in the first and second coils are taken as N_1 and

N_2. The currents in them are assumed to be I_1 and I_2, and their self inductances are L_{11} and L_{22} respectively.

Hence, L_{11} is written as $L_{11} = \frac{\psi_1}{I_1} = N_1 \cdot \frac{\phi_1}{I_1} = N_1^2 \cdot \frac{\mu_0 \mu_r a}{l}$.

The magnetic core is assumed to be same.

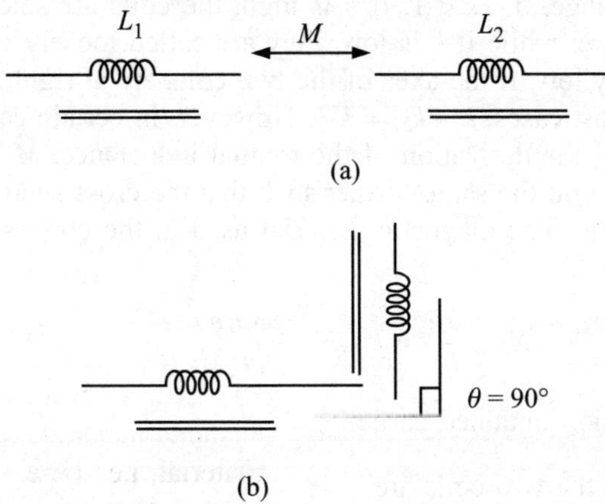

Fig. 9.8 *Mutual inductance between two coupled coils*

Mutual inductance

The mutual inductance between two coils in close proximity to each other is derived in the same manner as given for self inductance. Two coils may be wound on the same former, or on different ones, but kept close to each other (Fig. 9.8). These are called coupled coils. The symbols used are the same as given earlier, except that the symbol used for mutual inductance is M or M_{21}. This means, as we know, that if the current in the first coil or excitation is changed, then a voltage is induced in the second coil due to the change in flux caused. The induced voltage in the second coil is given by,

$$e_2 = M_{21} \cdot \frac{di_1}{dt}.$$

Similarly, voltage will be induced in the first coil by the change of current in the second coil, where, the symbol used is M_{12}. The induced voltage in the first coil is given by,

$$e_1 = M_{12} \cdot \frac{di_2}{dt}.$$

In most cases, $M = M_{21} = M_{12}$.

Coefficient of coupling

The flux, ϕ_{11} in the first coil is linked with the current in that coil, I_1, and its self inductance is given by $L_{11} = N_1 \cdot \frac{\phi_{11}}{I_1}$. The flux, ϕ_{21} linked to the second coil, which may be in the same core or in a different one as given earlier, due to the current in the first coil, is $\phi_{21} = k.\phi_{11}$, or $k = \dfrac{\phi_{21}}{\phi_{11}}$. The symbol, k is termed as the *coefficient of coupling* between the two coils. The value of k lies in the range, $0 < k < 1$. If k is high, the coils are said to be tightly coupled, may be on the same core, while if k is low, they are called loosely coupled. The value of k may be zero (0), or very low, if the axes of the two coils are at right angles to each other. In this case, $k = k_{21}$. In most cases, $k = k_{21} = k_{12}$. However, in certain cases, $k_{21} \neq k_{12}$.

Let us again come to the derivation of the mutual inductance, M. It is assumed that two coils, 1 and 2 are wound on the same former such that the cross-sectional area and length of the core remain the same. The magnetic material used in the core is also the same. Hence, the mutual inductance is

$$M_{21} = \frac{\psi_{21}}{I_1} = N_2 \cdot \frac{\phi_{21}}{I_1} = N_2 \cdot \frac{k\phi_{11}}{I_1} = kN_1N_2 \cdot \frac{\mu_0\mu_r a}{l},$$

substituting the expression obtained earlier.

The self inductances of the two coils are,

$$L_{11} = N_1^2 \cdot \frac{\mu_0\mu_r a}{l}, \text{ and } L_{22} = N_2^2 \cdot \frac{\mu_0\mu_r a}{l}.$$

If the symbols are changed to $L_1 = L_{11}$, $L_2 = L_{22}$ and $M = M_{21} = M_{12}$, it can be shown that

$$M = k\sqrt{L_1L_2}$$

The coils can be placed side by side as shown in Fig. 9.8(a). They can also be placed at an angle with each other, as shown in Fig. 9.8(b). Assume that θ is the angle between the two coils lying on a plane. When the two coils are placed together like in Fig. 9.8, nearly all the fluxes in the first coil, when it carries current, is linked with the second coil, and vice versa. The value of M, and also the coefficient of coupling, k are high. However, if they are placed at an angle, θ (Fig. 9.8(a)), only a part of the flux in the first coil, when it carries current, will be linked with the second coil, and vice versa. The value of M, and also the coefficient of coupling, $k(0 < k < 1)$ are smaller, as the angle, θ increases. As given earlier, the values of M and k are highest, with $k \approx 1$, at $\theta = 0°$. The values of M and k are zero (0), at $\theta = 90°$ (Fig. 9.8(b)). In this case, no flux in the first coil, when it carries current, is linked with the second coil, and vice versa.

9.3 Energy Stored in the Inductor

When the current in an inductor coil is changed, say, increased, a voltage is induced in it (as mentioned in the previous section), and is given by,

$$e = L\frac{di}{dt}.$$

Instantaneous power is $P = e.i = Li\frac{di}{dt}$.

The increase of energy in the inductor, when the current in the coil is increased from I_1 to I_2 in time, dt,

$$dW = P.dt = L.idi.$$

Hence, the energy stored in the inductor as the current in the coil increases from zero (0), to say, I is,

$$W = \int_0^I L \cdot idi = \frac{1}{2}L \cdot i^2\Big|_0^I = \frac{1}{2}L \cdot I^2.$$

Note that the energy is stored in the magnetic field in the core, which may be iron or air. The unit of stored energy is Joules (J). The core is now under stress because of the magnetic flux produced in it by the current in the inductor coil.

ENERGY STORED PER UNIT VOLUME IN THE MAGNETIC FIELD

As given earlier, the stored energy in the inductor coil is the energy stored in the magnetic field, when a current is passed through the coil. The formula for inductance is given by,

$$L = N^2 \cdot \frac{\mu_0 \mu_r a}{l}$$

where the parameters are defined earlier.

The coil is wound on a magnetic core, whose length is $l(l_i)$, and cross-sectional area is a. The volume of the magnetic core is $a.l$ (m^3). It can also be wound on a horse shoe magnet as shown in Fig. 9.9 and used for Example 9.6. A toroidal core with the winding around it (Fig. 8.9(a)), as shown in Section 8.3 (Chapter 8), while discussing magnetic circuits, can also be taken as an inductance coil.

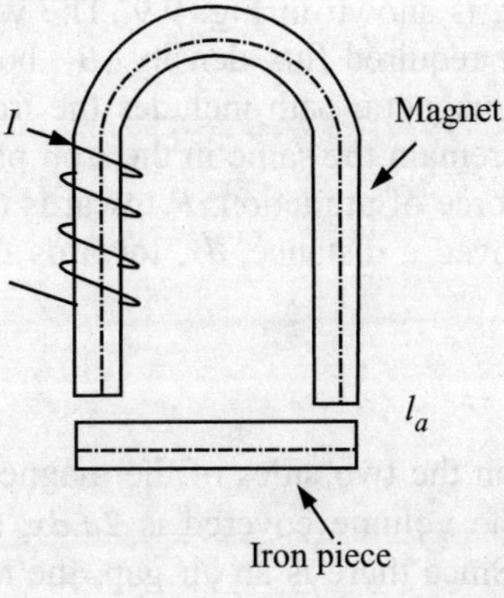

Fig. 9.9

If I is the current in the coil, the stored energy in the magnetic field is obtained by substituting the formula of the inductance,

$$W = \frac{1}{2}(NI)^2 \cdot \frac{\mu_0 \mu_r a}{l}.$$

Using Ampere's circuital law,

$$NI = \oint H.dl = H_i.l_i,$$

where H_i is assumed to be constant or uniform over the length of the magnetic path.

Substituting the expression in the above formula and assuming $l = l_i$, we get,

$$W = \frac{1}{2}(H_i \cdot l)^2 \cdot \frac{\mu_0 \mu_r a}{l} = \frac{1}{2}(H_i)^2 \cdot \mu_0 \mu_r \cdot (al).$$

Using flux density (B), instead of field intensity ($H = H_i$) and substituting

$$H = \frac{B}{\mu} = \frac{B}{\mu_0 \mu_r}$$

and total permeability of iron, $\mu = \mu_0 \mu_r$, the stored energy in the magnetic field is,

$$W = \frac{1}{2}\left(\frac{B}{\mu}\right)^2 \mu \cdot (al) = \frac{B^2}{2\mu} \cdot (al).$$

Energy stored per unit volume in the magnetic field $= \dfrac{B^2}{2\mu} = \dfrac{B^2}{2\mu_0 \mu_r}$.

Its unit is J/m^3, and the unit of B is Wb/m^2.

LIFTING POWER OF THE ELECTRO-MAGNET

A horse shoe magnet, with two air gaps, each of length, l_g and an iron piece, which is attracted by the electro-magnet, is shown in Fig. 9.9. The winding placed on the U-shaped magnet is excited such that the required flux density (B), both in the iron part and air gap, is obtained. The length of the magnetic path includes the iron piece. Fringing is neglected, so that the flux or flux density remain the same in the iron part and also in the air gaps. The iron piece now experiences a force of attraction, F towards the magnet.

Now, if the iron piece is moved a distance, dx, towards the magnet, work done is given by,

$$W = (F + F).dx = 2F.dx,$$

as there are two forces acting on the two sides of the magnet as shown.

As the iron piece is moved, the volume covered is $2a.dx$, as the cross-sectional area is a (m^2) and there are two sides. Since there is an air gap, the relative permeability is taken as $\mu_r = 1$. Hence the change in energy caused is $\dfrac{B^2}{2\mu_0} 2a.dx$.

Equating the two expressions, the force, F is

$$F = \frac{B^2.a}{2\mu_0}$$

There are two forces acting in two sides. Hence,

$$\text{Total force} = \quad 2F = \frac{B^2.a}{\mu_0}.$$

The unit of the force is N.

NOTE This force of attraction of an iron piece is expressed in terms of magnetic flux density (B). In the next chapter (in EXAMPLE 10.11, where we derive the force between two charged spherical shells), the force is expressed in terms of electric field intensity (E).

EXAMPLE 9.3 *An iron ring of mean diameter 20 cm and cross-sectional area 15 cm^2, is wound with 300 turns (Fig. 9.10(a)). The flux produced in the ring is 1.5 mWb. Assume the relative permeability of iron as 600. Determine (a) the exciting current, (b) the self inductance of the coil, and (c) the stored energy.*

Now, a saw cut of 1 mm is made in the core with flux remaining the same. Find the new values of the parameters as given earlier.

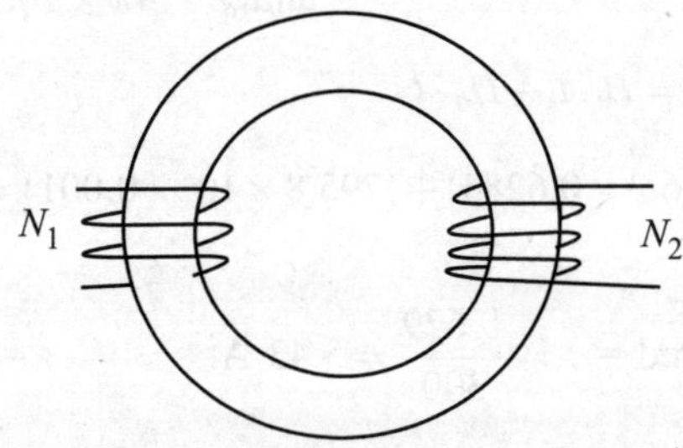

Fig. 9.10

SOLUTION The data for the iron ring, and also the coil wound on it (Fig. 9.10 (a)), are:

$$\text{Flux, } \phi = 1.5 \text{ mWb} = 1.5 \times 10^{-3} \text{ Wb}, \quad a = 15 \text{ cm}^2 = 15 \times 10^{-4} \text{ m}^2$$

$$d = 20 \text{ cm} = 0.2 \text{ m}, N = 300.$$

Relative permeability of iron, $\mu_{ri} = 600$.

Length of the magnetic path $= l_i = \pi.d = \pi \times 0.2 = 0.6283$ m $= 62.83$ cm.

$$\text{Flux density in the core} = B = \frac{\phi}{a} = \frac{1.5 \times 10^{-3}}{15 \times 10^{-4}} = 1.0 \text{ Wb/m}^2.$$

Magnetic field strength $= H_i = \dfrac{B}{\mu_0 \mu_{ri}} = \dfrac{1.0}{4\pi \times 10^{-7} \times 600} = 1{,}326.3$ A/m.

Ampere turns required $= NI = H_i \cdot l_i = 1{,}326.3 \times 0.6283 \approx 833.3$ A.

(a) Exciting current required $= I = \dfrac{833.3}{300} \approx 2.78$ A

(b) Self inductance of the coil $= L = N \cdot \dfrac{\phi}{I} = 300 \times \dfrac{1.5 \times 10^{-3}}{2.78} = 0.162$ H =162 mH.

(c) Stored energy in the coil $= W = \dfrac{1}{2} LI^2 = 0.5 \times 0.162 \times (2.78)^2 = 0.625$ J.

In the second case, when a saw cut of 1 mm is made in the core (Fig. 9.10(b)), the flux remains the same as given. All values computed earlier are the same in this condition.

Length of the air gap $= l_a = 1.0$ mm $= 0.001$ m.

Relative permeability, $\mu_{ra} = 1$.

Magnetic field strength in air gap $= H_i = \dfrac{B}{\mu_0 \mu_{ra}} = \dfrac{1.0}{4\pi \times 10^{-7} \times 1} = 795.8 \times 10^3$ A/m.

Ampere turns required $= NI = H_i \cdot l_i + H_a \cdot l_a$

$$= (1{,}326.3 \times 0.6283) + (795.8 \times 10^3 \times 0.001) = (833.3 + 795.8) \approx 1{,}629 \text{ A}$$

(a) Exciting current required $= I = \dfrac{1{,}629}{300} = 5.43$ A.

(b) Self inductance of the coil $= L = N \cdot \dfrac{\phi}{I} = 300 \times \dfrac{1.5 \times 10^{-3}}{5.43} = 0.083$ H $= 83$ mH.

(c) Stored energy in the coil $= W = \frac{1}{2} LI^2 = 0.5 \times 0.083 \times (5.43)^2 = 1.224$ J.

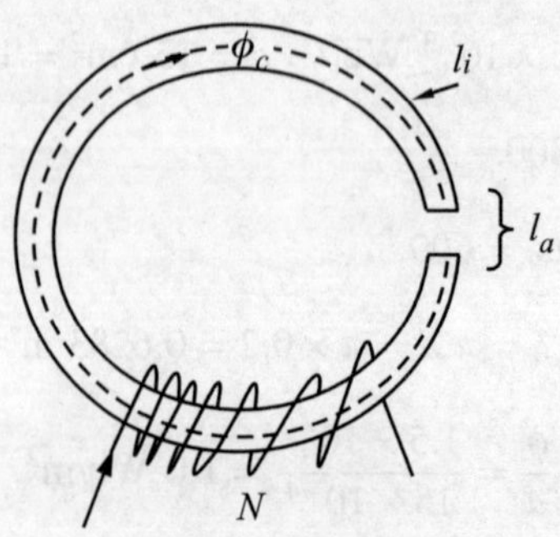

Fig. 9.10(b)

When an air gap is introduced in the magnetic circuit, the ampere turns required, and the exciting current in the coil increase. The self inductance of the coil decreases, and the stored energy increases.

An alternative procedure for finding total ampere turns is given below. It involves finding the reluctance as shown in Section 8.3 (Chapter 8).

The reluctance of the iron part

$$\Re_i = \frac{l_i}{\mu_0 \mu_{ri} a} = \frac{0.6283}{4\pi \times 10^{-7} \times 600.0 \times 15.0 \times 10^{-4}} = 555.5 \times 10^3 \text{ H}^{-1}.$$

The reluctance of the air gap $= \Re_a = \dfrac{l_a}{\mu_0 a} = \dfrac{0.001}{4\pi \times 10^{-7} \times 15.0 \times 10^{-4}} = 530.5 \times 10^3 \text{ H}^{-1}.$

Total reluctance $= \Re = [\Re_i + \Re_a] = [(555.5 + 530.5) \times 10^3] = 1.086 \times 10^3 \text{ H}^{-1}.$

Total ampere turns required for both the iron part and the air gap

$$NI = [(NI)_i + (NI)_a].\phi = [\{(555.5 + 530.5) \times 10^3\} \times 1.5 \times 10^{-3}]$$

$$= (833.0 + 796.0) = 1.629.0 \text{ A},$$ which is the same as obtained earlier.

EXAMPLE 9.4 *Two coils, 1 and 2, having 200 and 300 turns respectively, are wound side by side on a closed magnetic (iron) surface circuit, with cross-sectional area 100 cm², and mean length 150 cm (Fig. 9.8(a)). Assume that the relative permeability of iron is 1,000. Calculate (a) the self inductance of the coils, (b) the mutual inductance between the coils, and (c) the induced emf in coil 2, if the current in coil 1 changes from 0 to 6 A in 0.02 s (20 ms).*

SOLUTION The parameters of the coils (Fig. 9.8(a)) are:

$$a = 100 \text{ cm}^2 = 100 \times 10^{-4} = 0.01 \text{ m}^2, \qquad l = 150 \text{ cm} = 1.5 \text{ m},$$

$N_1 = 200, \quad N_2 = 300.$

Relative permeability, $\mu_r = 1{,}000$.

(a) Self inductance of coil 1 $= L_1 = (N_1)^2 \cdot \dfrac{\mu_0 \mu_r a}{l}$

$$= (200)^2 \times \frac{4\pi \times 10^{-7} \times 1{,}000 \times 0.01}{1.5} = 0.335 \text{ H} = 335 \text{ mH}.$$

Self inductance of coil 2 $= L_2 = (N_2)^2 \cdot \dfrac{\mu_0 \mu_r a}{l} = L_1 \cdot \left(\dfrac{N_2}{N_1}\right)^2 = 0.335 \times \left(\dfrac{300}{200}\right)^2$

$$= 0.7538 \text{ H} = 753.8 \approx 754 \text{ mH}$$

(b) Mutual inductance between two coils $= M = N_1 N_2 \cdot \dfrac{\mu_0 \mu_r a}{l} = L_1 \cdot \left(\dfrac{N_2}{N_1}\right)$

$$= 0.335 \times \frac{300}{200} = 0.5025 \text{ H} = 502.5 \text{ mH}.$$

Note that all parameters of the magnetic circuit including M are the same in all the three cases.

Also, the coefficient of the coupling is $k = 1.0(100\%)$. Mutual inductance can also be computed as given below.

$$M = k \cdot \sqrt{L_1 L_2} = 1.0 \times \sqrt{335 \times 754} = 502.5 \text{ mH},$$

which is the same as obtained earlier.

(c) As the current in coil 1 changes from $I_1' = 0$ to $I_1'' = 6$ A in $T = 0.02$ s = 20 ms,

$$\frac{di_1}{dt} = \frac{I_1'' - I_1'}{T} = \frac{6-0}{0.02} = 300 \text{ A/s},$$

as the slope is assumed to be constant over time, T, or the current in coil 1 changes linearly.

$$\text{Induced emf in coil 2} = e_2 = M \cdot \frac{di_1}{dt} = 502.5 \times 10^{-3} \times 300 \approx 151.0 \text{ V}$$

EXAMPLE 9.5 *Two identical coils, 1 and 2, of 250 turns each, lie in parallel planes, such that 60% of the flux produced by one coil, links with the other coil (Fig. 9.8(a)). A current of 6 A in coil 1 produces a flux of 0.06 mWb, in it. If the current in coil 1, changes from +4 A to − 4 A in 0.02 s, what is the induced emf in coil 2? Calculate the self inductance of the coils, and the mutual inductance between them.*

SOLUTION The parameters of the coils (Fig. 9.8(a)) are:

$$\phi_1 = 0.06 \text{ mWb} = 60\ \mu \text{ Wb}, \ I_1 = 6 \text{ A},$$

$$N_1 = 250, \quad N_2 = 250.$$

$$\text{The self inductance of coil } 1 = L_1 = N_1 \cdot \frac{\phi_1}{I_1} = 250 \times \frac{60 \times 10^{-6}}{6} = 2.5 \times 10^{-3} = 2.5 \text{ mH}.$$

As $N_2 = N_1$, the self inductance of coil 2 is the same as $L_2 = L_1 = 2.5$ mH. Coefficient of coupling, $k = 60\% = 0.6$, as 60% of the flux produced by one coil is linked with the other coil.

$$\text{Mutual inductance} = M = k\sqrt{L_1 L_2} = 0.6 \times \sqrt{2.5 \times 2.5} = 0.6 \times 2.5 = 1.5 \text{ mH}.$$

As the current in coil 1 changes from $I_1' = +4$ A to $I_1'' = -4$ A in $T = 0.02$ s = 20 ms,

$$\frac{di_1}{dt} = \frac{I_1' - I_1''}{T} = \frac{+4-(-4)}{0.02} = 400 \text{ A/s}.$$

The slope is assumed to be constant over time, T, or the current in coil 1 changes linearly.

$$\text{Induced emf in coil 2} = e_2 = M \cdot \frac{di_1}{dt} = 1.5 \times 10^{-3} \times 400 = 0.6 \text{ V}.$$

Alternatively, the mutual inductance can be computed as,

$$\frac{\phi_2}{\phi_1} = 60\% = 0.6,$$

or, $\phi_2 = 0.6 \cdot \phi_1 = 0.6 \times 0.06 = 0.036 \text{ mWb} = 36\,\mu\text{Wb}.$

This is because 60% of the flux produced by one coil is linked with the other coil.

Mutual inductance between the coils

$$M = N_2 \cdot \frac{\phi_2}{I_1} = 250 \times \frac{36 \times 10^{-6}}{6} = 1.5 \times 10^{-3} = 1.5 \text{ mH},$$

which is the same as obtained earlier.

EXAMPLE 9.6 *A horse shoe magnet is shown in Fig. 9.9, where there are two air gaps, each of length 1 mm, and an iron piece, which is attracted towards the magnet. The cross-sectional area of the magnet is 160 cm², and the length of the magnetic path, including the iron piece, is l_i = 80 cm. The flux produced in the air gap is 20 mWb. Neglect leakage and fringing. Find (a) the exciting ampere turns required, and (b) the force needed to separate the contact surface. The B–H curve of the magnetic material used is given in the table.*

Flux density (Wb/m²)	1.0	1.2	1.4
Field intensity (A/m)	900	1,230	2,100

SOLUTION A horse shoe (U-shaped) magnet, with an iron piece, is shown in Fig. 9.10. An exciting coil is wound on the magnet.

Flux, $\phi = 20 \text{ mWb} = 0.02 \text{ Wb}.$

Cross-sectional area = $a = 160 \text{ cm}^2 = 160 \times 10^{-4} \text{ m}^2.$

Length of the magnetic path = $l_i = 80 \text{ cm} = 0.8 \text{ m}.$

Length of the air gap = $2l_g = 2 \times 1 = 2 \text{ mm} = 0.002 \text{ m}.$

The length of the magnetic path remains nearly the same as the length of the air gap is small.

Flux density in the magnet, and in the air gap = $B = \frac{\phi}{a} = \frac{0.02}{160.0 \times 10^{-4}} = 1.25 \text{ Wb/m}^2.$

Fringing is neglected. Therefore, the flux and flux density are the same in both the magnet and the air gap.

The *B–H* curve for the magnetic material is plotted in Fig. 9.11. From this graph, the field intensity, H_i for $B = 1.25$ Wb/m² is found to be 1350 A/m.

An alternative procedure is given for finding field intensity using a part of the *B–H* curve. As the value of computed *B* lies between two points of the curve, 1.2 and 1.4 Wb/m², a straight can be drawn through the above points, whose slope is

$$\frac{2,100 - 1,230}{1.4 - 1.2} = \frac{870.0}{0.2} = 3,450.$$

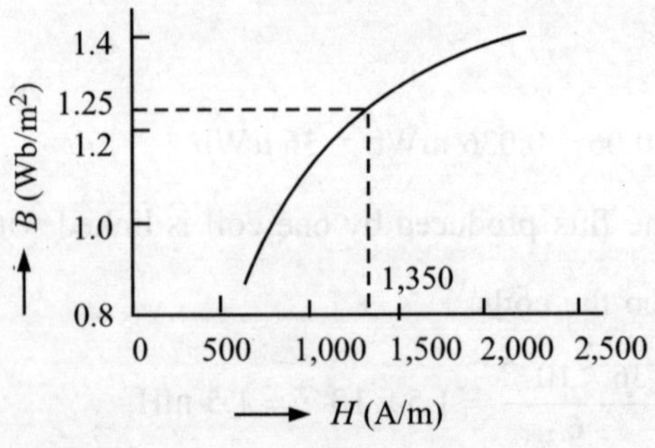

Fig. 9.11

The value of the field intensity in the magnetic material at $B = 1.25$ Wb/m², is obtained as

$H_i = 1{,}230.0 + [4{,}350.0 \times (1.25 - 1.2)] \approx 1{,}448.0$ A/m.

This value differs considerably from the earlier one as the curve is not linear, as assumed earlier.

The relative permeability of the magnetic material is computed as

$$\mu_{ri} = \frac{B}{\mu_0 H_i} = \frac{1.25}{4\pi \times 10^{-7} \times 1{,}350} = 737.0$$

The value of the field intensity in the air gap ($\mu_r = 1$), is

$$H_a = \frac{B}{\mu_0} = \frac{1.25}{4\pi \times 10^{-7}} = 994.7 \times 10^3 \text{ A/m}$$

(a) Neglecting leakage, and using Ampere's circuital law, the ampere turns required

$$= NI = H_i \cdot l_i + H_a \cdot l_a$$

$$= (1350 \times 0.8) + (994.7 \times 10^3 \times 0.002) = (1.080.0 + 1{,}989.4) \approx 3{,}070.0 \text{ A}$$

(b) Total force developed $= \dfrac{B^2 .a}{\mu_0} = \dfrac{(1.25)^2 \times 160.0 \times 10^{-4}}{4\pi \times 10^{-7}} \approx 19{,}894$ N

$$= \frac{19{,}894.0}{9.806} = 2{,}029 \text{ kg},$$

where, acceleration due to gravity $= g = 9.806$ m/s²

The procedure for finding total ampere turns required by first computing the reluctance is given in EXAMPLE 9.3.

9.4 Series and Parallel Connection of Inductances

SERIES CONNECTION

Two inductances, L_1 and L_2 are connected in series, and fed from a dc supply. A resistor is also connected in series to control the current (Fig. 9.12(a)). The current in the two inductors is the same. The voltage drops in the inductances, are

$$v_1 = L_1 \frac{di}{dt} \text{ and } v_2 = L_2 \frac{di}{dt},$$

according to the formula given earlier. Let the total voltage drop be v, which is equal to

$$v = v_1 + v_2 = L_1 \frac{di}{dt} + L_2 \frac{di}{dt} = (L_1 + L_2) \frac{di}{dt}.$$

If L_{eq} is the equivalent inductance, then the total voltage drop is,

$$v = L_{eq} \frac{di}{dt} = (L_1 + L_2) \frac{di}{dt}.$$

From this expression, $L_{eq} = L_1 + L_2$.

If three or more inductances are connected in series, the equivalent inductance can be obtained in the same manner. If, say, three inductances, L_1, L_2 and L_3 are connected in series, the equivalent inductance is $L_{eq} = L_1 + L_2 + L_3$.

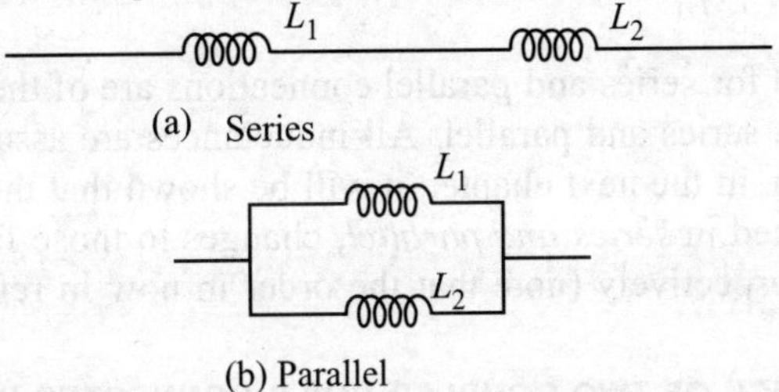

Fig. 9.12

PARALLEL CONNECTION

Two inductances, L_1 and L_2 are connected in parallel and fed from a dc supply. A resistor is connected in series to control the current (Fig. 9.12(b)). The voltage drops in the inductances are

$$v = L_1 \frac{di_1}{dt} = L_2 \frac{di_2}{dt},$$

where v is the voltage drop across the two inductances, and i_1 and i_2 are the current in the two inductances. The total current drawn from the supply is $i = i_1 + i_2$. From the first expression,

$$\frac{di_1}{dt} = \frac{v}{L_1}, \text{ and } \frac{di_2}{dt} = \frac{v}{L_2}.$$

Differentiating the expression for the total current, and substituting the above expressions, we get,

$$\frac{di}{dt} = \frac{di_1}{dt} + \frac{di_2}{dt} = \frac{v}{L_1} + \frac{v}{L_2}.$$

If the equivalent inductance is L_{eq}, the total voltage is given by,

$$v = L_{eq}\frac{di}{dt}, \quad \text{or,} \quad \frac{di}{dt} = \frac{v}{L_{eq}} = \frac{v}{L_1} + \frac{v}{L_2},$$

from which the equivalent inductance is obtained as,

$$\frac{1}{L_{eq}} = \frac{1}{L_1} + \frac{1}{L_2}, \quad \text{or,} \quad L_{eq} = \frac{L_1 \cdot L_2}{L_1 + L_2}.$$

If three or more inductances are connected in parallel, the equivalent inductance can be obtained in the same manner. If, say, three inductances, L_1, L_2 and L_3 are connected in parallel, the equivalent inductance is

$$\frac{1}{L_{eq}} = \frac{1}{L_1} + \frac{1}{L_2} + \frac{1}{L_3}$$

$$\text{or,} \quad L_{eq} = \frac{L_1 L_2 L_3}{L_1 L_2 + L_2 L_3 + L_3 L_1}.$$

The expressions derived for series and parallel connections are of the same form as those for resistances connected in series and parallel. All inductances are assumed to be lossless, i.e., with no resistance. Later, in the next chapter, it will be shown that the equivalent capacitance for capacitances connected in *series and parallel*, changes to those for resistances connected in *parallel and series* respectively (note that the order in now in reverse).

EQUIVALENT INDUCTANCE OF TWO COUPLED COILS CONNECTED IN SERIES

In circuit analysis, the dot convention is very important when describing inductance in a network. The convention is used to denote the voltage polarity of the mutual inductance of two components.

Dot convention

Two coils (Fig. 9.13(a)) are wound in the same way or direction, either on a single former, or on two formers, kept close to each other, so as to have a higher value of k. The terminals of two coils, 1 and 2 are A and B, and C and D.

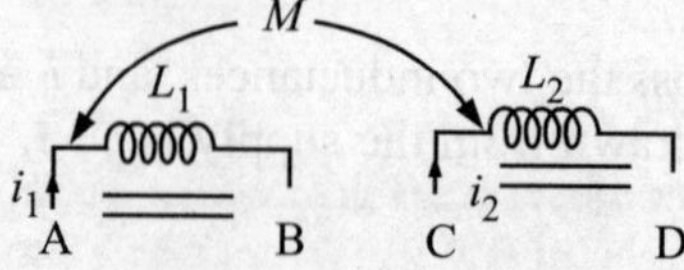

Fig. 9.13(a)

The voltage drop, or induced emf in coil 1 is

$$e_1' = L_1 \frac{di_1}{dt},$$

with the current in that coil as i_1. If the current in coil 2 is i_2, the magnitude of the voltage drop, or induced emf in coil 1 is

$$e_1'' = M \frac{di_2}{dt}.$$

Its sign indicates its polarity with respect to e_1'. In this case, the polarity is positive, i.e., this part is added to the first one, as the total voltage drop in coil 1 is

$$e_1 = e_1' + e_1'' = L_1 \frac{di_1}{dt} + M \frac{di_2}{dt}.$$

The polarity is shown in the circuit (Fig. 9.13(a)) by using dots at the points A and C, in the coils 1 and 2. This means that, if the current in coil 2 enters at the point C as shown, the induced voltage in coil 1 will have positive polarity at point A. As given in the equation, this voltage is added to the voltage drop due to change in coil 1. The sign of M is positive in this case.

Assuming the change in the current (i_2) of coil 2, the voltage drop in that coil is

$$e_2' = L_2 \frac{di_2}{dt}.$$

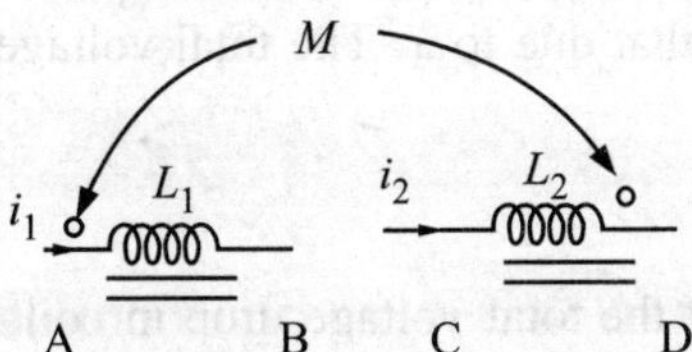

Fig. 9.13(b)

As is the case earlier, the voltage drop in coil 2 due to the change of current in coil 1 is

$$e_2'' = M \frac{di_1}{dt},$$

which is added to the earlier voltage drop such that the total voltage drop in coil 2 is

$$e_2 = e_2' + e_2'' = L_2 \frac{di_2}{dt} + M \frac{di_1}{dt}$$

Using dot convention, it can be observed that if the current in coil 1 enters at the point A as shown, the induced voltage in coil 2 will have positive polarity at point C, thus adding to the earlier voltage drop. This shows that the sign of M is positive when the currents in the two coils enter at the respective dot, i.e., A and C.

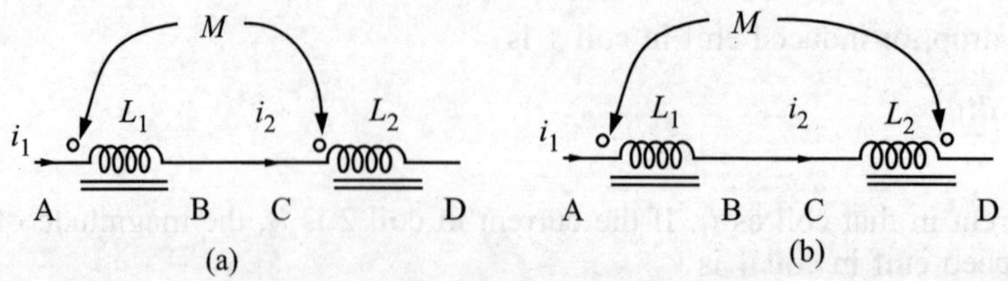

Fig. 9.14

If two coupled coils are connected in series (Fig. 9.14(a)), the currents in the coils are $i = i_1 = i_2$. The points B and C are connected. The total voltage drop now appears between A and D. Substituting this in the two equations, the total voltage drop will be the sum of the voltage drops in the two coils, which is,

$$e = L_{eq} \cdot \frac{di}{dt} = e_1 + e_2 = (L_1 + L_2 + 2M) \cdot \frac{di}{dt}.$$

The equivalent inductance for the two coupled coils connected in series aiding type, is

$$L_{eq} = (L_1 + L_2 + 2M).$$

The sign of M is positive in this case.

Consider another case where two coupled coils are connected in series (Fig. 9.14(b)), with B connected to D. The total voltage drop appears between A and C. The currents in the coils are the same as given earlier with $i = i_1 = i_2$. Using the dot convention as given here, the current in coil 2 enters at D, opposite to C (see the dot position). Hence, the voltage drop in coil 1 due to i_2 is opposite to that due to i_1. The total voltage drop in coil 1 now is

$$e_1 = v_{AB} = L_1 \frac{di_1}{dt} - M \frac{di_2}{dt}.$$

Similarly, it can be shown that the total voltage drop in coil 2 is

$$e_2 = v_{CD} = L_2 \frac{di_2}{dt} - M \frac{di_1}{dt}.$$

The reason for the negative sign has already been explained. Note that the two coupled coils are now connected in series opposition. The total voltage drop will be the sum of the voltage drops in the two coils, which is,

$$e = v_{AC} = v_{AB} + v_{DC} = L_{eq} \cdot \frac{di}{dt} = e_1 + e_2 = (L_1 + L_2 - 2M) \cdot \frac{di}{dt},$$

where $v_{DC} = v_{BC}$, since B is connected to D, as given earlier.

The equivalent inductance for the two coupled coils connected in series opposition type, is

$$L_{eq} = (L_1 + L_2 - 2M).$$

The sign of M is negative, as the current, i_1 enters at the dot (A), while the other current, i_2 leaves at the dot (C).

These two formulas may be combined as,

$$L_{\text{eq}} = L_1 + L_2 \pm 2M,$$

where the sign of M is to be chosen by the dot position given in the circuit—it shows, whether the coils are connected in series aiding, or series opposition.

Energy stored in two coupled coils

When the currents in two coils (Fig. 9.14) are changed, say, increased, the voltages induced in the two coils are given by,

$$e_1 = L_1 \frac{di_1}{dt} + M \frac{di_2}{dt}, \text{ and } e_2 = M \frac{di_1}{dt} + L_2 \frac{di_2}{dt}.$$

The symbols used are the same. Also note the *dot convention* given in the circuit (Fig. 9.14), which shows that the polarity is positive. Hence, M is taken as positive. The flux produced by the two currents aid each other here.

The instantaneous power in the two coils is given by

$$P = P_1 + P_2 = (e_1 i_1 + e_2 i_2) = L_1 \cdot i_1 \frac{di_1}{dt} + L_2 \cdot i_2 \frac{di_2}{dt} + M \cdot \left(i_1 \frac{di_2}{dt} + i_2 \frac{di_1}{dt} \right).$$

The increase of energy in the two coupled coils when the currents in the coils 1 and 2 are increased from I_1' to I_1'', and from I_2' to I_2'' respectively, in time, dt,

$$\begin{aligned} dW &= P \cdot dt = L_1 \cdot i_1 di_1 + L_2 \cdot i_2 di_2 + M \cdot (i_1 di_2 + i_2 di_1) \\ &= L_1 \cdot i_1 di_1 + L_2 \cdot i_2 di_2 + M \cdot d\,(i_1 \cdot i_2). \end{aligned}$$

Hence, the stored energy in the inductor, as the currents in the coils 1 and 2, increase from zero (0) to I_1, and 0 to I_2, respectively, is obtained as,

$$\begin{aligned} W &= \int_0^{I_1} L_1 \cdot i_1 di_1 + \int_0^{I_2} L_2 \cdot i_2 di_2 + \int_0^{I_1}\int_0^{I_2} M \cdot d\,(i_1 \cdot i_2) \\ &= \frac{1}{2} L_1 \cdot i_1^2 \Big|_0^{I_1} + \frac{1}{2} L_2 \cdot i_2^2 \Big|_0^{I_2} + M \cdot (i_1 \cdot i_2) \Big|_0^{I_1} \Big|_0^{I_2} = \frac{1}{2} L_1 I_1^2 + \frac{1}{2} L_2 I_2^2 + M I_1 I_2. \end{aligned}$$

Firstly, note the presence of an additional term involving the mutual inductance, M, between the two coupled coils. If two isolated inductors are used with no mutual inductance between them, then $M = 0$ and the additional term will not be there

Also note that the energy is stored in the magnetic field in either one core, or in two cores placed close to each other, over which the windings are placed. The core(s) may be iron or air.

EXAMPLE 9.7 Two *coupled coils are connected in series. The total inductance of the circuit are 0.8 mH and 0.2 mH, depending on the relative direction of the currents in the coils, as shown in Fig. 9.14(a) and Fig. 9.14(b) respectively. Find (a) the self inductances of two coils, (b) the mutual inductance between them, and (c) the coefficient of coupling. Assume that two coils are identical, with equal number of turns, and the magnetic circuit parameters are same for both of them.*

SOLUTION The circuits with two coupled coils connected in series are shown in Fig. 9.14(a)–(b). It is assumed that the self inductances of the two coils are L_1 and L_2, while the mutual inductance between them is M. As shown by the dot convention in Fig. 9.14, the total or equivalent inductance, L_{eq} of the circuits are given by

$$L_{eq} = L_1 + L_2 \pm 2M.$$

In the first case, the coils are connected in series aiding type according to dot convention (Fig. 9.14(a)), such that M is taken as +ve, while in the second case (Fig. 9.14(b)), the coils are connected in series opposition type, with M being taken as +ve. Hence, we get,

$$L_{eq1} = L_1 + L_2 + 2M = 0.8 \text{ mH, and } L_{eq2} = L_1 + L_2 - 2M = 0.2 \text{ mH}.$$

Please note that $L_{eq1} > L_{eq2}$, as M is +ve. Subtracting the first formula from the second one, $4M = (0.8 - 0.2) = 0.6$ mH,

or, mutual inductance, $M = \dfrac{0.6}{4} = 0.15$ mH.

According to assumptions, such as equal number of turns and the other conditions being the same as given before, the self inductance of the two coils are equal, i.e., $L_1 = L_2$.

Adding the two formulas given earlier, and substituting the above, we get,

$$4L_1 = (0.8 + 0.2) = 1.0 \text{ mH},$$

or, self inductance of the coils, $L_1 = L_2 = \dfrac{1.0}{4} = 0.25$ mH.

The coefficient of coupling, $k = \dfrac{M}{\sqrt{L_1 L_2}} = \dfrac{M}{L_1} = \dfrac{0.15}{0.25} = 0.6$ (60%), as $L_1 = L_2$.

EXAMPLE 9.8 *Two coupled coils have a self inductance of L_1 and L_2, and a mutual inductance of M. The coils are connected in parallel across A and B (Fig. 9.15(a)). Find the total or equivalent inductance of the combination.*

SOLUTION The circuit with two coupled coils connected in parallel across the points, A and B, is shown in Fig. 9.15(a). The sign of M is given by the position of dots. The circuit is connected to a dc supply and may be through a resistance (not shown). The currents in coils 1 and 2 are i_1 and i_2, with the total current being i, which is given by $i = i_1 + i_2$. As the currents enter at the dots (A) in two coils, the sign of M is positive.

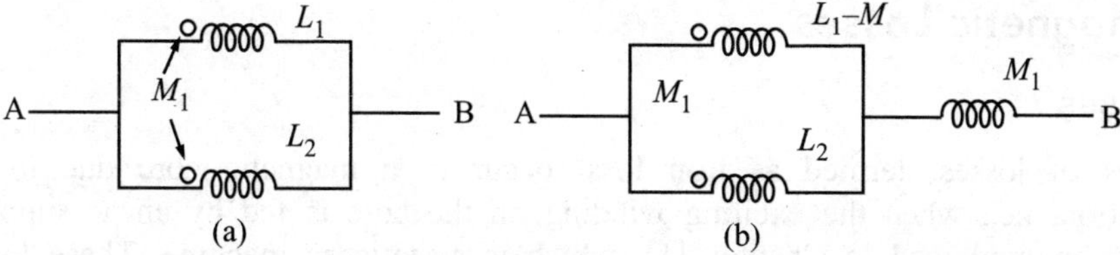

Fig. 9.15

The voltages across the two coils, are $v_{AB} = v_1 = v_2$, as they are connected in parallel. These are,

$$v_1 = L_1 \frac{di_1}{dt} + M \frac{di_2}{dt}, \text{ and } v_2 = M \frac{di_1}{dt} + L_2 \frac{di_2}{dt}.$$

The two equations are rewritten as

$$v_1 = (L_1 - M) \frac{di_1}{dt} + M \frac{d}{dt}(i_1 + i_2) = (L_1 - M) \frac{di_1}{dt} + M \frac{di}{dt}.$$

$$v_2 = (L_2 - M) \frac{di_2}{dt} + M \frac{d}{dt}(i_1 + i_2) = (L_2 - M) \frac{di_2}{dt} + M \frac{di}{dt}.$$

The equivalent circuit across A and B, according to these equations, is shown in Fig. 9.15(b). In the circuit, two inductances, $(L_1 - M)$ and $(L_2 - M)$ are connected in parallel, and the combination is in series with the inductance, M. The total or equivalent inductance across A and B, according to the dot convention shown in Fig. 9.15(b) is,

$$L_{eq} = L_{AB} = L_{AC} + L_{CB} = \frac{(L_1 - M) \cdot (L_2 - M)}{(L_1 - M) + (L_2 - M)} + M$$

$$= \frac{\left[L_1 L_2 - (L_1 + L_2) M + M^2\right] + \left[(L_1 + L_2) M - 2M^2\right]}{(L_1 + L_2 - 2M)} = \frac{L_1 L_2 - M^2}{L_1 + L_2 - 2M}.$$

If the dot position is changed, say, for coil 1, it is at A, and for coil 2 at B, the current in coil 1 enters at A, while the current in coil 2 enters at B (not shown). Hence, the sign of M is now negative. If $+M$ is substituted by $-M$ in the formula given earlier, the total or equivalent inductance is

$$L_{eq} = L_{AB} = \frac{L_1 L_2 - M^2}{L_1 + L_2 + 2M}.$$

These two formulas may be combined as

$$L_{eq} = \frac{L_1 L_2 - M^2}{L_1 + L_2 \mp 2M},$$

where the sign of M is to be chosen by the dot positions given in the circuit.

9.5 Magnetic Losses

Iron Losses

Two types of losses, termed as iron loss, occur in a magnetic core due to cyclical magnetisation, i.e., when the exciting winding on the core is fed by an ac supply, as in a transformer (explained in Chapter 11), which is a static ac machine. These losses are hysteresis and eddy current losses, which are explained briefly here.

Hysteresis loss

When a magnetic material is subjected to cyclical magnetisation, i.e., when the exciting winding on the magnetic material is fed by an ac supply, energy loss takes place. This is because the material is magnetised first in one direction (in the first cycle), and then in the opposite direction, as the current in the winding reverses in the second cycle. This process repeats every cycle. Losses occur in the magnetic material (in this case, iron) every cycle—this loss is termed as hysteresis.

The *B–H* curve of a magnetic material is explained in Chapter 8, when dc current is passed through the exciting winding. However, when ac voltage is applied, the curve is different—it becomes a hysteresis loop (Fig. 9.16) i.e., the relation between *B* and *H* is different. Starting from the negative maximum of the exciting current at point B of the loop, the curve of *B* vs *H*, follows a path as shown, when the exciting current increases from a –ve to a +ve value, till it reaches point A, which is the postive maximum of the exciting current. However, it first reaches zero (0) value of the intensity of the magnetic field, *H* at point C when the exciting current increases from a –ve value to zero (0). At this point, the value of the flux density, *B* still remains negative (–). If the current is increased from zero (0) to a positive value, *B* reaches zero (0) at point D, with *H* now becoming +ve. As the current is increased to the +ve maximum value, point A is reached. In the same way, if the current is decreased from the +ve maximum to the –ve maximum, the path (trajectory), AEFB is followed by the *B–H* curve, thus forming a loop in every cycle. The points, E and F correspond to C and D respectively.

Using the same symbols as used in Chapter 8, the strength of the magnetic field in the core, when the winding on it is excited, is given by,

$$H = \frac{N.i}{l}.$$

The emf induced in the coil is,

$$e = Na\frac{dB}{dt}.$$

The energy loss in the core (magnetic material) in Joules (J) is obtained as $dW = ei.dt = (al).H.dB$, where the volume of the core is $v = a.l$.

Now, $\int H.dB$, gives the area of the hysteresis loop, which gives the loss per cycle per volume.

The hysteresis loss (W) is found as $P_h = W_h.f = v.f.\int H.dB$.

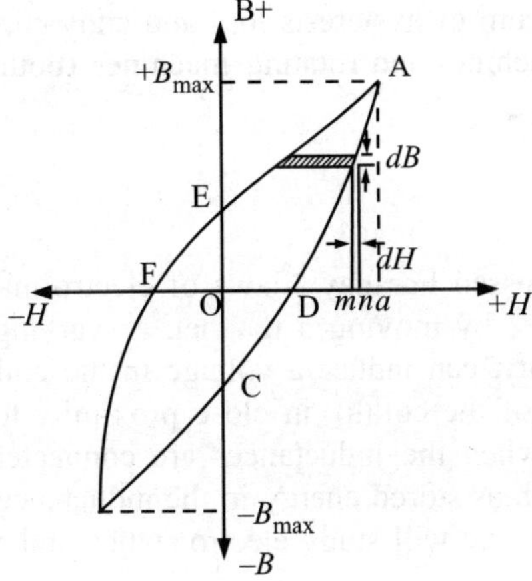

Fig. 9.16 *Hysteresis loops*

Eddy current loss

When a magnetic material is subjected to a changing magnetic field, eddy current loss also occurs, in addition to hysteresis loss. The changing flux induces voltage in the material (core) according to Faraday's law of electromagnetic induction. Since the material is conducting, these induced voltages circulate currents within the body of the material. The induced currents, termed as eddy currents, cause loss in the material.

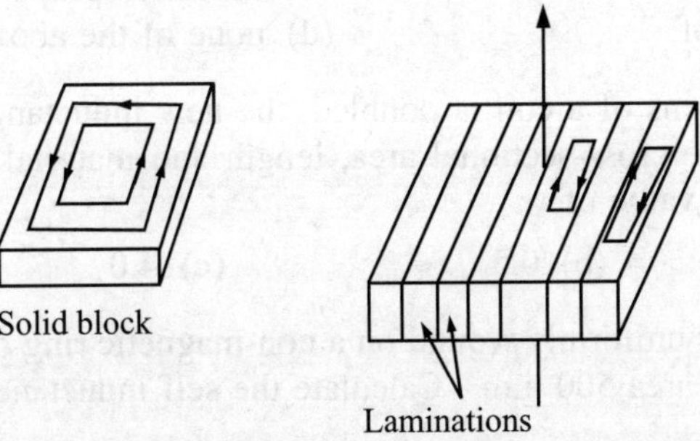

Fig. 9.17 *Eddy current loss*

Consider a solid block of iron subjected to a changing magnetic field (Fig. 9.17). The eddy current power loss in the block is given by i^2R, with i and R being the eddy current and resistance of the path. As the block has a continuous path, this loss is quite large. The obvious method to reduce the eddy current loss is to split the solid block into thin sheets called laminations (Fig. 9.17). This reduces the length of the current path, thus reducing its resistance. Further reduction can be achieved by using a magnetic material of high resistivity, so as to reduce the resistance, and hence the above loss. If the winding on the material (core) is fed by an ac supply of frequency, f, the formula for this loss (W) is $P_e = k_e.(B_m)^2 f^2.t^2.v$, where t is the thickness of the lamination (stamping) used.

The total iron loss is the sum of hysteresis loss and eddy current loss, which takes place in transformers (static ac machine) and rotating machines (both generators and motors) of both (dc and ac) types.

Conclusion

In this chapter, we first discussed Faraday's laws of electro-magnetic induction and show that the change of flux, either by moving a magnet, or varying the current in the coil, or another coil in close proximity, can induce a voltage in the coil. Then, the expressions for self and mutual inductances of the coil(s), in close proximity to each other, are presented. The equivalent inductance, when the inductances are connected in series and parallel, is determined. Other topics, such as stored energy in the inductance and magnetic loss are also discussed. In the next chapter, we will study electro statics and capacitance in detail.

Multiple Choice Type Questions

Tick the correct option(s), out of the four choices provided in each question.

9.1 The current in a coil is 1 A. The flux linkage due to it is found to be 8 Wb turn. Its self inductance in H is,

(a) 2.0 (b) 0.25 (c) 8.0 (d) 0.125

9.2 The emf induced in a coil is .... the rate of change in the flux linkages.

(a) directly proportional to (b) inversely proportional to
(c) independent of (d) none of the above

9.3 If the number of turns of a coil is doubled, the new inductance formed is x times the previous value. The cross-sectional area, length and material inside the core, remain the same. Find the value of x.

(a) 2.0 (b) 0.5 (c) 4.0 (d) 0.25

9.4 A 2,000-turn coil is uniformly wound on a non-magnetic ring of mean diameter 30 cm, and cross-sectional area 500 mm^2. Calculate the self inductance in mH of the toroidal coil so formed.

(a) 1.33 (b) 2.67 (c) 5.34 (d) 3.0

9.5 A conductor of length 10 cm, and carrying a current of 40 A, lies perpendicular to a magnetic field intensity of 1,000 A/m. Calculate the force in mN, acting on the conductor.

(a) 1.6 (b) 5.027 (c) 0.8 (d) 10.054

9.6 Two coils are wound on a single core whose cross-sectional area, length and the material remain the same. If the number of turns of the second coil is halved, while keeping that of the first one the same, the mutual inductance between the coils is found to be x times its previous value. Find the value of x.

(a) 2.0 (b) 0.25 (c) 4.0 (d) 0.5

9.7 The mutual inductance between two coils is 0.2 H. If the current in one coil changes at the rate of 5 A/s, then the emf induced in the other coil in V is,

(a) 2.0 (b) 0.05 (c) 1.0 (d) 0.04

9.8 Two coils, having 150 and 250 turns respectively, are wound side by side, on a closed iron circuit, with mean length of 200 cm and cross-sectional area of 100 cm^2. If the relative permeability of iron is assumed to be 1,000, find the mutual inductance between the coils in H.

(a) 0.47 (b) 0.5 (c) 0.118 (d) 0.236

9.9 The inductance of a coil is .... the reluctance of the magnetic path.

(a) directly proportional to (b) inversely proportional to
(c) independent of (d) none of the above

9.10 Two coils are wound on a single core having mutual inductance between them. Firstly, when two coils are connected in series with their fluxes aiding, the total inductance is 0.62 H. When the connection to one of the coils is reversed (they are still connected in series), the total inductance now changes to 0.18 H. Find the value of the mutual inductance in H.

(a) 0.2 (b) 0.4 (c) 0.22 (d) 0.11

9.11 A horse shoe magnet has two poles, each of area 5 cm^2. Find the force between the poles and the iron piece, in N, when the flux density at the contact surface is 0.8 Wb/m^2.

(a) 255.0 (b) 128.0 (c) 400.0 (d) 510.0

9.12 The self inductance of a coil is 0.2 H. If the current in the coil is 5 A, find the energy (J) stored in the magnetic field.

(a) 0.5 (b) 2.5 (c) 1.0 (d) 0.1

Answers

9.1 (c)	9.2 (a)	9.3 (c)	9.4 (b)	9.5 (b)	9.6 (d)
9.7 (c)	9.8 (d)	9.9 (b)	9.10 (d)	9.11 (a)	9.12 (b)

Problems

9.1 A conductor of length 75 cm moves with a uniform velocity of 4 m/s, perpendicular to a magnetic field of flux density 0.8 Wb/m^2 (Fig. P9.1). Calculate,

(a) The emf induced in the conductor.

(b) The mechanical force developed in the conductor, when it carries a current of 10 A.

(c) The power required to drive the conductor against the mechanical force.

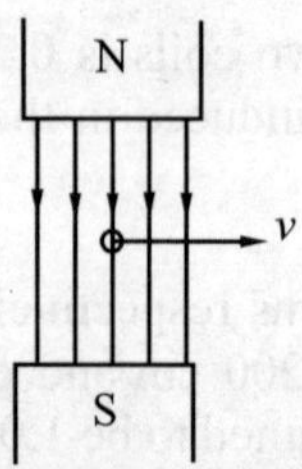

Fig. P9.1

9.2 A square coil of side 10 cm, having 100 turns, is rotated at a uniform speed of 500 rpm, about an axis at right angle to a uniform magnetic field of 0.5 Wb/m^2 (Fig. P9.2). Calculate the instantaneous value of induced emf, when the plane of the coil is
(a) at right angle ($\theta = 90°$) to the plane of the magnetic field,
(b) in the plane of the magnetic field ($\theta = 0°$), and
(c) at 45° with the direction of the magnetic field ($\theta = 45°$).

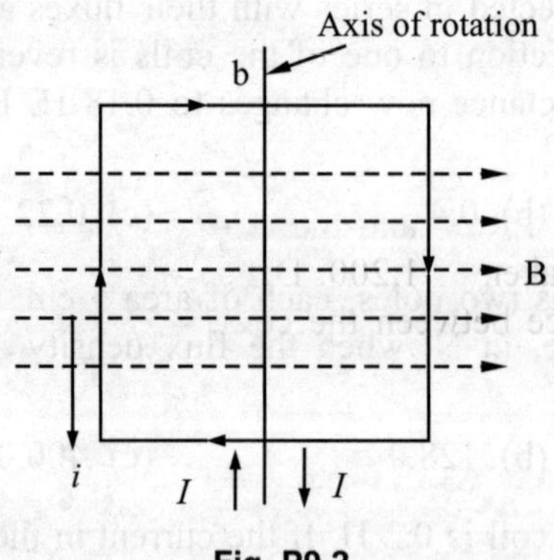

Fig. P9.2

9.3 A ring, made of steel has a rectangular cross-sectional area. Its inner and outer diameters are 15 cm and 20 cm respectively, the thickness being 2 cm. The ring has a winding of 800 turns, and when carrying a current of 2 A, a flux density of 1.2 Wb/m^2 is produced in the air gap. The ring is cut to form an air gap of 1 mm length (Fig. P9.3). Calculate,
(a) the relative permeability of the magnetic material,
(b) the inductance of the coil, and
(c) the average emf induced in it, when a current of 1 A is reversed in 0.05 s.

Also find the total reluctance of the above magnetic circuit. Neglect fringing and leakage effects.

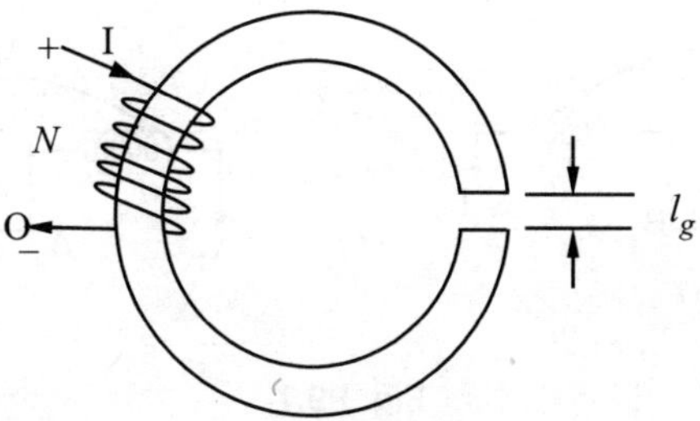

Fig. P9.3

9.4 An iron ring has a rectangular cross-sectional area of 25 cm^2, and a mean circumference of 100 cm. The ring has a winding of 250 turns. The exciting current is 2 A. Assume that the relative permeability of iron is 2,000. Find (a) the flux in the core, and (b) the self inductance of the winding (coil).

9.5 A coil has 600 turns, wound on an air cored toroidal core, which has a mean diameter of 25 cm. It also has a square cross-section with the side as 3 cm. If the exciting current is 2 A, determine (a) the flux in the core, and also (b) the self inductance of the coil.

9.6 Two coils, having 40 and 200 turns respectively, are wound side by side on a magnetic core. Its mean length is 1.0 m, and the cross-sectional area is 200 cm^2. The relative permeability of iron is taken as 1,200. Determine (a) the self inductance of two coils, (b) the mutual inductance between the coils, (c) the induced voltage in coil 2, if the current changes from 0 to 8 A in 0.1 s. If the coupled coils are connected in series aiding mode, as shown in Fig. P 9.6, find the total stored energy in the circuit, when the current in the coils is 8 A.

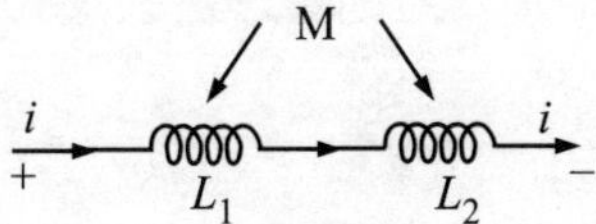

Fig. P9.6

9.7 Two coils, with terminals, A_1 and A_2, and B_1 and B_2 respectively, are placed side by side. Measured separately, the inductance of coils A and B, are 1.6 and 1.0 mH respectively. Point A_2 is connected to B_1 (Fig. P9.7a). The total inductance of two coils in series between A_1 and B_2 is 4.0 mH. What is the mutual inductance between the coils? Also find the coefficient of coupling.

Now, point A_2 is connected to B_2 (Fig. P9.7b). Determine the total inductance of two coils in series between A_1 and B_1. The dot convention is given in the two figures.

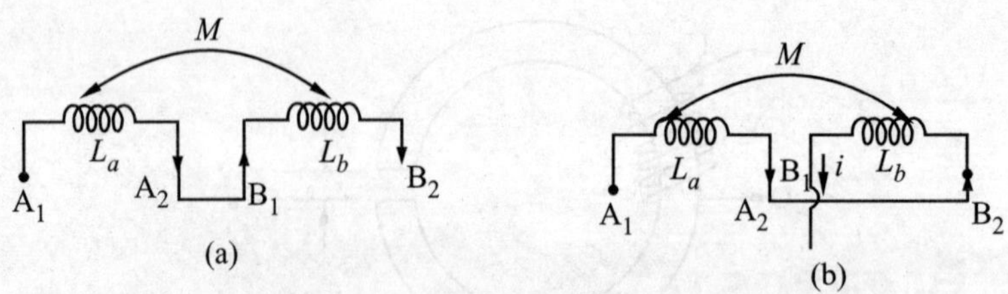

Fig. P9.7

9.8 A toroidal magnetic core, with a mean diameter of 30 cm, and a cross-sectional area of 5 cm^2, is uniformly wound with a coil of 800 turns. It carries a current of 3 A. The relative permeability of magnetic material is 600. Determine (a) the self inductance of the coil, and (b) the emf induced in the coil, when the current is reduced to zero (0) in 20 ms. Assume the slope to be linear.

CHAPTER 10

ELETROSTATICS AND CAPACITANCE

Introduction

In the previous chapter, we studied inductance of a coil. In this chapter, we will discuss capacitance in detail. Starting with Coulomb's laws, the electric field intensity, electric flux, potential, and Gauss's theorem, we go on to study different types of capacitors, along with series and parallel connections of capacitors. Then, capacitances of various shapes, like parallel plates, spheres (single and concentric) and also concentric cylinders, are discussed in detail.

10.1 Coulomb's Laws

FIRST AND SECOND LAWS

1. Like charges repel each other, and unlike charges attract each other. This means that if both charges are positive or negative, they will repel each other, while, if one charge is positive, and the other one is negative, they will attract each other. This is only a qualitative law.

2. The force between two point charges is proportional to the product of the charges, q_1 and q_2, and inversely proportional to the square of the distance, r between them (Fig. 10.1).

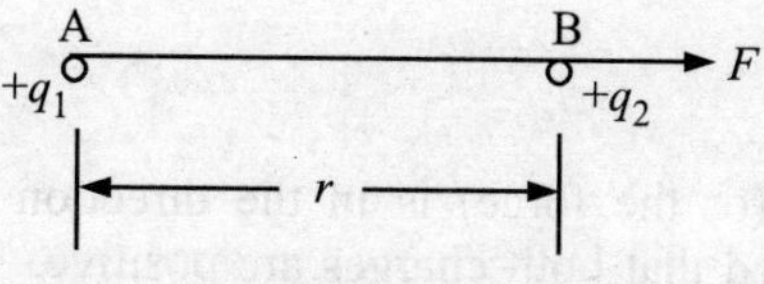

Fig. 10.1

This law, which is termed as a quantitative law, is given by,

$$F = \frac{q_1 q_2}{4\pi\varepsilon r^2}$$

where,

F = force in Newton

ε = $\varepsilon_0\varepsilon_r$ = total permittivity of the medium in F/m or Fm^{-1}

ε_0 = $8.854 \times 10^{-12} = \dfrac{1}{36\pi \times 10^9}$ = permittivity of the free space or air in F/m

ε_r = relative permittivity of the medium

Please note that the units of ε and ε_0 are the same (F/m), while ε_r is a constant which depends on the medium, with no units.

Two positive point charges, q_1 and q_2, are placed at A and B, with AB = r. The units of the two charges, q_1 and q_2, and the distance, r are in Coulomb (C) and m respectively. The direction of the force, F between them is determined by the first law. If both charges are positive, as given earlier, or negative, the force, F produced on the second charge, q_2 at B, is in the direction, AB (Fig. 10.1), as it is a repelling force. On the other hand, if q_1 is positive, and q_2 is negative, or vice versa, the force, F produced on the second charge, q_2 at B, is in the direction, BA, now being an attractive one. This is not shown here. It may be noted that in the above cases, the force is due to the first charge, q_1 (+ve/−ve, as the case may be), placed at point A.

10.2 Electric Field Intensity due to a Point Charge

An electric field is produced when a point charge, $+q$ is placed at O (Fig. 10.2). If a unit positive charge is placed at P, with OP = r, the repelling force (N) at P is given by,

$$F = \frac{q}{4\pi\varepsilon r^2}$$

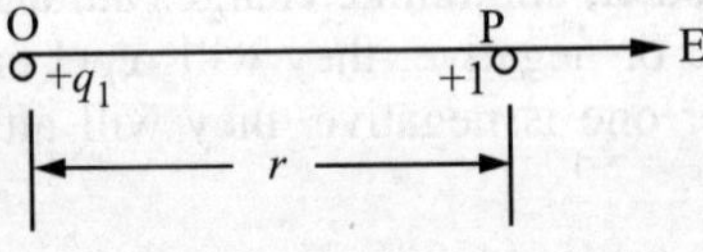

Fig. 10.2

The intensity of the electric field, E, the unit of which is N/C, or V/m, or force, F produced on the unit positive charge, at P, is,

$$E = \frac{q}{4\pi\varepsilon r^2}$$

The electric field intensity (or the force) is in the direction OP, as the above force is a repelling one. It may be noted that both charges are positive. The first unit (N/C), as given, is self-explanatory.

If the point charge is negative ($-q$), with its magnitude remaining the same, the electric field intensity at the point P, is given by

$$E = -\frac{q}{4\pi\varepsilon r^2},$$

which means that its magnitude remains the same, but its direction is opposite, i.e., in the direction, PO. The force produced on the unit positive charge at P, is an attracting one.

The electric field intensity, E, or the resultant force, F on a charge, q, due to a group of charges, q_1, q_2, etc. placed at different points, is computed, as given in the solved examples.

EXAMPLE 10.1 *A rectangle, ABGH is shown in Fig. 10.3(a), with its two sides, AB and BG as 10 cm and 5 cm respectively. Two point charges of +2μC and – 4μC, are placed in air at the points, A and B respectively. Find the resultant force on a charge of +1 μC, if it is placed at the points (i) G and (ii) H.*

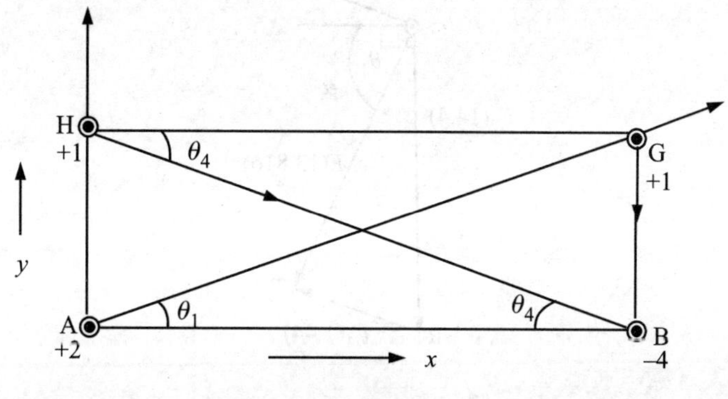

Fig. 10.3(a)

SOLUTION ABGH is a rectangle as shown in Fig. 10.3(a). The side, AB is taken as the x-axis, while the side, AH is the y-axis, as ∠ HAB = 90°.

AB = HG = 10 cm, BG = AH = 5 cm.

$$\varepsilon_0 = \frac{1}{36\pi \times 10^9} \text{ F/m}; \qquad 4\pi\varepsilon_0 = \frac{4\pi}{36\pi \times 10^9} = \frac{1}{9} \times 10^{-9}.$$

(i) $r_1 = \text{AG} = \sqrt{(\text{AB})^2 + (\text{BG})^2} = \sqrt{10^2 + 5^2} = \sqrt{125} = 11.18 \text{ cm} = 0.1118 \text{ m}$,

as ∠ ABG = 90°.

Force on charge $q' = +1\ \mu$C at point G due to charge, $q_1 = +2\ \mu$C at point A is repelling in the direction AG. Its magnitude is,

$$F_1 = \frac{q_1 q'}{4\pi\varepsilon_0\varepsilon_r r_1^2} = \frac{2 \times 10^{-6} \times 1 \times 10^{-6} \times 9 \times 10^9}{(0.1118)^2} = 1.44 \text{ N}.$$

$r_2 = \text{BG} = 5 \text{ cm} = 0.05 \text{ m}$.

Force on charge $q' = +1\ \mu$C at point G due to charge, $q_2 = -4\ \mu$C at point B is attracting in the direction GB, which is at right angles to AB (x-axis). Its magnitude is,

$$F_2 = \frac{q_2 q'}{4\pi\varepsilon_0\varepsilon_r r_2^2} = \frac{4 \times 10^{-6} \times 1 \times 10^{-6} \times 9 \times 10^9}{(0.05)^2} = 14.4 \text{ N}.$$

The two forces, F_1 and F_2, are to be added as vectors since each one has not only magnitude, but also a direction (Fig. 10.3(b)). Here, the origin is also taken as point G. GS = RT = F_1, and GR = F_2.

$$\theta_1 = \angle \text{BAG}, \qquad \tan\theta_1 = \tan\angle\text{BAG} = \frac{\text{BG}}{\text{AB}} = \frac{5}{10} = 0.5.$$

$$\theta_1 = \tan^{-1}(0.5) = 26.565°.$$

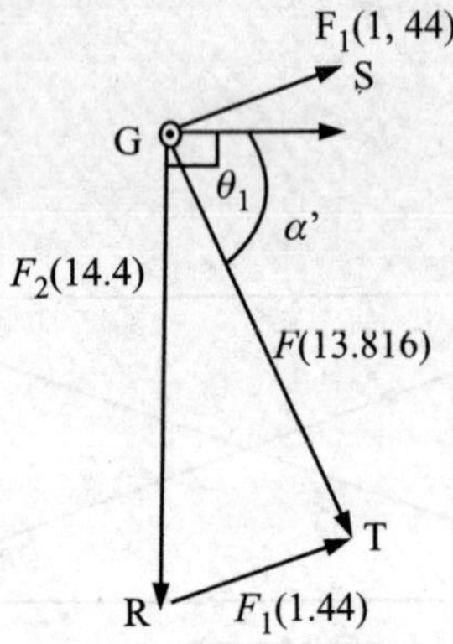

Fig.10.3(b)

The resultant force in the direction of the x-axis

$$= F_1\cos\theta_1 = 1.44 \times \cos 26.565° = 1.288 \text{ N}.$$

The resultant force in the direction of the y-axis

$$= F_1\sin\theta_1 - F_2 = 1.44 \times \sin 26.565° - 14.4 = 0.644 - 14.4 = -13.756 \text{ N}.$$

The magnitude of the resultant force = GT = $F'_G = \sqrt{(1.288)^2 + (13.756)^2}$

$$= 13.816 \approx 13.82 \text{ N}.$$

The angle of the resultant force with respect to the x-axis (AB), is found as,

$$\alpha' = \tan^{-1}\left(\frac{-13.756}{1.288}\right) = \tan^{-1}(-10.68) = -84.651°.$$

Alternatively, they can be added,

$$F'_G\angle\alpha' = F_1\angle\theta_1 + F_2\angle -90° = 1.44\angle 26.565° + 1.44\angle -90° = (1.288 + j0.644) - j14.4$$

$$= (1.288 - j13.756) = 13.816\angle -84.651°.$$

The values of F'_G and α' are the same as found earlier.

(b) r_3 = HA = 5 cm = 0.05 m.

Force on charge $q' = +1\ \mu$C at point H due to charge $q_1 = +2\ \mu$C at point A is in the direction, AH. Note that this is a repelling force. Its magnitude is

$$F_3 = \frac{q_1 q'}{4\pi\varepsilon_0\varepsilon_r r_3^2} = \frac{2\times10^{-6}\times1\times10^{-6}\times9\times10^9}{(0.05)^2} = 7.2\text{ N.}$$

$$r_4 = \text{BH} = \sqrt{(\text{BA})^2+(\text{AH})^2} = \sqrt{10^2+5^2} = \sqrt{125} = 11.18\text{ cm} = 0.1118\text{ m,}$$

as $\angle\text{BAH} = 90°$ (Fig. 10.3(a)).

Force on charge $q' = +1\ \mu$C at point H due to charge $q_2 = -4\ \mu$C at point B is in the direction, HB. Note that this is an attracting force. Its magnitude is

$$F_4 = \frac{q_2 q'}{4\pi\varepsilon_0\varepsilon_r r_4^2} = \frac{4\times10^{-6}\times1\times10^{-6}\times9\times10^9}{(0.1118)^2} = 2.88\text{ N.}$$

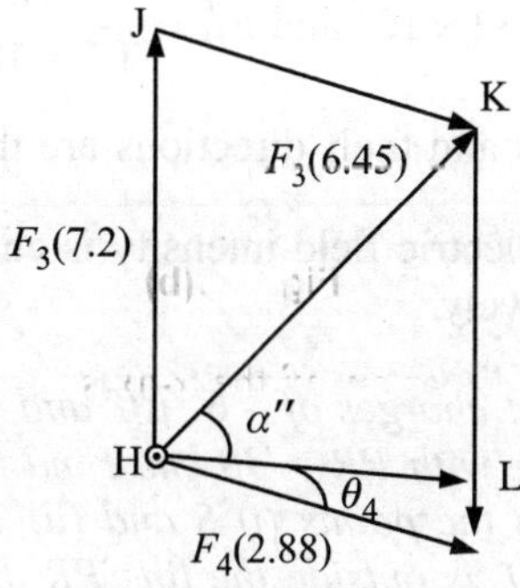

Fig.10.3(c)

As given earlier in the first example, two forces, F_3 and F_4, are to be added as vectors, since each one has not only magnitude, but also angle (Fig. 10.3(c)). Here, the origin is taken as the point H, to differentiate it from the first example. HJ = F_3 and HL = JK = F_4.

Again, firstly, the forces, F_3 and F_4 are resolved into x and y components.

$$\theta_4 = \angle\text{HBA},\ \tan\theta_4 = \tan\angle\text{HBA} = \frac{\text{AH}}{\text{AB}} = \frac{5}{10} = 0.5\ .$$

$$\theta_4 = \tan^{-1}(0.5) = 26.565°.$$

The resultant force in the direction of the x-axis

$$= F_4\cos\theta_4 = 2.88\times\cos 26.565° = 2.576\text{ N.}$$

The resultant force in the direction of the y-axis

$$= -F_4\sin\theta_4 + F_3 = -(2.88\times\sin 26.565°) + 7.2 = -1.288 + 7.2 = 5.912\text{ N.}$$

The magnitude of the resultant force $= F_H'' = \sqrt{(2.576)^2 + (5.912)^2} = 6.449 \approx 6.45$ N.

The angle of the resultant force with respect to the x-axis (AB), is found as,

$$\alpha'' = \tan^{-1}\left(\frac{5.912}{2.576}\right) = \tan^{-1}(2.295) = 66.46°.$$

Alternatively, they can be added,

$$F_H''\angle\alpha'' = F_3\angle 90° + F_4\angle - \theta_4 = 7.2\angle 90° + 2.88\angle - 26.565° = j7.2 + (2.576 - j1.288)$$

$$= (2.576 + j5.912) = 6.45\angle 66.46°.$$

The values of F_H'' and α'' are the same as found earlier.

The magnitudes of the resultant electric field intensity in the above two cases are,

$$E_G' = \frac{13.82}{1.0 \times 10^{-6}} = 13.82 \times 10^6 \text{ and } E_H'' = \frac{6.45}{1.0 \times 10^{-6}} = 6.45 \times 10^6.$$

Their units are N/C or V/m and their directions are the same as given earlier.

If the unit of the resultant electric field intensity is changed to kV/mm, the magnitudes are 13.82 and 6.45 respectively.

EXAMPLE 10.2 *Two point charges of – 6 μC and + 8 μC are placed in air at the points P and R respectively, with PR = 30 cm. Find the resultant force on the charge of +4 μC, if it is placed at the points (i) S and (ii) T. The point S is between P and R, with RS = 10 cm, while T is outside the line PR, being away from R with RT = 10 cm (Fig. 10.4(a)).*

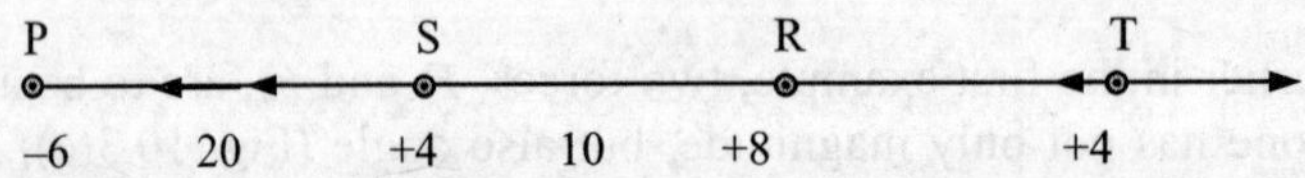

Fig.10.4(a)

SOLUTION: (i) PR = 30 cm = 0.3 m (Fig. 10.4(a)),

$$r_2 = \text{RS} = \text{SR} = 10 \text{ cm} = 0.1 \text{ m}.$$

Therefore, $r_1 = \text{PS} = \text{PR} - \text{SR} = 0.3 - 0.1 = 0.2 \text{ m} = 20 \text{ cm}.$

$$4\pi\varepsilon_0 = \frac{4\pi}{36\pi \times 10^9} = \frac{1}{9} \times 10^{-9}.$$

Force on charge $q' = +4$ μC at point S due to charge $q_2 = 8$ μC at point P is in the direction, SP. Note that this is an attracting force. Its magnitude is

$$F_1 = \frac{q_1 q'}{4\pi\varepsilon_0\varepsilon_r r_1^2} = \frac{6\times10^{-6}\times4\times10^{-6}\times9\times10^9}{(0.2)^2} = 5.4 \text{ N}.$$

Force on charge $q' = +1\ \mu$C at point S due to charge $q_2 = +8\ \mu$C at point R is in the direction RS or SP. Note that this is a repelling force. Its magnitude is,

$$F_2 = \frac{q_2 q'}{4\pi\varepsilon_0\varepsilon_r r_2^2} = \frac{8\times10^{-6}\times4\times10^{-6}\times9\times10^9}{(0.1)^2} = 28.8 \text{ N}.$$

As given, two forces, F_1 and F_2 are in the same direction (Fig. 10.4(b)), SP or RS, with the resultant force also being in the same direction.

The magnitude of the resultant force $= F' = F_1 + F_2 = 5.4 + 28.8 = 34.2$ N.

(ii) $r_4 =$ RT $= 10$ cm $= 0.1$ m (Fig. 10.4(a)).

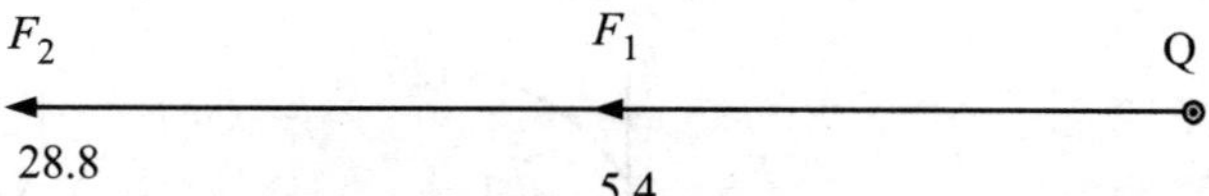

Fig.10.4(b)

Fig.10.4(c)

So, $r_3 =$ PT $=$ PR $+$ RT $= 0.3 + 0.1 = 0.4$ m $= 40$ cm.

Force on charge $q' = +4\ \mu$C at point T due to charge $q_1 = -6\ \mu$C at point P is in the direction TP, or TR. Note that this is an attracting force. Its magnitude is,

$$F_3 = \frac{q_1 q'}{4\pi\varepsilon_0\varepsilon_r r_3^2} = \frac{6\times10^{-6}\times4\times10^{-6}\times9\times10^9}{(0.4)^2} = 1.35 \text{ N}.$$

Force on charge $q' = +4\ \mu$C at point T due to charge $q_2 = +8\ \mu$C at point, R, in the direction, RT. Note that this is a repelling force. Its magnitude is,

$$F_4 = \frac{q_2 q'}{4\pi\varepsilon_0\varepsilon_r r_4^2} = \frac{8\times10^{-6}\times4\times10^{-6}\times9\times10^9}{(0.1)^2} = 28.8 \text{ N}.$$

The magnitude is the same as obtained earlier, as the distance, RT, between the two charges is the same, i.e., 10 cm.

As given, two forces, F_3 and F_4 are in the opposite direction (Fig. 10.4(b)), with the resultant force being the difference of the above two forces.

The magnitude of the resultant force $= F_T'' = F_4 - F_3 = 28.8 - 1.35 = 27.45$ N, with its direction being RT, or PT. It may be noted that this direction is opposite to that found in the earlier case (i).

As shown in EXAMPLE 10.1, the magnitudes of the resultant electric field intensity at points S and T due to two point charges, q_1 and q_2, placed at points P and R respectively, are,

$$E'_S = \frac{34.2}{4.0 \times 10^{-6}} = 8.55 \times 10^6, \text{ and } E''_T = \frac{27.45}{4.0 \times 10^{-6}} \simeq 6.863 \times 10^6.$$

Their units are N/C, or V/m. Their directions are the same as given earlier.

If the unit of resultant electric field intensity is changed to kV/mm, the magnitudes are 8.55 and 6.863 respectively.

EXAMPLE 10.3 *Three positive point charges of 2.0 μC each, are placed, in air, at the vertices of an equilateral triangle, ABD, whose sides are 15 cm each (Fig. 10.5(a)). Find the resultant force on one charge due to the other two charges.*

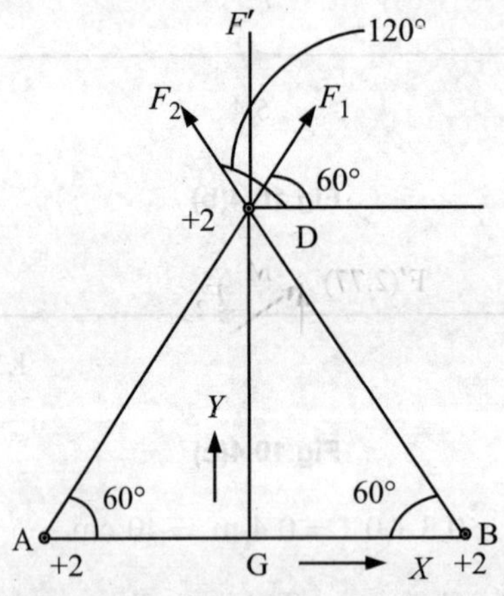

Fig.10.5(a)

SOLUTION ABC is an equilateral triangle (Fig. 10.5(a)), with the sides,

AB = AD = BD = 15 cm (0.15 m).

The angles are ∠ ABD = ∠ ADB = ∠DAB = 60°.

One charge of 2 μC is placed at point, D. The other two point charges, having the same value, are placed at points A and B.

$$r_1 = \text{AD} = 0.15 \text{ m}; \; 4\pi\varepsilon_0 = \frac{4\pi}{36\pi \times 10^9} = \frac{1}{9} \times 10^{-9}.$$

Force on charge $q_3 = +2.0\,\mu$C at point D, due to charge $q_1 = +2.0\,\mu$C at point A, is in the direction AD. Note that this is a repelling force. Its magnitude is,

$$F_1 = \frac{q_1 q_3}{4\pi\varepsilon_0\varepsilon_r r_1^2} = \frac{2.0 \times 10^{-6} \times 2.0 \times 10^{-6} \times 9 \times 10^9}{(0.15)^2} = 1.6 \text{ N}.$$

$r_2 = \text{BD} = 0.15$ m.

Force on charge $q_3 = +2.0\mu\text{C}$ at point D due to charge $q_2 = +2.0\mu\text{C}$ at point B is in the direction, BD. Note that this is also a repelling force. Its magnitude is,

$$F_2 = \frac{q_2 q_3}{4\pi\varepsilon_0\varepsilon_r r_2^2} = \frac{2.0\times10^{-6}\times2.0\times10^{-6}\times9\times10^{9}}{(0.15)^2} = 1.6 \text{ N}.$$

The two axes (x and y) are determined (or fixed) in this way (Fig. 10.5(a)). G is the mid-point of AB, with AG = GB = $(0.5\times15.0) = 7.5$ cm. Line GD is perpendicular to GB, as the triangle, ABD is equilateral. The point G is taken as the origin (0, 0). Line, GB is taken as the x-axis, while GD is the y-axis.

Two forces, F_1 and F_2, are to be added as vectors, since each has not only magnitude, but also a direction (Fig. 10.5(b)). In this case, the origin is also taken as the point, D. DH = F_1 and DJ = F_2. The resultant force is

DM = DH + HM, where HM = DJ.

Note that DM is along the y-axis.

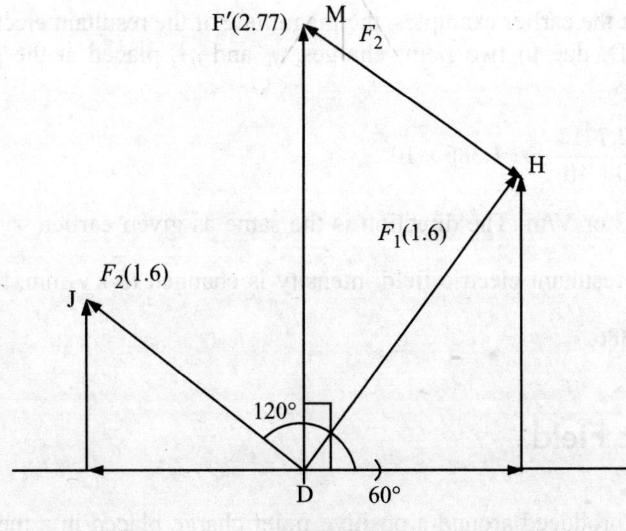

Fig.10.5(b)

The forces, F_1 and F_2 are resolved into x and y components.

As given earlier, $\theta_1 = \angle\text{DAB} = 60°$.

Force, F_2 is at an angle, $\theta_2 = (180° - 60°) = 120°$, with respect to the x-axis, GB, as

$\angle\text{DBA} = 60°$.

The resultant force in the direction of the x-axis (GB)

$$= F'_x = F_1 \cos\theta_1 + F_2 \cos\theta_2 = [(1.6 \times \cos 60°) + (1.6 \times \cos 120°)] = (0.8 - 0.8) = 0.0 \text{ N}.$$

The two x-axis components of F_1 and F_2 cancel each other.

The resultant force in the direction of the y-axis (GD)

$$= F'_D = F'_y = F_1 \sin\theta_1 + F_2 \sin\theta_2 = [(1.6 \times \sin 60°) + (1.6 \times \sin 120°)]$$

$$= (1.3856 + 1.3856) = 2.7712 \approx 2.77 \text{ N}.$$

The direction of this force is along GD (y-axis), as $F'_x = 0.0$.

Alternatively, they can be added,

$$F'_D \angle \alpha' = F_1 \angle 60° + F_2 \angle 120° = 1.6 \angle 60° + 1.6 \angle 120°$$

$$= (0.8 + j1.3856) + (-0.8 + j1.3856) = j2.7712 \approx j2.77 = 2.77 \angle 90°.$$

The values of F'_D and α' are the same as found earlier. The resultant force is in the direction of the y-axis (GD).

As given in the earlier examples, the magnitude of the resultant electric field intensity at the point, D, due to two point charges, q_1 and q_2, placed at the points, A and B respectively, is,

$$E'_D = \frac{2.7712}{2.0 \times 10^{-6}} \approx 1.386 \times 10^6.$$

Its unit is N/C or V/m. The direction is the same as given earlier.

If the unit of resultant electric field intensity is changed to kV/mm, the magnitude is,

$$E'_D = 1.386.$$

10.3 Electric Field

An electric field is produced around a positive point charge placed in a medium or air. The lines of force go out of this point charge (Fig. 10.6). If a point negative charge is placed, the lines of force enter this point charge (this is not shown in the figure). These lines are normal to the plane of the point charge, where the charge is placed—which means that the medium or air is stressed by the point charge. If two such point charges (one positive and the other negative) are placed at some distance, the lines of force come out from the positive charge and enter into the negative charge, thus closing the path, similar to the magnetic lines of force going from the North pole to the South pole, explained in Chapter 8.

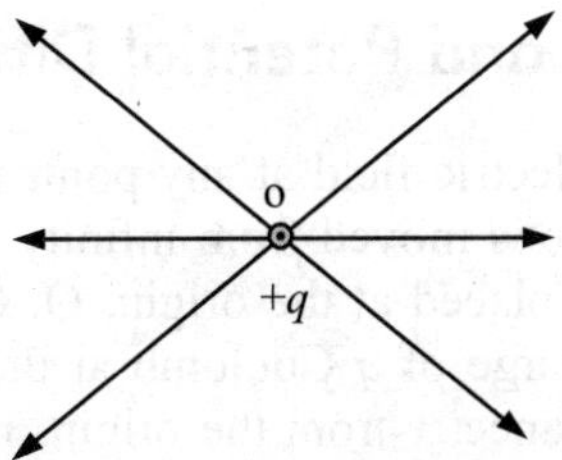

Fig.10.6

10.4 Electric Flux and Flux Density

Electric flux is measured by the number of lines of force in a given area, which may be small, in the medium. As given earlier, at least, one point charge (positive/negative) must be present. The symbol of electric flux is ψ. Since electric flux is equal to the charge contained (proved later), its unit can also be Coulomb (C).

Electric flux density is the electric flux divided by the small area around a point over which the electric flux is measured. This gives rise to uniform or constant electric flux density in that small area. The symbol of electric flux density is D; it can be measured by $D = \frac{\phi}{a}$, where a = area (small) in m^2. The unit of electric flux density is C/m^2.

Electric field intensity (E) has been discussed earlier. The electric flux density (D) is related to electric field intensity or field strength in the same manner as magnetic flux density (B) and magnetic field intensity, or field strength (H), explained in Chapter 8.

Electric flux density (D) = Total permittivity of the medium (ε) $\times$ Electric field intensity (H).

Total permittivity of the medium (ε) is expressed as the product of permittivity of free space or air (ε_0), and relative permittivity or dielectric constant (ε_r), of the medium, i.e.,

$$\varepsilon = \varepsilon_0 . \varepsilon_r, \text{ or } \varepsilon_r = \frac{\varepsilon}{\varepsilon_0}.$$

The relative permittivity or dielectric constant (ε_r) of free space or air is taken as 1.0. Two expressions are written mathematically as

$$D = \varepsilon . E = \varepsilon_0 \varepsilon_r . E, \text{ with } \varepsilon = \varepsilon_0 . \varepsilon_r, \text{ or } \quad E = \frac{D}{\varepsilon} = \frac{D}{\varepsilon_0 \varepsilon_r}$$

Hence, the intensity of the electric field in a medium is the product of the relative permittivity or dielectric constant (ε_r) and the intensity of the electric field in free space or air, if the electric flux density (D) produced is a same in both cases. The relative permittivity or dielectric constant (ε_r) is a property of the medium. The units of D, E and ε have been given earlier as, C/m^2, N/C or V/m and F/m.

10.5 Electric Potential and Potential Difference

The electric potential (V) of the electric field at any point is given by the amount of energy gained when a unit positive charge is moved from infinity to any point, say P, at a distance, r from the point charge, which is placed at the origin, O. Consider Fig. 10.7. Electric field is produced by placing a point charge of q Coulomb at the origin, O, as given earlier. The electric field intensity, E, at a distance, x from the origin, is

$$E = \frac{q}{4\pi\varepsilon x^2}$$

As it is a repelling force, its direction is away from the origin. If, now, it is moved by a distance, dx, toward the origin, the work done, or the energy gained is given by

$$dW = E.dx = \frac{q}{4\pi\varepsilon x^2}.dx.$$

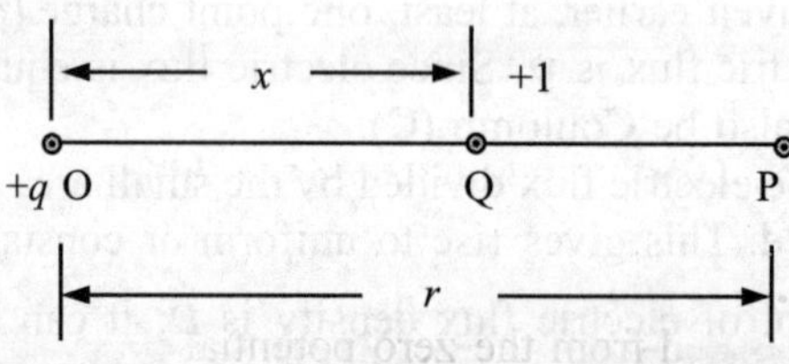

Fig.10.7

Now, if the above charge is moved from infinity to P, at a distance, r, from the origin, the work done or energy gained, is found as

$$W = \int_r^\infty \frac{q}{4\pi\varepsilon x^2}.dx = \frac{q}{4\pi\varepsilon x}\Big|_\infty^r = \frac{q}{4\pi\varepsilon r}.$$

The unit of energy or work done is J, or W.s. The unit of potential is the energy gained (J) divided by the unit charge (C), which is given by volt (V = J/C). This can be found easily, as J = W.s or N.m, and C = A.s (explained later). This value of W can be taken as *electric potential energy* in Joules, gained by a unit positive charge, when it is moved as shown earlier.

By substituting $r = \infty$, in the above equation, it can be shown that the potential at infinity is zero (0). Please note that the potential is a scalar quantity, unlike the electrical field intensity which has both magnitude as well as direction.

Potential difference is defined as a change of potential between any two points, say, R and S, at a distance of r_1 and r_2 from the origin, O, where a point charge (q) is placed (Fig. 10.8). This value, V_{RS} is obtained, when a unit positive charge is moved from R to S, the distance from the origin being r_1 and r_2, respectively. The value of the above potential difference is computed as

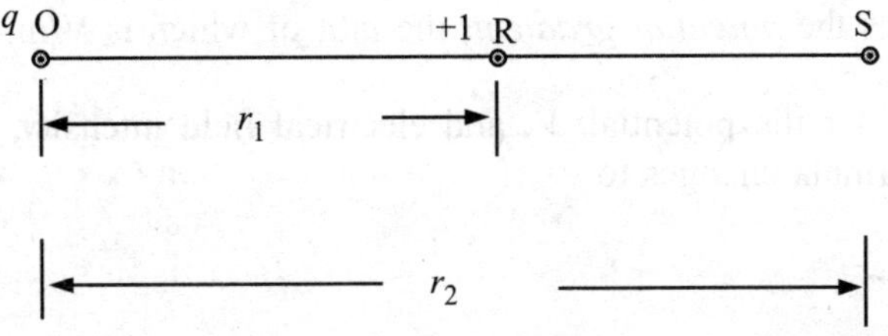

Fig.10.8

$$V_{RS} = V_R - V_S = \int_{r_1}^{r_2} \frac{q}{4\pi\varepsilon x^2}.dx = \left.\frac{q}{4\pi\varepsilon x}\right|_{r_2}^{r_1} = \frac{q}{4\pi\varepsilon r_1} - \frac{q}{4\pi\varepsilon r_2} = \frac{q}{4\pi\varepsilon}.\left(\frac{1}{r_2} - \frac{1}{r_1}\right).$$

This value is negative, as $r_2 > r_1$, which shows that $V_R < V_S$.

The potential difference can also be computed using the potential measured from the separate points to infinity. V_R is given by

$$V_R = \frac{q}{4\pi\varepsilon r_1},$$

which is the potential as measured from the zero potential point, at $r = \infty$. Similarly,

$$V_{SR} = V_S - V_R = \frac{q}{4\pi\varepsilon r_2} - \frac{q}{4\pi\varepsilon r_1} = \frac{q}{4\pi\varepsilon} \cdot \left(\frac{1}{r_2} - \frac{1}{r_1}\right),$$

which is the potential difference when a unit positive charge is moved from S to R. This is positive, as $V_{SR} = -V_{RS}$, as given in Chapter 2, where DC circuits was presented.

The potential and the field intensity at any point can be computed when more than one, say, two, point charges, are placed at different points. This has been illustrated earlier in the two solved examples (although, only the electric field intensity was calculated, not the potential).

10.6 Potential Gradient and Electric Field Intensity

As given earlier, the work done in moving a unit positive charge through a small distance, say dx, is given by

$$dW = E.dx,$$

where E = the force experienced by a unit positive charge placed at any point in a electric field, which is also the electric field intensity at that point.

dW can be taken as the change in potential (dV) at that point, which is equal to the work done on a unit positive charge. The electric field intensity, can now be written as

$$E = \frac{dV}{dx}$$

This can be termed as the *potential gradient*, the unit of which is V/m, being also equal to N/C, as given earlier.

If the expressions for the potential, V, and electrical field intensity, E, as found earlier, are used, then the formula changes to

$$E = -\frac{dV}{dx}$$

It is left to the reader to find out the reason for the negative sign in this formula. It is important to determine the direction of dx when the expression of the electric potential, V is found. From the above, it can be shown that,

$$V = -\int E.dx.$$

The negative sign for dx is used as the electric potential is the work done in moving a unit point charge for the distance, dx. The length, dx is taken as negative, as the charge is moved towards the origin, in the opposite direction to the positive direction of the x-axis.

10.7 Gauss' Theorem

STATEMENT: The total electric flux, ψ, coming out of from a closed surface is equal to the total charge, q, enclosed by the surface.

If the point charges enclosed by the surface are $q_1, q_2, q_3, \ldots, q_i, \ldots, q_n$, where these charges may be positive or negative, as the case may be, this law is expressed mathematically as,

$$\psi = \iint D \cdot ds = \sum_{i=1}^{n} q_i,$$

where D is the electric flux density (C/m^2), at the centre of the element represented by the outward vector, ds. The integration is over the entire closed surface.

If there is a continuous charge distribution of density, σ, instead of discrete charges, $\sum q_i$ is replaced by the volume integral of σ. Then, the above equation can be written as,

$$\psi = \iint D \cdot ds = \iiint \sigma \cdot dv.$$

DERIVATION: A positive charge, q is placed at a point, O inside a closed surface (Fig. 10.9). A point, P on the surface, with OP $= r$, is assumed. Now, a small area, ds is considered on the surface around P. It is assumed that the charge is placed in air or free space, with relative permittivity, $\varepsilon_r = 1$. The electric field intensity, E at P is,

$$E = \frac{q}{4\pi\varepsilon_0 r^2},$$

whose direction is towards OP.

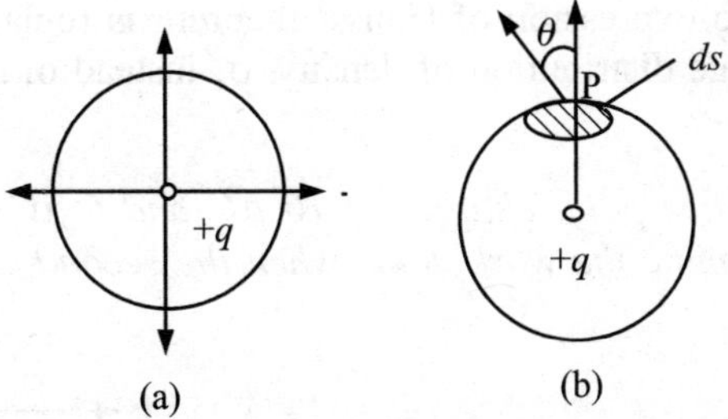

Fig.10.9 *Gauss' law*

The line, OP makes an angle, θ, with the line normal to the surface, ds. The relative permittivity of air or free space is $\varepsilon_r = 1$. The electric flux, $d\psi$ through ds, is given by,

$$d\psi = D.ds.\cos\theta = \varepsilon_0 E.ds.\cos\theta = \frac{q \cdot ds \cdot \cos\theta}{4\pi r^2} = \frac{q}{4\pi} \cdot d,$$

where, $d = \dfrac{ds \cdot \cos\theta}{r^2}$. d is the solid angle subtended by ds at O. Therefore, the total electric flux, ψ is,

$$\psi = \sum \frac{q}{4\pi} d = \frac{q}{4\pi}.4\pi = q,$$

where $\sum d$, represents the sum that is actually the total solid angle subtended by a closed surface at O. Obviously, the total solid angle is 4π.

The total electric flux, ψ due to the internal charge, q through the closed surface, is,

$$\psi = \frac{q}{4\pi}.\sum d = \frac{q}{4\pi}.4\pi = q$$

or, $\quad \psi = \iint D \cdot ds = q.$

Thus, Gauss' theorem is proved for a single internal charge, which can also be taken as q_1. Similarly, this can be proved for a second charge, q_2 placed at another point, S inside the closed surface.

This can be extended to the case where n number of charges, such as $q_1, q_2, q_3, \ldots, q_i, \ldots, q_n$, are placed at different points inside the closed surface. The theorem can then be expressed as,

$$\psi = \iint D \cdot ds = \sum_{i=1}^{n} q_i.$$

Similarly, if there is a continuous charge distribution of density, σ, inside the closed surface, then, the above equation can be written as,

$$\psi = \iint D \cdot ds = \sum_{i=1}^{n} q_i = \iiint \sigma \cdot dv.$$

The right-hand side of the expression of Gauss' theorem is replaced as shown above when there is a continuous charge distribution of density, σ, instead of n number of point charges, inside the closed surface.

EXAMPLE 10.4 *Two positive point charges of 10 μC and 6 μC, are placed 60 cm apart, in air (Fig. 10.10). Determine the work done, when the second charge is moved by 10 cm toward the first charge.*

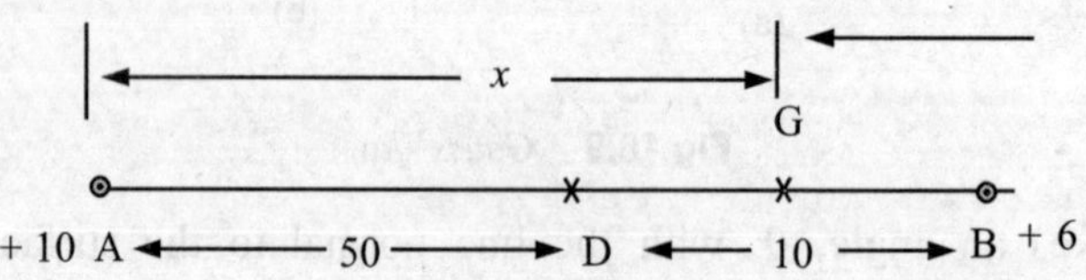

Fig.10.10

SOLUTION: All three points, A, B and D, given here, lie on a straight line (x-axis). Two charges, q_1 and q_2 are initially placed at the points, A and B (Fig. 10.10). Then, the second charge is moved to the point, D towards A, from B.

D is between A and B, such that DB = 10 cm (0.1 m).

$$r_1 = AB = 60 \text{ cm } (0.6 \text{ m}); \qquad r_2 = AD = AB - DB = 0.6 - 0.1 = 0.5 \text{ m}.$$

$$\varepsilon_0 = \frac{1}{36\pi \times 10^9} \text{ F/m}; \qquad 4\pi\varepsilon_0 = \frac{4\pi}{36\pi \times 10^9} = \frac{1}{9} \times 10^{-9}.$$

Let us use first principles to obtain the answer. The force between the two positive charges $q_1 = +10\,\mu$C and $q_2 = +6\,\mu$C, placed at A and an arbitrary point, G respectively, with AG $= x$, is given by,

$$F = \frac{q_1 q_2}{4\pi\varepsilon_0 x^2} = \frac{10 \times 10^{-6} \times 6 \times 10^{-6} \times 9 \times 10^9}{x^2} = \frac{0.54}{x^2} \text{ N}.$$

Note the following points:

(a) The point, G is chosen between D and B.
(b) The force is a repelling one in the direction, GA or DA.

The work done, when the second charge is moved through dx, towards A, is,

$$dW = -F \cdot dx = -0.54 \times \frac{dx}{x^2}.$$

Total work done when the second charge is moved from B to D, is,

$$W = \int_{r_1}^{r_2} F \cdot dx = -0.54 \times \int_{0.6}^{0.5} \frac{dx}{x^2} = \left.\frac{0.54}{x}\right|_{0.6}^{0.5} = 0.54 \times \left(\frac{1}{0.5} - \frac{1}{0.6}\right) = (1.08 - 0.9)$$

$$= 0.18 \text{ J or N.m (180 mJ)}.$$

Alternatively, the potential difference between the points, B and D, with the first charge placed at the point, A, is computed at first.

$$V_{DB} = V_D - V_B = \frac{q_1}{4\pi\varepsilon_0 r_2} - \frac{q_1}{4\pi\varepsilon_0 r_1} = \frac{q_1}{4\pi\varepsilon_0} \cdot \left(\frac{1}{r_2} - \frac{1}{r_1}\right)$$

$$= (10 \times 10^{-6} \times 9 \times 10^9) \times \left(\frac{1}{0.5} - \frac{1}{0.6}\right) = (90 \times 10^3) \times \left(\frac{1}{0.5} - \frac{1}{0.6}\right)$$

$$= (180 - 150) \times 10^3 = \quad 30 \text{ kV}.$$

This shows that the electric potential at D is greater than that at B, as the point D is nearer to A than B (Fig. 10.10).

The work done, or increase in energy, when the second charge is moved as given earlier, is found as,

$$W = q_2 \cdot V_{DB} = 6 \times 10^{-6} \times 30 \times 10^3 = 0.18 \text{ J},$$

which is the same as obtained earlier.

EXAMPLE 10.5 *A positive point charge of 0.02 μC, is placed at the point, A, in free space (Fig. 10.11). Calculate,*

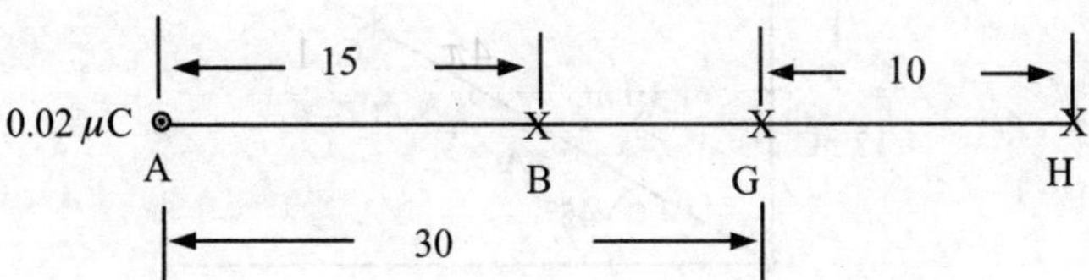

Fig.10.11

(a) the electric field intensity at the point, B, such that AB = 15 cm, and (b) the potential difference between the points, G and H, with AG = 30 cm, and GH = 10 cm.

SOLUTION All four points, A, B, G and H (Fig. 10.11), lie on a straight line, taken as the *x*-axis.

$$q = 0.02\ \mu\text{C} = 20 \text{ nC}.$$

(a) $r_1 = \text{AB} = 15 \text{ cm} = 0.15 \text{ m}.$

$$4\pi\varepsilon_0 = \frac{4\pi}{36\pi \times 10^9} = \frac{1}{9} \times 10^{-9}.$$

The electric field intensity at point B, is,

$$E = \frac{q}{4\pi\varepsilon_0 r_1^2} = \frac{20.0 \times 10^{-9} \times 9 \times 10^9}{(0.15)^2} = 8{,}000 \text{ V/m} = 8 \text{ kV/m}.$$

(b) $r_2 = \text{AG} = 30 \text{ cm} = 0.3 \text{ m}$; $\text{GH} = 10 \text{ cm} = 0.1 \text{ m}$.

$r_3 = \text{AH} = \text{AG} + \text{GH} = 30 + 10 = 40 \text{ cm} = 0.4 \text{ m}$.

The potential difference between the points, G and H, is computed as,

$$V_{GH} = V_G - V_H = \frac{q}{4\pi\varepsilon_0 r_2} - \frac{q}{4\pi\varepsilon_0 r_3} = \frac{q}{4\pi\varepsilon} \cdot \left(\frac{1}{r_2} - \frac{1}{r_3}\right)$$

$$= (20 \times 10^{-9} \times 9 \times 10^9) \times \left(\frac{1}{0.3} - \frac{1}{0.4}\right)$$

$$= 180 \times \left(\frac{1}{0.3} - \frac{1}{0.4}\right) = (600 - 450) = 150 \text{ V}.$$

This shows that the electric potential at G is greater than that at H, as the point, G is nearer to A than H (Fig. 10.11).

EXAMPLE 10.6 *Two point charges of + 12 nC and –10 nC, are placed in air, at the points, O (0, 0), and A (20, 20) respectively (Fig. 10.12). Find the potential difference between the points, G (40, 40), and H (50, 50). The scale on the two axes (x and y) is 1 unit =1 cm.*

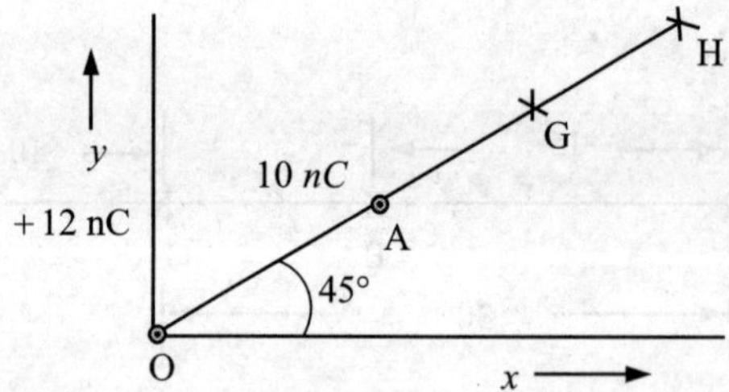

Fig.10.12

SOLUTION All four points, O, A, G and H lie on a straight line, with its slope ($\tan\theta = (20/20) = 1.0$) at an angle, $\theta = 45°$, with the x-axis (Fig. 10.12).

$$\text{OG} = d_1 = \sqrt{40^2 + 40^2} = 40\sqrt{2} \approx 56.57 \text{ m} = 0.5657 \text{ m}.$$

$$\text{OA} = \sqrt{20^2 + 20^2} = 20\sqrt{2} \approx 28.28 \text{ cm} = 0.2828 \text{ m}.$$

$$\text{AG} = \text{OG} - \text{OA} = d_2 = 0.5657 - 0.2828 = 0.2829 \text{ m} = 28.29 \text{ cm}.$$

$$4\pi\varepsilon_0 = \frac{4\pi}{36\pi \times 10^9} = \frac{1}{9} \times 10^{-9}.$$

The electric potential at the point, G, due to two charges, $q_1 = +12$ nC and $q_2 = -10$ nC, at the points, O and A respectively, is

$$V_G = \frac{q_1}{4\pi\varepsilon_0 d_1} + \frac{q_2}{4\pi\varepsilon_0 d_2} = \frac{1}{4\pi\varepsilon_0} \cdot \left(\frac{q_1}{d_1} + \frac{q_2}{d_2}\right) = (1 \times 10^{-9} \times 9 \times 10^9) \times \left(\frac{12.0}{0.5657} - \frac{10.0}{0.2828}\right)$$

$= 9.0 \times (21.21 - 35.36) = 190.9 - 318.2 = -127.3$ V.

$OH = d_3 = \sqrt{50^2 + 50^2} = 50\sqrt{2} = 70.71$ cm $= 0.7071$ m.

$AH = OH - OA = d_4 = 0.7071 - 0.2828 = 0.4243$ m $= 42.43$ cm.

The electric potential at the point, H, due to the same two charges, q_1 and q_2, is

$$V_H = \frac{q_1}{4\pi\varepsilon_0 d_3} + \frac{q_2}{4\pi\varepsilon_0 d_4} = \frac{1}{4\pi\varepsilon_0}\left(\frac{q_1}{d_3} + \frac{q_2}{d_4}\right) = (1 \times 10^{-9} \times 9 \times 10^9) \times \left(\frac{12.0}{0.7071} - \frac{10.0}{0.4243}\right)$$

$= 9.0 \times (16.971 - 23.568) = (152.74 - 212.11) \approx -59.4$ V.

The potential difference between the points, G and H $= V_{GH} = V_G - V_H$

$= -127.3 + 59.4 = -67.9$ V.

This shows that the electric potential at the point, G, is lower than that at H (Fig. 10.12).

Having now explored the electric field, we will discuss capacitances of different types, such as, parallel plates, concentric spheres and concentric cylinders etc.

10.8 Capacitance

Capacitance (the device is known as a capacitor) is produced between two conducting surfaces, when an insulating medium is inserted in the gap. The capacitor stores energy just like an inductor coil carrying current (Chapter 9), when it is charged to a certain voltage, by producing electric stress in the insulating medium between the conducting surfaces. One of the conducting surfaces is charged to a positive voltage, while the other surface gains a negative voltage. The capacitor is said to be charged. The charge stored, Q, on the capacitor is given by

$$Q = C.V,$$

where C is the value of the capacitor in Farad,

V is the voltage (potential) between the plates of the capacitor in volt, and

Q is the charge stored in the capacitor in Coulomb

The unit of capacitor is Farad, as given. Farad being a large unit, it is normally expressed in μF. The value of the capacitor (C) is the ratio of the charge stored (Q), divided by the voltage across it (V).

10.9 Types of Capacitors

Capacitors are classified in two ways:

- Different insulation (dielectric) mediums placed between two conducting surfaces.
- Types of conducting surfaces, like parallel plate, concentric cylinder, etc.

Firstly, these are classified according to different insulation (dielectric) mediums placed between two conducting surfaces. The choice of the dielectric medium depends on the following parameters:

1. The relative permittivity of the different dielectric material used, such as paper, mica, ceramic, etc. The value of the capacitor is normally determined by this parameter.
2. The magnitude of dielectric strength (potential gradient) that the above materials can withstand, which decides the maximum voltage to be applied across the capacitor.

Air capacitor: Normally, there are two sets of metal foils (aluminium or brass). The medium in between them is air, which has the lowest value of relative permittivity ($\varepsilon_r = 1$). Multi-plate capacitors, with alternate plates being connected to each other, are used to obtain higher values, as the capacitors between two adjacent plates, are connected in parallel.

Paper capacitor: A pair of elongated foils of metal (aluminium or copper or tin), inter-related with oil impregnated paper, form this capacitor. Multiple layers of foils with paper are available.

Mica capacitor: This consists of a series of aluminium or tin foils, separated by a very thin layer of mica sheets. Usually, multiple sheets are used, and alternate plates are connected to each other.

Ceramic capacitor: Discs of ceramic materials are used, with parallel facing surfaces being coated with silver.

Electrolytic capacitor: In this type, aluminum foils or cylinders are used as electrodes. Electrolytes, like porous paper, plastic, aluminium oxide, tantalum powder, etc. are placed between them. The thickness of the different electrolytes used is very small.

The range of values, applied voltage, and application to dc/ac circuits, for capacitors using different dielectric materials, are listed in Table 10.1.

Capacitors can also be classified according to the type or shape of the conducting plates:
(i) Parallel plate type
(ii) Spherical type (single or concentric spheres)
(iii) Concentric cylindrical type

Table 10.1 *Capacitors using different types of dielectric materials*

Capacitors	Values	Applied voltage	Applications to circuits
Air	up to 500 μF	low (100 V) to high (3 kV)	ac/dc circuits
Paper	small (in the range of pF)	high (100 V–10 kV)	ac/dc circuits
Mica	small (pF) to high (μF)	low (100–500 V)	ac circuits only
Ceramic	low values of pF to low values of μF	low (a few volts) to high (3 kV)	ac/dc circuits

There are several types of parallel plate capacitors, as described later. Also, it may be stated here that, one or two (in some cases, three or more) dielectric materials are used in the space between two parallel plates, or two concentric cylindrical surfaces (cables).

Before the expressions for the capacitance of the main types given here, in terms of their physical and other parameters are presented, an important topic in this connection, is taken up, which will be used later.

10.10 Series and Parallel Connection of Capacitors

SERIES CONNECTION

If two capacitors, C_1 and C_2 are connected in series across a supply voltage, V (Fig. 10.13), it can be shown that, in steady state, these are charged to two different voltages, V_1 and V_2 (one can assume a resistance connected in series with the above capacitors—see Chapter 3). As the capacitors are connected in series, the charging current is the same, resulting in the same charge, Q in them. So, the charge is related to the voltages in two capacitors as,

$$Q = C_1V_1 = C_2V_2.$$

The voltages across two capacitors are given by,

$$V_1 = \frac{Q}{C_1} \text{ and } V_2 = \frac{Q}{C_2}.$$

As they are connected in series, the supply voltage is $V = V_1 + V_2$.

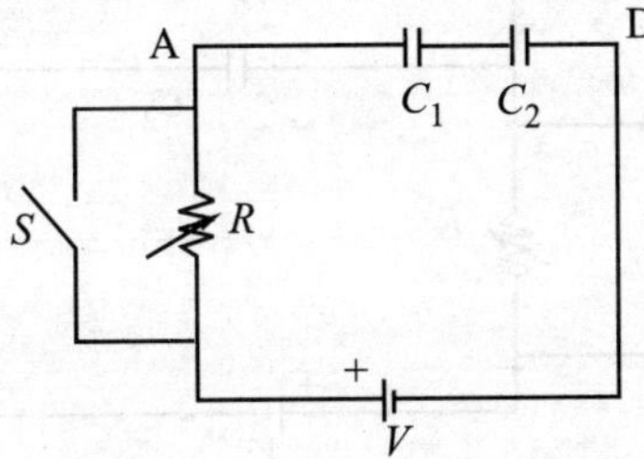

Fig.10.13

If C_{eq} is the equivalent capacitance of the above combination, the total charge can be written as

$$Q = C_{eq}V = C_1V_1 = C_2V_2.$$

Substituting these in the expression of the supply voltage, we get,

$$V = \frac{Q}{C_{eq}} = \frac{Q}{C_1} + \frac{Q}{C_2}$$

$$\text{or,} \quad \frac{1}{C_{eq}} = \frac{1}{C_1} + \frac{1}{C_2}.$$

Hence, the equivalent capacitance is

$$C_{eq} = \frac{C_1 \cdot C_2}{C_1 + C_2}.$$

This expression is of the same form when two resistances, R_1 and R_2 are connected in parallel.

If three or more capacitors are connected in series, the expression for the equivalent capacitance can be obtained in the same way as given before. The equivalent capacitance, when three capacitors, C_1, C_2 and C_3 are connected in series, is,

$$\frac{1}{C_{eq}} = \frac{1}{C_1} + \frac{1}{C_2} + \frac{1}{C_3},$$

from which the expression for C_{eq} is found as,

$$C_{eq} = \frac{C_1C_2C_3}{C_1C_2 + C_2C_3 + C_3C_1}$$

PARALLEL CONNECTION

If two capacitors, C_1 and C_2 are connected in parallel across a supply voltage, V (Fig. 10.14), it can be shown that, in steady state, these are charged to the supply voltage, V (one can assume a resistance connected in series with the above capacitors—see Chapter 3).

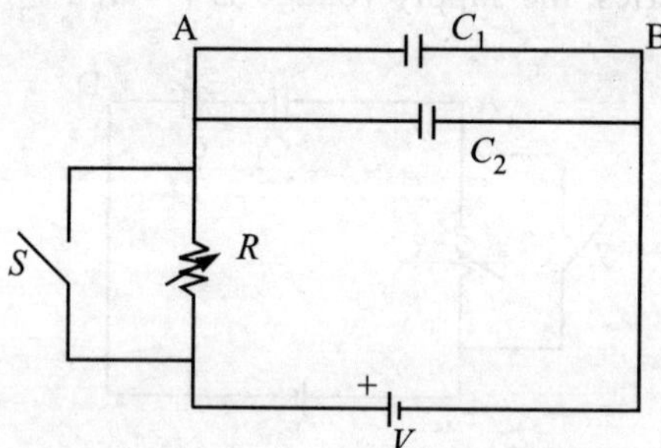

Fig.10.14

As the capacitors are connected in parallel, the charges in them are given by,

$$Q_1 = C_1V, \text{ and } Q_2 = C_2V.$$

The total charge in the two capacitors are given by,

$$Q = Q_1 + Q_2 = (C_1 + C_2)V,$$

substituting the two expressions obtained earlier.

If C_{eq} is the equivalent capacitance of the above combination, the total charge can be written as

$$Q = C_{eq}V = (C_1 + C_2)V.$$

The equivalent capacitance is given by,

$$C_{eq} = C_1 + C_2.$$

This expression is of the same form, when two resistances, R_1 and R_2 are connected in series.

If three or more capacitors are connected in parallel, the expression for the equivalent capacitance can be obtained in the same way as given before. The equivalent capacitance, when three capacitors, C_1, C_2 and C_3 are connected in parallel, is,

$$C_{eq} = C_1 + C_2 + C_3$$

If C is substituted by the inverse of the resistance, R, i.e., $\frac{1}{R}$, the expressions for two combination of resistances (series and parallel), can be obtained.

10.11 Energy Stored in a Capacitor

A capacitor, C is charged by connecting it to a dc voltage V (may be, through a resistance in series), as shown in Fig. 10.15. At steady state, the voltage across the capacitor is V, with its charge, $Q = CV$. As it charges, current is supplied to the capacitor from the source, the voltage across the capacitor is v, at time, t, the charge at that time is $q = Cv$. The current through the capacitor is

$$i = \frac{dq}{dt} = C\frac{dv}{dt} \quad \text{or} \quad i \cdot dt = dq = C \cdot dv.$$

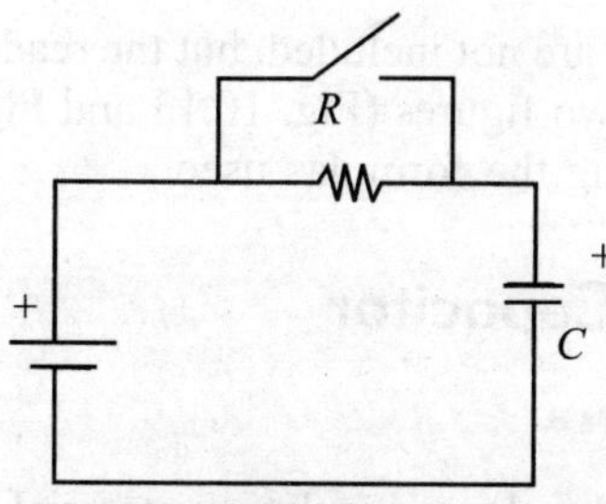

Fig.10.15

The power supplied from the source in time, dt

= increase of energy stored in the capacitor, C in time, dt

$= dW = P.dt = v.i.dt = v.dq = C.v.dv,$

where, $P = v.i$ is the instantaneous power drawn from the supply.

Hence, the energy stored in the capacitor, when the voltage across it is V, is given by,

$$W = \int_0^V C \cdot v \cdot dv = \frac{1}{2}C \cdot v^2\Big|_0^V = \frac{1}{2}C \cdot V^2 \text{ J.}$$

Note that the energy is stored in the electric field created in the dielectric medium, which is now under stress, because of the electric flux produced in it, as dc voltage is applied to the capacitor.

EXAMPLE 10.7 *Two capacitors, C_1 and C_2 are connected first in series, and then in parallel. If the equivalent capacitances are obtained as 2.4 μF and 10.0 μF, in that order, calculate the values of the two capacitors.*

SOLUTION: When two capacitors, C_1 and C_2 are connected in series, the equivalent capacitance is,

$$C_{eq1} = \frac{C_1 \cdot C_2}{C_1 + C_2} = 2.4\ (\mu\text{F}).$$

When two capacitors, C_1 and C_2 are connected in parallel, the equivalent capacitance is,

$$C_{eq2} = C_1 + C_2 = 10.0\ (\mu\text{F}).$$

Multiplying the two equations, we get,

$$C_{eq1} \cdot C_{eq2} = C_1 \cdot C_2 = (2.4 \times 10.0) = 24.$$

Solving the last two equations, the values of capacitances are computed as,

$$C_1 = 6\ \mu\text{F, and } C_2 = 4\ \mu\text{F}.$$

NOTE The steps, being simple, are not included, but the reader is expected to find the values. Also, no figures are given, but two figures (Fig. 10.13 and Fig. 10.14) can be referred. These figures are utilised while deriving the formulas used.

10.12 Parallel Plate Capacitor

UNIFORM DIELECTRIC MEDIUM

A parallel plate capacitor with the above insulating material is shown in Fig. 10.16.

The area of two plates is a m^2. The distance of separation is d m. Total permittivity of the dielectric is $\varepsilon = \varepsilon_0 . \varepsilon_r$ F/m.

If a charge of Q coulomb is given to one of the plates, then the electric flux through the dielectric medium is equal to $\psi = Q$, which can be obtained using Gauss's theorem. The flux density, D is assumed to be uniform or constant over the surface area, and is given by

$$D = \frac{\psi}{a} = \frac{Q}{a}.$$

As the total permittivity is ε, the electric field intensity, E is obtained as,

$$E = \frac{D}{\varepsilon} = \frac{Q}{\varepsilon_0 . \varepsilon_r . a}\ \text{V/m}.$$

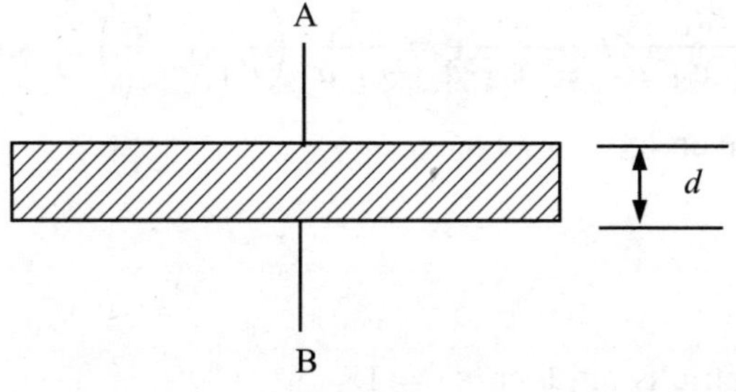

Fig.10.16

If the electric field intensity is assumed to be uniform or constant over the distance of separation, the voltage drop across two plates (V) is

$$V = E \cdot d = \frac{Q \cdot d}{\varepsilon_0 \varepsilon_r a}.$$

The capacitance, C in Farad, is

$$C = \frac{Q}{V} = \frac{\varepsilon \cdot a}{d} = \frac{\varepsilon_0 \cdot \varepsilon_r \cdot a}{d}.$$

The formula is nearly of the same form as derived for the self and mutual inductances of two coils in Chapter 9.

The dielectric medium is a composite one with two dielectric materials, of which one may be air, having different thickness

A parallel plate capacitor as given above is shown in Fig. 10.17.

The area of two plates is a. The thickness of the first dielectric is d_1. Total permittivity of the first dielectric is $\varepsilon_1 = \varepsilon_0.\varepsilon_{r1}$. The thickness of the second dielectric is d_2.

Total permittivity of the second dielectric is $\varepsilon_2 = \varepsilon_0.\varepsilon_{r2}$.

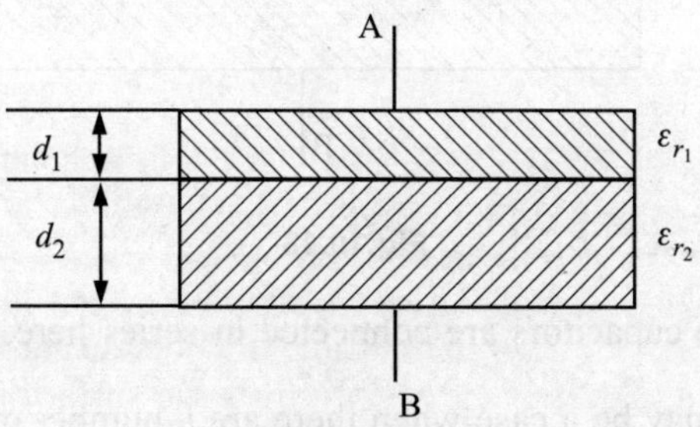

Fig.10.17

Using the expression found earlier, two capacitances, with two dielectric materials of different thicknesses, are obtained as,

$$C_1 = \frac{\varepsilon_1 \cdot a}{d_1} = \frac{\varepsilon_0 \cdot \varepsilon_{r1} \cdot a}{d_1}, \text{ and } C_2 = \frac{\varepsilon_2 \cdot a}{d_2} = \frac{\varepsilon_0 \cdot \varepsilon_{r2} \cdot a}{d_2}.$$

As the above two capacitors are connected in series, the equivalent capacitance of the combination, according to the formula derived in the previous section, is found as,

$$\frac{1}{C} = \frac{1}{C_1} + \frac{1}{C_2} = \frac{d_1}{\varepsilon_0 \cdot \varepsilon_{r1} \cdot a} + \frac{d_2}{\varepsilon_0 \cdot \varepsilon_{r2} \cdot a} = \frac{1}{\varepsilon_0 \cdot a} \cdot \left(\frac{d_1}{\varepsilon_{r1}} + \frac{d_2}{\varepsilon_{r2}} \right),$$

The capacitance, C is given by

$$C = \frac{\varepsilon_0 \cdot a}{\dfrac{d_1}{\varepsilon_{r1}} + \dfrac{d_2}{\varepsilon_{r2}}}.$$

If say, the second dielectric is air, then $\varepsilon_{r2} = 1$.

This formula can easily be extended to the case of a parallel plate capacitor, with *three or more dielectric materials,* having different thickness (Fig. 10.18). Let the thicknesses of *three layers* ($n = 3$) be d_1, d_2 and d_3, respectively, while the relative permittivities of the dielectric materials in the same order, are ε_{r1}, ε_{r2}, and ε_{r3}. In this case, the formula of the capacitance is obtained as

$$C = \frac{\varepsilon_0 \cdot a}{\dfrac{d_1}{\varepsilon_{r1}} + \dfrac{d_2}{\varepsilon_{r2}} + \dfrac{d_3}{\varepsilon_{r3}}} = \frac{\varepsilon_0 \cdot a}{\sum_{i=1}^{3} \dfrac{d_i}{\varepsilon_{ri}}}.$$

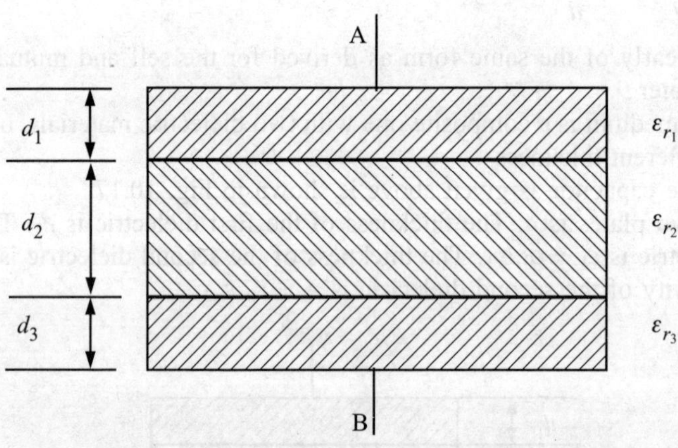

Fig.10.18

It may be noted that three capacitors are connected in series here.

SPECIAL CASE: There may be a case when there are n number of parallel plates, each one identical to other, placed in air ($\varepsilon_r = 1$), with alternate plates connected to the same polarity of the supply voltage as shown in Fig. 10.19. The total capacitance of this *multi-plate capacitor* is $(n-1)$ times the capacitance of only one type, as each one is connected in parallel with the others. If the area of each plate is a, and the distance of separation between the two plates is d, the capacitance of this combination, is given by,

$$C = \frac{\varepsilon_0 \cdot (n-1) \cdot a}{d}.$$

Fig.10.19

A few solved examples on the topics discussed here are given below.

EXAMPLE 10.8 *A 4 mm thick slab of an insulating material is inserted between the plates of a parallel plate capacitor (Fig. 10.17). It was found necessary to increase the separation between the plates by 2 mm, to maintain the capacitance to the same value, when no insulating material was present, i.e. the gap was filled with air. Find the dielectric strength of the slab material.*

SOLUTION $d = 4$ mm $= 0.004$ m; $\varepsilon_r = 1$ (for air).

Capacitance, $C_1 \propto \left(\frac{\varepsilon_r}{d}\right) = \left(\frac{1.0}{0.004}\right) = 250.$

Two dielectric materials are placed between the plates (Fig. 10.17).

$$d_1 = 4 \text{ mm} = 0.004 \text{ m}; \; d_2 = 2 \text{ mm} = 0.002 \text{ m}; \; \varepsilon_{r2} = 1 \text{ (for air)}.$$

The value of the inverse of the capacitance so formed, is given by

$$\frac{1}{C_2} \propto \left(\frac{d_1}{\varepsilon_{r1}} + \frac{d_2}{\varepsilon_{r2}}\right) = \left(\frac{0.004}{\varepsilon_{r1}} + \frac{0.002}{1.0}\right) = \left(\frac{0.004}{\varepsilon_{r1}} + 0.002\right),$$

As $C_1 = C_2$, $\frac{1}{C_1} = \frac{1}{C_2}$.

Hence, $\frac{1}{250} = 0.004 = \frac{0.004}{\varepsilon_{r1}} + 0.002$, as the proportionality constant is the same in both cases.

From the above, $\frac{0.004}{\varepsilon_{r1}} = 0.004 - 0.002 = 0.002.$

The dielectric constant of the material $= \varepsilon_{r1} = \frac{0.004}{0.002} = 2.$

EXAMPLE 10.9 *A multi-plate capacitor consists of 20 parallel plates with alternate plates connected to each other to form two terminals across which 500 V is applied (Fig. 10.19). The area of each plate is 80 cm². The air gap between any two plates is 5 mm. Calculate the energy stored in the capacitor.*

SOLUTION Number of plates $= n = 20$ (Fig. 10.19).

$$(n-1) = (20-1) = 19, d = 5 \text{ mm} = 0.005 \text{ m}.$$

$$a = 80 \text{ cm}^2 = 80 \times 10^{-4} \text{ m}^2; \ V = 500 \text{ V}.$$

$$\varepsilon_0 = 8.854 \times 10^{-12} \text{ F/m}; \ \varepsilon_r = 1 \text{ (for air)}.$$

Capacitance of a multi-plate capacitor

$$= C = \frac{\varepsilon_0 \cdot (n-1) \cdot a}{d} = \frac{8.854 \times 10^{-12} \times 19.0 \times 80 \times 10^{-4}}{0.005} = 269.2 \text{ pF}.$$

$$\text{Energy stored in the capacitor} = \frac{1}{2}CV^2 = \frac{269.2 \times 10^{-12} \times (500.0)^2}{2} = 33.65\mu\text{J}.$$

EXAMPLE 10.10 *The section of a composite parallel plate capacitor is shown in Fig. 10.20(a). Two plates are separated by 8 mm, which is termed as its thickness (t). It has three different dielectric materials, each having the same length, not shown in the figure. The first one has a thickness equal to half the total thickness (t), as measured from the top plate. Its width is the same as that of two plates. In the second half, there are two dielectric materials, whose thickness is the same as that of the first one, i.e., half the total thickness (t), placed between it (first) and the bottom plate. The width of the two insulating materials given above, are half of the first one, which means that their area is half the first one. The relative permittivities of the dielectric materials are 4, 2 and 3, respectively. Calculate the value of the capacitor per unit area so formed.*

SOLUTION The section of a composite parallel plate capacitor is shown in Fig. 10.20(a).

The length of the plates and the dielectric materials used are the same, its symbol being l (m). The width of the plates and the first material are the same, its symbol being b (m).

Hence, the area (cross-section) of the plates and the first material are the same, $a = l.b$ (m²).

Three capacitors are formed, as shown in Fig. 10.20(b). Two capacitors, C_2 and C_3 are connected in parallel. These are the lower ones. Their equivalent capacitance is

$$C' = C_2 + C_3.$$

This equivalent capacitance is connected in series with C_1, which is the top one. The combined capacitance is given by,

$$C'' = \frac{C_1 \cdot C'}{C_1 + C'} = \frac{C_1 \cdot (C_2 + C_3)}{C_1 + C_2 + C_3}.$$

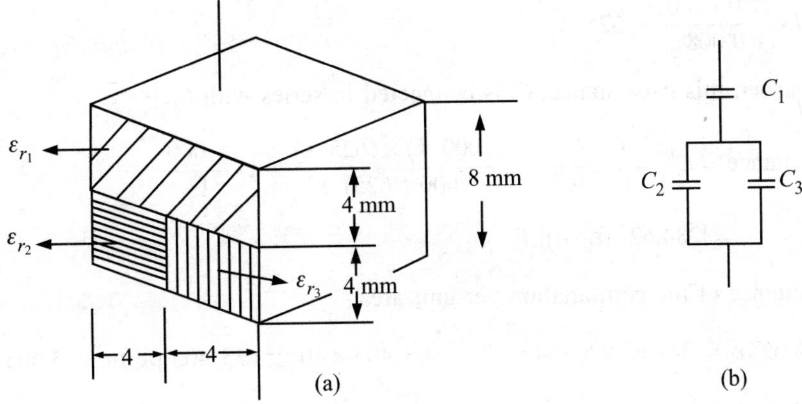

Fig.10.20

Total thickness $= t = 8$ mm $= 0.008$ m.

Thickness of the first dielectric material = half (50%) of the total thickness

$$= t_1 = 0.5 \cdot t = 0.5 \times 8.0 = 4 \text{ mm} = 0.004 \text{ m}.$$

Its area $= a_1 = a$, as given earlier.

Its relative permittivity $= \varepsilon_{r1} = 4$

The capacitance formed by the first dielectric placed below the top plate, is,

$$C_1 = \frac{\varepsilon_0 \cdot \varepsilon_{r1} \cdot a_1}{t_1} = \frac{\varepsilon_0 \cdot \varepsilon_{r1} \cdot a}{0.5 \cdot t} = k \cdot \frac{4.0}{0.004} = 1{,}000 \cdot k,$$

where the constant $k = \varepsilon_0 . a$.

The thickness of the second and third dielectric material

$$= t_2 = t_3 = 0.5 \cdot t = 4 \text{ mm} = 0.004 \text{ m},$$

which is the same as that of the first one.

The area of these dielectric materials are the same, being half the area of the plates, and is given by

$$a_2 = a_3 = 0.5 \cdot a.$$

Their relative permittivities are $\varepsilon_{r2} = 2$ and $\varepsilon_{r3} = 3$.

Hence, the equivalent capacitance of two capacitors connected in parallel is given by,

$$C' = C_2 + C_3 = \frac{\varepsilon_0 \cdot \varepsilon_{r2} \cdot (0.5 \cdot a)}{0.5 \cdot t} + \frac{\varepsilon_0 \cdot \varepsilon_{r3} \cdot (0.5 \cdot a)}{0.5 \cdot t} = \frac{\varepsilon_0 \cdot a \cdot (\varepsilon_{r2} + \varepsilon_{r3})}{t}$$

$$= k \cdot \frac{(2.0+3.0)}{0.008} = 625 \cdot k.$$

As given earlier, this capacitance, C' is connected in series with C_1.

$$\text{Total capacitance} = C'' = \frac{C_1 \cdot C''}{C_1 + C'} = \frac{(1,000 \cdot k) \times (625 \cdot k)}{(1,000 + 625) \cdot k} = k \cdot \frac{5,000}{13}$$

$$= [384.62 \cdot (\varepsilon_0 \cdot a)] \text{ F.}$$

Total capacitance of the combination per unit area

$$= 384.62 \cdot \varepsilon_0 = 384.62 \times 8.854 \times 10^{-12} = 3,405 \times 10^{-12} = 3,405 \text{ pF/m}^2 = 3.405 \text{ nF/m}^2.$$

If the unit area is taken as cm^2, then the total capacitance of the combination per unit area

$$= 3,405 \times 10^{-4} = 0.3405 \text{ pF/cm}^2 .$$

10.13 Capacitance of an Isolated Sphere

A perfectly insulated spherical conductor of radius, r (m), is placed in an insulating medium of relative permittivity, ε_r (Fig. 10.21).

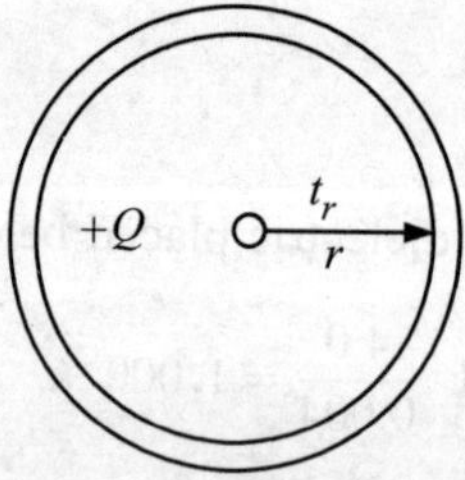

Fig.10.21 *Single spherical capacitor*

If a charge of Q coulomb is given to the spherical conductor, then the electric flux through the insulating medium is equal to $\psi = Q$, which can be obtained using Gauss's theorem. The flux density, D is assumed to be uniform or constant, over the surface area of the sphere, which is given by $4\pi r^2$. D can be determined by

$$D = \frac{\psi}{4\pi r^2} = \frac{Q}{4\pi r^2}.$$

As the total permittivity is ε, the electric field intensity, E is,

$$E = \frac{D}{\varepsilon} = \frac{Q}{\varepsilon_0 \cdot \varepsilon_r \cdot 4\pi r^2} \text{ V/m.}$$

The above expression is the same as that obtained when a charge of $+Q$, is placed at the centre of the sphere. The electric potential, V at the surface is given by,

$$V = \frac{Q}{4\pi\varepsilon_0\varepsilon_r r}.$$

This has been derived earlier, and is not repeated here.

The capacitance is,

$$C = \frac{Q}{V} = 4\pi\varepsilon_0\varepsilon_r r = \frac{4\pi\varepsilon_r r}{36\pi \times 10^9} = \frac{\varepsilon_r r}{9} \times 10^{-9} \text{ F}.$$

10.14 Capacitance of Concentric Spherical Conducting Plates

A capacitance is formed using two concentric spheres of inner radius, r_1 and outer radius, r_2, with dielectric material in between. This capacitor with the outer sphere earthed, is shown in Fig. 10.22. The space between two spheres is filled with a dielectric material having relative permittivity, ε_r.

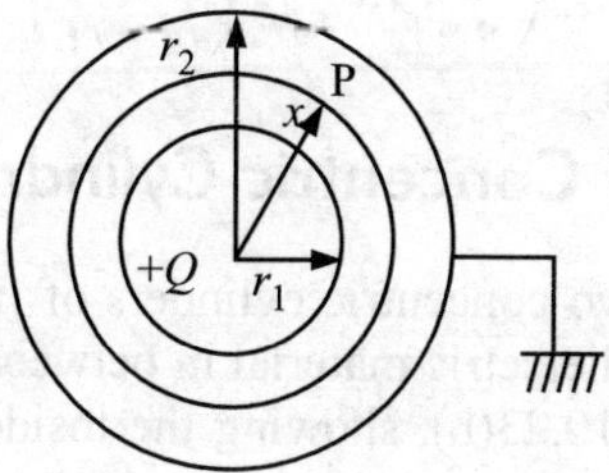

Fig.10.22 *Concentric spherical capacitor*

If the inner sphere is charged to $+Q$ Coulombs, a charge of $-Q$ Coulombs is induced in the inner surface of the outer sphere, and a charge of $+Q$ Coulombs appear on the outer surface of the inner sphere. The outer surface of the outer sphere is earthed, as explained earlier. Hence, no charge will appear there.

It can be stated that the electric flux in the dielectric medium between the two spheres is equal to $\psi = Q$, which can be obtained using Gauss's theorem. The flux density, D is assumed to be uniform or constant, over the surface area of an imaginary sphere with radius $= x$. The surface area is $4\pi x^2$. A point, P can be chosen on the surface of the imaginary sphere, with OP $= x$, if the centre of the sphere is taken as the origin, O. The electric flux density at the point, P is given by,

$$D = \frac{\psi}{4\pi x^2} = \frac{Q}{4\pi x^2}.$$

As the total permittivity is ε, the electric field intensity, E is obtained as,

$$E = \frac{D}{\varepsilon} = \frac{Q}{\varepsilon_0 \cdot \varepsilon_r \cdot 4\pi x^2} \text{ V/m, or N/C, where } r_2 > x > r_1.$$

This has also been discussed earlier.

The electric field intensity at the point P due to the charge on the outer sphere is zero. Hence, the potential difference between two concentric spheres,

V= Work done in bringing a unit positive from inner sphere, A to the outer one, B

$$= \int_{r_1}^{r_2} -E \cdot dx = \int_{r_1}^{r_2} \frac{Q}{4\pi\varepsilon_0\varepsilon_r} \cdot \left(-\frac{dx}{x^2}\right) = \left.\frac{Q}{4\pi\varepsilon_0\varepsilon_r x}\right|_{r_2}^{r_1} = \frac{Q}{4\pi\varepsilon_0\varepsilon_r} \cdot \left(\frac{1}{r_1} - \frac{1}{r_2}\right)$$

$$= \frac{Q}{4\pi\varepsilon_0\varepsilon_r x} \cdot \left(\frac{r_2 - r_1}{r_1 \cdot r_2}\right)$$

$$\varepsilon_0 = \frac{1}{36\pi \times 10^9} \text{ F/m; } 4\pi\varepsilon_0 = \frac{4\pi}{36\pi \times 10^9} = \frac{1}{9} \times 10^{-9}.$$

The capacitance is obtained as,

$$C = \frac{Q}{V} = 4\pi\varepsilon_0\varepsilon_r \cdot \left(\frac{r_1 \cdot r_2}{r_2 - r_1}\right) = \left(\frac{\varepsilon_r}{9} \times 10^{-9}\right) \cdot \left(\frac{r_1 \cdot r_2}{r_2 - r_1}\right) \text{ F.}$$

10.15 Capacitance of Concentric Cylindrical Conductors

A capacitance is formed using two concentric cylinders of finite length, l, with inner radius, r_1 and outer radius, r_2, having a dielectric material in between, which is a single core cable A plan (top view) is given in Fig. 10.23(b), showing the inside, from which the capacitance is derived, while Fig. 10.23(a) shows two concentric cylinders only, with the dielectric placed between them.

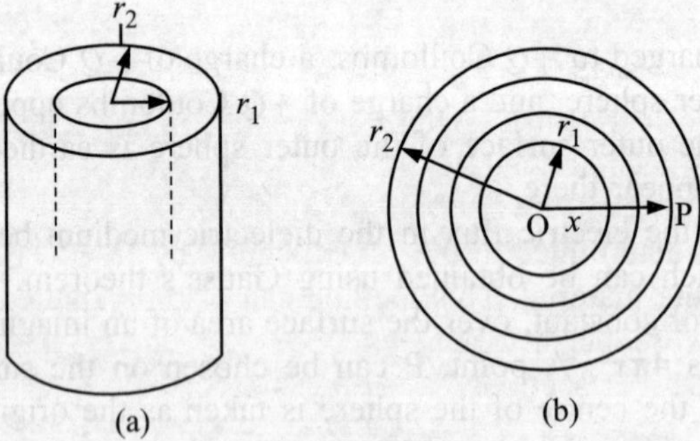

Fig.10.23 *Concentric cylindrical capacitor (cable)*

The charge per unit length in the cable on the outer surface of the inner cylinder is $+\sigma$, while that on the inner surface of the outer surface is $-\sigma$. Its unit is C/m. The relative permittivity of the dielectric material is ε_r.

It can be stated that the electric flux in the dielectric medium between the two cylinders, having unit length is equal to $\psi = \sigma$—this is obtained using Gauss's theorem. The flux density, D is assumed to be uniform or constant, over the surface area of an imaginary cylinder with radius $= x$. The surface area is $2\pi x.l = 2\pi x$. A point P can be chosen on the surface of the imaginary cylinder, with OP $= x$, which is the distance from the central line. The point, O is taken as the origin, which is the centre of the imaginary circle on the plane (Fig. 10.23).

The electric flux density at the point, P is given by,

$$D = \frac{\psi}{2\pi x} = \frac{\sigma}{2\pi x}.$$

As the total permittivity is ε, the electric field intensity, E at that point is obtained as,

$$E = \frac{D}{\varepsilon} = \frac{\sigma}{\varepsilon_0 \cdot \varepsilon_r \cdot 2\pi x} \text{ V/m, or N/C, where } r_2 > x > r_1.$$

Note that in both cases, D or E is based on the charge per unit length.

The potential difference between two concentric cylinders is

$$V = \int_{r_1}^{r_2} E \cdot dx = \int_{r_1}^{r_2} \left(\frac{\sigma}{2\pi\, \varepsilon_0\, \varepsilon_r} \right) \cdot \frac{dx}{x} = \frac{\sigma}{2\pi\, \varepsilon_0\, \varepsilon_r} \cdot \log_e x \Big|_{r_1}^{r_2} = \frac{\sigma}{2\pi\, \varepsilon_0\, \varepsilon_r} \cdot \log_e \frac{r_2}{r_1}.$$

$$2\pi\varepsilon_0 = \frac{2\pi}{36\pi \times 10^9} = \frac{1}{18} \times 10^{-9}.$$

The capacitance per unit length, is obtained as,

$$C = \frac{\sigma}{V} = \frac{2\pi\varepsilon_0\varepsilon_r}{\log_e\left(\frac{r_2}{r_1}\right)} = \frac{\varepsilon_r}{18 \times \log_e\left(\frac{r_2}{r_1}\right)} \times 10^{-9} \text{ F/m}.$$

Thus, the capacitance of concentric cylindrical conductors, or a single core cable, with one insulating material, has been derived. If two insulating materials, with variable thickness, and different relative pemittivities, are used in the space between the inner and outer cylinders, the capacitance can be determined in the same manner as given earlier for the parallel plate capacitor with composite material. This method can also be applied for the capacitors of an isolated sphere, and two concentric spheres, both with one insulating material, in the same manner, as derived in earlier sections, when the space is filled with composite material.

POTENTIAL GRADIENT IN A CABLE (CONCENTRIC CYLINDERS)

It is important to determine the potential gradient in the insulation used in a cable. In a composite cable (for example, a single core), with at least two dielectric materials, this, i.e. determination of potential gradient, is more vital. However, the topic is outside the scope of this book.

The electrical field intensity or potential gradient at a point P (Fig. 10.23) is,

$$E = \frac{dV}{dx} = \frac{\sigma}{\varepsilon_0 \cdot \varepsilon_r \cdot 2\pi x} \text{ N/C , or V/m.}$$

The same symbol is used for both, and the units as given, are also the same. The units can be obtained by equating the mechanical unit with the electrical one.

The potential difference is found as (derived earlier),

$$V = \frac{\sigma}{2\pi \varepsilon_0 \varepsilon_r} \cdot \log_e \frac{r_2}{r_1}.$$

From the above, the charge per unit length, σ is given by,

$$\sigma = \frac{2\pi\varepsilon_0 \varepsilon_r \cdot V}{\log_e \left(\frac{r_2}{r_1}\right)}.$$

Substituting this formula in the expression of potential gradient (or electric field intensity), we get,

$$E = \frac{dV}{dx} = \frac{V}{x.\log_e \left(\frac{r_2}{r_1}\right)} \text{ V/m.}$$

Hence, the potential gradient is inversely proportional to x. It is maximum at $x = r_1$.

$$E_{\max} = \frac{V}{r_1 \cdot \log_e \left(\frac{r_2}{r_1}\right)} \text{ V/m.}$$

It is minimum at $x = r_2$.

$$E_{\min} = \frac{V}{r_2 \cdot \log_e \left(\frac{r_2}{r_1}\right)} \text{ V/m.}$$

The potential gradient is also termed as dielectric stress.

10.16 Capacitance of Parallel Conductors

Two parallel conductors, for example, two wires, are placed in air ($\varepsilon_r = 1$), as shown in Fig. 10.24. The distance between them is d, which is large compared to the radius of the two conductors, r. If a positive charge of $+\sigma$ C/m is given to the conductor A, a negative charge of $-\sigma$ C/m is induced in the conductor, B. A point P is taken at a distance x from the conductor A. The electric field intensity at that point due to the charge in the conductor A is

$$E_1 = \frac{\sigma}{\varepsilon_0 \cdot 2\pi x}.$$

Its direction is towards B.

The electric field intensity at that point due to the induced charge in conductor B is

$$E_2 = \frac{\sigma}{\varepsilon_0 \cdot 2\pi(d-x)}.$$

Its direction is also towards B.

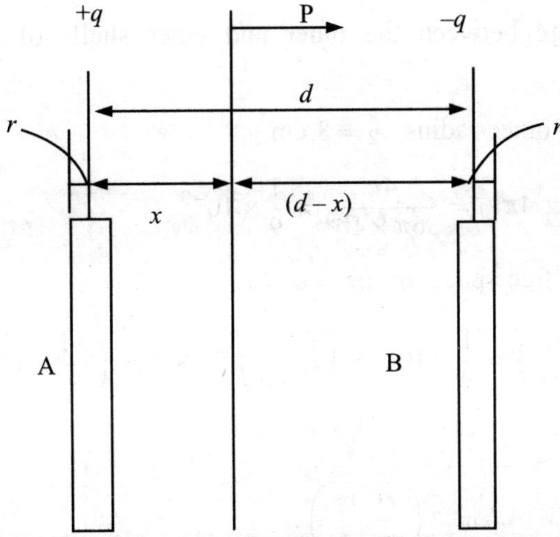

Fig.10.24

The resultant electric field is obtained as,

$$E = E_1 + E_2 = \frac{\sigma}{\varepsilon_0 \cdot 2\pi x} + \frac{\sigma}{\varepsilon_0 \cdot 2\pi(d-x)} = \frac{\sigma}{2\pi\varepsilon_0} \cdot \left(\frac{1}{x} + \frac{1}{d-x}\right).$$

Note that the direction remains the same as given earlier.

Hence, the potential difference is

$$= \int_r^{d-r} E \cdot dx = \int_r^{d-r} \frac{\sigma}{2\pi\varepsilon_0} \cdot \left(\frac{1}{x} + \frac{1}{d-x}\right) dx = \left(\frac{\sigma}{2\pi\varepsilon_0}\right) \cdot \left(\log_e x + \log_e(d-x)\right)\Big|_r^{d-r}$$

$$= \left(\frac{\sigma}{\pi\varepsilon_0}\right) \cdot \log_e\left(\frac{d-r}{r}\right).$$

It may be noted that the potential difference is the same due to each of the two conductors.

The capacitance of the two parallel conductors, per unit length, is found as,

$$C = \frac{\sigma}{V} = \frac{\pi\varepsilon_0}{\log_e\left(\frac{d-r}{r}\right)} = \frac{\frac{\pi}{36\pi \times 10^9}}{\log_e\left(\frac{d-r}{r}\right)} = \frac{\frac{1}{36} \times 10^{-9}}{\log_e\left(\frac{d-r}{r}\right)} \text{ F/m.}$$

A few examples are given below.

EXAMPLE 10.11 *A capacitor has two concentric thin spherical shells with inner and outer radii of 6 cm and 8 cm respectively (Fig. 10.22). The outer shell is earthed, and a charge is given to the inner shell such that its voltage is 250 V. Find the capacitance. Also determine the maximum dielectric stress, deriving its formula.*

SOLUTION The voltage between the inner and outer shells of a concentric spherical capacitor is $V = 250$ V.

Inner radius, $r_1 = 6$ cm; outer radius, $r_2 = 8$ cm.

$$\varepsilon_0 = \frac{1}{36\pi \times 10^9}\text{F/m};\ 4\pi\varepsilon_0 = \frac{4\pi}{36\pi \times 10^9} = \frac{1}{9} \times 10^{-9}.$$

Relative permittivity of free space, or air $= \varepsilon_r = 1$.

$$C = 4\pi\varepsilon_0\varepsilon_r \cdot \left(\frac{r_1 \cdot r_2}{r_2 - r_1}\right) = \frac{1}{9} \times 10^{-9} \times \left(\frac{6 \times 8}{8-6}\right) \times 10^{-2} = \left(\frac{24}{9} \times 10^{-9} \times 10^{-2}\right)$$

$$= 26.667 \text{ pF.}$$

The charge, $Q = V \cdot C = V \cdot 4\pi\varepsilon_0\varepsilon_r \cdot \left(\dfrac{r_1 \cdot r_2}{r_2 - r_1}\right).$

The dielectric stress, or electric field intensity at any distance, x, from the centre, as obtained earlier, is

$$E_{\max} = \frac{Q}{4\pi\varepsilon_0\varepsilon_r x^2} = \frac{V \cdot 4\pi\varepsilon_0\varepsilon_r}{4\pi\varepsilon_0\varepsilon_r x^2} \cdot \left(\frac{r_1 \cdot r_2}{r_2 - r_1}\right) = \frac{V}{x^2} \cdot \left(\frac{r_1 \cdot r_2}{r_2 - r_1}\right).$$

The maximum value of dielectric stress, or electric field intensity will be at $x = r_1$, and is given by,

$$E_{\max} = \frac{V}{r_1^2} \cdot \left(\frac{r_1 \cdot r_2}{r_2 - r_1}\right) = \frac{V \cdot r_2}{r_1 \cdot (r_2 - r_1)}.$$

Substituting the values, we get,

$$E_{\max} = \frac{250 \times 8}{6 \times (8-6) \times 10^{-2}} = 16.167 \text{ kV/m.}$$

EXAMPLE 10.12 *A concentric cylindrical capacitor (cable) has inner and outer diameters of 2.0 cm and 6.8 cm respectively (Fig. 10.23). An insulating material, used in the space between the cylinders, has a relative permittivity of 3. It is covered with lead sheath. If the voltage applied is 10 kV, calculate the capacitance per unit length (m). Also determine the maximum and minimum dielectric stresses.*

SOLUTION The voltage applied to a concentric cylindrical capacitor (cable) is,

$V = 10$ kV.

Inner diameter, $d_1 = 2.0$ cm; Outer diameter, $d_2 = 6.8$ cm.

$$\text{Thickness} = \frac{1}{2}(d_2 - d_1) = 0.5 \times (6.8 - 2.0) = 3.4 - 1.0 = 2.4 \text{ cm}.$$

$$2\pi\varepsilon_0 = \frac{2\pi}{36\pi \times 10^9} = \frac{1}{18} \times 10^{-9}.$$

Relative permittivity, $\varepsilon_r = 3.0$.

Capacitance per unit length

$$= C = \frac{2\pi\varepsilon_0\varepsilon_r}{\log_e\left(\frac{r_2}{r_1}\right)} = \frac{3}{18 \times \log_e\left(\frac{d_2}{d_1}\right)} \times 10^{-9} = \frac{1}{6 \times \log_e\left(\frac{6.8}{2.0}\right)} \times 10^{-9} = \frac{1}{6.0 \times 1.2238} \times 10^{-9}$$

$$= 136.2 \times 10^{-12} = 136.2 \text{ pF/m}.$$

The maximum value of dielectric stress, or electric field intensity is at $x = r_1$,

$$E_{max} = \frac{V}{r_1 \cdot \log_e\left(\frac{r_2}{r_1}\right)} = \frac{2V}{d_1 \cdot \log_e\left(\frac{d_2}{d_1}\right)} = \frac{2 \times 10 \times 10^3}{2.0 \times 10^{-2} \times \log_e\left(\frac{6.8}{2.0}\right)}$$

$$= \frac{1{,}000}{1.2238} = 817 \text{ kV/m} = 0.817 \text{ kV/mm}.$$

The minimum value of dielectric stress, or electric field intensity is at $x = r_2$,

$$E_{min} = \frac{V}{r_2 \cdot \log_e\left(\frac{r_2}{r_1}\right)} = \frac{2V}{d_2 \cdot \log_e\left(\frac{d_2}{d_1}\right)} = \frac{2 \times 10 \times 10^3}{6.8 \times 10^{-2} \times \log_e\left(\frac{6.8}{2.0}\right)}$$

$$= \frac{294.1}{1.2238} = 240.3 \text{ kV/m} \approx 0.24 \text{ kV/mm}.$$

Conclusion

In this chapter, both electrostatics and capacitance have been discussed in detail. Starting with electrostatics, the terms pertaining to electric field produced by point charge(s), viz., the electric field density, electric flux, potential, including Coulomb's laws and Gauss' theorem, have been explained. Then, different types of capacitors, such as parallel plate, spheres (single and concentric) and concentric cylinders were presented. The insulation used in these has also been taken up. Series and parallel connection of capacitors, and stored energy in them were finally discussed. Numerical examples were included to illustrate the different topics presented. In the next few chapters, we will learn about electric machines, starting with single-phase transformers in Chapter 11.

Multiple Choice Type Questions

Tick the correct option(s), out of the four choices provided in each question.

10.1 Two positive point charges, 9.0 μC and 1.0 μC, are placed at the points, A and B on a horizontal plane, with AB = 40 cm. A third charge, Q is placed at a point D between A and B, with AD = x, such that no force due to the two charges acts on it. Find the value of x in cm.

(a) 30 (b) 40 (c) 20 (d) 0

10.2 If the potential at point, P due to a positive point charge, Q placed at A, such that AP = 30 cm, is to be increased to three times its value at R due to the same charge, what should be the distance, AR in cm?

(a) 20 (b) 10 (c) 40 (d) 50

10.3 The capacitance of a parallel plate capacitor depends upon

(a) the dielectric used between the plates. (b) the separation between the plates. (c) the potential difference between the plates (d) the thickness of the plates

10.4 Two point charges of −4.0 μC and 6.0 μC, are placed in air at points, A and B, with AB = 25 cm. The electric potential at point, P, between A and B due to two point charges, with AP = x, is zero (0). Find the value of x in cm.

(a) 0 (b) 15 (c) 25 (d) 10

10.5 Two point charges of +1.0 μC and −3.0 μC, are placed at two points, A and B with AB = 40 cm. If a point, D is chosen outside the line, AB, such that BD = 20 cm, find the resultant force in N, on a charge of +2.0 μC placed at D.

(a) 1.3 (along the line, BA) (b) 1.4 (along the line, BA)
(c) 1.5 (along the line, AB) (d) 1.7 (along the line, AB)

10.6 The thickness of a parallel plate capacitor is doubled. The cross-sectional area and the dielectric placed in between the parallel plates remain the same. The new capacitor so formed has x times the value of the previous one. Find the value of x.

(a) 1.0 (b) 0.5 (c) 2.0 (d) 4.0

10.7 A capacitor consists of two parallel plates separated by a distance, d. The cross-sectional area is divided into two equal parts, having the same length, l and half its width, $\frac{w}{2}$, each filled with different dielectric materials of relative permittivities, 2 and 4 respectively. A capacitor is now assumed to be filled with a dielectric of the same relative permittivity, ε_r, with the distance and total cross-sectional area remaining the same. If the new capacitor so formed has the same value as the earlier one, find the value of ε_r.

(a) 3.0 (b) 6.0 (c) 1.5 (d) 4.0

10.8 A multi-plate parallel capacitor has 6 fixed plates connected in parallel, interleaved with 5 similar (variable) plates, each having an effective cross-sectional area of 100 cm^2. The gap between the adjacent plates is 1 mm. The capacitor is immersed in oil of relative permittivity, 4. Calculate its capacitance in pF.

(a) 7.08 (b) 1.77 (c) 3.54 (d) none of the above

10.9 Determine the capacitance in pF, of a spherical capacitor, if the inner and outer diameters are 0.2 m and 0.3 m respectively, with the space between them filled with a liquid having a dielectric constant of 12.

(a) 600 (b) 400 (c) 250 (d) 600

10.10 The relation between absolute permittivity of free space, ε_0 (F/m), absolute permeability of free space, μ_0 (H/m), and the velocity of light, c (m/s), is given by,

(a) $\mu_0.\varepsilon_0 = c^2$ (b) $\frac{\mu_0}{\varepsilon_0} = c$

(c) $\frac{\varepsilon_0}{\mu_0} = c$ (d) $\frac{1}{\mu_0.\varepsilon_0} = c^2$

10.11 If the voltage across a capacitor of 50 μF is 100 V, find the energy (J) stored in it.

(a) 0.375 (b) 0.5 (c) 0.25 (d) 0.125

10.12 Four capacitors, each of 12 μF, are connected in the circuit shown in Fig.MCQ10.12. What is the equivalent capacitance in μF across A and B?

(a) 8.0 (b) 12.0 (c) 4.8 (d) 6.0

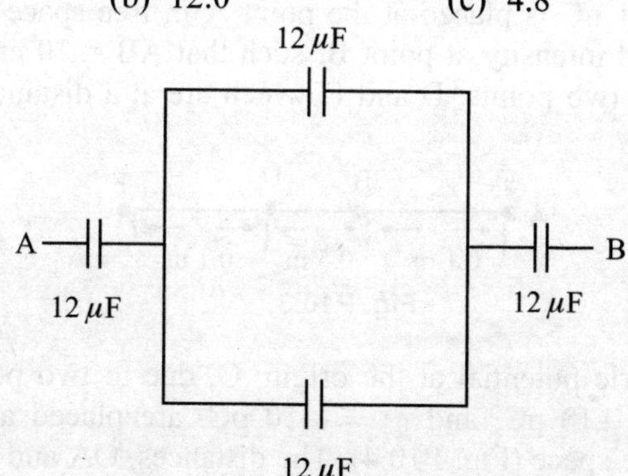

Fig. MCQ10.12

Answers

10.1 (a)	10.2 (b)	10.3 (a),(b)	10.4 (d)	10.5 (a)	10.6 (b)
10.7 (a)	10.8 (c)	10.9 (b)	10.10(d)	10.11(c)	10.12(c)

Problems

10.1 Three point charges of $+0.02\ \mu$C, $-0.03\ \mu$C and $+0.04\ \mu$C, are placed in air at the corners of an equilateral triangle, having each side, 2 m long (Fig. P10.1). Determine the resultant field intensity at the mid-point of the side, with charges of $0.02\ \mu$C, and $0.04\ \mu$C.

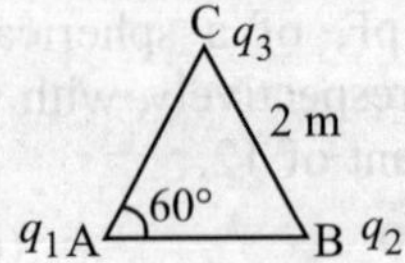

Fig. P10.1

10.2 Three point charges, $9q$, q and Q are placed at points, A, B and D respectively (Fig. P10.2). All points lie on a straight line, AB, whose length is r. The point, D is chosen on the line, AB, such that the length of AD is double that of BD. What is the value of Q, such that the net force on charge, q placed at D, is zero (0)?

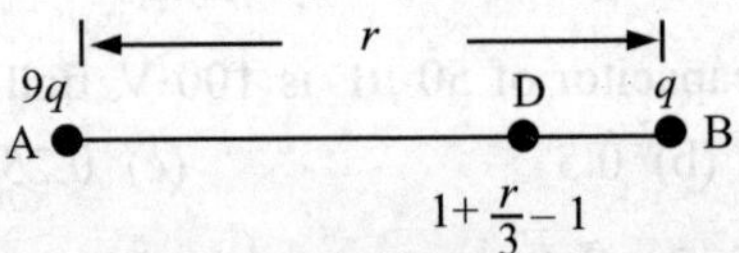

Fig. P10.2

10.3 A point charge of 1 nC is placed at the point A in free space (Fig. P10.3). Calculate (a) the electric field intensity at point B, such that AB = 10 cm, and (b) the potential difference between two points, D and F, which are at a distance of 15 cm and 25 cm from A respectively.

Fig. P10.3

10.4 Calculate the electric potential at the origin, O, due to two point charges, q_1 and q_2. The charges, $q_1 = +10$ pC, and $q_2 = -50$ pC, are placed at the points, A and B respectively, in free space (Fig. P10.4). The distances, OA and OB are 20 cm and -20 cm respectively, with the points, A and B on opposite sides of the point O, but on the same line, AOB.

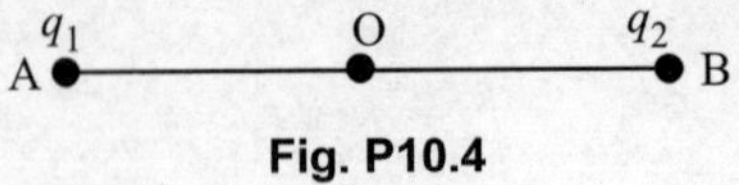

Fig. P10.4

10.5 Two capacitors, C_1 and C_2 are connected in series across a 100 V dc supply (Fig. P10.5). The voltage drops in the capacitors, C_1 and C_2 are 40 V and 60 V respectively. Now, a capacitor, $C_3 = 1\ \mu$F is connected across C_1 and the combination is connected to the same supply (Fig. P10.5). The respective voltage drops in the capacitors, C_1 and C_2 are 30 V and 70 V. Determine the values of the two capacitors, C_1 and C_2 in μF.

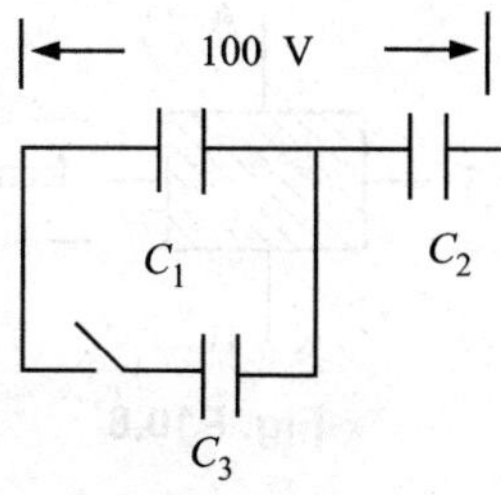

Fig. P10.5

10.6 A composite parallel plate capacitor is made of two different slabs (material) of permittivity, 2 and 3, and thickness 2.0 mm and 4.0 mm respectively, separated by a conducting foil, P (Fig. P10.6). A voltage of 140 V is applied across the combination. Find the voltage of the conducting foil.

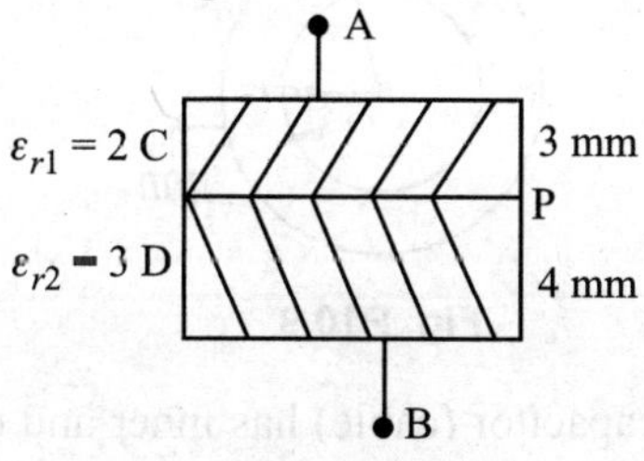

Fig. P10.6

10.7 An air capacitor has two parallel plates of cross-sectional area 1,500 cm^2 placed 5 mm apart. Determine the value of the capacitor.

If a dielectric slab (material) having the same cross-sectional area thickness, 2 mm, and relative permittivity 3, is now placed on the bottom plate in between two plates (Fig. P10.7), calculate the new value of the capacitor so formed.

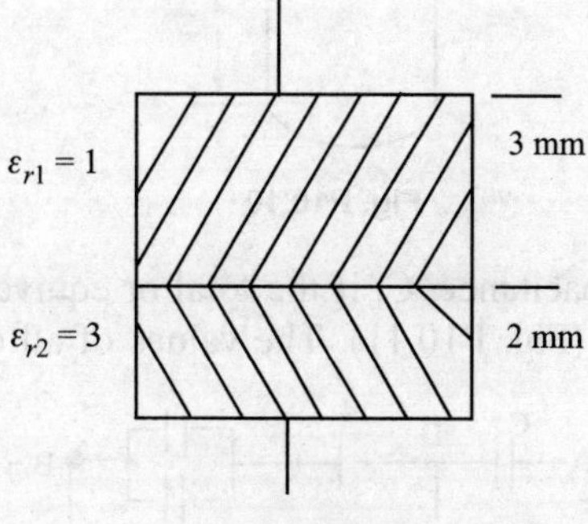

Fig.P10.7

10.8 A parallel plate capacitor is charged to a voltage of 5 kV (Fig. P10.8). The cross-sectional area of each plate is 120 sq. cm, and the distance between the plates is 10 mm. The relative permittivity of the insulating material used is 3. Calculate the force between the plates, and the energy stored per sq. mm. of plate. Derive the formula for the force developed, using first principles.

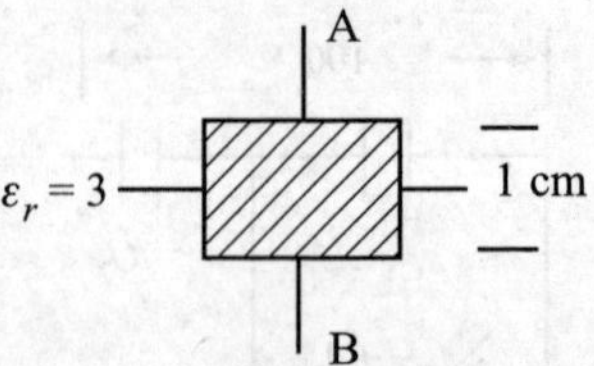

Fig. P10.8

10.9 A capacitor has two concentric thin spherical shells with inner and outer radii of 8 cm and 10 cm respectively (Fig. P10.9). The outer shell is earthed, and a charge is given to the inner shell, such that its voltage is 250 V. Find the capacitance. Also determine the maximum and minimum dielectric stress.

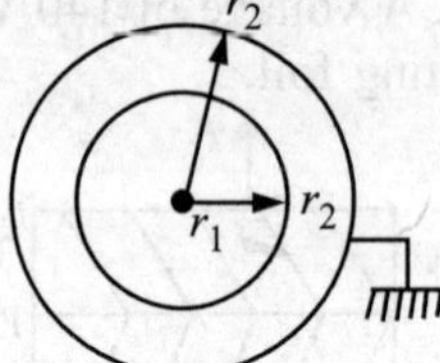

Fig. P10.9

10.10 A concentric cylindrical capacitor (cable) has inner and outer diameters of 2.5 cm and 5.5 cm respectively (Fig. P10.10). An insulating material (impregnated paper) used in the space, has a relative permittivity of 4. It is covered with lead sheath. If the voltage applied is 1,000 V, calculate the capacitance per unit length (m). Also determine the maximum and minimum dielectric stresses.

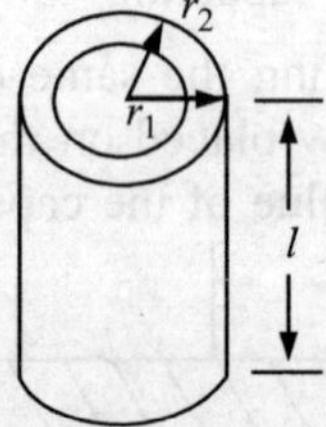

Fig. P10.10

10.11 Find the value of the capacitance, C, if the total or equivalent capacitance between the points, A and B is 2 μF (Fig. P10.11). The values of all capacitances are given in μF.

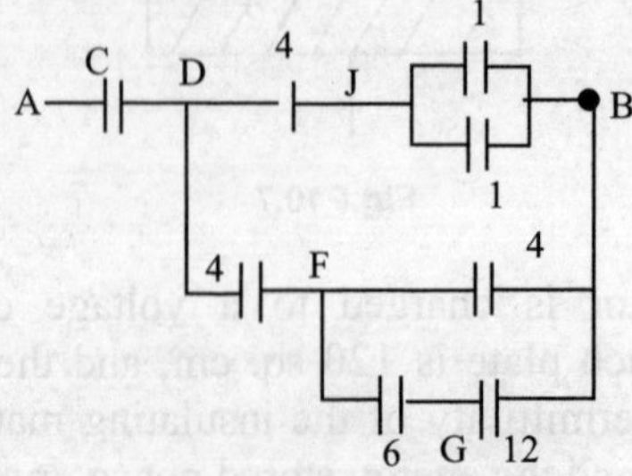

Fig. P10.11

CHAPTER 11

TRANSFORMERS

To start with, we will discuss the first electrical machine—the *transformer*. The transformer is a static machine, unlike the rotating machines (dc/ac) which will be described in later chapters. The input to this machine is an ac voltage normally at line frequency (50 Hz); the output voltage is also an ac one with the same frequency. The input and output voltages depend on applications, and in most cases, these are different, though in a small number of cases, these voltages can be equal. Starting with types of construction, the emf equation, operation on no-load and load, and equivalent circuit of a single-phase transformer are presented. This is followed by a discussion on how to calculate regulation and efficiency, along with losses, and tests (open circuit and short circuit).

11.1 Construction

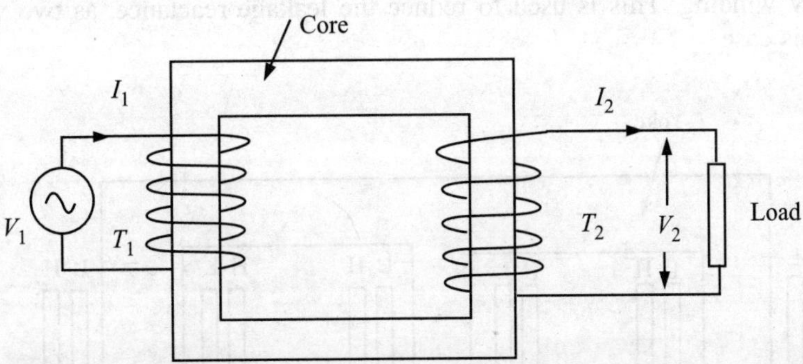

Fig. 11.1(a) *A single-phase transformer*

The schematic diagram of a single-phase two-winding transformer is shown in Fig. 11.1(a). Two closely coupled coils, normally of different number of turns, T_1 and T_2, are wound on a magnetic core. The coils must be placed close to each other so that the coupling coefficient is high, which in turn decreases the leakage reactance (the details of two types of transformers normally used, are given later, along with figures). The magnetic core used is laminated, so as to reduce eddy current losses. As stated earlier, the primary winding is fed from an ac supply, while the output voltage appears at secondary terminals which is connected to the load. The load may be resistive, inductive, and also capacitive (in some cases). The circuit symbol of a two-winding transformer is given in Fig. 11.1(b).

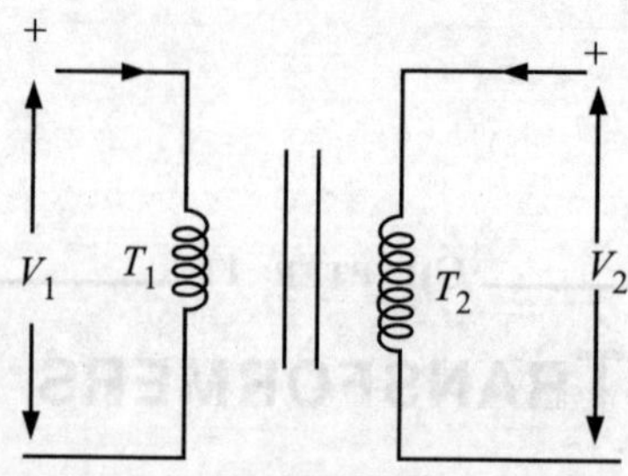

Fig. 11.1(b) *Circuit symbol*

CORE AND SHELL TYPES

Two types of construction are generally used for transformers. Normally, the *core type* is used for three-phase transformers—here, three pairs of windings, i.e., coils (primary and secondary) are used (Fig. 11.2(a)). The low voltage (LV) one, which has the lower number of turns, is kept near the core, the core being earthed, while the high voltage winding, having higher number of turns, is put on top of the LV winding. There are three limbs or cores, joined by a yoke to complete the magnetic circuit. The leakage reactance decreases as the two windings are close to each other, i.e., tightly coupled.

The other type of construction—the *shell* type—is used in single-phase transformers. Two windings, LV and HV, are sectionalised, divided into a number of sections (Fig. 11.2(b)). The sectionalised windings are placed on a central limb, with two outer limbs used to complete the magnetic path. Each section of LV winding placed close to the core is followed by a section of HV winding. This is used to reduce the leakage reactance, as two windings are coupled in this case.

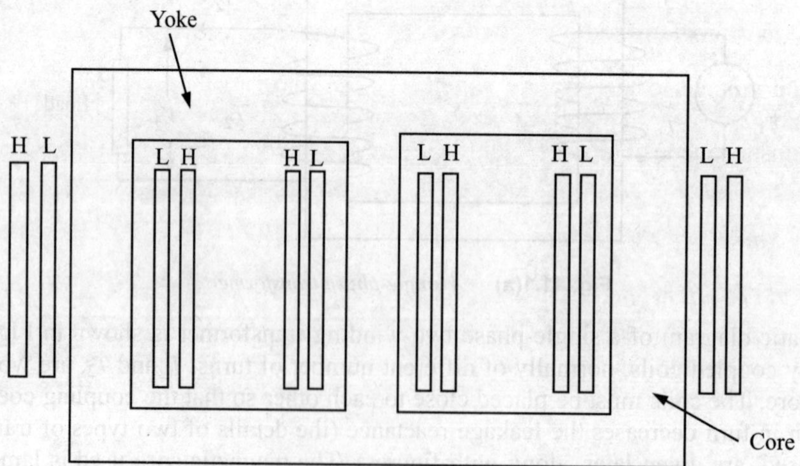

Fig. 11.2(a) *Three-phase transformer (core type)*

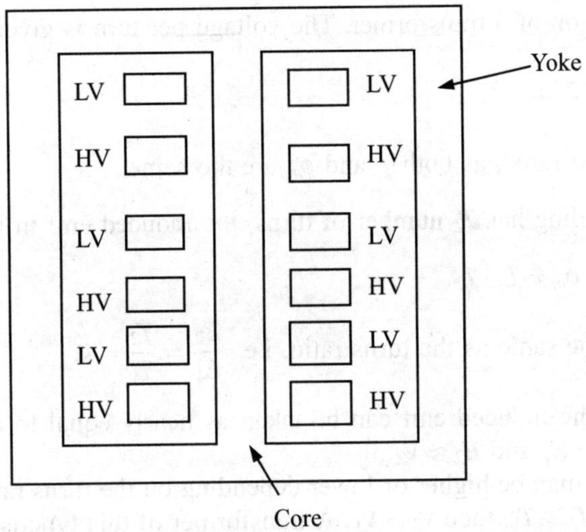

Fig. 11.2(b) *Single-phase transformer (shell type)*

11.2 EMF Equation

The flux in the core (Fig. 11.1(a)) is sinusoidal, and its value at any instant, is given by,

$$\phi = \phi_m \sin \omega t = \phi_m \sin 2\pi f t,$$

where, ϕ = flux (Wb),
f = frequency in Hz (normally assumed as 50.0),
$\omega = 2\pi f$ = angular frequency in rad/s, and
$\theta = \omega t$ = angle in rad.

The applied voltage to the primary winding is sinusoidal. Hence, the flux is sinusoidal as given earlier.

The instantaneous emf, e_1 (V) in the primary winding of turns (T_1) is.

$$e_1 = -\frac{d}{dt}(T_1 \cdot \phi) = -T_1 \cdot \frac{d\phi}{dt} = -T_1 \cdot \frac{d}{dt}(\phi_m \sin \omega t) = -T_1 \cdot \omega \cdot \phi_m \cos \omega t$$

$$= T_1 \cdot \omega \cdot \phi_m \sin\left(\omega t - \frac{\pi}{2}\right) = E_{m1} \sin\left(\omega t - \frac{\pi}{2}\right),$$

where the maximum value of the induced emf $= E_{m1} = T_1.\omega.\phi_m$. For a sine wave, the rms value of the induced emf,

$$E_1 = \frac{E_{m1}}{\sqrt{2}}$$

or, $$E_1 = \frac{1}{\sqrt{2}} T_1.\omega \cdot \phi_m = \frac{1}{\sqrt{2}} T_1 \cdot 2\pi \cdot f \cdot \phi_m = \sqrt{2}\pi \cdot T_1 \cdot f \cdot \phi_m = 4.44 \cdot T_1 \cdot f \cdot \phi_m.$$

This is the emf equation of a transformer. The voltage per turn is given by,

$$E_t = \frac{E_1}{T_1} = 4.44 \cdot f \cdot \phi_m.$$

This value remains the same, as both f and ϕ_m are the same.

As the secondary winding has T_2 number of turns, the induced emf in it is obtained as,

$$E_2 = 4.44 \cdot T_2 \cdot f \cdot \phi_m = E_t \cdot T_2.$$

The voltage ratio is the same as the turns ratio, i.e., $\frac{E_2}{E_1} = \frac{T_2}{T_1} = k.$

For the transformer, the induced emf can be taken as nearly equal to the terminal voltage, which means that $V_1 \approx E_1$ and $E_2 \approx V_2$..

The output voltage may be higher or lower depending on the turns ratio, defined earlier. If k is greater than 1 or $T_2 > T_1$, then $V_2 > V_1$. A transformer of this type is termed as a step-up transformer. If k is lower than 1 or $T_2 < T_1$, then $V_2 < V_1$. In this case, the transformer is a step-down transformer. If k is equal to 1 or $T_2 = T_1$, then $V_2 = V_1$, i.e., both input and output voltages are equal. In certain cases, this type is used for isolation.

If the net cross-section (m^2) of the core is a_i, and maximum flux density (Wb/m^2) is B_m, then maximum flux is $\phi_m = B_m . a_i$.

RATING AND IDEAL TRANSFORMERS

The *rating* of a transformer is expressed in VA as the load power factor (pf) varies from lagging to leading, in the range $1.0 >$ pf > 0.0. As stated earlier, the load connected across secondary terminals may be resistive, inductive or capacitive. If the rated primary voltage is V_1, the rated (full-load) current in the primary winding is

$$I_1 = \frac{S}{V_1}, \text{ where } S \text{ is the VA rating.}$$

Similarly, if the rated secondary voltage is V_2, the rated (full-load) current in the secondary winding is $I_2 = \frac{S}{V_2}$.

Hence, $S = V_1 . I_1 = V_2 . I_2$, or, $k = \frac{V_2}{V_1} = \frac{T_2}{T_1} = \frac{I_1}{I_2}$.

The ampere-turns in the two windings is the same, i.e.,

$$T_1 . I_1 = T_2 . I_2.$$

In the above equation, no-load current, being small is neglected.

Normally, S is given in kVA, as the rating of a transformer is large. Therefore, the current in the primary winding is

$$I_1 = \frac{S \times 10^3}{V_1}$$

An *ideal transformer* is one in which the losses (iron and copper), magnetising current, and leakage reactance, are neglected, as these are low. Hence, no-load current, which also includes its loss component and total impedance, including total resistance is neglected.

For this transformer, the primary and secondary voltage and current ratings are related as

$$V_1 = \frac{V_2}{k}, \text{ and } I_1 = k \cdot I_2,$$

where the turns ratio $= k = \dfrac{T_2}{T_1}$.

It may be stated that $V_1 = E_1$ and $E_2 = V_2$, and the ampere-turns in the two windings are equal because no-load current is neglected as mentioned earlier.

EXAMPLE 11.1 *A 50 kVA, 3,300/220 V, 50 Hz, single-phase transformer is built on a core, having an effective cross-section of 150 cm². It has 80 turns in the low voltage winding. Calculate, (a) maximum flux density, (b) number of turns in the high voltage winding and (c) full load currents in both the low and high voltage windings.*

SOLUTION: kVA = 50; $V_1 = 3,300$ V; $V_2 = 220$ V; $f = 50$ Hz.

Effective cross-section of the core $= a_i = 150\ \text{cm}^2 = 150 \times 10^{-4}\ \text{m}^2$.

Number of turns in the secondary winding (2) $= T_2 = 80$.

$$V_2 \approx E_2 = 4.44 f . B_m . a_i . T_2 = 220\ \text{V}.$$

(a) Maximum flux density

$$B_m = \frac{V_2}{4.44 f a_i T_2} = \frac{220.0}{4.44 \times 50.0 \times 150.0 \times 10^{-4} \times 80.0} = \frac{220.0}{266.4} \approx 0.826\ \text{Wb/m}^2.$$

(b) $\dfrac{V_1}{V_2} \approx \dfrac{E_1}{E_2} = \dfrac{T_1}{T_2}$.

Number of turns in the high voltage winding (1) $= T_1 = T_2 \cdot \dfrac{V_1}{V_2} = 80.0 \times \dfrac{3,300}{220} = 1,200$.

(c) Full-load current in the high voltage winding (1) $= I_1 = \dfrac{S}{V_1} = \dfrac{50.0 \times 10^3}{3,300} = 15.152$ A.

Full-load current in the low voltage winding (2) $= I_2 = \dfrac{S}{V_2} = \dfrac{50.0 \times 10^3}{220} = 227.3$ A.

NOTES

1. The voltage per turn is the same in both the windings, as

$$e_t = \frac{V_1}{T_1} = \frac{V_2}{T_2} = \frac{220.0}{80.0} = 2.75\ \text{V}.$$

2. The ampere-turns in both the windings are equal, as, $I_1 T_1 = I_2 T_2 = 227.3 \times 80 = 18,184$ AT or A,

$$\text{or,} \quad I_1 = I_2 \cdot \frac{T_2}{T_1} = 15.152 \times \frac{1,200}{80.0} = 227.3\,.$$

EXAMPLE 11.2 *A 50 Hz, single-phase transformer has 550 and 40 turns in the high and low voltage windings respectively. The input voltage on the HV side (primary) is 6,600 V. The full-load current in the secondary (LV) winding is 220 A. Find (a) the output voltage on the LV side (secondary) and (b) the full-load current in the primary (HV) winding. Also determine the kVA rating.*

SOLUTION $V_1 = 6,600$ V; $f = 50$ Hz.

No. of turns in the primary winding (1) $= T_1 = 550$.

No. of turns in the secondary winding (2) $= T_2 = 40$.

Full-load current in the secondary winding (LV) = 220 A.

$$\text{The voltage per turn} = e_t = \frac{V_1}{T_1} = \frac{V_2}{T_2} = \frac{6,600.0}{550.0} = 12.0.$$

(a) Output voltage in the secondary winding (LV) $= V_2 = e_t T_2 = 12.0 \times 40 = 480$ V.

This voltage is also calculated as $V_2 = V_1 \cdot \dfrac{T_2}{T_1} = 6,600.0 \times \dfrac{40}{550} = 480$ V.

(b) Full-load current in the primary winding (HV)

$$= I_1 = I_2 \cdot \frac{T_2}{T_1} = 220.0 \times \frac{40}{550.0} = 16.0 \text{ A},$$

as ampere-turns in the two windings are nearly equal.

$$\text{kVA rating} = S = (V_1 I_1) \times 10^{-3} = (V_2 I_2) \times 10^{-3} = 480.0 \times 220.0 \times 10^{-3} = 105.6.$$

This value can be checked, using the data for the HV side.

11.3 Operation on No-load and Load

We will first consider the operation at no-load. The primary winding is connected to the supply (ac), while the secondary winding is not connected to the load, i.e., it is open-circuited (Fig. 11.3(a)). Hence, there is no current in the secondary winding, although no-load current (I_0) is drawn from the supply, as the transformer is not an ideal one. It may be noted that the ampere-turns is given by $T_1 I_0$, which is small.

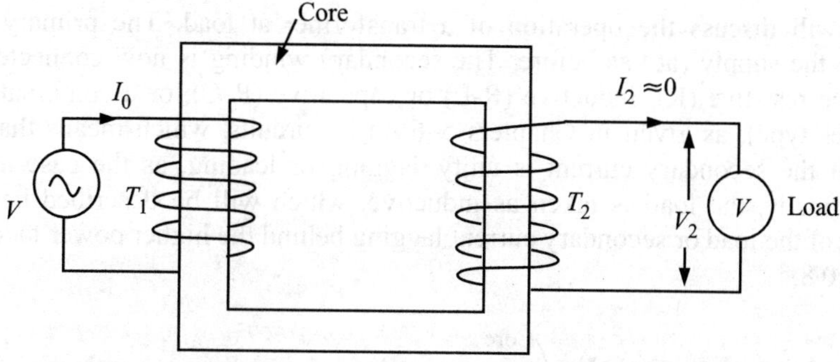

Fig. 11.3(a) *A single-phase transformer with load not connected to the secondary winding (open circuit)*

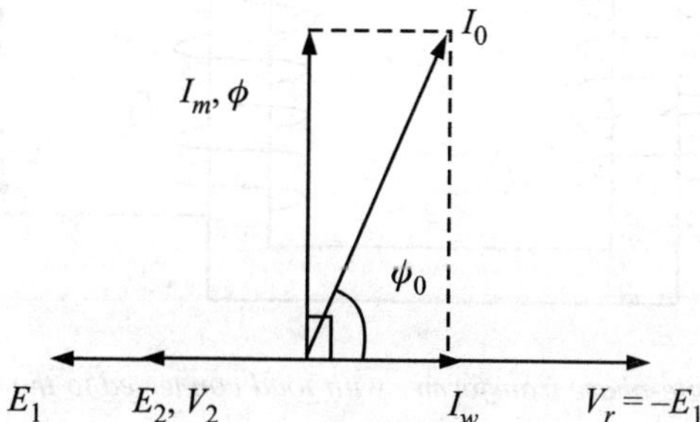

Fig. 11.3(b) *Phasor diagram at no-load*

The current is resolved into two components—first being the magnetising one (I_m), which produces the flux in the core. The other is the loss component (I_w), in phase with the supply (input) voltage, as there is iron loss in the core, even on no-load. The magnetising component is in phase with the maximum flux. As derived earlier for the emf equation, the flux is taken as $\phi = \phi_m \sin \omega t$, and the induced voltage in the primary winding is,

$$e_1 = E_{m1} \sin\left(\omega t - \frac{\pi}{2}\right) = E_{m1} \sin(\omega t - 90°).$$

The rms value of the induced voltage is $E_1 = \dfrac{E_{m1}}{\sqrt{2}}$.

The magnetising component is in phase with the flux, ϕ, which is leading the induced voltage, E_1 by 90°. Hence, this component lags behind the applied voltage, V_1 by 90°.

As $I_m > I_w$, the angle of lag between I_0 and V_1 is large, close to 90°. Therefore, power factor at no-load is low.

The relation between the supply (input) voltage and the induced voltage is,

$$V_1 = -E_1 + I_0 \cdot (r_1 + jx_1)$$

where, r_1 and x_1 are the resistance and leakage reactance of the primary winding respectively. As no-load current, I_0 is small, V_1 can be taken as approximately equal to E_1 (in magnitude only). The phasor diagram at no-load is shown in Fig. 11.3(b).

Now we will discuss the operation of a transformer at load. The primary winding is connected to the supply (ac) as before. The secondary winding is now connected to a load, which may be resistive (R), inductive (R-L) or capacitive (R-C), or a combination of them (mainly series type), as given in Chapters 5–6 on ac circuits, which means that the power factor (pf) of the secondary current is unity, lagging or leading, as the case may be (Fig. 11.4(a)). Normally, the load is taken as inductive, which will be described here, with the power factor of the load or secondary current lagging behind the higher power factor, normally greater than 0.8.

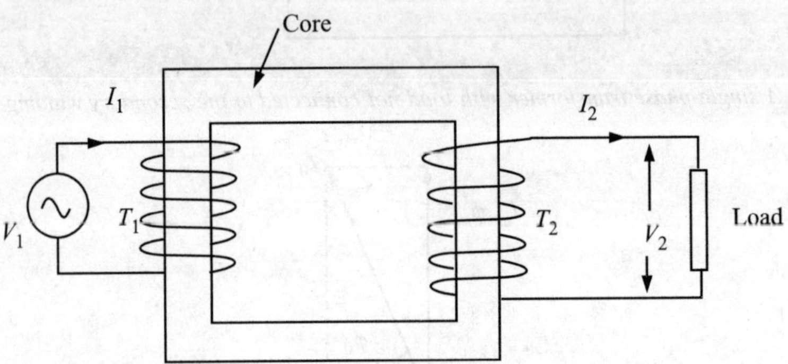

Fig. 11.4(a) *A single-phase transformer with load connected to the secondary winding*

As presented earlier, at no-load, the primary current is small, given by I_0. When the secondary winding is connected to a load (inductive), the current, I_2 starts flowing in it, drawing a corresponding increase in primary current, I_1. As will be shown later, the two currents, I_1 and I_2 oppose each other, the phase difference between them being nearly 180°. Hence, the total ampere-turns at load is the same as that at no-load, as the flux in the transformer remains the same,

i.e., $T_1 I_1 + T_2 I_2 = T_1 I_0$, or $I_1 = I_0 - \frac{T_2}{T_1} I_2 = I_0 - I_2'$.

$$I_2' = \frac{T_2}{T_1} I_2 = kI_2$$

is the secondary current referred to in the primary winding, where the turns ratio is

$$k = \frac{T_2}{T_1} = \frac{I_1}{I_2}.$$

The relation between the output voltage and the induced emf in the secondary side is

$$E_2 = V_2 - I_2 \cdot (r_2 + jx_2),$$

where r_2 and x_2 are the resistance and leakage reactance of the secondary winding respectively. The relation between the input voltage and the induced emf in the primary side is

$$V_1 = -E_1 + I_1 \cdot (r_1 + jx_1),$$

where the relation between the two induced emfs is given by,

$$\frac{E_2}{E_1} = \frac{T_2}{T_1} = k, \quad \text{or,} \quad E_2' = \frac{T_2}{T_1}E_1 = kE_1,$$

which is the induced voltage in the secondary winding referred to as primary. Similarly, we can write,

$$V_2' = \frac{T_2}{T_1}V_1 = kV_1,$$

which is the terminal voltage in the secondary side referred to as primary. The phasor diagram at load is shown in Fig. 11.4(b).

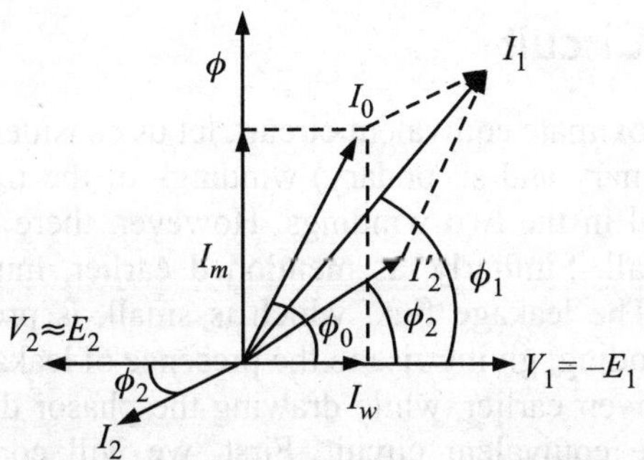

Fig. 11.4(b) *Phasor diagram at load*

EXAMPLE 11.3 *A 50 Hz, single-phase transformer has 600 and 150 turns in the primary and secondary windings respectively. The load current in the secondary winding is 100 A at 0.8 power factor (lag). The no-load current is 2 A at 0.2 pf (lag), with reference to the primary voltage. Find the primary current drawn from the supply.*

SOLUTION $T_1 = 600, T_2 = 150, f = 50$ Hz.

The phasor diagram of the currents is shown in Fig. 11.4(b). The primary voltage is taken as reference.

No-load current $= |I_0| = 2.0$ A, power factor $= \cos\phi_0 = 0.2$ (lag).

$$\phi_0 = \cos^{-1}(0.2) = 78.46°, \sin\phi_0 = \sin 78.46° \approx 0.98.$$

No-load current $= I_0\angle -\varphi_0 = (I_w - jI_m) = 2.0\angle -78.46° = (0.4 - j1.96)$ A .

Loss component $= I_w = I_0.\cos\phi_0 = 2.0 \times 0.2 = 0.4$ A.

Magnetising component $= I_m = I_0.\sin\phi_0 \approx 2.0 \times 0.98 = 1.96$ A.

Secondary current $= |I_2| = 100.0$ A, power factor $= \cos\phi_2 = 0.8$ (lag).

$\phi_2 = \cos^{-1}(0.8) = 36.87°$.

Secondary current in terms of primary $= |I_2'| = |I_2| \cdot \dfrac{T_2}{T_1} = 100.0 \times \dfrac{150}{600} = 25.0$ A.

Secondary current in terms of primary with direction $= I_2' \angle -\phi_2 = 25.0 \angle -36.87° = (20.0 - j15.0)$ A.

Primary current $= I_1 \angle -\phi_1 = (I_0 \angle -\phi_0 + I_2' \angle -\phi_2)$

$$= [(0.4 - j1.96) + (20.0 - j15.0)] = (20.4 - j16.96) = 26.53 \angle -39.74° \text{ A.}$$

Power factor = $\cos\phi_1 = \cos 39.74° \approx 0.77$ (lag).

11.4 Equivalent Circuit

Before we derive the approximate equivalent circuit, let us consider the resistance and leakage reactance of the two (primary and secondary) windings of the transformer. Copper is used as the conducting material in the two windings. However, there will be some resistance in the windings, though small. Similarly, as mentioned earlier, mutual flux, linking the two windings, is quite high. The leakage flux, which is small, is present, as it links only one (primary or secondary) winding, giving rise to the presence of leakage inductance or reactance in the two windings, as given earlier, while drawing the phasor diagram.

Now, let us derive the equivalent circuit. First, we will consider the parallel part of the approximate equivalent circuit (Fig. 11.5(a)). The no-load current, I_0 consists of two components—the loss component and the magnetising component. The loss component, I_w gives rise to the loss in the core (iron) at line frequency, which is in phase with the supply voltage. It is represented in the equivalent circuit by resistance, R_0 or R_m, connected in parallel with the input or supply voltage, V_1. According to the phasor diagram of no-load, where the no-load current lags behind the input voltage by an angle, ϕ_0, the loss component is obtained as $I_w = I_0 \cos\phi_0$, which is in phase with the supply voltage, V_1. The resistance in the parallel circuit is

$$R_0 = \frac{V_1}{I_w} = \frac{V_1}{I_0 \cos\phi_0}.$$

Similarly, the magnetising component, I_m, which creates flux, ϕ in the core (iron) at line frequency, is in the same quadrature as the supply voltage. It is represented in the equivalent circuit by the magnetising reactance, X_0 or X_m, connected also in parallel with the input or supply voltage, V_1. The magnetising component is found as $I_m = I_0 \sin\phi_0$, which is at 90° (quadrature) with the supply voltage, V_1. The magnetising reactance in the parallel circuit is

$$X_0 = \frac{V_1}{I_m} = \frac{V_1}{I_0 \sin\phi_0}.$$

This is the equivalent circuit at no-load.

The series part is represented by the impedance (z_1) of the primary winding, consisting of its resistance, r_1 and leakage reactance, x_1 followed by an ideal transformer, with the same input/output voltage, and also the same number of turns in the primary/secondary windings. On the output side of the ideal transformer is the impedance (z_2) of the secondary winding, consisting of its resistance, r_2 and leakage reactance, x_2, which is placed between the output and the load impedance, connected to the output terminals. The approximate equivalent circuit, consisting of both parallel and series parts, is shown in Fig. 11.5(a).

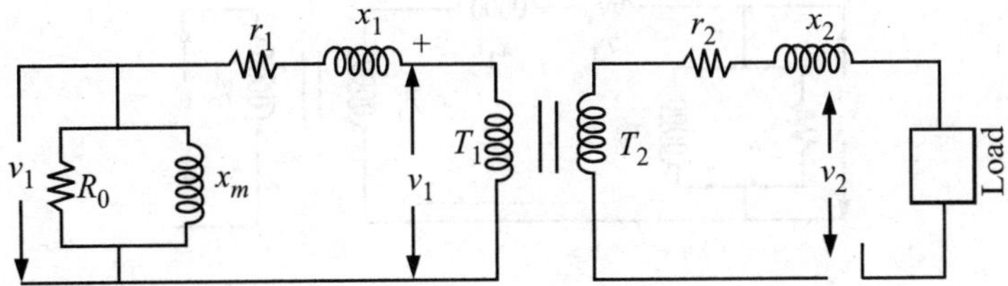

Fig. 11.5(a) *Approximate equivalent circuit, with an ideal transformer between the primary and secondary windings, which is connected to the load*

The impedance of the secondary winding, i.e., its resistance and leakage reactance, can be obtained in terms of the impedance of the primary winding by the relation,

$$z_2' = (r_2' + jx_2') = \frac{z_2}{k^2} = \frac{(r_2 + jx_2)}{k^2}.$$

For transferring the resistance part, the copper loss can be equated as,

$$(I_2)^2 r_2 = (I_2')^2 r_2' = (kI_2)^2 r_2',$$

from which the relation, $r_2' = \frac{r_2}{k^2}$ is found. Similarly, it can shown that $x_2' = \frac{x_2}{k^2}$. Hence, eliminating the ideal transformer, the equivalent circuit is drawn as in Fig. 11.5(b), where total resistance and leakage reactance of the transformer in terms of the primary side is given by

$$r_t = (r_1 + r_2') = \left(r_1 + \frac{r_2}{k^2}\right), \text{ and } x_t = (x_1 + x_2') = \left(x_1 + \frac{x_2}{k^2}\right),$$

with its impedance as $z_t = (r_t + jx_t)$.

11.5 Regulation

The regulation of a transformer is defined as the difference in the secondary voltage on no-load from the same voltage at a load, in terms of the secondary voltage on no-load. The ideal transformer in the equivalent circuit (Fig. 11.5(b)) is transferred to the primary side, with a corresponding change in the impedance (i.e., resistance and leakage reactance) as

$$r_t' = (r_1' + r_2) = [(k^2 r_1) + r_2], \quad \text{and} \quad x_t' = (x_1' + x_2) = [(k^2 x_1) + x_2],$$

where k is the turns ratio. These values can be obtained in the same way as given earlier. The secondary voltage at no-load and at load are E_2 and V_2 respectively. This equivalent circuit is shown in Fig. 11.5(b). The expressions used here can be obtained in the same way as given earlier.

Mathematically, the percentage voltage regulation is $\frac{E_2 - V_2}{E_2} \times 100.$

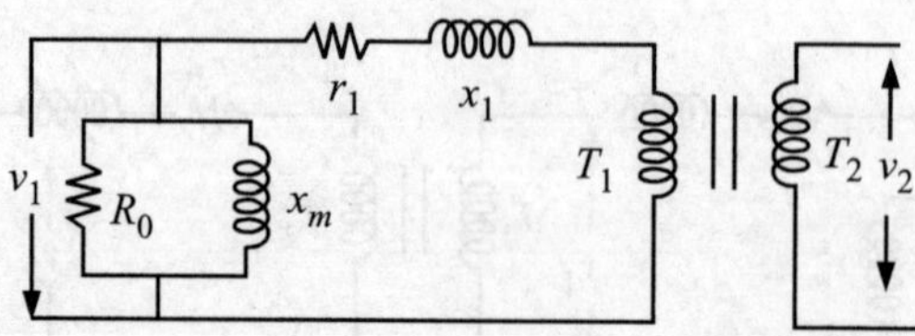

Fig. 11.5(b) *Approximate equivalent circuit with reference to the primary side, when the ideal transformer is placed at the end, between the primary and secondary ones*

It can also be defined as the difference in the secondary voltage on no-load from the same voltage at a load, in terms of secondary voltage on load, when the load is switched off. The secondary voltage on load is assumed to be a rated value. Hence, the percentage voltage regulation is

$$\frac{E_2 - V_2}{V_2} \times 100.$$

Assuming that the load is inductive, i.e., the current, I_2 is lagging the voltage, V_2 by an angle, ϕ, the phasor diagram is shown in Fig. 11.6(a).

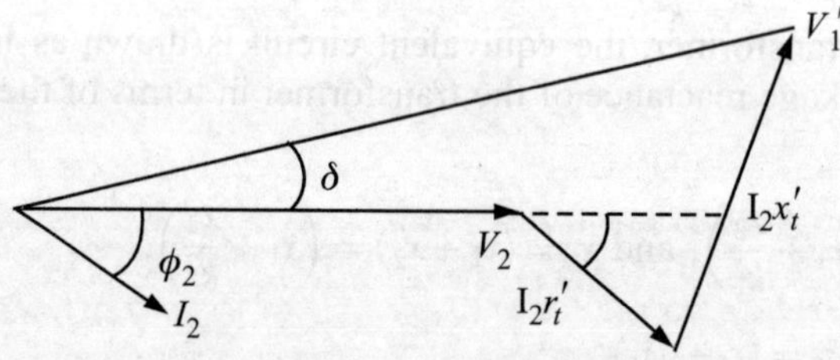

Fig. 11.6(a) *Resistive (upf) load*

If the projection of the voltage phasor, E_2, which leads V_2 by an angle, δ, on the voltage axis (x-axis) is taken, then

$$E_2 = V_2 + I_2 \cdot (r'_t \cos\phi + x'_t \sin\phi).$$

The projection of the voltage phasor, E_2, on the y-axis, which is perpendicular to the voltage axis, is taken as small, the values of r'_t and x'_t being small. The voltage drops, along with the angle, δ, are small.

The percentage voltage regulation is

$$\frac{E_2 - V_2}{V_2} \times 100 = \frac{I_2 \cdot (r_t' \cos\phi + x_t' \sin\phi)}{V_2} \times 100.$$

If the current, I_2 is the full-load current, per unit (pu) values of resistance and leakage reactance are obtained as $\varepsilon_r = \frac{I_2.r_t'}{V_2}$ and $\varepsilon_x = \frac{I_2.x_t'}{V_2}$ respectively.

The voltage regulation in per unit is written as

$$\varepsilon = (\varepsilon_r \cos\phi + \varepsilon_x \sin\phi).$$

Note that for an inductive load, the phase angle, ϕ is taken as positive. The regulation is positive in this case, as $|E_2| > |V_2|$, i.e., the magnitude of the voltage at load is lower than the magnitude of the voltage at no-load.

If the load is taken as resistive, i.e., the current, I_2 is in phase with the voltage, V_2 (the phase angle, $\phi = 0°$), the phasor diagram is shown in Fig. 11.6(b).

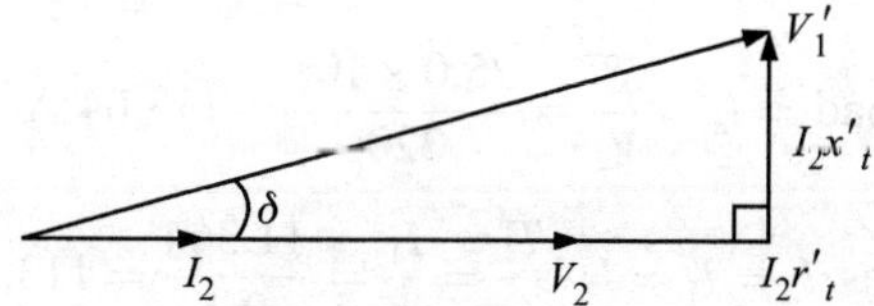

Fig. 11.6(b) *Inductive (lagging) load*

The percentage voltage regulation in this case, is given by $\frac{I_2.r_t'}{V_2} \times 100$.

Here, also the regulation is positive.

If the load is taken as capacitive, i.e., the current, I_2 is leading the voltage, V_2 by an angle, ϕ, the phasor diagram is shown in Fig. 11.6(c).

The percentage voltage regulation in this case, is given by $\frac{I_2 \cdot (r_t' \cos\phi + x_t' \sin\phi)}{V_2} \times 100$.

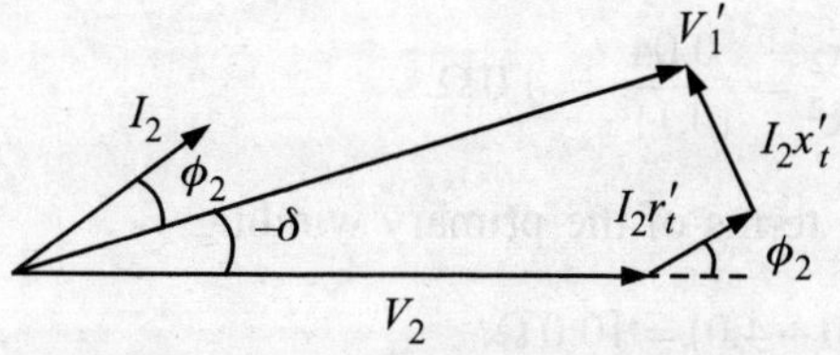

Fig. 11.6(c) *Capacitive (leading)*

Note that for this load, the phase angle, ϕ is taken as negative. The regulation is negative in this case, as $|E_2| < |V_2|$, i.e., the magnitude of the voltage at load is greater than the magnitude of the voltage at no-load.

In both cases of resistive and capacitive load, it can be stated that the angle, δ is small, with the projection of the voltage phasor, E_2, on the *y*-axis, also being small.

EXAMPLE 11.4 *A 25 kVA, 2,200/220 V, 50 Hz, single-phase transformer has a primary resistance of 1.5 Ω, and leakage reactance of 6.0 Ω. The secondary resistance and leakage reactance are 0.01 Ω, and 0.04 Ω respectively. Determine the percentage regulation, at half full load, with power factor of 0.8 lag and lead.*

SOLUTION $V_1 = 2,200$ V, $V_2 = 220$ V, $f = 50$ Hz.

$$r_1 = 1.5\,\Omega,\ x_1 = 6.0\,\Omega.\ r_2 = 0.01\,\Omega,\quad x_2 = 0.04\,\Omega.\ \text{kVA} = S = 25.0.$$

Primary current at full load $= I_1 = \dfrac{S}{V_1} = \dfrac{25.0\times 10^3}{2,200} = 11.364$ A.

Turns ratio $= k = \dfrac{T_2}{T_1} = \dfrac{V_2}{V_1} = \dfrac{220}{2,200} = 0.1.$

Secondary current at full load $= I_2 = \dfrac{S}{V_2} = \dfrac{25.0\times 10^3}{220} = 113.64$ A.

This current can be found as $I_2 = I_1' = I_1\cdot\dfrac{T_1}{T_2} = \dfrac{I_1}{k} = \dfrac{11.364}{0.1} = 113.64$ A.

Similarly, the secondary voltage $= V_2 = V_1' = V_1\cdot\dfrac{T_2}{T_1} = k.V_1 = 0.1\times 2,200 = 220$ V.

Resistance of the secondary winding, in terms of the primary winding

$$= r_2' = r_2.\left(\frac{T_1}{T_2}\right)^2 = \frac{r_2}{k^2} = \frac{0.01}{(0.1)^2} = 1.0\,\Omega.$$

Total resistance, in terms of the primary winding $= r_t = (r_1 + r_2') = (1.5 + 1.0) = 2.5\,\Omega.$

Leakage reactance of the secondary winding, in terms of the primary winding

$$= x_2' = x_2\cdot\left(\frac{T_1}{T_2}\right)^2 = \frac{x_2}{k^2} = \frac{0.04}{(0.1)^2} = 4.0\,\Omega.$$

Total leakage reactance, in terms of the primary winding

$$= x_t = (x_1 + x_2') = (6.0 + 4.0) = 10.0\,\Omega.$$

Primary current at half full load (50%) $= I_1 = x\cdot(I_1)_{fl} = 0.5\times 11.364 \approx 5.68$ A.

The value of *x* (which we have used here without explanation), the per unit (pu) or percentage of full load, will be discussed in the next section.

Power factor $= \cos\phi = 0.8$ (lag) or, $\phi = \cos^{-1}(0.8) = 36.87°$ (+ve).

$$\sin\phi = \sin(36.87°) = 0.6\ (+\text{ve}).$$

Regulation at half full load, with power factor of 0.8 lag

$$= \frac{I_1 \cdot (r_t \cos\phi + x_t \sin\phi)}{V_1} = \frac{5.68 \times [(2.5 \times 0.8) + (10.0 \times 0.6)]}{2,200} = \frac{5.68 \times 8.0}{2,200} \approx 0.0207 = 2.07\%.$$

Power factor $= \cos\phi = 0.8$ (lead) or, $\phi = \cos^{-1}(0.8) = 36.87°$ (−ve).

$\sin\phi = \sin(-36.87°) = -0.6$ (−ve).

Regulation at half full load, with power factor of 0.8 lead

$$= \frac{I_1 \cdot (r_t \cos\phi + x_t \sin\phi)}{V_1} = \frac{5.68 \times [(2.5 \times 0.8) - (10.0 \times 0.6)]}{2,200} = \frac{-5.68 \times 4.0}{2,200}$$

$\approx -0.01033 = -1.033\%.$

An alternative method of calculating regulation, using percentage or per unit (pu) values of resistance and leakage reactance of transformers is presented here. Let us first compute the above values in terms of the secondary side.

Total resistance, in terms of the secondary side

$$= r_t' = (r_1' + r_2) = [(k^2 r_1) + r_2] = [\{(0.1)^2 \times 1.5\} + 0.01] = 0.025\ \Omega.$$

This resistance can also be computed as $r_t' = (k^2 r_t) = [(0.1)^2 \times 2.5] = 0.025\ \Omega$.

Total leakage reactance, in terms of the secondary side

$$= x_t' = (x_1' + r_2) = [(k^2 x_1) + x_2] = [\{(0.1)^2 \times 6.0\} + 0.04] = 0.1\ \Omega.$$

pu resistance is the voltage drop in the total resistance at full load in terms of the rated voltage of the secondary winding, and is given by,

$$\varepsilon_r = \frac{r_t' \cdot I_2}{V_2} = \frac{0.025 \times 113.64}{220.0} \approx 0.0129 = 1.29\%.$$

pu reactance is $\varepsilon_x = \dfrac{x_t' \cdot I_2}{V_2} = \dfrac{0.1 \times 113.64}{220.0} \approx 0.0516 = 5.16\%.$

The secondary current at half full load (50%) $= I_2 = x \cdot (I_2)_{fl} = 0.5 \times 113.64 \approx 56.8$ A,

or 50% of the full load current, i.e., $x = 50\% = 0.5$.

Power factor $= \cos\phi = 0.8$ (lag) or, $\phi = \cos^{-1}(0.8) = 36.87°$ (+ve).

$\sin\phi = \sin(36.87°) = 0.6$ (+ve).

Regulation at half full load, with power factor of 0.8 lag

$$= [x \cdot (\varepsilon_r \cos\phi + \varepsilon_t \sin\phi)] = [0.5 \times \{(0.0129 \times 0.8) + (0.0516 \times 0.6)\}]$$

$= (0.5 \times 0.0413) = 0.02065 = 2.065\%$.

Power factor $= \cos\phi = 0.8$ (lead) or, $\phi = \cos^{-1}(0.8) = 36.87°(-ve)$.

$\sin\phi = \sin(-36.87°) = -0.6$ (–ve).

Regulation at half full load, with power factor of 0.8 lead

$= [x \cdot (\varepsilon_r \cos\phi + \varepsilon_t \sin\phi)] = [0.5 \times \{(0.0129 \times 0.8) - (0.0516 \times 0.6)\}]$

$= -(0.5 \times 0.002064) = -0.001032 = -0.1032\%$.

The voltage regulation as computed is nearly the same as obtained earlier. The values of phase angles are repeated here for convenience.

Thus, the voltage regulation at any load current with known power factor (lag/lead) can be determined by the procedure given here. The power factor may be unity (1.0)—upf— for resistive load.

11.6 Losses and Efficiency

There are two types of losses—the iron loss, which occurs in the magnetic core, and the copper loss in the two windings (primary and secondary). Iron loss, P_i is taken as constant, as it is related to the maximum value of flux (ϕ_m) or flux density (B_m). The voltage, V_1, is applied to the primary winding, the relation being given as,

$$V_1 \approx E_1 = 4.44 f T_1 \phi_m = 4.44 f T_1 B_m a_i.$$

The symbols used are given earlier. As V_1 is constant, the flux density is also constant, as are the parameters, like frequency, number of turns, and net cross-sectional area of the core. The iron or core loss is the sum of two losses, viz., hysteresis and eddy current.

Hysteresis loss, $P_h = k_h f B_m^x$, while eddy current, $P_e = k_e f^2 B_m^2 t^2$,

t being the thickness (mm) of the stamping used in the core. As the parameters, including t, are constant, iron loss, $P_i = P_h + P_e$, remains constant.

Copper loss is variable in electrical machines, both static and rotating, as it depends on the load current in the two windings.

Copper loss at full load, $(P_{cu})_{fl} = (I_2)^2 \cdot r_t' = (I_2)^2 \cdot [(k^2 r_1) + r_2]$,

where I_2 is the secondary current also at full load. The secondary current at other loads is $x.I_2$, where x is pu of full load, being in the range, $0.0 < x < 1.0$. At full load, $x = 1.0$, and at no-load, $x \approx 0.0$. The copper loss at other loads is $P_{cu} = x^2 \cdot (P_{cu})_{fl}$.

Note that there is no other loss in the transformer, as it is a static machine. The two losses (iron and copper) are of the order of nearly 1%. Hence, the efficiency of a transformer is about 97–97.5%, i.e., very high, compared to rotating machines.

Total losses, at any other load = Iron loss + Copper loss $= P_i + P_{cu} = P_i + x^2 \cdot (P_{cu})_{fl}$.

VA at full load $= S = V_2.I_2$.

VA at other loads $= x.S = x.V_2.I_2$.

Output power at that load $= P_o = x.S.\cos\phi = x.V_2.I_2.\cos\phi$,

assuming any power factor (pf), be it lagging, unity or loading, depending upon the type of load at the output. Note that the power factor is positive.

Input power at that load = Output + Losses (total)

$= P_{in} = x.V_2.I_2.\cos\phi + P_i + x^2.(P_{cu})_{fl}$.

$$\text{Efficiency} = \text{Output/Input} = \eta = \frac{P_0}{P_{in}} = \frac{x \cdot V_2 \cdot I_2 \cdot \cos\phi}{x \cdot V_2 \cdot I_2 \cdot \cos\phi + P_i + x^2 \cdot (P_{cu})_{fl}}.$$

Efficiency increases first with increase in output, reaches maximum value at a particular output, and then decreases with increase in output, i.e., it varies, as x is changed.

Maximum efficiency occurs when $\dfrac{d\eta}{dx} = 0$.

If we take the inverse of efficiency, then also $\dfrac{d}{dx}\left(\dfrac{1}{\eta}\right) = 0$, as its value will be minimum.

$$\frac{1}{\eta} = 1 + \frac{1}{V_2 \cdot I_2 \cdot \cos\phi} \cdot \left(\frac{P_i}{x} + x \cdot (P_{cu})_{fl}\right).$$

$$\text{Hence, } \frac{d}{dx}\left(\frac{1}{\eta}\right) \propto \left(-\frac{P_i}{x^2} + (P_{cu})_{fl}\right) = 0,$$

from which the value of x is obtained as $x = \sqrt{\dfrac{P_i}{(P_{cu})_{fl}}}$.

At this load, iron loss = copper load $= P_i = x^2 \cdot (P_{cu})_{fl}$.

Substituting this value of x in the expression for efficiency, maximum efficiency (η_m) can be obtained. Numerical examples are given at the end of this chapter.

For a distribution transformer, the value of x is chosen in the range 0.6–0.7 (60–70%), as the output of this transformer varies with the time of the day. However, for a power (unit) transformer, the value of $x = 1.0(100\%)$, as it is always on full load.

EXAMPLE 11.5 *A 10 kVA, 50 Hz, single-phase transformer has an efficiency of 95% at both full load, and at half full load. Find the efficiency at 75% of full load. Also calculate the load at which maximum efficiency occurs, and its value. Assume power factor as 0.8, in all cases.*

SOLUTION kVA $= S = 10.0, f = 50$ Hz.

Power factor $= 0.8(\cos\phi)$, in all cases.

Percentage of full load $= x_1 = 1.0(100\%)$.

Output power at full load $= P_{01} = x_1.S.\cos\phi = 1.0 \times 10.0 \times 0.8 = 8.0$ kW$= 8,000$ W.

Efficiency $= \eta_1 = \dfrac{P_{01}}{P_1} = \dfrac{8.0}{P_1} = 0.95 = 95.0\%$.

Input power = Output power + losses $= P_1 = \dfrac{P_{01}}{\eta_1} = \dfrac{8.0}{0.95} = 8.421$ kW $= 8,421$ W.

Losses (total) = Iron loss + Copper loss (at full load)

$$= (P_i + x_1^2.P_{C1}) = (P_i + P_{C1}) = (P_1 - P_{01}) = (8,421 - 8,000) = 421 \text{ W}. \qquad \text{(i)}$$

Percentage of full load $= x_2 = 0.5(50\%)$, i.e., half full load.

Output power $= P_{02} = x_2.S.\cos\phi = 0.5 \times 10.0 \times 0.8 = 4.0$ kW $= 4,000$ W.

Efficiency $= \eta_2 = \dfrac{P_{02}}{P_2} = \dfrac{4.0}{P_2} = 0.95 = 95.0\%$.

Input power = Output power + losses $= P_2 = \dfrac{P_{02}}{\eta_2} = \dfrac{4.0}{0.95} = 4.211$ kW $= 4,211$ W.

Losses (total) = Iron loss + Copper loss (at half full load)

$$= (P_i + P_{C2}) = (P_i + x_2^2 \cdot P_{C1}) = [P_i + (0.5)^2 \cdot P_{C1}] = (P_2 - P_{02}) = (4,211 - 4,000) = 211 \text{ W (ii)}$$

Subtracting (i) from (ii), we get,

$$0.75.P_{C1} = 421 - 211 = 210$$

or, copper loss (at full load) $= P_{C1} = \dfrac{210.0}{0.75} = 280$ W.

From (i), Iron loss = losses (total) − copper loss (at full load) $= 421 - 280 = 141$ W.

At 75% full load, $x_3 = 0.75 = 75\%$,

Output power $= P_{03} = x_3.S.\cos\phi = 0.75 \times 10.0 \times 0.8 = 6$ kW.

Losses (total) = Iron loss + Copper loss (at 75% full load)

$$= (P_i + P_{C3}) = (P_i + x_3^2 \cdot P_{C1}) = [P_i + (0.75)^2 \cdot P_{C1}] = (P_3 - P_{03})$$

$$= [141.0 + (0.5625 \times 280)] \approx (141.0 + 158.0)$$

$$= 299.0 \approx 300.0 \text{ W} = 0.3 \text{ kW}.$$

Input power = output power + total loss $= P_3 = 6.0 + 0.3 = 6.3$ kW.

$$\text{Efficiency} = \eta_3 = \frac{P_{03}}{P_3} = \frac{600}{6.3} = 0.9524 = 95.24\%.$$

Percentage of full load, at which maximum efficiency is obtained

$$= x_4 = \sqrt{\frac{P_i}{P_{C1}}} = \sqrt{\frac{141.0}{280.0}} \approx 0.71 = 71.0\%.$$

At this load, copper loss = iron loss = 141 W.

Losses (total) $= 2 \times 141 = 282$ W $= 0.282$ kW.

Output power $= P_{04} = x_4.S.\cos\phi = 0.71 \times 10.0 \times 0.8 = 5.68$ kW.

Input power = output power + total loss $= P_4 = 5.68 + 0.282 = 5.962$ kW.

$$\text{Efficiency} = \eta_4 = \frac{P_{04}}{P_4} = \frac{5.68}{5.962} = 0.9527 = 95.27\%.$$

EXAMPLE 11.6 *The maximum efficiency of a 25 kVA, 50 Hz, single-phase transformer is 97.5%, and occurs at 80% of the full load at 0.9 power factor. If the leakage impedance of the transformer is 4.5%, find the voltage regulation at rated load, 0.8 power factor leading.*

SOLUTION kVA $= S = 25.0, f = 50$ Hz.

Percentage of full load $= x = 0.8 (80\%)$.

Power factor $= \cos\phi = 0.9$.

Output power at full load $= P_0 = x.S.\cos\phi = 0.8 \times 25.0 \times 0.9 = 18.0$ kW.

$$\text{Efficiency} = \eta_m = \frac{P_0}{P_1} = \frac{18.0}{P_1} = 0.975 = 97.5\%.$$

$$\text{Input power} = \text{Output power} + \text{losses} = P_1 = \frac{P_0}{\eta_m} = \frac{18.0}{0.975} = 18.462 \text{ kW}.$$

$$\text{Losses (total)} = (\text{Iron loss} + \text{Copper loss}) = (P_i + x^2.P_C) = [P_i + (0.8)^2.P_C]$$

$$= [P_i + (0.64.P_C)] = (P_1 - P_0) = (18.462 - 18.0) = 0.462 \text{ kW} = 462 \text{ W}.$$

As maximum efficiency occurs at this load, iron loss $= P_i = 0.64 \times P_C = \dfrac{462}{2} = 231$ W

$$\text{or, copper loss at full load} = P_C = \frac{231.0}{0.64} \approx 361.0 \text{ W}.$$

$$\text{Per unit (pu) resistance} = \varepsilon_r = \frac{P_C}{S} = \frac{361.0}{25 \times 10^3} = 0.0144.$$

Per unit (pu) impedance $= \varepsilon_z = 0.045 = 4.5\%$.

$$\text{Per unit (pu) reactance} = \varepsilon_x = \sqrt{(\varepsilon_z)^2 - (\varepsilon_4)^2} = \sqrt{(0.045)^2 - (0.0144)^2} = 0.0426 = 4.26\%.$$

Power factor $= \cos\phi = 0.8$ (lead) or, $\phi = \cos^{-1}(0.8) = 36.87°$ (−ve).

$\sin\phi = \sin(-36.87°) = -0.6$ (−ve).

Regulation at full (rated) load, with power factor of 0.8 lead

$$= (\varepsilon_r \cos\phi - \varepsilon_x \sin\phi) = [(0.0144 \times 0.8) - (0.0426 \times 0.6)] = -0.014 = -1.4\%.$$

11.7 Testing

An electrical machine, say a transformer, is tested first, before it is used in any system. Though direct testing, is preferred—mainly to check whether the temperature rise is within the specified limits, with the machine delivering full load output—it is not often used. This is because the energy consumed is as per its full load rating and arranging the dissipation of the energy is a costly affair, both in terms of capital, and energy or power consumed. Two types of indirect tests, viz., open circuit (OC) and short circuit (SC), are normally conducted, and the data obtained is used to predict the efficiency and regulation at any load.

Open Circuit (OC) Test

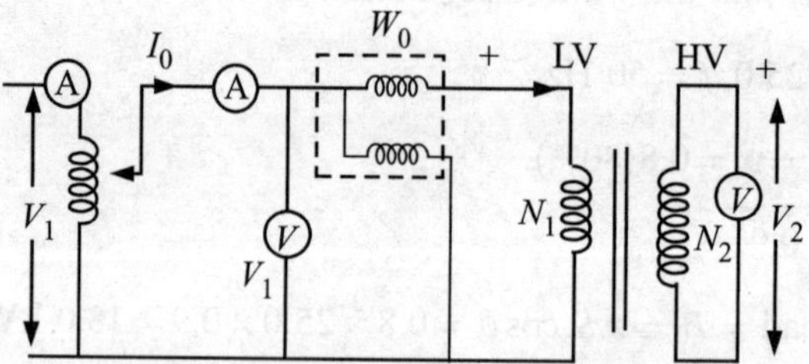

Fig. 11.7(a) *Open circuit test for a transformer*

The low voltage (LV) winding of the single-phase transformer is connected to a rated voltage as specified, with the high voltage (HV) winding on open circuit, i.e., at no load. This is, because of easy availability of this voltage, and also metering, which is required to conduct this test. A voltmeter and an ammeter are connected to measure the input voltage and current respectively, while a wattmeter is also connected to measure the input power at no load (Fig. 11.7(a)). Another voltmeter of higher rating is needed to measure the output voltage. The readings are V_1, I_0 and W_0, in the order specified, while the output voltage is V_2. This voltage along with the earlier one is used to check the rating, i.e., the voltages of LV (1) and HV (2) windings, and also the turns ratio, given as

$$k = \frac{T_2}{T_1} = \frac{V_2}{V_1}$$

As rated voltage is applied to the LV winding, the flux in the core is rated. The magnetising component is higher in magnitude than the loss component of the no-load current, which is small, about 5–8% of the rated current in the LV winding. The power factor is low, which is computed as $\cos\phi_0 = \frac{W_0}{V_1 \cdot I_0}$, ϕ_0 is the phase angle between V_1 and I_0. The no-load current lags behind the voltage applied across the LV winding, with the magnetising component ($I_m = I_0 \cdot \sin\phi_0$) being in phase with the flux, but in quadrature with the applied voltage. The loss component ($I_m = I_0 \cdot \cos\phi_0$) is in phase with the applied voltage (Fig. 11.3(a)). The computation of the parameters of the parallel branch in the equivalent circuit (Fig. 11.5(a)) has been given earlier, and is not repeated here.

SHORT CIRCUIT (SC) TEST

The high voltage (HV) winding of the single-phase transformer is connected to the supply, with the low voltage (LV) winding short-circuited (Fig. 11.7(b)). The input voltage needed is quite low in this case, i.e., about 4–6% of the rated voltage of HV winding. The metering scheme, or common measuring instruments used is nearly the same as for the open circuit test. As the input voltage required for passing full load currents both in HV and LV windings is limited to about 4–6% of the rated voltage of the HV winding, a voltmeter of similar range can be used. The same is the case in respect of the ammeter and wattmeter, except that the current is high depending on the kVA rating of the transformer. If needed, a current transformer (CT) of proper rating may be used, the secondary current of which is the same as the rating of the ammeter and current coil of the wattmeter. If an ammeter is to be connected to the LV side, a current transformer is needed to measure the rated current. The turns ratio given by

$$k = \frac{T_2}{T_1} \approx \frac{I_1}{I_2},$$

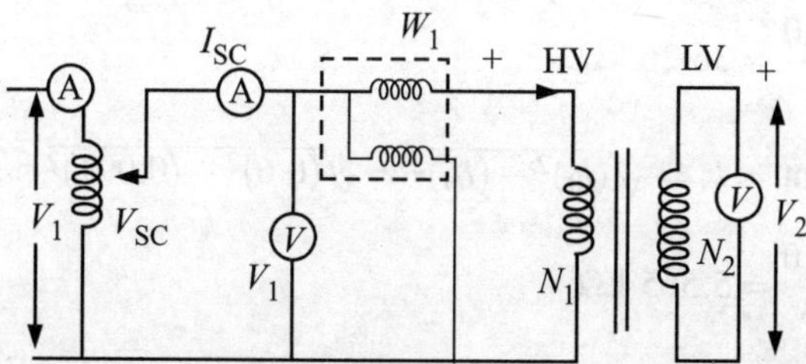

Fig. 11.7(b) *Short circuit*

can be checked from the readings of the ammeters. As low voltage is applied to the HV winding, the flux in the core is small, so that the iron loss, being also small, can be neglected when compared to the copper loss at rated currents. The power factor of the current in the HV winding is found as

$$\cos\phi_{sc} = \frac{W_{sc}}{V_2 \cdot I_2} = \frac{r_t'}{z_t'}.$$

ϕ_{sc} is the phase angle between V_2 and I_2, r_t', x_t' and z_t' are the resistance, leakage reactance and impedance of the transformer referred to the HV winding. It must be noted that the voltage, V_2, as given here is the voltage applied to the HV winding, not its rated value. The magnitude of the impedance is obtained as $|z_t'| = \frac{V_2}{I_2}$, I_2 being the rated current. Now, it is possible to substitute the parameters of the series branch in the equivalent circuit (Fig. 11.5(a)) given earlier.

ADDITIONAL SOLVED EXAMPLES

EXAMPLE 11.7 *The following results were obtained in tests on a 40 kVA, 3,300/220 V, 50 Hz, single-phase transformer:*
Open circuit test: Primary voltage = 3,300 V, primary current = 0.6 A and input power 300 W.
Short circuit test: Reduced voltage on the primary side (150 V) to give rated secondary current; primary current is 12.1 A and input power 450 W.
Calculate the parameters of the equivalent circuit, referred to as the primary side. Determine the percentage regulation and efficiency at full load, with power factor of 0.707 lag.

SOLUTION $V_1 = 3{,}300$ V, $V_2 = 220$ V, $f = 50$ Hz,

kVA = $S = 40.0$.

From open circuit tests, iron losses = $W_{oc} = 300$ W.

Primary current = $I_0 = 0.6$ A.

$$\text{Core loss component} = I_w = \frac{W_{oc}}{V_1} = \frac{300}{3,300} = 0.091 \text{ A.}$$

$$R_0 = \frac{V_1}{I_w} = \frac{3,300.0}{0.091} = 36.26 \text{ k}\Omega.$$

$$\text{Magnetising component} = I_m = \sqrt{(I_0)^2 - (I_m)^2} = \sqrt{(0.6)^2 - (0.091)^2} \approx 0.593 \text{ A.}$$

$$X_0 = \frac{V_1}{I_m} = \frac{3,300.0}{0.593} = 5.565 \text{ k}\Omega$$

$$\text{Power factor for this test (OC)} = \cos\phi_0 = \frac{I_c}{I_0} = \frac{0.091}{0.6} = 0.152 \text{ (lag).}$$

$$\phi_0 = \cos^{-1}(0.152) = 81.26^\circ \text{ (+ve).}$$

$$\text{This value can be obtained as } \cos\phi_0 = \frac{W_0}{V_1 I_0} = \frac{300.0}{3,300.0 \times 0.6} \approx 0.152 \text{ (lag).}$$

$$\text{Primary current at full load} = I_1 = \frac{S}{V_1} = \frac{40.0 \times 10^3}{3,300} \approx 12.1 \text{ A.}$$

From the short circuit test,

Copper losses on full load = $W_{sc} = (P_c)_{fl} = 450$ W.

Applied voltage on the primary side = 150 V

The current for this test is the same as the primary current on full load.

Equivalent impedance, in terms of the primary winding $= z' = \dfrac{V_{sc}}{I_1} = \dfrac{150.0}{12.1} \approx 12.38\ \Omega.$

Equivalent resistance, in terms of the primary winding $= r' = \dfrac{W_{sc}}{(I_1)^2} = \dfrac{450.0}{(12.1)^2} \approx 3.063\ \Omega.$

Equivalent reactance, in terms of the primary winding

$$= x' = \sqrt{(z')^2 - (r')^2} = \sqrt{(12.38)^2 - (3.063)^2} \approx 12.0 \quad \Omega.$$

Power factor for this test (SC) $= \cos\phi_{sc} = \dfrac{r'}{z'} = \dfrac{3.063}{12.38} = 0.247$ (lag).

$$\phi_{sc} = \cos^{-1}(0.247) = 75.68^\circ \text{ (+ve)}.$$

This value can be obtained as $\cos\phi_{sc} = \dfrac{W_{sc}}{V_{sc} I_1} = \dfrac{450.0}{150.0 \times 12.1} \approx 0.247$ (lag).

The equivalent circuit referred to primary is given in Fig. 11.8.

Power factor $= \cos\phi = 0.707$ (lag) or, $\phi = \cos^{-1}(0.707) \approx 45^\circ$(+ve).

$$\sin\phi = \sin(45^\circ) = 0.707 \text{ (+ve)}.$$

Regulation at half full load, with power factor of 0.707 lag

$$= \frac{I_1 \cdot (r\cos\phi + x\sin\phi)}{V_1} = \frac{12.1 \times [(3.063 \times 0.707) + (12.0 \times 0.707)]}{3,300}$$

$$= \frac{12.1 \times 10.65}{3,300} \approx 0.0391 = 3.91\%.$$

Total loss at full load $= P_i + (P_c)_{fl} = 300 + 450 = 750$ W $= 0.75$ kW.

Power factor $= 0.707(\cos\phi)$.

Output power $= P_0 = S.\cos\phi = 40.0 \times 0.707 = 28.28$ kW.

Input power = output power + total loss $= P_1 = 28.28 + 0.75 = 29.03$ kW.

Efficiency at full load $= \eta_{fl} = \dfrac{P_0}{P_1} = \dfrac{28.28}{29.03} = 0.9742 = 97.42\%.$

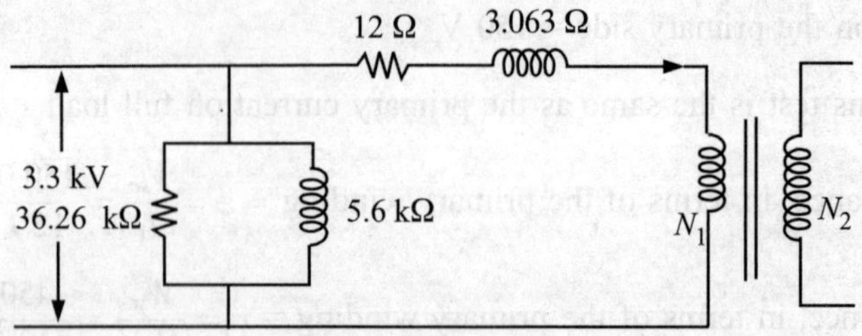

Fig. 11.8 *Short circuit*

EXAMPLE 11.8 *A 50 kVA, 11,000/220 V, 50 Hz, single-phase transformer requires 350 V at the HV winding to circulate full-load current, with the LV winding short-circuited. The input power is 600 W. Find the percentage regulation and the secondary (LV) terminal voltage on full load at 0.8 power factor lag.*

SOLUTION $V_1 = 11,000$ V (HV), $V_2 = 220$ V (LV), $f = 50$ Hz, kVA $= S = 50.0$.

Primary current at full load $= I_1 = \dfrac{S}{V_1} = \dfrac{50.0 \times 10^3}{11.0 \times 10^3} = 4.545$ A.

Secondary current at full load $= I_2 = \dfrac{S}{V_2} = \dfrac{50.0 \times 10^3}{220} = 227.3$ A.

From the short circuit test, copper losses on full load $= W_{sc} = P_c = 600$ W.

Applied voltage on the primary side = 350 V.

The current for this test is the same as the primary current on full load.

Equivalent impedance, in terms of the primary winding $= z' = \dfrac{V_{sc}}{I_1} = \dfrac{350.0}{4.545} \approx 77.0\,\Omega$.

Equivalent resistance, in terms of the primary winding $= r' = \dfrac{W_{sc}}{(I_1)^2} = \dfrac{600.0}{(4.545)^2} \approx 29.05\Omega$.

Equivalent reactance, in terms of the primary winding

$$= x' = \sqrt{(z')^2 - (r')^2} = \sqrt{(77.0)^2 - (29.05)^2} \approx 71.3\,\Omega.$$

Power factor for this test (SC) $= \cos\phi_{sc} = \dfrac{r'}{z'} = \dfrac{29.05}{77.0} = 0.377$ (lag).

$$\phi_{sc} = \cos^{-1}(0.377) = 67.85° \text{ (+ve)}.$$

This value can be obtained as

$$\cos\phi_{sc} = \frac{W_{sc}}{V_{sc}I_1} = \frac{600.0}{350.0 \times 4.545} \approx 0.377 \text{ (lag)}.$$

Power factor $= \cos\phi = 0.8$ (lag) or, $\phi = \cos^{-1}(0.8) = 36.87°(+\text{ve})$.

$\sin\phi = \sin(36.87°) = 0.6\ (+\text{ve})$).

Regulation at full load, with power factor of 0.8 lag

$$= \frac{I_1.(r\cos\phi + x\sin\phi)}{V_1} = \frac{4.545 \times [(29.05 \times 0.8) + (71.3 \times 0.6)]}{11{,}000} = \frac{4.545 \times 66.0}{11{,}000}$$

$$\approx 0.0273 = 2.73\%.$$

The secondary (LV) terminal voltage $= [220.0 \times (1.0 - 0.0273)] \approx 214.0$ V.

Conclusion

In this chapter, we studied the transformer, a static electrical machine. Both the input and output voltages are ac, at normally, line frequency (50 Hz). The voltages are dependent on the number of turns in the primary (input) and secondary (output) windings. The topics presented are types of construction, the emf equation, operation on no-load and load and equivalent circuit of a single-phase transformer. Then, regulation and efficiency at various loads with power factor (lagging, unity and leading) are described. Two types of losses (iron and copper), and tests (open and short circuit), are also discussed. In the next chapter, the first rotating machine—a dc one—which can be operated both as a generator and a motor, is presented.

Multiple Choice Type Questions

Tick the correct option(s), out of the four choices provided in each question.

11.1 A single-phase transformer is supplied at 100 V, 50 Hz. The flux density in the core is 1.0 Wb/m^2. If the number of primary turns and cross-sectional area of the core remain the same, with the transformer now being fed at 120 V, 60 Hz supply, find the new value of the flux density in Wb/m^2.

(a) 1.2 (b) 0.833 (c) 1.0 (d) 0.5

11.2 The thickness of the stamping in a transformer is doubled, with the cross-sectional area of the core remaining the same. The flux density in the core and the supply frequency, remain the same. If the new eddy current loss under this condition is x times the previous value, then x is given by,

(a) 8 (b) 4 (c) 2 (d) 1

11.3 In a step-down transformer, which of the following is correct?

(a) The secondary winding is a high voltage one.
(b) The secondary winding is a low voltage one.
(c) The primary winding is a high voltage one.
(d) The primary winding is a low voltage one.

11.4 The magnetising component of the primary current in a single-phase transformer rated at 110/220 V is 3 A. Its core loss component is neglected. If the secondary current on the resistive load is 2 A, find the magnitude of the primary current.

(a) 4 A (b) 3 A (c) 7 A (d) 5 A

11.5 The regulation of a single-phase transformer,

(a) decreases with lagging power factor (pf).
(b) is constant at unity power factor.
(c) increases with lagging power factor.
(d) none of the above

11.6 The percentage resistance and leakage reactance of a single-phase transformer are 2% and 6% respectively. If it is on full load at a lagging power factor of 0.8, what is the regulation?

(a) 6.4% (b) 2% (c) 6% (d) 5%

11.7 In the previous problem (#11.6), with all the conditions remaining the same, if the power factor is changed only to 0.8 leading, what is the new value of regulation?

(a) 0% (b) –6% (c) –3.2% (d) –2%

11.8 The percentage resistance and leakage reactance of a single-phase transformer are 3% and 4% respectively. What is its percentage impedance?

(a) 4% (b) 3% (c) 7% (d) 5%

11.9 The core loss and full load copper loss of a single-phase transformer are 300 W and 400 W respectively. Find the load in pu, at which both the core and copper losses are equal.

(a) 1.0 (b) 0.866 (c) 0.5 (d) 0.75

11.10 The efficiency of a single-phase transformer,

(a) increases with load.
(b) decreases with load.
(c) is independent of load.
(d) is maximum for a certain value of load.

11.11 The efficiency of a transformer at full load, is in the range of

(a) 97% (b) 85% (c) 90% (d) none of (a)–(b)

11.12 For a single-phase transformer, tick the correct statement(s).

(a) The core loss is higher than the copper loss at full load.
(b) The core loss is lower than the copper loss at full load
(c) The copper loss is proportional to the load current.
(d) The copper loss is proportional to the square of the load current.

11.13 A single-phase transformer has a turns ratio of 1:2, and is connected to a pure resistive load as shown in Fig. MCQ11.13. The magnetizing current drawn is 1 A, and the secondary current is 1 A. If core losses and leakage reactances are neglected, the primary current is [GATE]

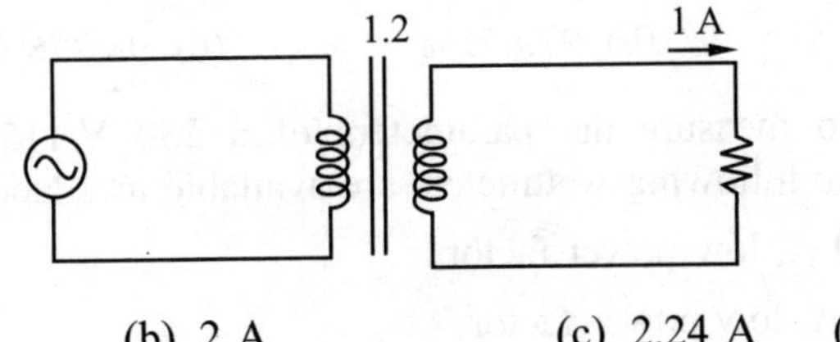

Fig. MCQ11.13

(a) 1.41 A (b) 2 A (c) 2.24 A (d) 3 A

11.14 For a single-phase, two winding transformer, the supply frequency and voltage are both increased by 10%. The percentage changes in the hysteresis loss and eddy current loss, respectively, are [GATE]

(a) 10 and 21 (b) −10 and 21 (c) 21 and 10 (d) −21 and 10

11.15 Assuming an ideal transformer, the Thevenin's equivalent voltage and impedance as seen from the terminals x and y for the circuit in Fig. MCQ11.5 are

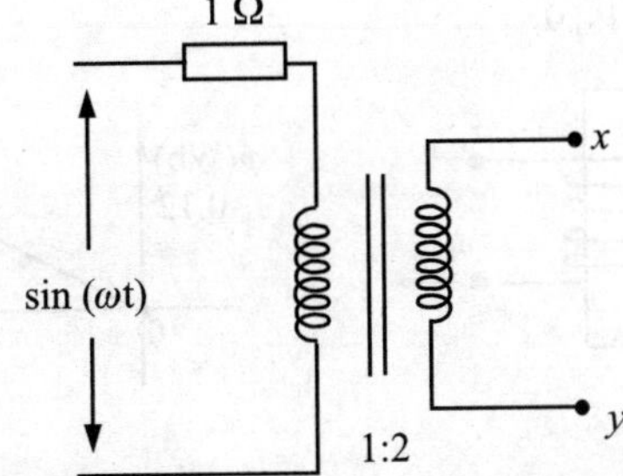

Fig. MCQ11.15

(a) $2\sin(\omega t), 4\,\Omega$ (b) $1\sin(\omega t), 1\,\Omega$ (c) $1\sin(\omega t), 2\,\Omega$ (d) $2\sin(\omega t), 0.5\,\Omega$

11.16 A single-phase air core transformer, fed from a rated sinusoidal supply, is operating at no load. The steady state magnetizing current drawn by the transformer from the supply will have the waveform [GATE]

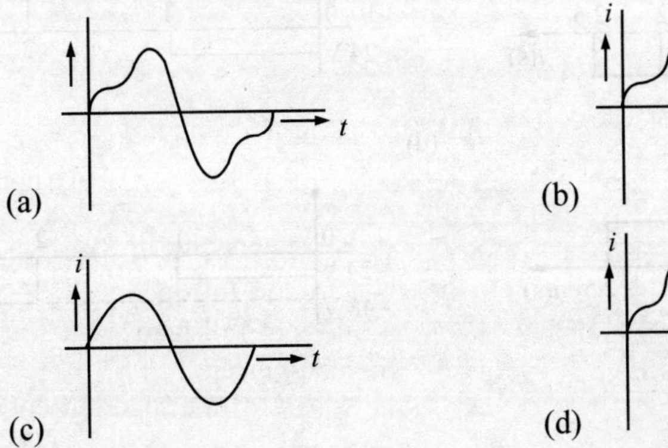

11.17 A single-phase, 50 kVA, 250 V/500 V two winding transformer has an efficiency of 95% at full load, unity power factor. If it is re-configured as a 500 V/750 V auto-transformer, its efficiency at the new rated load at unity power factor will be [GATE]

(a) 95.752% (b) 97.851% (c) 98.276% (d) 99.241%

11.18 It is desired to measure the parameters of a 230 V/115 V, 2 kVA, single-phase transformer. The following wattmeters are available in a laboratory:

W_1 : 250 V, 10 A, low power factor

W_2 : 250 V, 5 A, low power factor

W_3 : 150 V, 10 A, high power factor

W_4 : 150 V, 5 A, high power factor

The wattmeters used in the open circuit test and short circuit test of the transformer will respectively be [GATE]

(a) W_1 and W_2 (b) W_2 and W_4 (c) W_1 and W_4 (d) W_2 and W_3

11.19 The core of a two-winding transformer is subjected to a magnetic flux variation as indicated in Fig. MCQ11.19. [GATE]

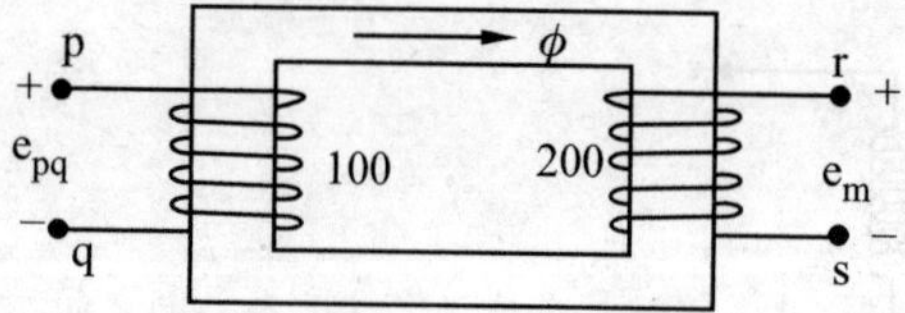

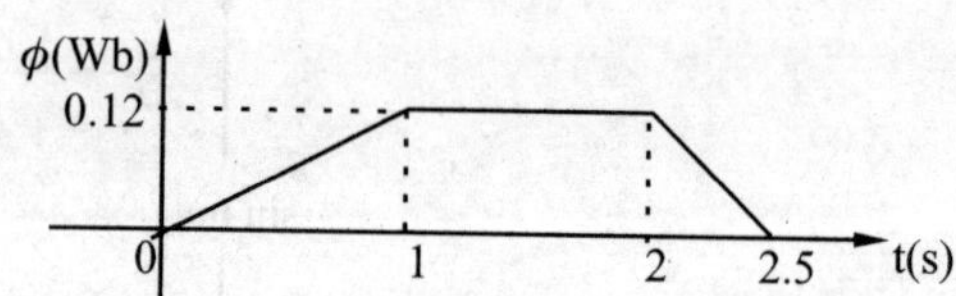

Fig. MCQ.11.19

The induced emf (e_{rs}) in the secondary winding as a function of time will be of the form

(a)

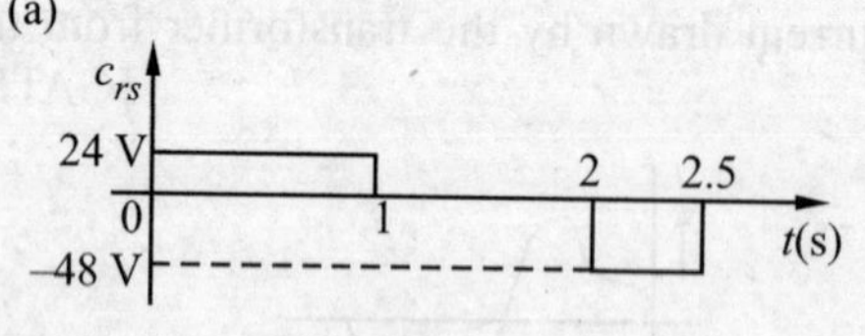

(b)

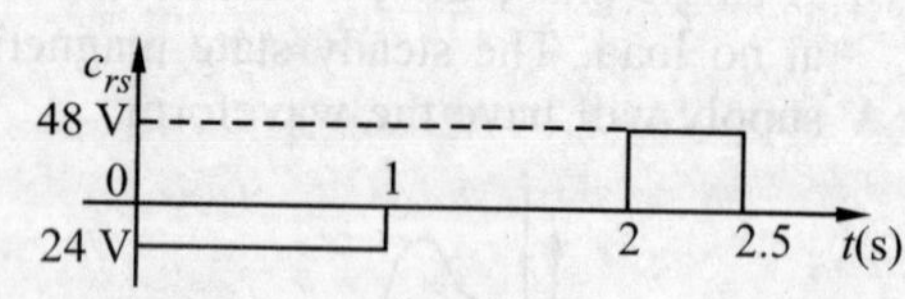

(c)

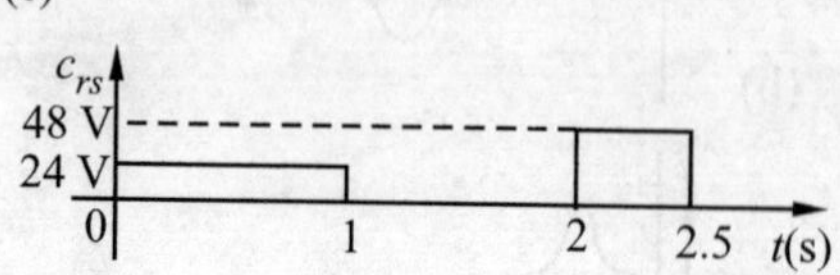

(d)

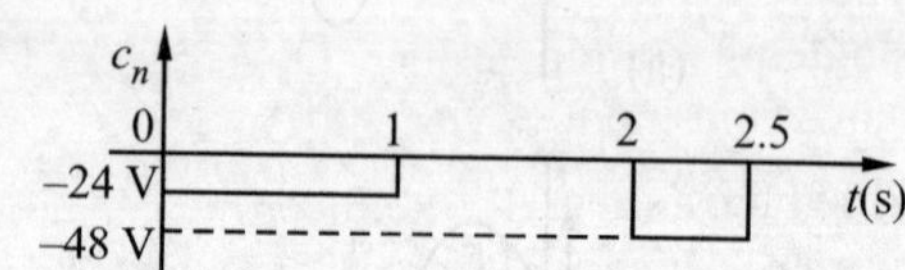

Answers

11.1	(c)	11.2	(b)	11.3	(b), (c)	11.4	(d)	11.5	(c)	11.6	(a)
11.7	(c)	11.8	(d)	11.9	(b)	11.10	(a), (d)	11.11	(a)	11.12	(b), (d)
11.13	(a)	11.14	(c)	11.15	(a)	11.16	(c)	11.17	(c)	11.18	(d)
11.19	(a)										

Problems

11.1 A 50 kVA, 3,300/220 V, 50 Hz, single-phase transformer is built on a core, having an effective cross-section of 150 cm^2. It has 80 turns in the low voltage winding. Calculate (a) maximum flux density, (b) no. of turns in the high voltage winding and (c) full load currents in both low and high voltage windings.

11.2 A 50 Hz, single-phase transformer has 550 and 40 turns in the high and low voltage windings respectively. The input voltage on the HV side (primary) is 6,600 V. The full load current in the secondary (LV) winding is 220 A. Find (a) the output voltage on the LV side (secondary) and (b) the full load current in the primary (HV) winding. Also determine the kVA rating.

11.3 The maximum flux density in the core of a 3,000/250 V, 50 Hz, single-phase transformer is 1.0 Wb/m^2. If the emf per turn is 6 V, calculate the number of turns in the primary (HV) and secondary (LV) windings and the effective cross-section of the core.

11.4 A 50 Hz, single-phase transformer has 600 and 150 turns in the primary and secondary windings respectively. The load current in the secondary winding is 100 A at 0.8 power factor (lag). The no-load current is 2 A at 0.2 pf (lag), with reference to the primary voltage. Find the primary current drawn from the supply.

11.5 A 100 kVA, 1,100/220 V, 50 Hz, single-phase transformer has a primary resistance of 0.1 Ω and leakage reactance of 0.3 Ω. The secondary resistance and leakage reactance are 0.0005 Ω and 0.0012 Ω respectively. Determine the percentage regulation, at full load, with power factor of 0.8 lag and unity (1.0).

11.6 The following results were obtained in tests on a 40 kVA, 3,300/220 V, 50 Hz, single-phase transformer: Open circuit test: Primary voltage 3,300 V, primary current 0.6 A and input power 300 W

Short circuit test: Reduced voltage on primary (150 V) to give rated secondary current, primary current 12.1 A and input power 450 W. Calculate the parameters of the equivalent circuit referred to the primary side. Determine the percentage regulation and efficiency at full load, with power factor of 0.707 lag.

11.7 A 50 kVA, 11,000/220 V, 50 Hz, single-phase transformer requires 350 V at HV winding to circulate full-load current, with LV winding short-circuited. The input power is 600 W. Find the percentage regulation and the secondary (LV) terminal voltage on full load at 0.8 power factor lag.

11.8 The maximum efficiency of a 25 kVA, 50 Hz, single-phase transformer is 97.5%, and occurs at 80% of full load at 0.9 power factor. If the leakage impedance of the transformer is 4.5%, find the voltage regulation at rated load, 0.8 power factor lagging.

11.9 A 10 kVA, 50 Hz, single-phase transformer has efficiency of 95% at both full load, and at half full load. Find the efficiency at 75% of full load. Also calculate the load, at which maximum efficiency occurs, and its value. Assume power factor as 0.8, in all cases.

11.10 A 500 kVA, 50 Hz, single-phase transformer is tested first on open circuit, and then, on short circuit. A power of 3.5 kW is drawn when rated (full load) voltage is applied to the secondary (LV) winding, with the primary (HV) winding being open-circuited. A power of 6.0 kW is drawn when rated (full load) current flows in the secondary (LV) winding, it being short-circuited, with low voltage applied to the primary (HV) winding. Find its efficiency at full load, with power factor of 0.7. Also calculate the load at which maximum efficiency occurs, and its value.

CHAPTER 12

DC MACHINES

DC machines are rotating machines which can be operated both as a generator and a motor. When it is run as a generator, using a prime mover at a rated speed, an output voltage—a dc one—is developed across the output terminals. A resistive load is connected across the output, so that current flows through the load. When it is run as a motor, the input terminals are connected to a dc supply. The motor now draws current from the supply. Torque is developed, driving the mechanic load coupled to the shaft. Hence, it is different from a transformer—a static machine—where both input and output voltages are ac, normally of line frequency, as presented in Chapter 11.

The dc machine has some special features—an armature with commutaor and brushes, which rotates at rated speed. The field winding, where flux is formed by passing current through it, is in the stator. There are two types of armature windings—lap and wave. These features will be explained in detail in the next section.

Next, we will discuss the different types of connections used for dc machines, which will result in different types of machines. Then, starting with the emf equation, generator operation is taken up, followed by open circuit (no-load) and load characteristics of different types of dc generators, such as shunt and compound. Next, we will discuss what will happen if the machine is run as a motor, the armature being connected to a dc supply. The speed–torque characteristics of two types of dc motors, viz., shunt and series, are discussed. Two other aspects—methods of starting and speed control of these motors, are also presented. Lastly, efficiency and various types of losses are described.

12.1 Construction

The components of a dc machine, whether it is working as a generator or a motor, are shown in Fig. 12.1. There are two main parts in all types of rotating electrical machines, including a dc machine, viz., a stator and a rotor.

The stator includes the poles along with excitation windings (the types of connections are given later), and the yoke needed to complete the magnetic path, so that flux is produced in the air gap between the stator and the rotor surface. In the rotor, the armature winding, again of different types, is the main part. It is connected to the commutator, an integral part of a dc machine. Brushes are placed on the commutator, which is used, either to obtain a dc output voltage in a generator to be fed to a resistive load, or to feed the current from a dc supply, in a motor. The rotor assembly, i.e., armature and commutator, must be placed on a bearing, such that it runs at a rated speed with low friction.

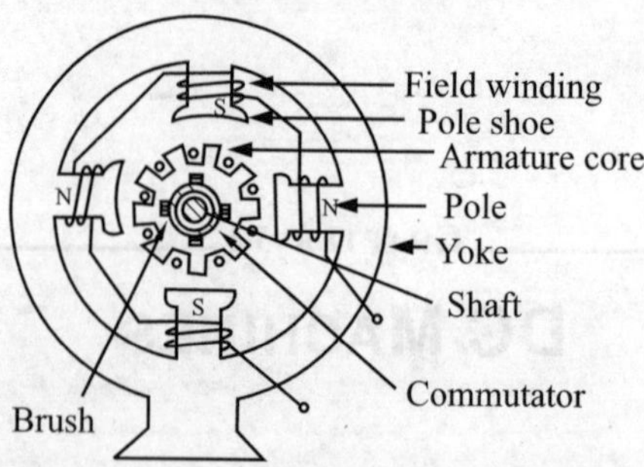

Fig. 12.1 *Parts of a dc machine*

To explain further, the generator converts mechanical energy taken from the prime mover, into electrical energy, obtained as a dc output at the brushes and fed to resistive load. For motor operation, electrical energy, i.e., the current fed to the armature via the brushes from a dc input voltage, is converted into mechanical energy, which is used to drive the load, coupled to the rotor.

ARMATURE WINDING

There are two types of armature windings, viz., lap and wave. In a *lap-connected* winding (Fig. 12.2(a)), one end of a coil, having a certain number of turns, is connected to a commutator segment, and the next coil end to an adjacent one, and so on. This gives rise to a number of parallel paths (a) in the winding, which is equal to the number of poles (p). It is suitable for low voltage and high current machines, as the number of parallel paths increases with the increase in number of poles.

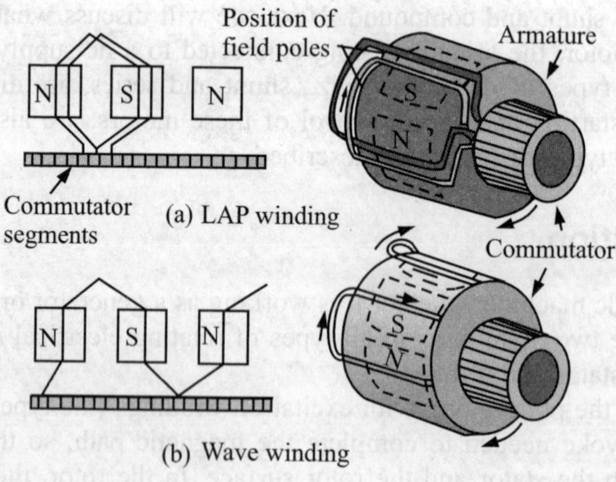

Fig. 12.2 *Types of armature winding*

In a *wave-connected* winding, one end of a coil (Fig.12.2(b)), having a certain number of turns, is connected to a commutator segment, and the next coil end to another segment far away from the previous one, and so on. The number of parallel paths (a) in this winding is two, irrespective of the number of poles (p). Wave-connected windings are used for high voltage and low current machines, as the number of parallel paths is only two, the conductor current being the same.

12.2 EMF Equation for DC Machines

Let us take a dc generator. Flux in the air gap is formed when the field winding is excited or current flows in it. As the armature is rotated, the conductors cut the flux at a rate depending on the speed of rotation. Hence, an emf is induced in the armature winding, which appears across the two brushes. For a dc motor, this will be described later, when the torque equation is derived.

The symbols used are as follows:

ϕ=Flux per pole (Wb)
Z=No. of armature conductors in series
p=No. of poles
a=No. of parallel paths
N=rpm

As the armature conductors rotate one revolution, the flux cut (Wb) is given by $p.\phi$. The revolution per second, is $\frac{N}{60}$. Therefore, the time required for one revolution in seconds (s) is $t = \frac{60}{N}$. The generated or induced emf in one conductor is

$$e = \frac{p \cdot \phi}{t} = p \cdot \phi \times \frac{N}{60},$$

according to Faraday's law. The emf is proportional to the rate at which flux changes. This rate of change of flux is taken as uniform or constant, assuming the machine speed to remain constant. As there are a number of parallel paths, the armature conductors in series per parallel path is $\frac{Z}{a}$. The generated emf (V) of a dc machine is given by

$$E_g = e \cdot \frac{Z}{a} = \phi \cdot Z \cdot \frac{p}{a} \cdot \frac{N}{60} = k' \cdot \phi \cdot N,$$

where, $k' = \frac{Z}{60} \cdot \frac{p}{a}$.

The angular speed in rad/s is

$$\omega = 2\pi \cdot \frac{N}{60} \quad \text{or,} \quad \frac{\omega}{2\pi} = \frac{N}{60} = n,$$

n being speed in rps.

Hence, if E_g is expressed in terms of angular speed, ω, we get, by substituting the relation given earlier,

$$E_g = \phi \cdot z \cdot \frac{p}{a} \cdot \frac{\omega}{2\pi} = k \cdot \phi \cdot \omega,$$

where, $k = \dfrac{Z}{2\pi} \cdot \dfrac{p}{a}$.

EXAMPLE 12.1 *Calculate the useful flux per pole in a 4-pole dc generator, to give an induced emf of 250 V at 1,200 rpm, when the armature is (a) lap-connected, and (b) wave-connected. The generator has 51 slots with 24 conductors/slot.*

SOLUTION $E_g = 250$ V, $N = 1,200$ rpm.

No. of slots = 51, No. of conductors/slot = 24.

No. of armature conductors $= Z = 51 \times 24 = 1,224$.

No. of poles $= p = 4$.

Lap-connected winding

No. of parallel paths, $a = p = 4$.

$$\text{The constant} = k' = \frac{Z}{60} \cdot \frac{p}{a} = \frac{1,224}{60} \times \frac{4}{4} = 20.4$$

$$\text{Generated voltage} = E_g = V = k' \cdot \phi \cdot N = 20.4 \times \phi \times 1,200 = \phi \times 24.48 \times 10^3 = 250.0 \text{ V.}$$

$$\text{Flux per pole} = \phi = \frac{250.0}{24.48 \times 10^3} = 10.21 \text{ mWb.}$$

Alternatively, the flux per pole can be found by the procedure given below.

$$\text{Angular speed} = \omega = 2\pi.\frac{N}{60} = \frac{2\pi \times 1,200}{60} = 125.664 \text{ rad/s.}$$

$$\text{The emf constant} = k = \frac{Z}{2\pi}.\frac{p}{a} = \frac{1,224}{2\pi} \times \frac{4}{4} = 194.81.$$

$$\text{Generated voltage} = E_g = V = k.\phi.\omega = 194.81 \times \phi \times 125.664 = \phi \times 24.48 \times 10^3 = 250.0 \text{ V.}$$

$$\text{Flux per pole} = \phi = \frac{250.0}{24.48 \times 10^3} = 10.21 \text{ mWb.}$$

Wave-connected winding

No. of parallel paths $= a = 2$.

$$\text{The constant} = k' = \frac{Z}{60}.\frac{p}{a} = \frac{1,224}{60} \times \frac{4}{2} = 40.8.$$

$$\text{Generated voltage} = E_g = V = k'.\phi.N = 40.8 \times \phi \times 1,200 = \phi \times 48.96 \times 10^3 = 250.0 \text{ V.}$$

$$\text{Flux per pole} = \phi = \frac{250.0}{48.96 \times 10^3} = 5.11 \text{ mWb.}$$

EXAMPLE 12.2 *An 8-pole, lap-connected, armature has 960 conductors, a flux/pole of 40 mWb, and a speed of 800 rpm. Find the emf generated in the armature. Determine the speed at which the armature is to be driven to maintain a voltage of 400 V, if it is wave-connected.*

SOLUTION $N_1 = 800$ rpm, No. of poles $= p = 8$.

No. of armature conductors $= Z = 960$.

Flux per pole $= \phi = 40$ mWb $= 0.04$ Wb.

Lap-connected

No. of parallel paths $a = p = 8$.

The constant $= k' = \frac{Z}{60} \cdot \frac{p}{a} = \frac{960}{60} \times \frac{8}{8} = 16.0.$

Generated voltage $= E_g = V = k'.\phi.N = 16.0 \times 0.04 \times 800 = 512.0$ V.

Wave-connected

No. of parallel paths $= a = 2$

The constant $= k' = \frac{Z}{60} \cdot \frac{p}{a} = \frac{960}{60} \times \frac{8}{2} = 64.0.$

Generated voltage $= E_g = V = k'.\phi.N_2 = 64.0 \times 0.04 \times N_2 = 2.56 \times N_2 = 400.0$ V.

Speed in rpm $= N_2 = \frac{400.0}{2.56} \approx 156.0.$

Alternatively, the speed in rpm can be found by the procedure given in Example 12.1.

The emf constant $= k = \frac{Z}{2\pi} \cdot \frac{p}{a} = \frac{960}{2\pi} \times \frac{8}{2} = 611.155.$

Generated voltage $= E_g = V = k.\phi.\omega = 611.155 \times 0.04 \times \omega_2 = \omega_2 \times 24.446 = 400.0$ V.

Angular speed $= \omega_2 = \frac{400.0}{24.446} = 16.362$ rad/s.

The speed in rpm $= N_2 = \frac{60}{2\pi} \times \omega_2 = \frac{60}{2\pi} \times 16.362 \approx 156.0.$

12.3 Types of Connections Used for Field Windings in DC Machines

Different types of field winding connections are used for dc machines. The most commonly used ones are the shunt and series connections. We will describe them in detail later, while discussing the characteristics of the machines. Figure 12.3(a) shows a separately-excited dc generator where the shunt field is supplied from a separate dc supply. Shunt field winding is connected in parallel with the armature terminals (Fig. 12.3(b)). To vary the field current, an additional resistance, R' is connected in series with the field winding.

The output voltage (rated) from the armature terminals of the dc generator is normally fed to a resistive load. The resistance of the shunt field winding is high, as the current is low and the number of turns required is high. This is used to provide the number of ampere-turns needed for the rated flux in the air gap. The length of the wires used in the winding is large, being proportional to the number of turns and also inversely proportional to the cross-sectional area, which is low, as the current in it is low, resulting in its resistance being high, as given earlier. We must remember two things—the first is that the current density in the wires is to be kept the same so that the copper losses are within limits. The second is that the length of the mean turn of the winding remains the same as the pole dimensions are also the same.

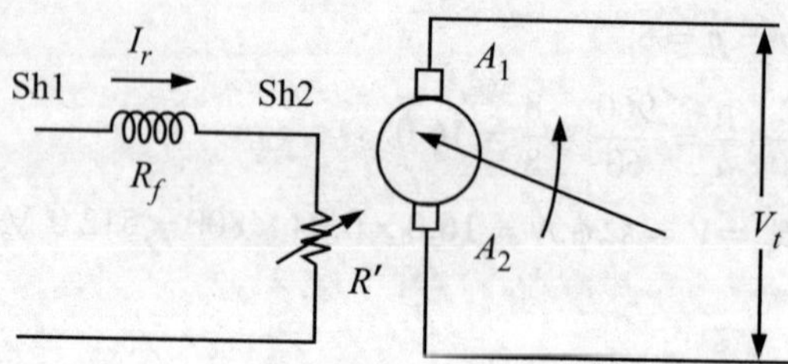

Fig. 12.3(a) *Connections for dc separately exicited generators: Shunt field winding*

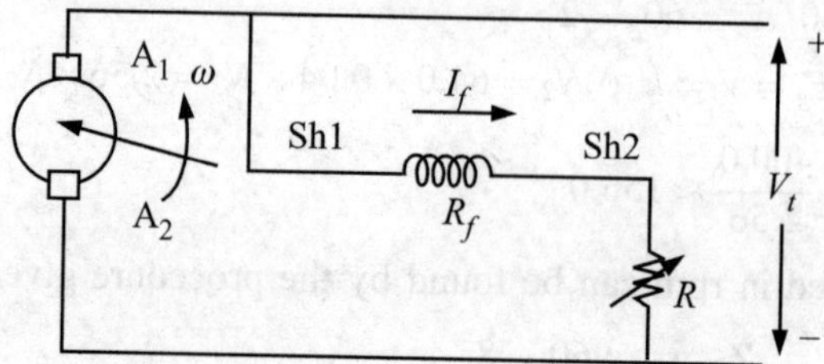

Fig. 12.3(b) *Connections for dc separately exicited generators: Parallel field winding*

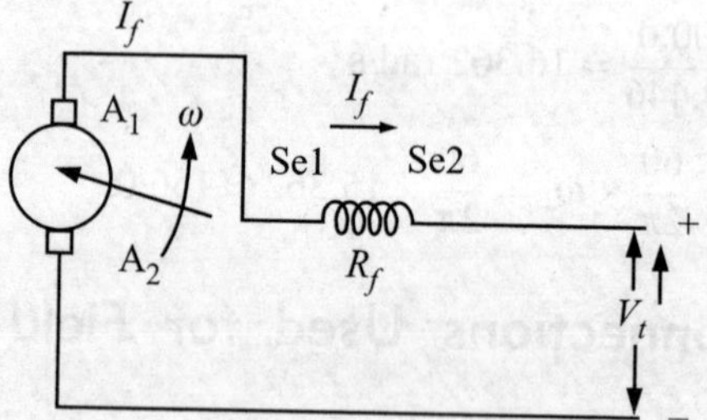

Fig. 12.3(c) *Connections for dc separately exicited generators: Series field winding*

The series type dc machine is connected normally in series with the armature terminals of the dc generator (Fig. 12.3(c)). The resistance of this winding is low, being nearly equal to the armature resistance. The current is high, being the same as the armature current and the number of turns required is low. This is used to provide the number of ampere-turns needed for creating the flux in the air gap. The length of the wires required here is low, being proportional to the number of turns and also inversely proportional to the cross-sectional area, which is high, as the current in it is large, resulting in its resistance being low, as given earlier. As for the shunt connection, current density in the wires and the length of mean turn of the winding is to be kept the same.

Table 12.1 *Classification according to the method of excitation in a dc generator*

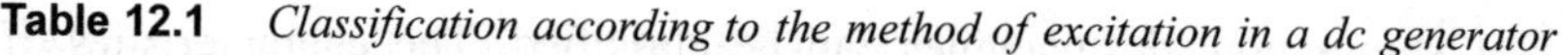

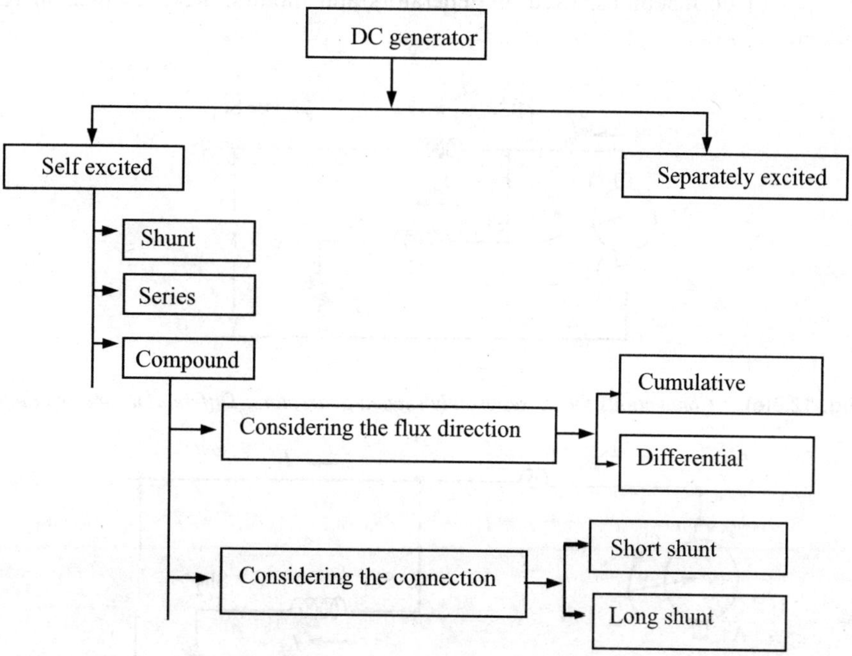

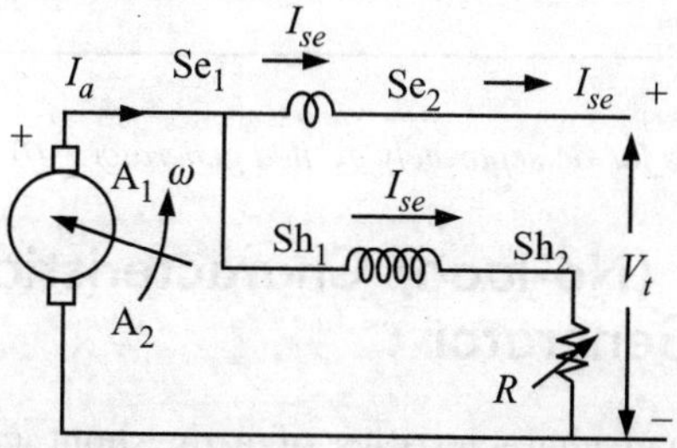

- DC generator
  - Self excited
    - Shunt
    - Series
    - Compound
      - Considering the flux direction
        - Cumulative
        - Differential
      - Considering the connection
        - Short shunt
        - Long shunt
  - Separately excited

Fig. 12.3(d) *Connections for dc separately excited generators: Cumulative field winding*

Another dc machine, termed as the dc compound generator is also used, where both windings—shunt and series—are placed together on the pole, the latter being on top of the former. These coils or windings can be taken as mutually coupled coils. The dot convention for mutual inductance given in any standard textbook is also used here to indicate whether the flux produced by one coil aids or opposes the flux in another coil. Cumulative (Fig. 12.3(d)) and differential (Fig. 12.3(e)) windings are also used for dc compound generators. There can be another classification where firstly, the shunt coil is connected in parallel with the armature terminals (Fig. 12.3(d)). This is known as short shunt connection. The series one is now in series with the resistive load. For long shunt connection, the shunt coil is connected in parallel with the output terminals (Fig. 12.3(f)). As usual, the output voltage is fed to a resistive load. In this case, the series one is in series with the armature terminals, in this case. Table 12.1 shows how dc machines can be classified. The characteristics of different types of dc machines, used as generators and motors, are presented in respective sections later.

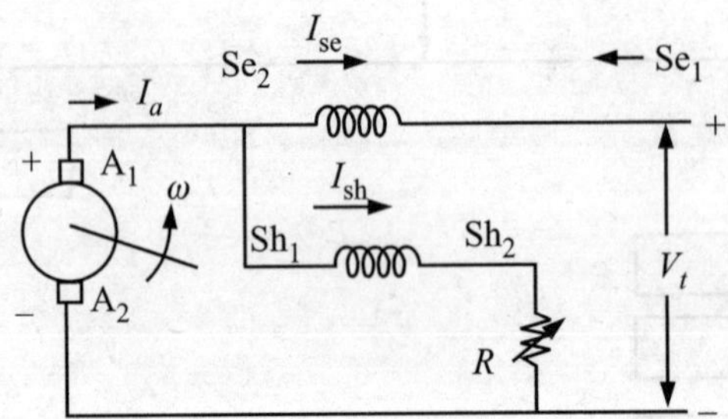

Fig. 12.3(e) *Connections for dc separately excited generators: Differential field winding*

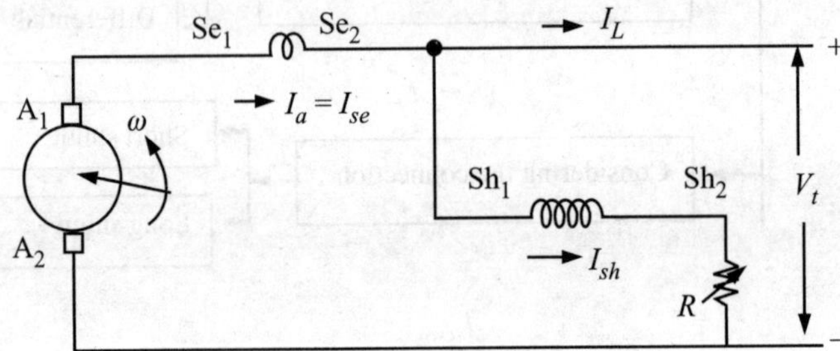

Fig. 12.3(f) *Connections for dc separately excited generators: (f) long shunt field winding*

12.4 Open Circuit (No-load) Characteristic of a DC Shunt or Self-excited Generator

Before going into the no-load characteristics of a dc shunt generator, the operation of a separately-excited generator driven by a prime mover at a rated speed is taken up. Normally, a shunt winding having high resistance is connected to a separate dc supply, in series with a variable resistance to control the field current (as shown earlier in Fig. 12.3(a)). Alternatively, the field winding can also be fed through a potential divider, the input of which is the same dc supply (Fig. 12.4). The advantage is that the field current can be varied from zero (0) to slightly more than the rated value. In the earlier case, the field can be reduced only to a low value, not zero. When the generator is run at a rated speed, with the field current also at the rated value, the generated or induced voltage is rated, as given earlier by the emf equation, with no resistive load connected to the armature terminals. When the field current is zero (0), a small voltage will be generated due to the residual flux present. As the field current is increased, the induced emf increases, but not linearly.

Only in a small region, will the emf increase linearly—it saturates with increase in field current. The no-load characteristic ($E_g - I_f$), at say, rated speed, is plotted in Fig. 12.5. The graph resembles a non-linear $B - H$ characteristic of the magnetic material used in the magnetic circuit of the generator. This is because the induced emf is proportional to flux (ϕ), or flux density (B), the pole area remaining the same, and the field current is related to magnetic field intensity (H), due to excitation or ampere-turns, number of turns of the field winding being constant.

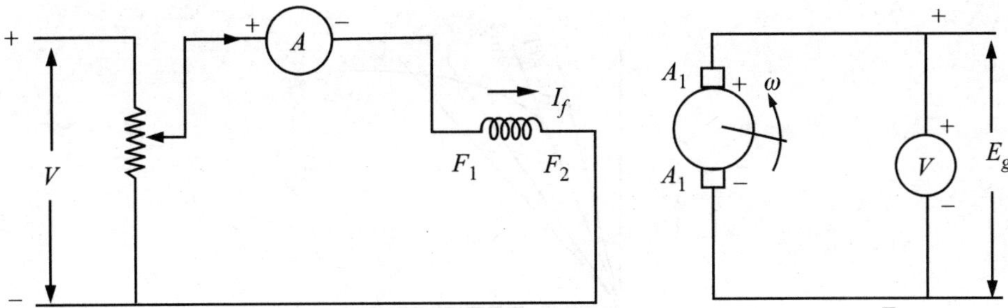

Fig. 12.4 *Separately-excited dc generator*

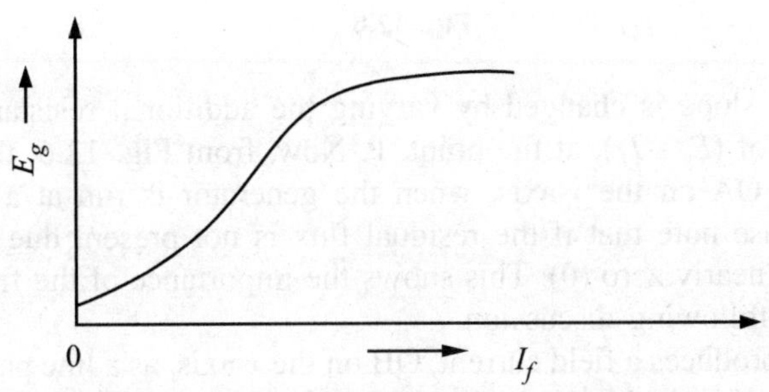

Fig. 12.5

We will now discuss how the voltage is built up in a dc shunt or self-excited generator run at rated speed. The shunt winding, as shown earlier in Fig. 12.3(a), is connected across the armature terminals. Note that the connection of the shunt winding remains the same, and the shunt generator is not connected to a load, i.e., it is an open circuit. Three conditions are to be satisfied. These are: (a) the presence of residual flux, (b) the field winding connection or its polarity is such that the flux created by it, aids or adds to the residual flux and (c) the total field circuit is lower than a critical value, at a particular speed, which may be rated. Let us now take the conditions one by one. The $(E_g - I_f)$ characteristic given earlier, is repeated here as Fig. 12.6. The current in the shunt winding is

$$I_f = \frac{V_t}{R_f + R'} = \frac{V_t}{R},$$

where the terminal voltage, V_t = induced or generated voltage, E_g,

shunt winding resistance = R_f, and additional resistance in the field circuit = R', with total resistance $= R = R_f + R'$.

The graph of the equation for the field circuit is

$$V_t = (R_f + R') \cdot I_f = R \cdot I_f,$$

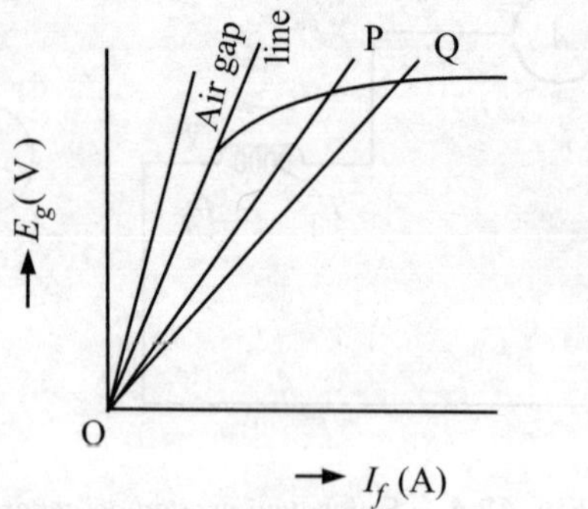

Fig. 12.6

which is linear. The slope is changed by varying the additional resistance of R'. This line intersects the graph of $(E_g - I_f)$, at the point, P. Now, from Fig. 12.6, there is residual flux and induced emf is OA on the y-axis, when the generator is run at a certain speed, near the rated value. Please note that if the residual flux is not present due to some reason, the induced emf will be nearly zero (0). This shows the importance of the first condition, which is confirmed by the following discussion.

The voltage, OA produces a field current, OB on the x-axis, as a line parallel to the x-axis is drawn from the point A, which cuts the line, OP at B (Fig.12.12). The current (OB) depends on the total resistance in the field circuit. At this point, let us take the second condition given earlier, into account, which states that the field winding connection or its polarity is such that the flux created by it, aids or adds to the residual flux. Only if this condition is satisfied, will the induced voltage increase, its value being OC, when a line parallel to the y-axis is drawn from the point, B, intersecting OCC at C. If the above condition is not satisfied, because of the wrong polarity used in the connection, the residual flux or emf will be reduced to nearly zero (0), thereby producing no emf, also violating the first condition. After satisfying the second condition, the induced voltage will build up as shown in Fig. 12.7, the points are D, E, F, and so on. Ultimately, the point, O is reached, which gives the induced emf and also the current in the shunt field, when the generator is run at the speed at which the OCC is drawn.

If a line is drawn from the point, P, tangential to OCC, this line is termed as the air gap line (Fig. 12.6); the slope of this line gives the value of critical resistance at the speed for which OCC is drawn. It may be noted here that for voltage build-up, the field resistance must be less than the critical resistance. If the total resistance of the shunt field circuit is more than the critical resistance, a small voltage, which is nearly zero (0), appears across the armature terminals (Fig. 12.6). If the field circuit resistance is increased by inserting an additional resistance, R', in series with the field resistance, the slope of the line changes, and the induced voltage is decreased. Lastly, if the air gap line is fixed and the value of the resistance is known, then the speed, for which OCC is drawn, is termed as the critical speed for this resistance.

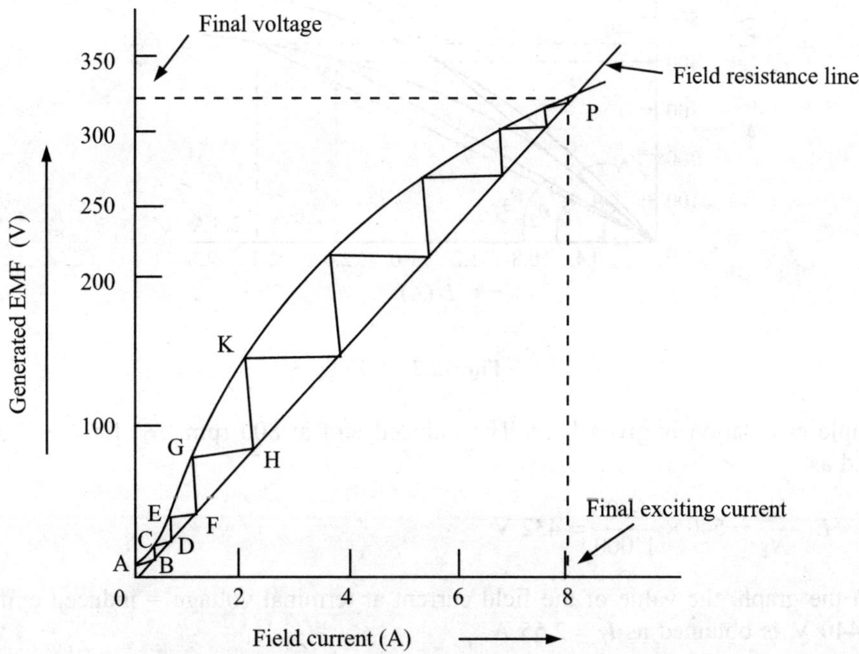

Fig. 12.7 *Graph for voltage build-up in a dc shunt generator*

EXAMPLE 12.3 *The open circuit characteristic of a dc separately-excited generator (Fig. 12.8) at 1,000 rpm is given in the table. The generator is now run as a shunt generator at a speed of 800 rpm. (a) What should be the field circuit resistance to generate a voltage of 440 V? (b) Find the voltage, when the field circuit resistance is 200 Ω. (c) Find the critical speed corresponding to the above field circuit resistance of 200 Ω. (d) Find the critical field circuit resistance at the speed of 800 rpm.*

Field current (A)	0.4	0.8	1.2	1.6	2.0	2.4	2.8
Induced EMF (V)	140	275	375	440	500	540	570

SOLUTION The open circuit characteristic (OCC) of a dc separately-excited generator (Fig. 12.8) at 800 rpm is given in the following table.

Field current (A)	0.4	0.8	1.2	1.6	2.0	2.4	2.8
Induced EMF (V)	112	220	300	352	400	432	456

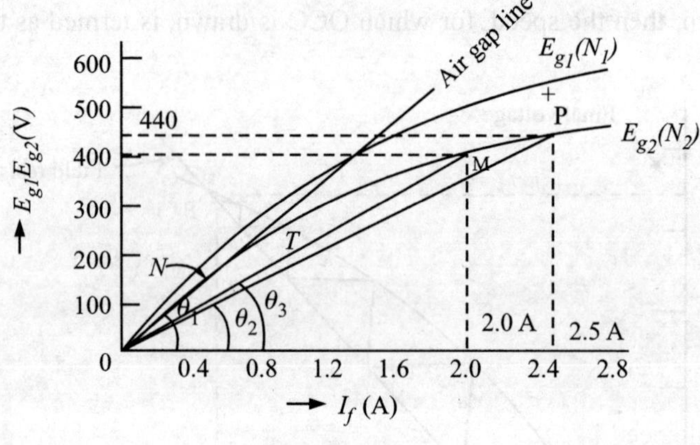

Fig. 12.8

One sample calculation is given here. The induced emf at 800 rpm (N_2) for $I_f = 2.5$ A, is calculated as

$$E_2 = E_1 \cdot \frac{N_2}{N_1} = 540 \times \frac{800}{1,000} = 432 \text{ V}$$

(a) From the graph, the value of the field current at terminal voltage = induced emf (at no load) = 440 V, is obtained as $I_f = 2.55$ A.

The value can be found by the procedure as stated. The voltage, 440 V at point P, lies between the points, Q (2.4, 432.0) and R (2.8, 456.0). The line, QR is assumed to be linear, with its slope (Ω), being

$$\frac{456.0 - 432.0}{2.8 - 2.4} = \frac{24.0}{0.4} = 60.0 \quad \Omega.$$

The value of the field current at point P (440.0 V) is given by

$$I_f = 2.4 + \frac{440.0 - 432.0}{60.8} = 2.4 + \frac{8.0}{60.0} = 2.533 \approx 2.53 \text{ A}.$$

The total resistance of the field circuit is $R = \dfrac{440.0}{2.53} \approx 174.0 \quad \Omega.$

The new value of the field current is $I'_f = \dfrac{440.0}{174.0} \approx 2.53$ A.

The induced emf at this value is $E_g = 432.0 + (0.13 \times 60.0) = 432.0 + 7.8 = 439.8$ V, which can be taken approximately as 440.0 V. This is to check the value found earlier.

(b) A straight line, whose slope is 200 Ω, is drawn with points O (0, 0) and T (1.0, 200.0), which cuts the OOC at S. The value of the induced emf = 400 V, with field current = 2 A. The slope of the line is

$$\frac{400.0 - 0.0}{2.0 - 0.0} = 200.0 \quad \Omega.$$

Note that this is one the points, (2.0, 400.0).

(d) To find the critical resistance at the speed of 800 rpm, a straight line, which is tangent to OCC, is drawn. This is known as the air gap line, whose slope is

$$\frac{112.0 - 0.0}{0.4 - 0.0} = 280.0 \quad \Omega,$$

the value of critical resistance.

(c) To find the critical speed, with total field circuit resistance of 200 Ω as given in (b), the value of the critical speed is

$$800 \times \frac{200.0}{280.0} \approx 571 \approx 570 \text{ rpm}.$$

If the OCC is drawn at a speed of 571 rpm, the slope of the tangent or air gap line is obtained as 200 Ω, which is the critical resistance at that speed. The ratio of the two resistances, multiplied by the speed of 800 rpm, gives the value of critical speed in rpm.

12.5 Load Characteristics of Different Types of DC Generators

Three types of dc generators are taken up.

DC SEPARATELY EXCITED GENERATOR

This is shown in Fig. 12.3(a), with the armature output connected to a resistive load. The direction of the currents is shown there, with the field current (I_f) assumed constant. It can be stated that the induced emf, $E_g =$ constant, as both flux or field current and generator speed is constant. In all cases, the speed is assumed to be constant when driven by a prime mover. The equation for terminal or output voltage is $V_t = E_g - I_a r_a$. r_a is the armature resistance, which is small, and $I_a = I_L$, which are currents in the armature and load respectively. Load characteristic of this generator is shown in Fig. 12.9. The voltage drop at the rated armature current is about 5% or lower than this, in terms of rated value.

DC SHUNT GENERATOR

This is shown in Fig. 12.4, where the output is connected to the load. The load characteristic is described briefly. The direction of the currents is shown there. Both field current (I_f), and total field circuit resistance, including the additional resistance, are assumed to be constant. The equations are:

$$E_g \propto \phi, \text{ or } E_g \propto I_f, \text{ with } \phi \propto I_f, \text{ i.e., in the linear region.}$$

$$V_t = E_g - I_a r_a.$$

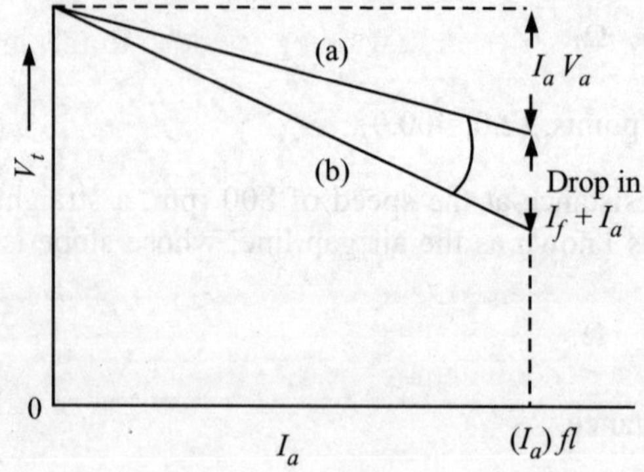

Fig. 12.9

$$I_a = I_L + I_f, \text{ with } I_f = \frac{V_t}{R_f + R'} = \frac{V_t}{R}.$$

The field current decreases as output voltage decreases due to voltage drop in the armature resistance. There is a small drop in armature current. However, this decrease is low as the rated value of armature current is much higher that the field current. Only a part of the load characteristic is shown and also explained.

DC COMPOUND GENERATOR

A dc compound generator with load is shown in Fig. 12.3(f). It is a short shunt type with the shunt field winding connected across the armature terminals. Please note here that the flux produced by the series field winding is added to the flux produced by the field produced in the air gap.

The induced emf, $E_g \propto \phi$, where the flux is related to total ampere-turns, which is the sum of the ampere-turns produced by the two windings—shunt and series—and is

$$\phi \propto (N_f I_f + N_{se} I_{se}).$$

Other equations are,

$$V_t = V_t' - I_{se} r_{se}, \quad \text{where} \quad V_t' = E_g - I_a r_a.$$

$$I_a = I_L + I_f, \text{ with } I_f = \frac{V_t'}{R_f + R'} = \frac{V_t'}{R}.$$

In this type of dc generator, the ampere-turns produced by the shunt field winding provide the main flux, as in the case of other dc shunt generators. Note that here the ampere-turns and also the flux in the air gap decrease. Similarly, the terminal voltage and also the current in the shunt field winding decrease, as given earlier. However, in the case of a dc cumulatively compound generator, the total ampere-turns and also the flux in the air gap increase, as the ampere-turns produced by the series field winding is added to the ampere-turns produced by the shunt field winding. As the load current increases, the currents in both the armature and series field winding increase, irrespective of the connection (short shunt or long shunt), which is given in Fig. 12.3(f) and 12.3(e). The amount of increase in the flux or ampere-turns is controlled by the number of turns in the series field winding, according to the formula given. As the flux increases, the induced emf, E_g increases, thus, increasing the terminal voltage, V_t to some extent, which also increases the current in the shunt field winding.

Depending on the number of turns in the series field winding, there can be three types of load characteristics—over compounding, level compounding, and under compounding. In the first case (over compounding), the terminal voltage, V_t on load, is more than the terminal voltage, V_t on no-load (Fig. 12.10), as the number of turns in the series field winding is high, resulting in the increase of the flux or ampere-turns on load, which increases the flux in the air gap resulting in increase in the induced emf, E_g, according to the formula given. If the number of turns in the series field winding is reduced, the terminal voltage, V_t on the load can be made equal to the terminal voltage, V_t on no-load, which is the second case (level compounding). This is, because, here, the increase in flux in the air gap at load is quite small, as the number of turns, N_{se} is reduced. Here, though the induced emf, E_g increases to some extent, it is nullified by the increase in voltage drop across the armature winding. If the number of turns in the series field winding is reduced further, the terminal voltage, V_t on the load is lower than the terminal voltage, V_t on no-load, which is the third or last case (under compounding). Moreover, the flux in the air gap decreases at load, due to the reduction in the number of turns, N_{se}. The reasons are the same as in the second case.

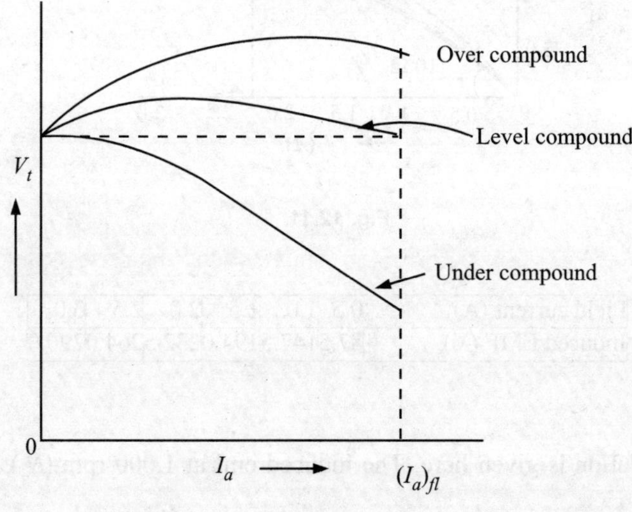

Fig. 12.10

The load characteristics of two types of dc generators are not taken up here, as they are not used much these days. The first one is the dc compound generator of differential type, where the total ampere-turns and also the flux in the air gap decrease at a rapid rate, as the ampere-turns, produced by the series field winding, is subtracted from the ampere-turns produced by the shunt field winding. As the load current increases, the rate of decrease is very rapid, resulting in low terminal voltage. The second one is the series type, where only the series field winding is present. The induced emf is proportional to the armature or load current, and is low, compared to the induced emf or terminal voltage in the earlier cases.

EXAMPLE 12.4 *The open circuit characteristic of a dc separately-excited generator at 800 rpm is given in the table. The armature resistance is 0.1 Ω. The generator is run as a shunt generator at 1,000 rpm. Determine the total resistance of the field circuit for obtaining an emf of 250 V at no-load. Calculate also the terminal voltage and field current when the armature current is 80 A.*

Field current (A)	0.5	1.0	1.5	2.0	2.5	3.0
Induced EMF (V)	70	118	155	186	211	232

SOLUTION: The open circuit characteristic (OCC) of a dc separately-excited generator at 1,000 rpm is given in the following table, and also plotted in Fig. 12.11.

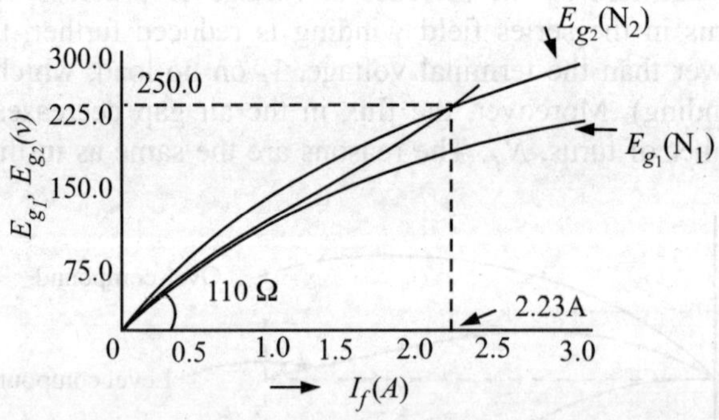

Fig. 12.11

Field current (A)	0.5	1.0	1.5	2.0	2.5	3.0
Induced EMF (V)	87.5	147.5	194.0	232.5	264.0	290.0

One sample calculation is given here. The induced emf at 1,000 rpm (N_2) for $I_f = 2.5$ A, is calculated as

$$E_2 = E_1 . \frac{N_2}{N_1} = 211 \times \frac{1000}{800} = 263.75 \approx 264.0 \text{ V}.$$

From the graph, the value of the field current at terminal voltage = induced emf (at no-load), obtained as $I_f = 2.275$ A.

The value can be found by the procedure stated before. The voltage 250 V at point P lies between the points, Q (2.0, 232.5) and R (2.5, 264.0). The line, QR is assumed to be linear, with its slope (Ω), being

$$\frac{263.75-232.5}{2.5-2.0}=\frac{31.25}{0.5}=62.5 \quad \Omega.$$

The value of the field current at point P (250.0 V) is given by

$$I_f=2.0+\frac{250.0-232.5}{62.5}=2.0+\frac{17.5}{62.5}=2.28 \text{ A}.$$

The total resistance of the field circuit is $R=\dfrac{250.0}{2.28}=109.65\ \Omega.$

This value is approximated as 110.0 Ω.

The new value of the field current is $I'_f=\dfrac{250.0}{110.0}=2.273$ A.

The induced emf at this value is $E_g=232.5+(0.273\times 62.5)\approx 232.5+17.1=249.6$ V, which can be taken approximately as 250.0 V. This is to check the value found earlier.

The armature current at load, $I_a=80$ A.

The armature resistance, $r_a=0.1\ \Omega$.

At load, the terminal voltage is given by,

$$V_t=E_g-(I_a.r_a)=250.0-(80.0\times 0.1)=250.0-8.0=242.0 \text{ V}.$$

The value of the field current at this terminal voltage is $\dfrac{242.0}{110.0}=2.2$ A.

The induced emf at this value is $E_g=232.5+(0.2\times 62.5)=232.5+12.5=245.0$ V.

The terminal voltage is $V_t=245.0-8.0=237.0$ V.

One more step is included here, before going into the final results.

The value of the field current is $\dfrac{237.0}{110.0}=2.155$ A.

The induced emf at this value is $E_g=232.5+(0.155\times 62.5)\approx 232.5+9.7\approx 242.0$ V.

The terminal voltage is $V_t=242.0-8.0=234.0$ V.

Lastly, the terminal voltage at the load current of 80 A is found as $V_t=231.0$ V.

The value of the field current is $\dfrac{231.0}{110.0}=2.1$ A.

The induced emf at this value of field current is,

$$E_g=232.5+(0.1\times 62.5)=232.5+6.25=238.75\approx 239.0 \text{ V}.$$

Now, the terminal voltage is $V_t = 239.0 - 8.0 = 231.0$ V, which is the same as found earlier.

The steps are simple and are not repeated here, but left to the students as an exercise.

12.6 Torque Equation for a DC Machine

Normally, torque equation is derived for a dc machine when operated as a motor. In this case, a rated voltage is fed to the armature terminals while the shunt field winding may be connected to the same supply, as shown in Fig. 12.12. A dc series motor is shown in Fig. 12.13, where a series field winding is used.

The variables used include:

I_a = Armature current fed from a dc supply (A).

I_c = Conductor current (A) = $\dfrac{I_a}{a}$.

D = Diameter of armature (m).

L = Length of armature (m).

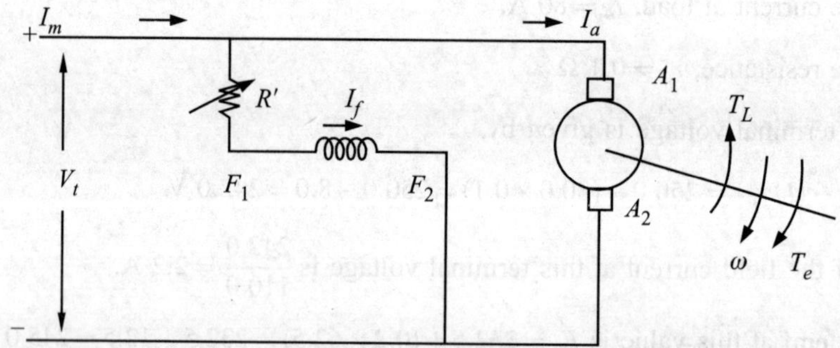

Fig. 12.12 *DC shunt motor*

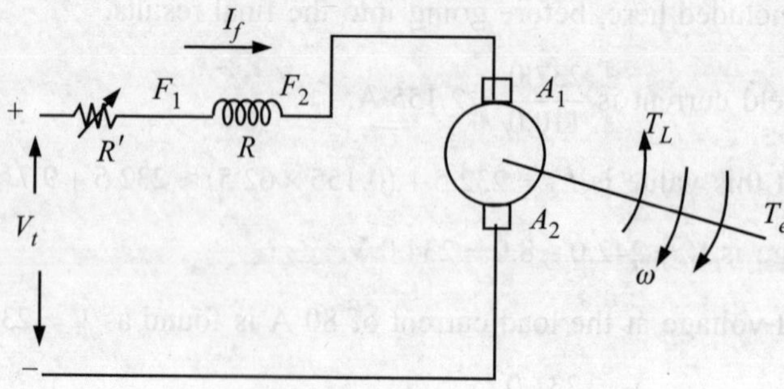

Fig. 12.13 *DC series motor*

The symbols, p and a, refer to the number of poles and number of parallel paths respectively. The number of parallel paths depends on the type of windings used in the dc machine.

Area under all poles (p) in m^2 = circumference $\times$ length = πDL.

Total flux for all poles (Wb) $= p\phi$.

Average flux density under the poles (Wb/m^2) $= B_{av} = \dfrac{p\phi}{\pi DL}$.

Force produced in one conductor (N) $= F = B_{av} I_c L$.

Torque produced in one conductor (N.m) $= T = B_{av} I_c L \dfrac{D}{2}$.

Substituting the values as obtained earlier, we get

$$T = \frac{p\phi}{\pi DL} \cdot \frac{I_a}{a} . L . \frac{D}{2} = \frac{1}{2\pi} \cdot \frac{p}{a} \cdot \phi I_a.$$

As the number of conductors, Z, are connected in series in the armature winding,

$$\text{Total torque developed} = T_e = \frac{Z}{2\pi} . \frac{p}{a} . \phi I_a = k \phi I_a,$$

where, torque constant = emf constant $= k = \dfrac{Z}{2\pi} \cdot \dfrac{p}{a}$.

Torque (T_e) is expressed in N.m, and k in N m/A, or V. s/rad.

The expression for average flux density can also be obtained in this way.

Pole pitch (m) $= \tau = \dfrac{\pi D}{p}$.

Area for a pole (m^2) = pole pitch $\times$ length $= \dfrac{\pi DL}{p}$.

Average flux density under the poles (Wb/m^2) $= B_{av} = \phi . \dfrac{p}{\pi DL} = \dfrac{p\phi}{\pi DL}$.

The equation of the induced emf can be derived in the following way, using the expressions derived earlier.

Induced emf in one conductor (V) $= e = B_{av} L v$.

Linear velocity (m/s) $= v = \omega . \dfrac{D}{2} = \dfrac{\pi DN}{60}$.

Angular speed (rad/s) $= \omega = \dfrac{2\pi}{60} \cdot N$, N being machine speed in rpm.

Substituting the values as obtained earlier, we get

$$= \frac{p\phi}{\pi DL} \cdot L \cdot \omega \cdot \frac{D}{2} = \frac{p}{2\pi} \cdot \phi \, \omega.$$

As the number of conductors in series is Z and the number of parallel paths is a, the number of conductors in series per parallel path is $\frac{Z}{a}$.

$$\text{Induced emf} = E = e \cdot \frac{Z}{a} = \frac{Z}{2\pi} \cdot \frac{p}{a} \cdot \phi\,\omega = k\phi\,\omega.$$

This is the same expression as derived earlier. The expression can also be obtained in terms of speed, by substituting the expression for ω.

BACK (INDUCED) EMF IN A DC MOTOR

As the armature of a dc shunt motor rotates, the conductors cut the magnetic flux in the air gap produced by the current in the shunt field winding, and induced or back emf develops in the armature winding, as in a dc generator, given earlier. It is termed as back emf as it opposes the applied voltage. The voltage equation in the armature circuit is $V_t \approx E_b$, neglecting the voltage drop in the armature winding, as it is small.

Neglecting losses, the input power (W) $= V_t.I_a \approx E_b.I_a$, is equal to output power $= \omega.T_e$, from which the mechanical torque developed is obtained as

$$T_e = \frac{E_b I_a}{\omega}.$$

Substituting the expression for back emf which is nearly equal to the induced emf,

$$\text{Mechanical torque developed } (N.m) = T_e = \frac{k\phi\,\omega.I_a}{\omega} = k\phi\,I_a.$$

This is the same as derived earlier.

EXAMPLE 12.5 *A 4-pole dc shunt motor takes 60 A, when it is run using a 250 V supply. The armature winding resistance is 0.2 Ω. The winding is wave-connected, with 700 conductors. Flux/pole is 20 mWb. Find the motor speed, and gross torque developed. Assume that voltage drop per brush is 1 V.*

SOLUTION The circuit of a dc shunt motor is shown in Fig. 12.12.

$$r_a = 0.2\,\Omega, V_t = 250 \text{ V}.$$

Armature current $= I_a = 60$ A.

Back emf $= E_b = V_t - (I_a \cdot r_a) -$ brush drop $= 250.0 - (60.0 \times 0.2) - (2 \times 1) = 236.0$ V.

$$P = 4, a = 2 \text{ (wave)}.$$

$$\phi = 20 \text{ mWb} = 0.02 \text{ Wb}; Z = 700.$$

$$\text{Back emf or torque constant} = k = \frac{\phi.Z}{2\pi}.\frac{p}{a} = \frac{0.02 \times 700}{2\pi} \times \frac{4}{2} = 4.456 \text{ V.s/rad or N.m/A}.$$

Hence, back emf $= E_b = k.\omega = 4.456.\omega = 236.0$ V.

Angular speed of motor $= \omega = \dfrac{2\pi}{60} \cdot N = \dfrac{E_b}{k} = \dfrac{236.0}{4.456} = 52.96$ rad/s.

Motor speed $= N = \dfrac{60}{2\pi} \cdot \omega = 9.55 \times 52.96 = 506$ rpm.

Gross torque developed $= T_e = k.I_a = 4.456 \times 60.0 = 267.4$ N.m.

Gross output power $= P_o = E_b.I_a = 236.0 \times 60.0 = 14,160$ W $=14.16$ kW.

The torque is also found as $T_e = \frac{P_o}{\omega} = \frac{E_b.I_a}{\omega} = \frac{14,160}{52.96} = 267.4$ N. m.

12.7 Load Characteristics of DC Motors

The load characteristics of two main types of dc motors, viz. shunt and series, are described in this section.

DC SHUNT MOTOR

A dc shunt motor is shown in Fig. 12.12. It is assumed that the mechanical load is connected to the shaft. The direction of the currents is also given.

The equations for the electrical part are:

$$V_t = E_b + I_a r_a, \quad \text{or,} \quad E_b = V_t - I_a \cdot r_a$$

$E_b = k \cdot \phi \cdot \omega$, or, $E_b = k.I_f.\omega$, with $\phi \propto I_f$, i.e., in the linear region.

$$I_a = I_m - I_f, \text{ with } I_f = \frac{V_t}{R_f + R'} = \frac{V_t}{R}.$$

The symbols used are:

ω is the angular speed (rad/s) of the motor $= \omega = \dfrac{2\pi}{60} \cdot N$.

N is the motor speed in rpm $= N = \dfrac{60}{2\pi} \cdot \omega$.

I_m is the current (A), fed to the dc shunt motor from the supply.

The torque balance equation for the mechanical part is:

$$T_e = T_L = k.\phi.I_a,$$

where T_L is the torque of the mechanical load connected to the shaft.

The variations of motor speed with load torque or input/armature current are termed as load characteristics. In this case, the shunt field current is kept constant at a value at which near rated speed is obtained with rated voltage fed to the motor. Load torque, which is equal to the developed torque at constant speed, is also proportional to the armature current. The angular speed of the motor is

$$\omega = \frac{E_b}{k.\phi} = \frac{V_t - I_a r_a}{k.\phi}.$$

The speed in rpm can be computed by the formula given. We find that the speed is nearly constant, with a small drop (Fig. 12.14(a)), as the armature current changes with load. The armature voltage drop at rated current is about 8–10% of the rated voltage. The developed or load torque in proportional to the armature current.

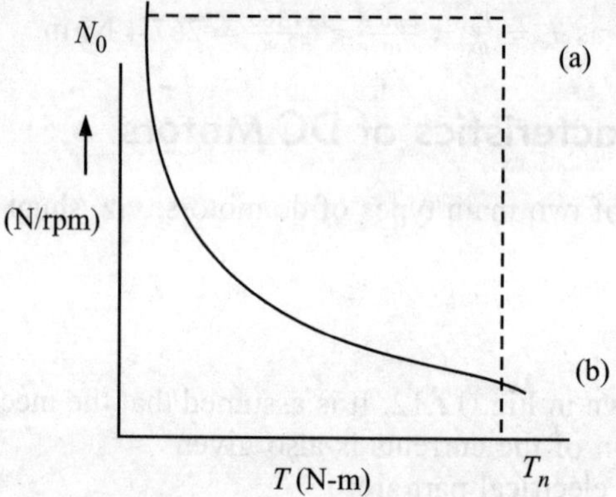

Fig. 12.14 *Speed–torque characteristics of dc (a) shunt and (b) series*

DC SERIES MOTOR

A dc series motor is shown in Fig. 12.13. In a series motor, the field winding is connected in series with the armature, with the current being the same in the armature and series field winding, i.e., $I_a = I_{se}$.
The equations for the electrical part are:

$$V_t = E_b + I_a(r_a + r_{se}), \quad \text{or,} \quad E_b = V_t - I_a.(r_a + r_{se}).$$

$$E_b = k.\phi.\omega, \quad \text{or,} \quad E_b = k.I_{se}.\omega = k.I_a.\omega,$$

with $\phi \propto I_{se} = I_a$, i.e., in the linear region.

The torque balance equation for the mechanical part is:

$$T_e = T_L = k.\phi.I_a = k.(I_a)^2,$$

using the relation given earlier.

The angular speed of the series motor is

$$\omega = \frac{E_b}{k \cdot \phi} = \frac{V_t - I_a r_a}{k.I_a},$$

which shows that the speed varies with load torque (Fig. 12.14(b)). The load torque, which is equal to the developed torque at constant speed, is proportional to square of the armature current. The back emf at full load is nearly equal to terminal voltage, with the voltage drop in armature and series field windings being small, as given in the previous case. The speed is, thus, inversely proportional to the input current, making this motor a variable speed one, unlike the shunt motor, where speed is nearly constant. This shows that the speed decreases at a high rate, when the load torque is increased. It may be mentioned here that it is dangerous to start this motor at no-load, or with small load torque, as the speed will be very high. Normally, the motor is directly coupled to the load here. The series motor is much used, unlike dc series generators, as the load characteristics of this motor makes it quite suitable for all types of traction, such as electric vehicles/trains, for cranes/hoists, and also for pumps/compressors. The load characteristics of these drives match that of a dc series motor, and also in most cases, the series motor is directly coupled to load, which is essential.

Other types of dc motors are not described, as they are not much used, except the cumulatively compound one, where flux may slightly increase with increase of current in armature, and also series field winding. The speed decreases to a value lower than that in the case of shunt motor. Now, we will discuss, with emphasis on shunt motors, the methods, used for starting (starters) and speed control of two main types of dc motors.

12.8 Starters for DC Shunt Motors

The circuit diagram is shown in Fig. 12.15. The armature current is

$$I_a = \frac{V_t - E_b}{r_a},$$

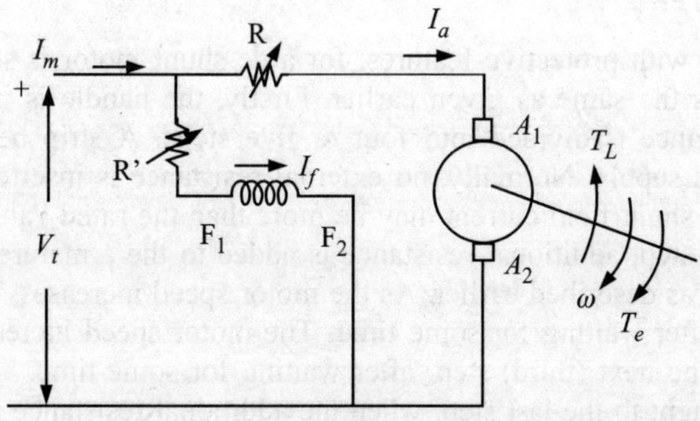

Fig. 12.15

when no resistance is added in the armature current. The formula has been derived earlier. At starting, angular speed, ω (or N) = 0. Therefore, back emf, $E_b = 0$ as $E_b \propto \omega$. The armature current,

$$I_a = \frac{V_t}{r_a},$$

in this case, is very high, as r_a is small. To control the armature current, I_a at $2.I_{fl}$ (twice the full load current), an additional resistance, R is added to armature resistance, the formula being changed to

$$I_a = 2 \cdot I_{fl} = \frac{V_t}{r_a + R}.$$

The value of R is found using this formula. It may be noted that no resistance is inserted in series with the field winding, which is connected to the rated supply voltage. As current flows in the armature, torque is developed with $T_e \propto I_a$, and the motor accelerates. The speed and back emf increases, and the armature current,

$$I_a = \frac{V_t - E_b}{r_a},$$

decreases. When the armature current reaches the rated (fl) value, the additional resistance, R in series with the armature is decreased to a value such that the armature current,

$$I_a = \frac{V_t - E_b}{r_a + R},$$

again increases to twice the rated (fl) value. The motor speed increases, and the additional resistance is decreased to increase the armature current again. This process is followed in four or five steps, till the additional resistance is completely cut out and the motor reaches the rated speed. As stated earlier, the field current may be more than the rated value, with the developed torque, $T_e \propto \phi I_a = I_f I_a$, increasing, leading to higher acceleration.

THREE–POINT STARTER

A three-point starter, with protective features, for a dc shunt motor is shown in Fig. 12.16. The principle used is the same as given earlier. Firstly, the handle is in the OFF position. The additional resistance is divided into four or five steps. A strip below it connects the field winding to a dc supply. Normally, no external resistance is inserted in series with the field winding, so the shunt field current may be more than the rated value. When the handle is moved to the first step, additional resistance is added to the armature to keep the current within certain limits, as described earlier. As the motor speed increases, the handle is moved to the second step, after waiting for some time. The motor speed increases further, and the handle is moved to the next (third) step, after waiting for some time. The process goes on, till the handle is brought to the last step, when the additional resistance is fully cut, with the armature directly connected to a dc supply. The motor now runs at nearly rated (fl) speed.

The protective features, included in this starter, are now explained. The first one is an electromagnet, which is placed near the last step of the handle. Its coil is connected in series with the shunt field winding. There is a spring, which is used to bring the handle back to the OFF position. As the handle reaches the last step, the electromagnet attracts it by a force which is termed as no-volt release. When the supply goes off, or the field current decreases to a low a value, the electromagnet no longer attracts the handle. The handle now moves to the OFF position. If the handle is not brought to the OFF position, but kept at the last step, next time, when the supply is switched on, the motor will start with no additional resistance in the armature. In that case, very high current will flow in the armature, which may result in damage to it, though the supply fuse will act under this condition. Thus, this no-volt release performs an essential function—it helps the motor to start in the manner desired.

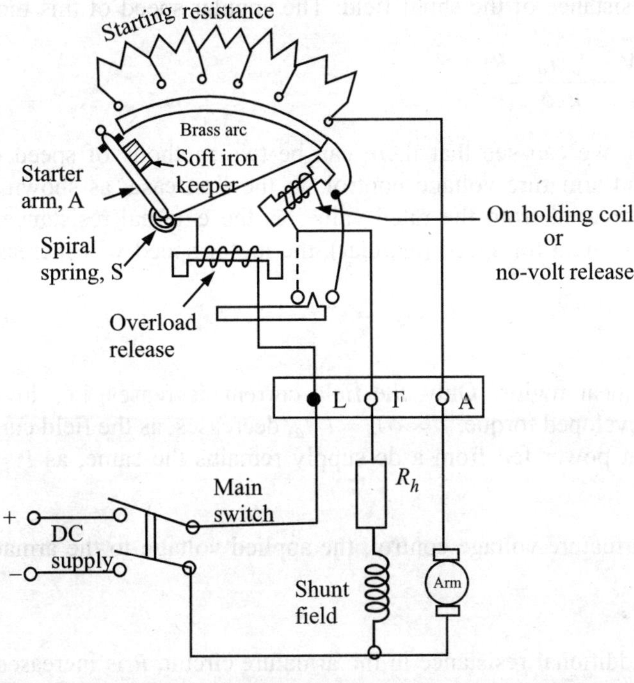

Fig. 12.16

The second electromagnet is termed as overload release, the coil of which is connected in series with the armature winding. This electromagnet tries to attract an iron piece which is held by a spring attached to it. When the armature current exceeds a certain value, say around 125% of the rated (fl) current, the electromagnet attracts the iron piece against the spring. This short circuits two points of the no-volt release coil, and the handle goes back to the OFF position, thereby protecting the motor against overload. This is another protective feature in the starter.

Additional resistance may be used in the motor circuit to start the motor, as was used in the case of a dc shunt motor. However, a discussion on this additional resistance is not included in this book, as it is beyond the scope of the syllabus.

12.9 Methods Used for Speed Control of DC Motors

The speed control methods of two types of dc motors are presented.

The circuit required in the case of a *dc shunt motor* is shown in Fig. 12.12, which is nearly similar to that used for a starter (Fig. 12.15). Here, also an additional resistance R is added to the armature resistance, to control the speed. Similarly, an external resistance, R, is inserted in the field current to decrease the current in the shunt field winding, according to the formula,

$$I_f = \frac{V_t}{R_f + R'} = \frac{V_t}{R},$$

where R_f is the resistance of the shunt field. The angular speed of this motor is given by

$$\omega = \frac{E_b}{k \cdot \phi} = \frac{V_t - I_a\,(r_a + R)}{k \cdot \phi}.$$

From this formula, we can see that there can be two methods of speed control, viz, field current control, and armature voltage control. In the first case, as shown earlier, the shunt field current is decreased from the rated value, as the external resistance, R' is increased. According to the formula for speed (angular), the motor speed will increase, as

$$\omega \propto \frac{1}{\phi} = \frac{1}{I_f},$$

as $\phi \propto I_f$, in the linear region. Only, the field current decreases, i.e., lower than the rated speed. Also, the developed torque, $T_e \propto \phi\, I_a = I_f\, I_a$, decreases, as the field current is decreased. However, the input power fed from a dc supply remains the same, as $P = V_t I_a$, both terms being constant.

In the case of armature voltage control, the applied voltage to the armature,

$$V_a = V_t - I_a R$$

decreases, as the additional resistance in the armature circuit, R is increased, with back emf,

$$E_b = V_a - I_a . r_a,$$

also decreasing. Hence, according to the formula, motor speed decreases below the rated value, as

$$\omega \propto E_b \approx V_a,$$

with the field current remaining constant. Here, the developed torque,

$$T_e \propto \phi\, I_a = I_f\, I_a,$$

remains constant, as both terms remain the same, but the input power, $P = V_a I_a$, decreases, as the applied voltage to the armature decreases, with the armature current remaining the same. It may be noted that the energy loss increases, as R is increased, with I_a being the same, thus, reducing the efficiency of this scheme. Now-a-days, the input voltage to the armature is decreased using phase-controlled rectifiers (converters) fed from a constant voltage ac supply, with the system efficiency being quite high in this case.

The circuit required for a *dc series motor* is shown in Fig. 12.17. Two points are to be noted. As in the previous case, the additional resistance, R is connected in series with the motor, with the field winding here being connected in series with the armature. The second point is that a small resistance, R_d is connected in parallel with the series field, so as to reduce the current in it, such that,

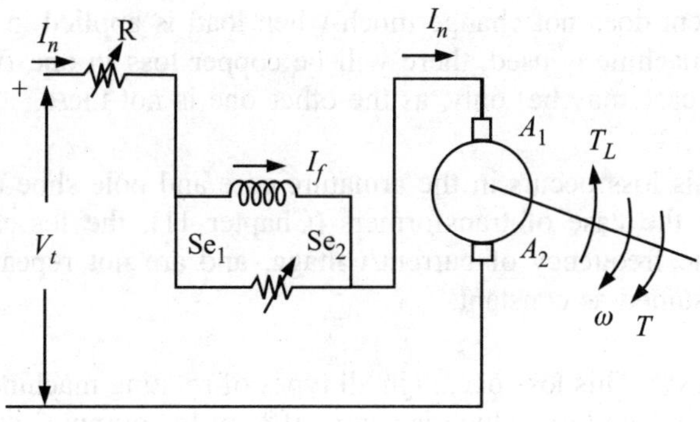

Fig. 12.17

$$I_{se} = \frac{R_d}{r_{se} + R_d} \cdot I_a = a \cdot I_a,$$

where $a < 1$, and also $I_{se} < I_a$. As the diverter resistance, R_d is changed, the series field current is decreased, with a part of the armature current flowing in the diverter. According to the formula given earlier, speed, which is inversely proportional to the series field current, is thus increased, as shown earlier. The developed torque, $T_e \propto \phi\, I_a = I_{se}\, I_a$, decreases, as the field current is decreased. However, the input power fed from the dc supply remains the same, as $P = V_t I_a$, both terms being constant.

The rest remains the same as for dc shunt motors. Hence, it is not repeated here.

Only thing not discussed here, in the case of both types of dc motors is the rating of the additional resistance used in series, which are termed as starting and controller resistances for starting and speed control respectively, as again it is beyond the scope of the syllabus. The speed control methods for other types of dc motors are not described, as these are not much used. In the case of dc cumulatively compound motors, the same scheme as is used for dc shunt motors can be used, as the main excitation there is provided by the shunt field winding, as given earlier. In certain cases, the field winding of a dc shunt motor is separately supplied from another dc supply, with the armature fed as usual. The dc shunt motor may be termed as separately excited, but the speed control methods remain the same as given earlier.

12.10 Losses and Efficiency of DC Machines

There are three types of losses in this rotating machine—copper, iron and mechanical.

(a) *Copper losses*: Taking a dc compound motor as an example, copper losses (I^2R) occur in two types of field windings—shunt and series, and also in the armature winding. The losses in armature and shunt and series field windings are given by ($I_a^2 r_a$), ($I_f^2 R_f = V_t I_f$) and ($I_{se}^2 r_{se}$) respectively. Two points may be noted in this connection. The current in a series field winding may or may not be the same as the armature current, as this depends upon the type of connection used—short shunt or long shunt. The copper losses in the armature and series field windings are termed as variable loss, as it depends on the load current, while the losses in the shunt field winding is generally taken as constant, as the field current does not change much when load is applied in both cases. If a dc shunt or series machine is used, there will be copper loss in one field winding (shunt or series, as the case may be) only, as the other one is not there.

(b) *Iron losses*: This loss occurs in the armature core and pole shoe (not discussed). As given earlier, in the case of transformers (Chapter 11), the losses are of two types depending on the frequency of current/voltage, and are not repeated here. Iron loss, being low, is assumed as constant.

(c) *Mechanical losses*: This loss occurs in all types of rotating machines. It is also known as friction and windage loss, which is nearly 10% of the output. Like the iron loss, this one is taken as constant, irrespective of the machine speed.

EFFICIENCY

Now, let us discuss the procedure of determining the efficiency of a dc shunt machine, used as a generator and a motor.

Generator:

Output = $V_t I_L$,

where I_L is the load current.

There are two types of losses in the dc machine (generator)—constant and variable.

Constant loss = Mechanical loss + iron (core) loss + shunt field copper loss

Variable loss = Armature copper loss

Total losses = Constant loss + variable loss

Input = Output + losses

Efficiency = Output/Input = Output/(Output + losses)

Motor:

Input = $V_t I_m$, where I_m is the input current.

Losses in the two machines remain same.

Output = Input – losses

Efficiency = Output/Input = (Input – losses)/Input

The efficiency of rotating machines at full load, whether it is working as a generator or a motor, is of the order of about 85%, with a major part of the losses being mechanical losses, which is high. The expressions for different types of losses have been given earlier. The armature and field currents, as required, are found from the circuit diagrams for dc shunt generator and motor, which are given earlier, and not repeated here.

ADDITIONAL EXAMPLES

EXAMPLE 12.6 *A 4-pole, long shunt dc compound generator delivers 10 kW to a load, at 250 V. The armature has 50 slots, with 6 conductors per slot. The flux/pole is 60 mWb. The winding is lap connected. The generator has shunt field, armature, and series field winding resistance of 250 Ω, 0.1 Ω, and 0.15 Ω respectively. Find the speed of the generator.*

SOLUTION The circuit diagram of a long shunt dc compound generator is shown in Fig. 12.3(f).

$$r_a = 0.1\,\Omega; R_{sh} = 250\,\Omega;\; r_{se} = 0.15\,\Omega.$$

$$V_t = 250\text{ V}.$$

Output power $= P_o = V_t.I_L = 10$ kW.

Load current $= I_L = \dfrac{P_o}{V_t} = \dfrac{10\times 10^3}{250} = 40$ A.

Shunt field current $= I_{sh} = \dfrac{V_t}{R_{sh}} = \dfrac{250.0}{250.0} = 1.0$ A.

Armature current $= I_a = (I_L + I_{sh}) = (40.0 + 1.0) = 41.0$ A.

Generated emf $= E_g = V_t + [I_a.(r_a + r_{se})] = 250.0 + [(41.0\times(0.1+0.15)] = 260.25$ V,

as $I_{se} = I_L$ in the circuit (Fig. 12.3(f)).

$p = a = 4$ (lap-connected).

No. of slots = 50, conductors/slot = 12.

No. of conductors $= Z = 50\times 6 = 300$.

Flux/pole = 0.06 Wb = 60 mWb.

$$E_g = \frac{\phi\cdot Z\cdot N}{60}\cdot\frac{p}{a} = \left(\frac{0.06\times 300.0}{60}\times\frac{4}{4}\right)\cdot N = 0.3\cdot N = 260.25\text{ V}.$$

Speed $= N = \dfrac{260.25}{0.3} \approx 868.0$ rpm.

EXAMPLE 12.7 *A 400 V dc shunt motor runs at 800 rpm, and draws an armature current of 40 A, from a 400 V supply. The armature resistance is 0.3 Ω. If the flux is reduced by 80%, and the load torque remains the same, find the speed under this condition.*

SOLUTION $r_a = 0.3\ \Omega$, $V_t = 400$ V, $N_1 = 800$ rpm.

Armature current $= I_1 = 40$ A.

Back emf at this speed (N_1)

$$= E_{b1} = V_t - (I_{a1} \cdot r_a) = 400.0 - (40.0 \times 0.3) = 388.0 \text{ V}.$$

The ratio of the fluxes is given as $\dfrac{\phi_2}{\phi_1} = 0.8 = 80\%$.

As load torque remains the same as the developed torque,

$$\frac{T_{e2}}{T_{e1}} = \frac{\phi_2 \cdot I_{a2}}{\phi_1 \cdot I_{a1}} = 0.8 \cdot \frac{I_{a2}}{40.0} = 1.0$$

or, $$I_{a2} = \frac{40.0}{0.8} = 50.0 \text{ A}.$$

Back emf at this speed (N_2)

$$= E_{b2} = E_{b1} \cdot \frac{\phi_2 \cdot N_2}{\phi_1 \cdot N_1} = V_t - (I_{a2} \cdot r_a) = 400.0 - (50.0 \times 0.3) = 385.0 \text{ V}$$

or, $$N_2 = N_1 \cdot \frac{E_{b2} \cdot \phi_2}{E_{b1} \cdot \phi_1} = \frac{800.0}{0.8} \times \frac{385.0}{388.0} = 992.0 \text{ rpm}.$$

EXAMPLE 12.8 *A 250 V dc shunt motor runs at 750 rpm, and draws an armature current of 40 A, from a 250 V supply. The armature and field winding resistances are 0.25 Ω, and 125 Ω respectively. Find the resistance to be added to the field winding to increase speed to 1,200 rpm, at an armature current of 60 A. Assume that the flux is proportional to the field current.*

SOLUTION The circuit diagram of a dc shunt motor is shown in Fig. 12.12.

$$r_a = 0.25\ \Omega, R_f = 125\ \Omega, V_t = 250 \text{ V}.$$

Armature current $= I_{a1} = 40$ A, $N_1 = 750$ rpm.

Field current $= I_{f1} = \dfrac{V_t}{R_f} = \dfrac{250}{125} = 2.0$ A.

Back emf at this speed (N_1)

$$= E_{b1} = k \cdot I_{f1} \cdot N_1 = V_t - (I_{a1} \cdot r_a) = 250.0 - (40.0 \times 0.25) = 240.0 \text{ V}.$$

Note that $\phi_1 \propto I_{f1}$ (flux is proportional to field current).

$$k = \frac{E_{b1}}{I_{f1} \cdot N_1} = \frac{240.0}{2.0 \times 750.0} = 0.16.$$

Armature current, in this case $= I_{a2} = 60$ A, $N_2 = 1{,}200$ rpm.

Back emf at this speed (N_1)

$$= E_{b2} = k \cdot I_{f2} \cdot N_2 = V_t - (I_{a2} \cdot r_a) = 250.0 - (60.0 \times 0.25) = 235.0 \text{ V}.$$

Hence, field current $= I_{f2} = \dfrac{E_{b2}}{k \cdot N_2} = \dfrac{235.0}{0.16 \times 1{,}200.0} = 1.224$ A.

Note that the external resistance, R is added in series with the field winding (Fig. 12.12).

$$R_f + R = \frac{V_t}{I_{f2}} = \frac{250.0}{1.224} = 204.25\ \Omega$$

or, $R = 204.25 - 125.0 = 79.25\ \Omega$.

The ratio of two torques are found as

$$\frac{T_{e2}}{T_{e1}} = \frac{I_{f2}.I_{a2}}{I_{f1}.I_{a1}} = \frac{1.224 \times 60.0}{2.0 \times 40.0} = 0.918.$$

Hence, the developed torque is slightly reduced.

EXAMPLE 12.9 *A 220 V dc shunt motor runs at 500 rpm. A current of 30 A, including field current, is drawn from a 220 V supply, under this condition. The armature and field winding resistances are 0.5 Ω, and 200 Ω respectively. What resistance is to be inserted in series with the armature in order to reduce the speed to 300 rpm, the armature current remaining the same?*

SOLUTION $r_a = 0.5\ \Omega, R_f = 200\ \Omega, V_t = 220$ V.

Motor current $= I_m = 30$ A, $N_1 = 500$ rpm.

Field current $= I_f = \dfrac{V_t}{R_f} = \dfrac{220}{200} = 1.1$ A.

Armature current $= I_a = I_m - I_f = 30.0 - 1.1 = 28.9$ A.

Back emf at this speed (N_1)

$$= E_{b1} = k \cdot N_1 = V_t - (I_a \cdot r_a) = 220.0 - (28.9 \times 0.5) = 205.55 \text{ V}.$$

Both armature and field currents remain the same. The flux also is the same. A resistance, R is inserted in series with the armature to reduce the speed to $N_2 = 300$ rpm.

Back emf at this speed (N_2)

$$= E_{b2} = k \cdot N_2 = E_{b1} \cdot \frac{N_2}{N_2} = V_t - [I_a \cdot (r_a + R)] = 205.55 \times \frac{300}{500} = 123.33 \text{ V}$$

or, $E_{b2} = V_t - [I_a \cdot (r_a + R)] = E_{b1} - (I_a \cdot R) = 230.0 - [28.9 \times (0.5 + R)]$

$$= 205.55 - (28.9 \times R) = 123.33 \text{ V}.$$

The additional resistance to be inserted in series with the armature

$$= R = \frac{E_{b1} - E_{b2}}{I_a} = \frac{205.55 - 123.33}{28.9} = 2.845 \quad \Omega.$$

EXAMPLE 12.10 *A dc series motor runs at 600 rpm, and draws a current of 60 A, being supplied from a 230 V supply. The total resistance, including the series field, is 0.2 Ω. Calculate the resistance, R connected in series with the motor, so as to reduce the speed to 300 rpm, the load torque being half the previous value. Assume flux as proportional to current.*

SOLUTION The circuit of a dc series motor is shown in Fig. 12.13.

$$r = (r_a + r_{se}) = 0.2\,\Omega, V_t = 230 \text{ V}.$$

Motor current $= I_1 = 60$ A, $N_1 = 600$ rpm.

Back emf at this speed (N_1)

$$= E_{b1} = k \cdot I_1 \cdot N_1 = V_t - (I_1 \cdot r) = 230.0 - (60.0 \times 0.2) = 218.0 \text{ V}.$$

Note that $\phi_1 \propto I_1$ (flux is proportional to the series field current). As this is a series motor, the armature and series fields are the same as the motor current.

$$k = \frac{E_{b1}}{I_1 \cdot N_1} = \frac{218.0}{60.0 \times 600.0} = 6.056 \times 10^{-3}.$$

Torque developed $= T_{e1} = T_{L1} = k'.(I_1)^2$.

As the torque is now half the previous value, $T_{e2} = T_{L2} = \dfrac{T_{e1}}{2} = k' \cdot (I_2)^2$.

From the two expressions, we get, $\left(\dfrac{I_2}{I_1}\right)^2 = \dfrac{1}{2}$

or, $I_2 = \dfrac{I_1}{\sqrt{2}} = \dfrac{60.0}{\sqrt{2}} = 42.43$ A.

$$N_2 = 300 \text{ rpm}.$$

Back emf at this speed (N_2)

$$= E_{b2} = k \cdot I_2 \cdot N_2 = V_t - [I_2 \cdot (r + R)] = 6.056 \times 10^{-3} \times 42.43 \times 300.0 = 77.1 \text{ V}.$$

E_{b2} is also found as

$$E_{b2} = E_{b1} \cdot \frac{\phi_2.N_2}{\phi_1.N_1} = E_{b1} \cdot \frac{I_2.N_2}{I_1.N_1} = 218.0 \times \frac{42.43 \times 300.0}{60.0 \times 600.0} = \frac{218.0}{2\sqrt{2}} = 77.1 \text{ V}.$$

From this expression, we get,

$$(r+R) = (0.2+R) = \frac{V_t - E_{b2}}{I_1} = \frac{230.0 - 77.1}{42.43} \approx 3.6 \quad \Omega.$$

The additional resistance to be inserted in series with the motor

$$= R = 3.6 - 0.2 = 3.4\,\Omega.$$

EXAMPLE 12.11 *A 230 V dc shunt motor runs at 800 rpm, and the power drawn is 1,600 W, under no-load condition. The armature and field winding resistances are* 0.2 Ω *and* 200 Ω *respectively. Find the efficiency (a) when it is run as a motor, taking 80 A at 230 V, and (b) when it is run as a generator, supplying 80 A at 230 V.*

SOLUTION The circuits of a dc shunt machine, running as a motor and a generator are shown in Fig. 12.18(a) and 12.18(b) respectively.

$$r_a = 0.2\,\Omega, R_f = 200\,\Omega, V_t = 230 \text{ V}.$$

Input power at no-load $= P_{i0} = V_t.I_{m0} = 1,600$ W.

$$\text{Motor current} = I_{m0} = \frac{P_{i0}}{V_t} = \frac{1,600}{230} = 6.96 \text{ A}.$$

$$\text{Field current} = I_f = \frac{V_t}{R_f} = \frac{230}{200} = 1.15 \text{ A}.$$

Field copper loss $= P_{Cf} = V_t \cdot I_f = 230.0 \times 1.15 = 264.5$ W, which is constant.

Armature current $= I_{a0} = I_{m0} - I_f = 6.96 - 1.15 = 5.81$ A.

Armature copper loss $= P_{Ca0} = (I_{a0})^2 \cdot r_a = (5.81)^2 \times 0.2 = 6.75$ W, which is very low.

As there is no output, constant loss = input – total copper loss

$$= 1,600.0 - (6.75 + 264.5) = 1,328.75 \text{ W}.$$

This loss includes iron and mechanical (friction and windage) losses, which depends on the speed of the rotating machine.

If we add field copper loss, constant loss = 1,328.75 + 264.5 = 1,593.25 W.

(a) Motor (Fig. 12.18(a))

Motor (input) current $= I_{m1} = 80$ A, $N = 800$ rpm.

Armature current $= I_{a1} = I_{m1} - I_f = 80.0 - 1.15 = 78.85$ A.

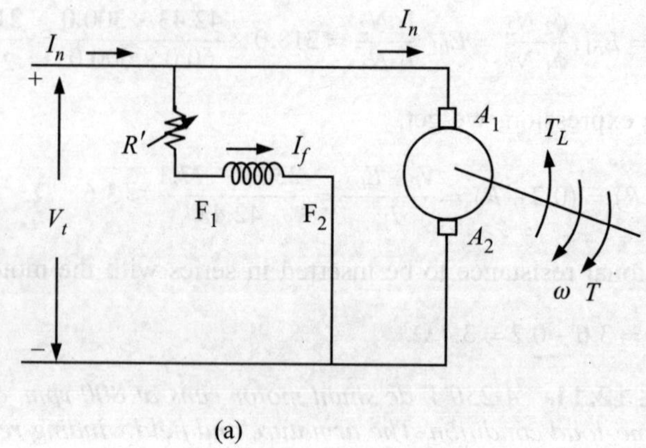

(a)

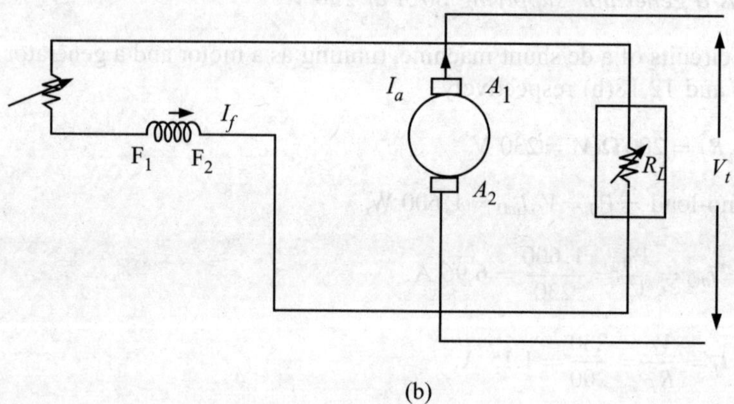

(b)

Fig. 12.18

Input power $= P_{i1} = V_t \cdot I_{m1} = 230.0 \times 80.0 = 18,400$ W $= 18.4$ kW.

Armature copper loss $= P_{Ca1} = (I_{a1})^2 \cdot r_a = (78.85)^2 \times 0.2 = 1,243.5$ W.

Total losses = armature copper loss + constant losses (including field copper loss)

$$= 1,243.5 + 1,593.75 = 2,837 \text{ W} = 2.837 \text{ kW}.$$

Output power = input power − total losses $= P_{01} = 18.4 - 2.837 = 15.563$ kW.

Motor efficiency $= \eta_m = \dfrac{P_{01}}{P_{i1}} = \dfrac{15.563}{18.4} = 0.8458 = 84.58\%$.

(b) Generator (Fig. 12.18(b))

Load (output) current $= I_{L2} = 80$ A, $N = 800$ rpm.

Armature current $= I_{a2} = I_{L2} + I_f = 80.0 + 1.15 = 81.15$ A.

Output power $= P_{02} = V_t \cdot I_{L2} = 230.0 \times 80.0 = 18{,}400$ W = 18.4 kW.

Armature copper loss $= P_{Ca2} = (I_{a2})^2 \cdot r_a = (81.15)^2 \times 0.2 = 1{,}317.1$ W.

Total losses = armature copper loss + constant losses (including field copper loss)

$= 1{,}317.1 + 1{,}593.75 = 2{,}911$ W $= 2.911$ kW.

Input power = input power + total losses $= P_{i2} = 18.4 + 2.911 = 21.311$ kW.

$$\text{Generator efficiency} = \eta_g = \frac{P_{02}}{P_i} = \frac{18.4}{21.311} = 0.8634 = 86.34\%.$$

Conclusion

In this chapter, the second one on electrical machines, a rotating machine—a dc machine, which can operate, both as a generator and a motor—is presented. This is in contrast to the earlier machine the — transformer an ac machine, and static one. When a dc machine is run as generator at rated speed, dc voltage is obtained at the output. The machine can also be run as a motor, producing torque output driving a load, when a rated voltage (dc) is supplied to it. Starting with the construction, some special features, like poles with field winding(s), armature winding, and commutator are described. The field winding connections used for dc machines resulting in different types are then described. The emf equation, and both open circuit (no-load) and load characteristics of dc generators, such as shunt and compound, are discussed in detail. The torque equation of a dc motor, the load characteristics, speed control and starting methods of different types of dc motors, such as shunt and series, are explained, emphasizing their uses. Lastly, the losses and efficiency of dc machines, running as a generator and a motor, are discussed. Note that the efficiency of dc rotating machines is quite lower than that of a transformer (static one) due to rotational losses. The details of an ac rotating machine—three-phase induction motor—which is widely used in industry, are described in the next chapter, the third in a series of four chapters on electrical machines.

Multiple Choice Type Questions

Tick the correct option(s), out of the four choices provided in each question.

12.1 For a dc shunt generator running at 1,250 rpm, the generated emf is 100 V. If it is run at 1,000 rpm, with all its parameters, including the field current, being kept constant, the generated emf in V is,

(a) 125 (b) 100 (c) 80 (d) 50

12.2 For a dc shunt generator running at 1,000 rpm, the generated emf is 100 V at a field current of 2 A. If its speed and all parameters are kept constant, the generated emf in V, at a field current of 1.5 A, is (assume the flux to be proportional to field current),

(a) 133.3 (b) 100 (c) 50 (d) 75

12.3 A dc shunt generator, which is rotated at a rated speed, fails to excite. Choose the correct reason(s).

(a) There is no residual magnetism.
(b) The total resistance of the field circuit is lower that the critical resistance at rated speed.
(c) The total resistance of the field circuit is higher that the critical resistance at rated speed.
(d) None of the above.

12.4 For a dc shunt generator running at 1,500 rpm, the output voltage is 100 V at a load current of 10 A. The armature resistance is 1 Ω. If its speed is changed to 1,000 rpm, and all parameters, including shunt field and load currents are kept constant, find its output voltage in V.

(a) 50 (b) 66.7 (c) 150 (d) 100

12.5 For a dc cumulatively coupled compound generator running at rated speed, the output voltage is 150 V, at a load current of 10 A. The shunt field current is 1 A, and the number of turns in it is 300. The copper losses in armature, shunt and series resistance are neglected. All parameters, including its speed, are kept constant and the number of turns of the series field is 2. Assume the flux to be proportional to total field excitation. Find its output voltage at no-load in V.

(a) 160 (b) 150 (c) 170 (d) 100

12.6 For a dc cumulatively coupled compound generator running at rated speed, the output voltage at full load is greater than that at no-load. State whether the generator can be termed as,

(a) under compounded. (b) level compounded.
(c) over compounded. (d) none of (a)–(c)

12.7 A dc shunt machine is run as a generator and a motor with the output and input voltages, both being rated. The machine runs at the same speed. The field current in both cases is 2 A. The output and input currents are both 30 A respectively. If all parameters, including speed, are kept constant, the armature currents (A) in both cases (generator and motor) are,

(a) 30 and 30 (b) 32 and 30 (c) 30 and 28 (d) 32 and 28

12.8 For a dc series motor, which is being fed from a rated supply, the torque developed,

(a) is proportional to the square of the input current.
(b) is proportional to the input current.
(c) is constant at load.
(d) none of (a)–(c)

12.9 For a dc shunt motor which is fed from a rated supply with the additional resistance in the field circuit being kept constant,

(a) the torque is proportional to the square of the armature current.
(b) the torque is proportional to the armature current.
(c) the speed is nearly constant at all loads.

(d) the speed is variable as the load is varied.

12.10 A dc shunt motor, being fed from a 200 V supply, runs at 750 rpm. Its field current is 2 A. The armature copper losses are neglected when the machine runs as a motor or a generator. If all parameters, including supply voltage, are kept constant, the motor speed in rpm, at a field current of 1.5 A, is (the flux being proportional to the field current),

(a) 1,000 (b) 750 (c) 400 (d) 560

12.11 For a dc series motor, which is being fed from a rated supply, the speed is,

(a) constant at load.
(b) inversely proportional to the input current.
(c) inversely proportional to the torque developed.
(d) inversely proportional to the square root of torque developed.

12.12 A dc shunt motor is fed from a 200 V supply, its rated current being 25 A. The armature resistance is 1 Ω. Find the value of additional resistance (Ω), to be inserted in series with the armature at starting, such that the current drawn from the supply is twice the value of rated current. Neglect the value of shunt field current.

(a) 1 (b) 2 (c) 4 (d) 3

12.13 For a dc motor, the core losses in the armature mainly,

(a) increase with speed. (b) decrease with speed.
(c) are constant at all speeds. (d) none of the above.

12.14 The efficiency of a dc shunt motor at full load, is in the range of,

(a) 97% (b) 85% (c) 90% (d) 70%

Answers

12.1 (c) 12.2 (d) 12.3 (a),(c) 12.4 (a) 12.5 (b) 12.6 (c) 12.7 (d)
12.8 (a), (c) 12.9 (b) 12.10(a) 12.11(b), (d) 12.12(d) 12.13(a) 12.14(b)

Problems

12.1 Calculate the useful flux per pole in a 4-pole dc generator, to give an induced emf of 250 V at 1,200 rpm, when the armature is (a) lap-connected and (b) wave-connected. The number of slots in the machine is 51, and it has 24 conductors/slot.

12.2 An 8-pole, lap-connected, armature has 960 conductors, a flux/pole of 40 mWb, and a speed of 800 rpm. Find the emf generated in the armature. Determine the speed at which the armature is to be driven to maintain a voltage of 400 V, if it is wave-connected.

12.3 The armature of an 8-pole, dc machine has a wave-winding containing 664 conductors. Calculate the generated emf, when the flux per pole is 0.08 Wb, and the speed is 210 rpm. At what speed should the armature be driven to generate 500 V, if the flux per pole is changed to 0.06 Wb?

12.4 An 8-pole, lap-connected, armature of a dc machine has 40 slots, with 12 conductors per slot. The flux/pole is 50 mWb. Determine the speed at which the armature is to be driven to generate a voltage of 500 V.

12.5 The open circuit characteristic of a dc separately-excited generator at 1,000 rpm is given here. The generator is run as a shunt generator at the same speed, i.e., 1,000 rpm, and has a field resistance of 90 Ω. Find (a) the open-circuit voltage and field current, (b) the critical resistance at this speed and (c) the resistance to be inserted in the field circuit to induce a voltage of 105 V on open circuit.

Field current (A)	0.2	0.4	0.6	0.8	1.0	1.2	1.4	1.6
Induced EMF (V)	30	55	75	90	100	110	115	120

12.6 The open circuit characteristic of a dc separately-excited generator at 1,000 rpm is given here. The generator is now run as a shunt generator at a speed of 800 rpm. (a) What should be the field circuit resistance to generate a voltage of 440 V? (b) Find the voltage, when the field circuit resistance is 200 Ω. (c) Find the critical speed corresponding to the above field circuit resistance of 200 Ω. (d) Find the critical field circuit resistance at the speed of 800 rpm.

Field current (A)	0.4	0.8	1.2	1.6	2.0	2.4	2.8
Induced EMF (V)	140	275	375	440	500	540	570

12.7 A 6-pole, dc shunt generator delivers 50 A at 110 V. The armature winding is lap-connected with 360 conductors. The generator has a shunt field, and armature winding resistance of 55 Ω and 0.2 Ω respectively. Determine the flux per pole, if the machine is run at a rated speed of 1,800 rpm.

12.8 A 4-pole, 220 V dc shunt generator supplies power to parallel connected hundred 220 V, 40 W lamps. The generator has shunt field, and armature winding resistance of 220 Ω and 1 Ω respectively. Calculate the generated emf if the load voltage is to be maintained at 220 V. Assume voltage drop per brush to be 1 V.

12.9 A dc shunt generator has an induced emf of 126 V, while its terminal voltage is 120 V. Determine the armature current, if the armature and field winding resistances are 0.025 Ω and 30 Ω respectively. Neglect armature reaction and also brush drop.

12.10 A short shunt dc compound generator delivers 100 A to a load at 250 V. The generator has shunt field, armature, and series field winding resistance of 125 Ω, 0.1 Ω, and 0.15 Ω respectively. Find the generated emf. Assume voltage drop per brush to be 1 V.

12.11 A 4-pole, long shunt dc compound generator delivers 10 kW to a load, at 250 V. The armature has 50 slots, with 6 conductors per slot. The winding is lap-connected. The generator has shunt field, armature, and series field winding resistance of 250 Ω, 0.1 Ω, and 0.15 Ω respectively. Find the speed of the generator.

12.12 A 4-pole, dc shunt motor takes 20 A, when it is run from a 220 V supply. The armature and field winding resistances are 0.06 Ω and 100 Ω respectively. The winding is lap-connected, with 300 conductors. Flux/pole is 20 mWb. Find the motor speed, and gross torque developed.

12.13 A 4-pole dc shunt motor takes 60 A, when it is run from a 250 V supply. The armature winding resistance is 0.2 Ω. The winding is wave-connected with 700 conductors. Flux/pole is 20 mWb. Find the motor speed and gross torque developed. Assume voltage drop per brush to be 1 V.

12.14 The armature of a 4-pole dc shunt motor has 60 slots, with 20 conductors per slot. The winding is lap-connected. Flux/pole is 25 mWb. If the armature current is 40 A, find the gross torque developed. Also calculate its speed, when input voltage is 300 V. The armature winding resistance is 0.25 Ω.

12.15 A 10 HP, 230 V dc shunt motor takes an armature current of 6 A from a 230 V supply at no-load. It runs at 1,200 rpm. The armature resistance is 0.25 Ω. Calculate the speed and gross torque developed, when the armature takes 30 A with the same flux.

12.16 A 220 V dc shunt motor runs at 500 rpm. A current of 30 A, including field current, is drawn from a 220 V supply, under this condition. The armature and field winding resistances are 0.5 Ω, and 200 Ω respectively. What resistance is to be inserted in series with the armature, in order to reduce the speed to 300 rpm, the armature current remaining the same?

12.17 A 250 V dc shunt motor runs at 750 rpm, and draws an armature current of 40 A, from a 250 V supply. The armature and field winding resistances are 0.25 Ω and 100 Ω respectively. Find the resistance to be added to the field winding to increase the speed to 1,200 rpm at an armature current of 60 A, with the load torque remaining constant. Assume flux as proportional to field current.

12.18 A 400 V dc shunt motor runs at 800 rpm and draws an armature current of 40 A, from a 400 V supply. The armature resistance is 0.3 Ω. If the flux is reduced by 80%, and the load torque remains the same, find the speed under this condition.

12.19 A dc series motor runs at 600 rpm, and draws a current of 60 A, being supplied from a 230 V supply. The total resistance, including the series field, is 0.2 Ω. Calculate the resistance, R connected in series with the motor, so as to reduce the speed to 300 rpm, the load torque being half the previous value. Assume flux as proportional to current.

12.20 A dc shunt motor develops 25 HP at 200 V, while consuming 22 kW of power. The armature and field winding resistances are 0.06 Ω, and 80 Ω respectively. Find the efficiency and input power, when the output is 10 HP.

12.21 A 230 V dc shunt motor runs at 800 rpm, and the power drawn is 1,600 W, under no-load conditions. The armature and field winding resistances are 0.2 Ω, and 200 Ω respectively. Find the efficiency (a) when it is run as a motor, taking 80 A at 230 V and (b) when it is run as a generator, supplying 80 A at 230 V.

CHAPTER 13

THREE-PHASE INDUCTION MOTORS

The three-phase induction motor is the third of the series of four electrical machines that we will discuss in our book. It is a rotating machine, run from an ac supply which works only as a motor unlike the dc machine discussed in Chapter 12 which can be used both as a generator and a motor. It is widely used in industry in various ratings, as it is cheaper than other motors. We will first describe the construction of two types of this motor, i.e., cage and slip-ring rotors. The first one, being rugged and maintenance-free, is generally used. As we will study, a rotating magnetic field is produced when a three-phase balanced supply is given normally to the stator, having three balanced windings, of equal number of turns, and spaced at 120° electrical degree. Then, the principle of operation, followed by the equivalent circuit, which is nearly the same as that of the transformer given in Chapter 11, is presented. The torque–speed (slip) characteristics are discussed, along with the various starters used for these two types of motors. Speed control methods are briefly given.

13.1 Production of Rotating Magnetic Field in a Three-phase Induction Motor

A two-pole, three-phase balanced winding in the stator of an induction motor is shown in Fig.13.1. For a three-phase balanced winding, the number of turns in the three windings is equal. The angle between the adjacent phases, say R and Y, is 120°. Note that for a two-pole stator, the mechanical angle is same as the electrical angle. The angle between the phases Y and B is also 120°. The winding in each of the phases is concentrated in a slot having the required number of turns. The two (forward and return) parts of the winding in each phase, say R, are referred to as R and R′ respectively. A similar situation arises for the two other phases, Y and B.

A three-phase balanced voltage, with a phase sequence of R–Y–B, is applied to the above winding. In a balanced voltage, the magnitude of the voltage in each phase, assumed to be in star in this case, is equal to the phase angle of the voltage between the adjacent phases, say R and Y (here, 120°). This is shown in the sinusoidal voltage waveforms of the three phases, R, Y and B (Fig. 13.2(a)).

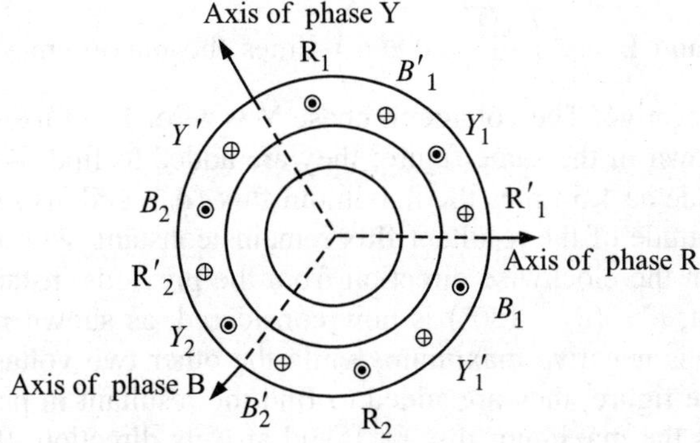

Fig. 13.1 *Three-phase two pole stator winding of an induction motor*

Let us first consider what happens at the time instant t_1, of the voltage waveforms. At this instant (Fig. 13.2(b)), the voltage in the R phase is a positive maximum ($\theta_1 = 90°$), while the two other voltages in the phases, Y and B, are half of the maximum value, and also negative. The three waveforms are represented by the following equations:

$$E_{RN} = E_m \sin\theta; E_{YN} = E_m \sin(\theta - 120°); E_{BN} = E_m \sin(\theta + 120°),$$

where, $\theta = \omega t$ (radians)

f = Supply frequency (Hz or cycles/s)

$\omega = 2\pi f$ = Angular frequency (rad./s) .

The voltage waveforms in the three windings are shown in Fig. 13.2(a). The fluxes due to the currents in the windings are shown in the same figure; they are added to find the resultant in phasor form, with the magnitude as 1.5 times the maximum value of the flux (ϕ_m), and also its direction.

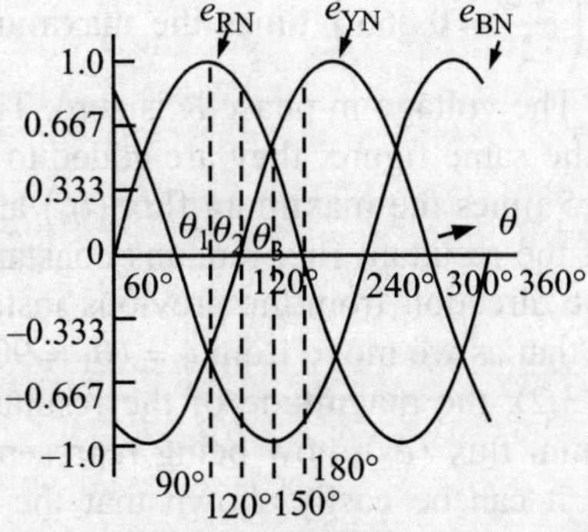

Fig. 13.2(a)

Now, let us now shift to the instant, $t_2 = (\theta_2 = 120°)$ as shown in Fig. 13.2(c). The voltages in the two phases, R and B are $\left(\frac{\sqrt{3}}{2} = 0.866\right)$ times the maximum value, with R phase positive and B phase negative. The voltage in phase Y is zero. The fluxes due to the currents in the windings are shown in the same figure; they are added to find the resultant in phasor form, with the magnitude as 1.5 times the maximum flux (ϕ_m), and also its direction. It may be noted that the magnitude of the resultant flux remains constant, with its direction shifting by 30° = (120° – 90°) in the clockwise direction from the previous instant.

Similarly, the instant, $t_3 = (\theta_3 = 150°)$ is now considered, as shown in Fig. 13.2(d). The voltage in the B phase is negative maximum, while the other two voltages in the phases, R and Y are half the same figure; they are added to find the resultant in phasor form, with the magnitude as 1.5 times the maximum flux (ϕ_m), and also its direction. It may be noted that the magnitude of the resultant flux remains constant, with its direction shifting by another 30° in the clockwise direction from the previous instant.

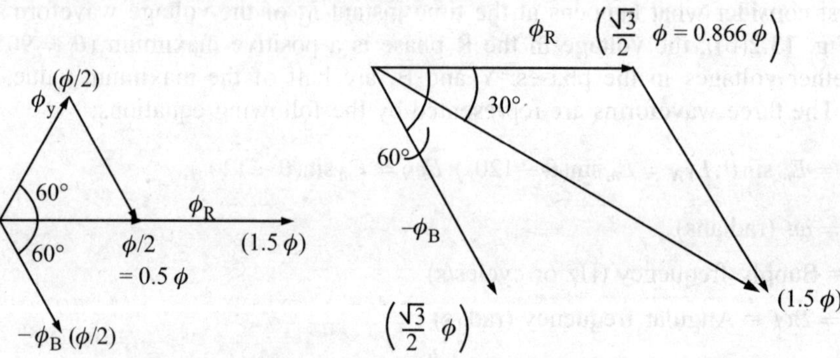

Fig. 13.2(b) $\theta_1 = 90°$ **Fig. 13.2(c)** $\theta_2 = 120°$

The instant, $t_4 = (\theta_4) = 180°$ is now taken up, as shown in Fig. 13.2(d). The voltages in the two phases, Y and B are $\left(\frac{\sqrt{3}}{2} = 0.866\right)$ times the maximum value, with the Y phase positive and B phase negative. The voltage in phase R is zero. The fluxes due to the currents in the windings are shown in the same figure; they are added to find the resultant in phasor form, with the magnitude as 1.5 times the maximum flux (ϕ_m) and also its direction. It may be noted that the magnitude of the resultant flux remains constant with its direction shifting by another 30° in the clockwise direction from the previous instant.

A study of the above shows that as we move from $t_1 = (\theta_1 = 90°)$ to $t_4 = (\theta_4 = 180°)$ along the voltage waveforms (Fig. 13.2), the magnitude of the resultant flux remains constant at the value 1.5 times the maximum flux (ϕ_m), now being represented by M. If another point, say, $t_5 = (\theta_5 = 210°)$ is taken, it can be easily shown that the magnitude of the resultant flux at that instant is 1.5M. If any other arbitrary point, $t(\theta)$ on the waveform is taken, the magnitude of the resultant flux at that instant also remains the same at 1.5M. It can also be seen that the axis of the resultant flux moves through 90°, as the angle changes from 90° to 180°, i.e., by the same angle of 90°. Hence, if we move through one cycle of the waveform (360°, electrical), the axis of the resultant flux also moves through 360° (2-pole stator), i.e., one complete revolution. As mentioned before, for a two-pole configuration, the mechanical and electrical angles are the same. Hence, the speed of the rotating magnetic field for this case is 50 rev./s (rps), or 3,000 rev./m (rpm), as the supply frequency is 50 Hz or cycles/s, with its magnitude, i.e., the resultant remaining the same.

It has been shown here that if a balanced three-phase voltage is supplied to the balanced three-phase windings in the stator of an induction motor, the resultant flux remains constant in magnitude, but rotates at a synchronous speed, which is related to the supply frequency and number of poles, for which the winding (stator) has been designed. This resultant flux is termed as a rotating magnetic field formed in the air gap of the motor, if a proper rotor is placed inside the stator.

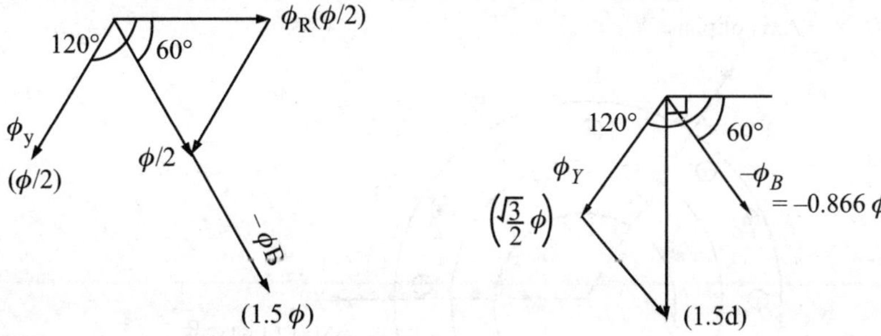

Fig. 13.2(d) $\theta_3 = 150°$ **Fig. 13.2(e)** $\theta_4 = 180°$

Four-pole Stator

A four-pole stator ($p = 4$) with balanced three-phase winding (Fig. 13.3) is taken as an example. The windings for each of the three phases are in two parts, with the mechanical angle between the start of the adjacent windings being 60° only, whereas, the electrical angle remains the same at 120°. As there are two pairs of poles, there are electrically two cycles, i.e., 720°, for one complete revolution, with an N–S pair for each cycle of 360°. However, the mechanical angle is only 360°. If we move through one cycle of the waveform, by 360° (electrical), the axis of the resultant flux in this case moves through a mechanical angle of 180°, i.e., one pole pair of 360° (electrical), or half a revolution only. As stated earlier, for the resultant flux axis to make one complete revolution (360°, mechanical), two cycles of the waveform 720° (electrical), are required, as the number of poles (p) is four (4). 1° mechanical in this case is equal to $\left(\frac{p}{2} = 2°\right)$ electrical. Therefore, for the supply frequency of f (50 Hz), the speed of the rotating magnetic field is $n_s = 25$ rps, or is $N_s = 1,500$ rpm. The relation between the synchronous speed, i.e., the speed of the rotating magnetic field, in rpm, and the supply frequency in Hz, is given by

$$N_s = \frac{(60 \times 2f)}{p} = \frac{120.f}{p}.$$

Reversal of the Direction of the Rotating Magnetic Field

The direction of the rotating magnetic field is reversed by changing the phase sequence to R-B-Y, i.e., changing the connection of any two of the three phases, and keeping the third one the same. If the three waveforms of the above phase sequence are drawn, it can be shown that the rotating magnetic field will move in the reverse direction, from phase R to phase Y, which is reverse of the earlier one shown in Fig. 13.1. Similarly, in the three waveforms shown in Fig. 13.2, the two waveforms of the phases Y and B will change, with the reference phase R (assumed in this case) remaining the same, as the phase sequence is changed or reversed. In the next section, we will discuss the construction of the three-phase induction motor.

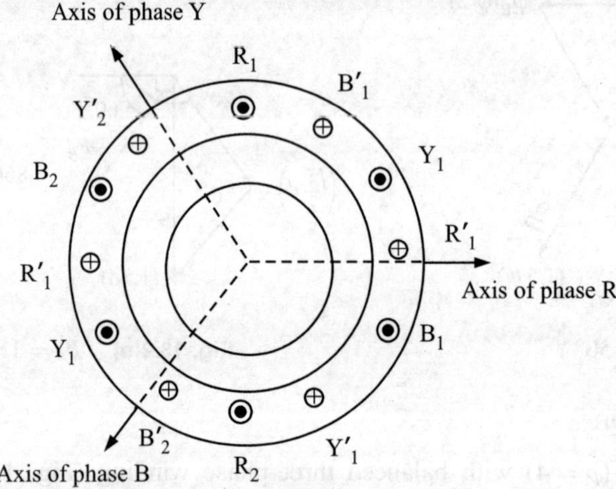

Fig. 13.3

13.2 Construction of a Three-phase Induction Motor

Like all rotating machines, the three-phase induction motor consists of two parts—the stator and the rotor. In the stator (Fig. 13.4), the winding used is a balanced three-phase one, which means that the number of turns in each phase, connected in star/delta, is equal. The windings of the three phases are placed 120°(electrical) apart. The mechanical angle between the adjacent phases is $\left[\frac{(2 \times 120°)}{p}\right]$, where p is the number of poles. For a four-pole ($p = 4$) stator, the mechanical angle between the winding of the adjacent phases, is

$$\left[\frac{(2 \times 120°)}{4}\right] = \frac{120°}{2} = 60°,$$

as shown in Fig. 13.3. Three conductors, mostly multi-turn, are placed in slots, which may be closed, or semi-closed, to keep the leakage inductance low. The start and return parts of the winding are placed nearly 180° or $(180° - \beta)$ apart. The angle of short chording (β) is nearly equal to 30°, or close to that value. Short chording results in reducing the amount of copper used for the winding, as the length of the conductor needed for the overhang part is reduced. There are also other advantages. The section of the stampings used for both stator and rotor is shown in Fig. 13.5. It is necessary to place the core below the teeth to reduce the reluctance of the magnetic path, which carries the flux in the motor (machine). The stator is normally kept inside a support.

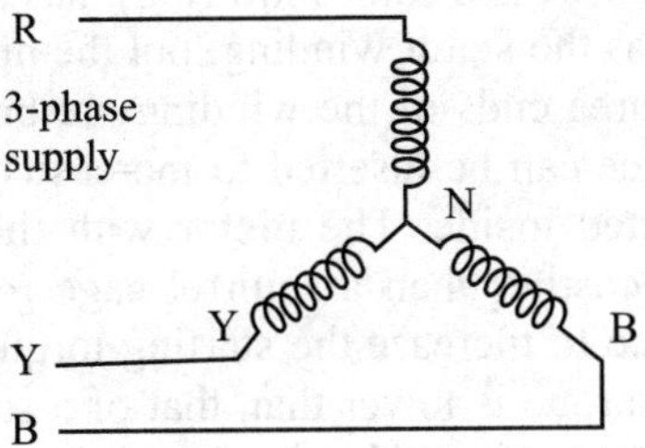

Fig. 13.4 *Schematic diagram of the stator windings in a three-phase induction motor*

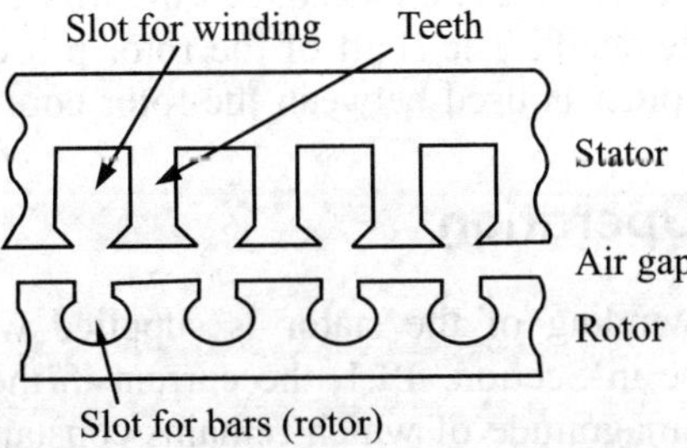

Fig. 13.5 *Section for stampings of stator and rotor in a three-phase induction machine (not to scale)*

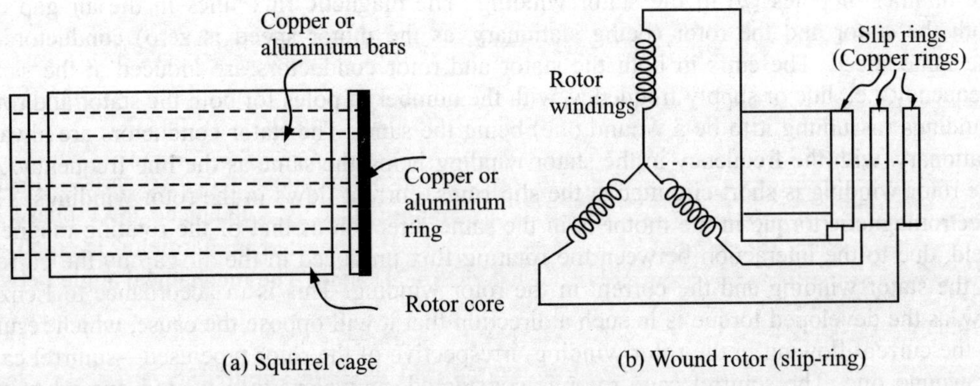

Fig. 13.6 *Two types of rotors for three-phase induction motors*

There are two types of rotors used in three-phase induction motors, viz., squirrel cage and wound (slip-ring) rotor one. The squirrel cage rotor (Fig. 13.6(a)) is mainly used, as it is cheap, rugged and needs little or no maintenance. It consists of copper bars placed in the slots of the rotor, short circuited at the two ends by end rings and brazed with the bars. This type of rotor is similar to a wound (slip-ring) one, with the advantage that it may be used for a stator with different number of poles. It is a universal one, and takes up any number of poles of the stator windings. The only disadvantage is that the equivalent resistance of the rotor is constant. Hence, at the design stage, the value is so chosen as to obtain a certain starting torque; the slip at full load torque is also kept within limits as needed.

The other type of rotor, i.e., a wound rotor (slip-ring), has a balanced three-phase winding (Fig. 13.6(b)), being the same as the stator winding, but the number of turns used depends on the voltage in the rotor. The three ends of the winding are brought at the three slip-rings, at which points, external resistance can be inserted to increase the starting torque requirement. The other three ends are shorted inside. The motor with the additional starting resistance is costlier—the rotor itself is costlier than a squirrel cage rotor of same power rating, and the additional starting resistance to increase the starting torque as required incurs more cost. However, the slip at full load torque is lower than that of a squirrel cage rotor with identical rating, when no additional resistance is used, with direct short-circuiting at the three slip-ring terminals. In both types of rotors, below the teeth, in which the bars of a squirrel cage rotor or the conductors of the rotor winding are placed, lies the iron core which carries the flux as is the case of the core in the stator. The shaft of the rotor passes below the rotor core. For rotors of large diameter, a spider is used between the rotor core and the shaft.

13.3 Principle of Operation

The balanced three-phase winding of the stator is supplied with a balanced three-phase voltage. As explained before in Section. 13.1, the current in the stator winding produces a rotating magnetic field, the magnitude of which remains constant. The axis of the magnetic field rotates at a synchronous speed $\left(n_s = \dfrac{2.f}{p}\right)$, a function of the supply frequency (f) and the number of poles (p) in the stator winding. The magnetic flux lines in the air gap cut both the stator and the rotor (being stationary, as the motor speed is zero) conductors at the same speed. The emfs in both the stator and rotor conductors are induced at the same frequency, i.e., line or supply frequency, with the number of poles for both the stator and rotor windings (assuming it to be a wound one) being the same. The stator conductors are always stationary, with the frequency in the stator winding being the same as the line frequency. As the rotor winding is short-circuited at the slip-rings, current flows in the rotor windings. The electromagnetic torque in the motor is in the same direction as that of the rotating magnetic field, due to the interaction between the rotating flux produced in the air gap by the current in the stator winding and the current in the rotor winding. This is in accordance to Lenz's law, as the developed torque is in such a direction that it will oppose the cause, which results in the current flowing in the rotor winding, irrespective of the rotor type used—squirrel cage or wound one. The squirrel cage rotor is considered equivalent to a wound one when the bars are short-circuited by two end-rings.

The induced emf and the current in the rotor are due to the relative velocity between the rotor conductors and the rotating flux in the air gap, which is maximum when the rotor is stationary ($n_r = 0.0$). As the rotor starts rotating in the same direction as that of the rotating magnetic field due to the production of the torque, the relative velocity decreases. The induced emf and current in the rotor also has lower values. If the rotor speed is equal to that of the rotating magnetic field, which is termed as synchronous speed, and also in the same direction, the relative velocity is zero, which causes both the induced emf and current in the rotor to be reduced to zero. Under this condition, torque will not be produced. Hence, for production of positive (motoring) torque, the rotor speed must always be lower than the synchronous speed. The rotor speed is never equal to the synchronous speed in an induction machine. It is determined by the mechanical load on the shaft and the total rotor losses, mainly copper. The difference between the synchronous speed and the rotor speed, expressed as a ratio of the synchronous speed, is termed as 'slip' in an induction machine. Hence, slip (s) in per unit (pu) is

$$s = \frac{n_s - n_r}{n_s} = \left(1 - \frac{n_r}{n_s}\right) \text{ or, } n_r = (1 - s)n_s,$$

where n_s and n_r are the synchronous speed and the rotor speed respectively in (rev./s) or shortly (rps).

In terms of $N_s = (60 \times n_s)$ and $N_r = (60 \times n_r)$, both in (rev./m) or shortly (rpm), slip is given by,

$$s = \frac{N_s - N_r}{N_s} = \left(1 - \frac{N_r}{N_s}\right) \text{ or, } N_r = (1 - s)N_s.$$

If the slip is expressed in %, then $s = \left[\left\{\frac{N_s - N_r}{N_s}\right\} \times 100\right]$.

Normally, for torques varying from no-load ($\approx$ zero) to full load value, the slip is proportional to torque. The slip at full load is 4–5% (0.04–0.05).

We can provide an alternative explanation for the production of torque in a three-phase induction motor by using the two rules (right-hand rule and left-hand rule) of Fleming. The stator and rotor, along with the air gap, is shown in Fig. 13.7. Both the stator and rotor is shown here as surfaces, but without the slots. Also shown is the path of the flux in the air gap. This is for a section, which is under the North Pole, as the flux lines move from the stator to the rotor. The rotor conductor shown in the figure is at rest, i.e., zero speed (standstill). The rotating magnetic field moves past the conductor at synchronous speed in the clockwise direction. Thus, there is a relative movement between the flux and the rotor conductor. Now, if the magnetic field, which is rotating, is assumed to be at standstill as shown in Fig. 13.7(b), the conductor will move in the direction shown. Hence, an emf is induced in the rotor conductor according to Faraday's law, due to change in flux linkage. The direction of the induced emf as shown in the figure can be determined using Fleming's right-hand rule.

As described earlier, the rotor bars in the squirrel cage rotor are short-circuited via end-rings. Similarly, in the wound rotor, the rotor windings are normally short-circuited externally via slip-rings. In both cases, as emf is induced in the rotor conductor (bar), current flows there, as it is short-circuited. The flux in the air gap, due to the current in the rotor conductor is shown in Fig. 13.7(c). The flux pattern in the air gap due to the magnetic fields produced by the stator windings and the current carrying rotor conductor is shown in Fig. 13.7(d). The flux lines bend as shown there. The property of flux lines is to travel via the shortest path as shown in Fig. 13.7(a). If the flux lines try to move to form a straight line, then the rotor conductor has to move in the direction of the rotating magnetic field, but not at the same speed, as explained earlier. The current carrying rotor conductor and the direction of flux are shown in Fig. 13.7(e). It is known that force is produced on a conductor carrying current, when it is placed in a magnetic field. The direction of the force on the rotor conductor is obtained by using Fleming's left-hand rule—it is the same as that of the rotating magnetic field. Thus, the rotor experiences a motoring torque in the same direction as that of the rotating magnetic field.

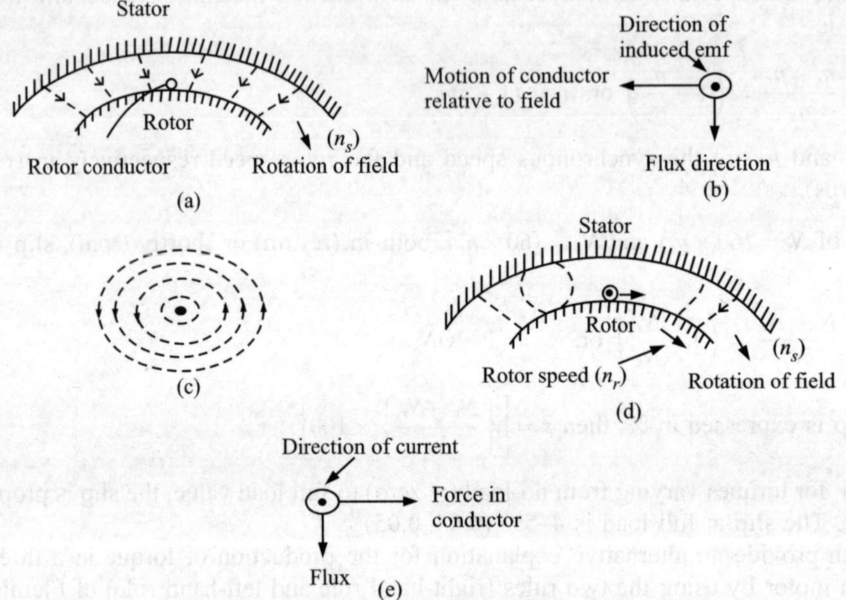

Fig. 13.7 *Production of torque in a three-phase induction motor*

FREQUENCY OF THE INDUCED EMF AND CURRENT IN THE ROTOR

As given earlier, both the induced emf and the current in the rotor are due to the relative velocity between the rotor conductors and the rotating flux in the air gap, the speed of which is the synchronous speed $\left(N_s = \dfrac{120 \times f}{p}\right)$. The rotor speed is

$$N_r = (1 - s)N_s$$

The frequency of the induced emf and current in the rotor is

$$f_r = p(n_s - n_r) = s(p.n_s) = s.f.$$

For normal values of slip, the above frequency is small. Taking an example, with full load slip as 4% (0.04) and supply (line) frequency as 50 Hz, the frequency (Hz) of the rotor induced emf and current is $f_r = (0.04 \times 50.0) = 2.0$, which is very small, whereas the frequency (f) of the stator induced emf and current is 50 Hz, i.e., line frequency. At standstill, i.e., when the rotor is stationary ($n_r = 0.0$), the rotor frequency is same as the line frequency with slip [$s = 1.0(100\%)$].

INDUCTION MOTOR (MACHINE) AS A GENERALIZED TRANSFORMER

The induction motor is considered analogous to the transformer, presented in Chapter 11. The currents in the balanced three-phase windings of the stator produce a resultant mmf (magneto-motive force), rotating at the synchronous speed n_s relative to the stator conductors, which are stationary. Similarly, the currents in the balanced three-phase windings of the rotor (taken as a wound one) produce another resultant mmf, rotating at the speed of $[(n_s - n_r) = s.n_s]$ relative to rotor conductors in the same direction as that of the rotor, which rotates at a speed, $[n_r = (1-s)n_s]$. Hence the speed of the resultant rotor mmf relative to the stator conductors is $[(s.n_s + n_r) = n_s]$. The axes of the resultant rotor mmf rotate at the same speed as those of the resultant stator mmf, so that the mmfs are stationary relative to each other. Consequently, the three-phase induction motor can be regarded as analogous to a transformer. The difference between the induction motor and the transformer is the presence of a large air gap in the induction motor (IM). Moreover, in IMs, the frequency in the rotor is different from that in the stator. Therefore, all induction motors can be termed as generalized transformers, where the voltage and frequency in the rotor is different from those in the stator.

13.4 Equivalent Circuit of a Three-phase Induction Motor

INDUCED EMF PER PHASE IN THE MOTOR WINDINGS

The induced emf per phase in the stator windings is nearly equal to the input voltage per phase, and is given by,

$$V_s \approx E_s = 2.22.k_{ws}.f.\phi.Z_s = 4.44.k_{ws}.f.\phi.T_s$$

where,

V_s =Input voltage (V) per phase to the stator winding
E_s =Induced emf (V) per phase in the stator winding
ϕ =Flux (Wb) per pole
f =Frequency (Hz or cycles per seconds) of the input voltage
Z_s =$2.T_s$ = No. of stator turns or coils in series per phase
k_{ws}=$k_{ds}.k_{ps}$ = Stator winding factor
k_{ds}=Distribution factor of stator winding
k_{ps}=Pitch factor of stator winding

The above formula, not derived here, is a standard one available in any text book. It is similar to the formula for the induced emf in a transformer, derived in the earlier chapter. As the winding in this case is stationary, with the rotating magnetic field rotating at a speed of

$$n_s = \frac{f}{\left(\frac{p}{2}\right)} = \frac{2.f}{p} \text{ (also termed as the synchronous speed),}$$

the frequency of the induced emf in the stator winding is the same as that of the input frequency

$\left[f = \left(\frac{p}{2}\right).n_s\right]$,

where,

n_s=Speed of the rotating magnetic field (rev./s or rps)
p =No. of poles, for which the stator winding of IM is designed

The induced emf per phase at standstill $\{(\omega_r = 0.0)$ or $(s = 1.0)\}$ in the rotor (wound or slip-ring one), is given by

$$E_r = 2.22.k_{wr}.f_r.\phi.Z_r = 4.44.k_{wr}.f_r.\phi.T_r$$

where,

E_r = Induced emf (V) per phase, in the rotor winding at standstill ($s = 1.0$)
$f_r = s.f$ = Frequency (Hz or cycles per seconds) in the rotor winding at standstill ($s = 1.0$)
$Z_r = 2.T_t$ = No. of rotor conductors in series per phase
$k_{wr} = k_{dr}.k_{pr}$ = Rotor winding factor
k_{dr} = Distribution factor of rotor winding
k_{pr} = Pitch factor of rotor winding

It may be noted that for the wound type (slip-ring) rotor, the number of poles in the rotor winding must be the same as those in the stator winding. The squirrel cage rotor can be considered as equivalent to a wound one.

The frequency of the induced emf in the rotor winding with the rotor rotating at any speed, is proportional to the relative speed of the speed of the rotating magnetic field and that of the rotor, and is given by

$$f_r = \left(\frac{p}{2}\right).(n_s - n_r) = \left(\frac{p}{2}\right).s.n_s = s.f,$$

where,

$n_r = (1 - s).n_s$ = Speed of the rotor (no. of revolutions per seconds)

$s = \frac{(n_s - n_r)}{n_s} = \left(1 - \frac{n_r}{n_s}\right)$ = Slip in IM (motor).

The ratios of the induced emfs per phase in the stator and rotor windings, with the rotor at standstill, is given by,

$$\frac{E_s}{E_r} \approx \frac{T_s}{T_r} = \frac{Z_s}{Z_r}.$$

Assume that the winding factors in both the windings are the same. It may be noted that the induced voltage per phase in the winding is proportional to the effective turns per phase of the winding ($= k_w.T$). The formula is similar to that obtained in a transformer, where the induced voltage per turn $\left(E_c = \frac{E}{T}\right)$ is the same in the two (primary/secondary) windings.

The induced emf in the rotor winding with the motor running at any speed is proportional to the rotor frequency ($f_r = s.f$), and given by

$$s.E_r = 2.22.k_{wr}.s.f.\phi.Z_r.$$

The current per phase in the rotor winding is

$$I_2 = \frac{sE_r}{\sqrt{r_2^2 + (sx_2)^2}}.$$

EQUIVALENT CIRCUIT

The equivalent circuit of the rotor circuit, per phase is shown in Fig. 13.8. The first circuit (i) has the input voltage to the rotor as $E_2 = s.E_r$. The current per phase in the rotor winding can be written as,

$$I_2 = \frac{s.E_r}{\sqrt{(r_2)^2 + (s.x_2)^2}} = \frac{E_r}{\sqrt{\left(\frac{r_2}{s}\right)^2 + (x_2)^2}},$$

dividing both the numerator and the denominator of the previous equation by slip (s). The input voltage to the rotor is now constant as E_r, being independent of slip, according to the second circuit (ii), which is equivalent to the first one. The total resistance changes to $\left(\frac{r_2}{s}\right)$, with the leakage reactance, x_2 remaining the same. If the rotor resistance, r_2 is subtracted from the previous value, the fictitious resistance is obtained as $\left(r_2.\frac{(1-s)}{s}\right)$. Only the rotor part is given here. The stator part can be added as given earlier for the transformer in Chapter 11.

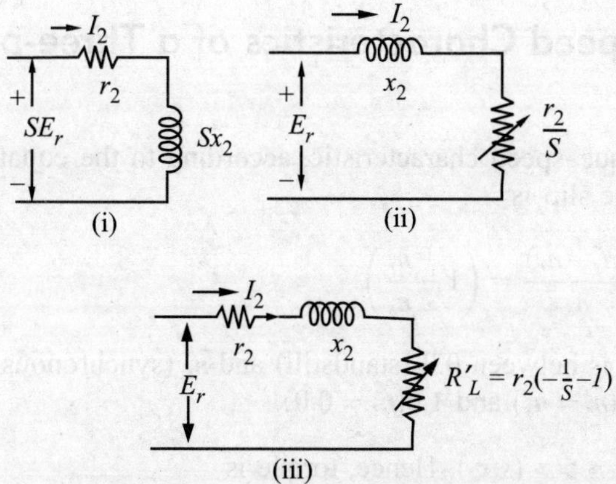

Fig. 13.8 *Equivalent circuit of the rotor per phase of the three-phase induction motor*

GROSS TORQUE DEVELOPED

The output power (gross) developed (Watt) is the loss in the fictitious resistance in the equivalent circuit (Fig. 13.8), which is,

$$P_0 = 3.(I_2)^2\left[r_2.\frac{(1-s)}{s}\right] = \frac{3.(s.E_r)^2.\left[r_2.\frac{(1-s)}{s}\right]}{[(r_2)^2+(s.x_2)^2]} = \frac{3.(E_r)^2 r_2 s(1-s)}{[(r_2)^2+(s.x_2)^2]}.$$

The motor speed in rps is $n_r = (1-s)n_s$.

The motor speed (angular) in radians/seconds is $\omega_r = (1-s)\omega_s$.

The gross torque developed in N.m is

$$T_0 = \frac{P_0}{\omega_r} = \frac{3.(E_r)^2.r_2.s.(1-s)}{(1-s)\omega_s[(r_2)^2+(s.x_2)^2]} = \frac{3.(E_r)^2.r_r.s}{2\pi n_s[(r_2)^2+(s.x_2)^2]}.$$

The synchronous speed (angular) is $\omega_s = 2\pi n_s$.

The input power to the rotor (or the power transferred from the stator via air gap) is the loss in the total resistance $\frac{r_2}{s}$ which is

$$P_i = 3.(I_2)^2\left(\frac{r_2}{s}\right) = \frac{3.(s.E_r)^2.[.\frac{r_2}{s}]}{[(r_2)^2+(s.x_2)^2]} = \frac{3.(E_r)^2 r_2 s}{[(r_2)^2+(s.x_2)^2]}$$

The relationship between the input power and the gross torque developed is $P_i = T_0.\omega_s$. Hence, the input power is also called the torque in synchronous watts, or the torque is

$$T_0 = \frac{P_0}{\omega_s}$$

13.5 Torque–Speed Characteristics of a Three-phase Induction Motor

The torque–slip or torque–speed characteristic, according to the equation derived earlier, is shown in Fig. 13.9. The slip is

$$s = \frac{(\omega_s - \omega_r)}{\omega_s} = \frac{(n_s - n_r)}{n_s} = \left(1 - \frac{n_r}{n_s}\right).$$

The range of speed, n_r is between 0.0 (standstill) and n_s (synchronous speed). The range of the slip is between $0.0(n_r = n_s)$ and $1.0(n_r = 0.0)$.

For low values of slip, $r_2 >> (s.x_2)$. Hence, torque is

$$T_0 = \frac{3.(E_r)^2.r_2..s}{\omega_s.(r_2)^2} = \frac{3.(E_r)^2}{\omega_s}.\frac{s}{r_2}.$$

This shows that $T_0 \propto s$, the characteristic being linear. The following points may be noted. The output torque developed is zero (0.0) at $s = 0.0$, if the motor is rotated at synchronous speed ($n_s = n_r$). Moreover, the slip at full load (output torque $= (T_0)_{fl}$) is normally 4–5% ($s_{fl} = 0.04 - 0.05$) the speed of the induction motor at full load, being 95–96% of the synchronous speed

$$(n_r)_{fl} = (1 - s_{fl})n_s = (0.95 - 0.96)n_s.$$

For large values of slip, $r_2 \ll (s.x_2)$. Hence torque is

$$T_0 = \frac{3.(E_r)^2.r_2.s}{\omega_s.(s.x_2)^2} = \frac{3.(E_r)^2}{\omega_s} \frac{r_2}{s(x_2)^2}.$$

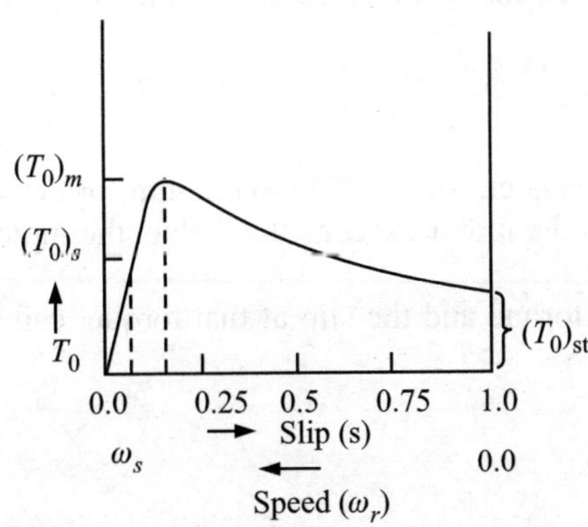

Fig. 13.9 *Torque–slip (speed) characteristic of a three-phase induction motor*

This shows that $T_0 \propto \frac{1}{s}$, the characteristic being hyperbolic. The starting torque ($s = 1.0$, or, $n_r = 0.0$) developed along with starting current, is discussed later.

Hence, starting from a low value of slip ($s > 0.0$), when torque is proportional to slip, torque becomes inversely proportional to slip for large values of slip ($s < 1.0$), both being derived earlier. In the characteristic shown, it may be observed that torque reaches a maximum value, which can be obtained in the following way. The relation between torque and slip is

$$T_0 = \frac{K.r_2.s}{[(r_2)^2 + (s.x_2)^2]}, \text{ where, } K = \frac{3.(E_r)^2}{\omega_s}$$

$$\text{or, } \frac{1}{T_0} = \frac{[(r_2)^2 + (s.x_2)^2]}{K.r_2.s} = \frac{1}{K}\left(\frac{r_2}{s} + \frac{s.(x_2)^2}{r_2}\right).$$

To determine the maximum value of torque (T_0) in terms of slip, the minimum value of its inverse $\left(\frac{1}{T_0}\right)$ need be determined from the relation,

$$\frac{d}{ds}\left(\frac{1}{T_0}\right)=\frac{1}{K}\left(-\frac{r_2}{s^2}+\frac{(x_2)^2}{r_2}\right)=0$$

from which, $s^2=\left(\frac{r_2}{x_2}\right)^2$ or, $s=\frac{r_2}{x_2}$.

Please note that for motoring conditions as shown earlier, slip, s is positive (+ve), as ($n_r < n_s$). At this slip, $(s=s_m)$, $r_2=s_m.x_2$. This may be termed as slip at maximum torque. The motor speed is $[(n_r)_m=(1-s_m).n_s]$. This value of slip is small for normal wound rotor (or slip-ring) induction motors, without any additional resistance inserted in the rotor circuit. The value is higher in the case of a squirrel cage induction motor.

Substituting the value of s, the maximum value of torque is

$$(T_0)_m=\frac{K}{2.x_2}=\frac{3.(E_r)^2}{\omega_s}\frac{1}{2.x_2},$$

which shows that it is independent of r_2 . The maximum torque is also termed as pull-out torque. If the load torque on the motor exceeds this value, the motor will stall, i.e., will come to a standstill condition.

The values of maximum torque and the slip at that torque, can also be obtained by using

$$\frac{d}{ds}(T_0)=0.0,$$

which is not shown here.

The ratio of torque at any slip to the maximum torque is given by,

$$\frac{T_0}{(T_0)_m}=\frac{(r_2.s).(2.x_2)}{[(r_2)^2+(s.x_2)^2]}.$$

Dividing both the numerator and denominator by $(x_2)^2$, and then substituting $s_m=\frac{r_2}{x_2}$, the above ratio is obtained as,

$$\frac{T_0}{(T_0)_m}=\frac{2.s.\left(\frac{r_2}{x_2}\right)}{\left[\left(\frac{r_2}{x_2}\right)^2+s^2\right]}=\frac{2.s.s_m}{[(s_m)^2+s^2]}=\frac{2}{\left(\frac{s}{s_m}+\frac{s_m}{s}\right)}.$$

It may be observed from the torque–slip characteristic (Fig. 13.9), that the output torque developed increases if the slip increases from 0.0 to s_m, or the motor speed decreases from n_s to $(n_r)_m$. This ensures stable operation of the induction motor in this region $(0.0 < s < s_m)$, for constant load torque. However, the output torque developed decreases, if the slip increases from s_m to 1.0, or the motor speed decreases from $(n_r)_m$ to 0.0. This results in the unstable operation of the induction machine in the region $(s_m < s < 1.0)$, for constant load torque. However, for fan type loads with the torque as $(T_L \propto (n_r)^2)$, stable operation of the induction motor is achieved in this region $(s_m < s < 1.0)$.

STARTING CURRENT AND TORQUE

The starting current (rotor) is,

$$(I_2)_{st} = \frac{E_r}{\sqrt{(r_2)^2 + (x_2)^2}},$$

as slip at starting ($n_r = 0.0$) is 1.0, which is the same at standstill (or stalling condition). The magnitude of the induced voltage per phase in the stator winding is nearly the same as the input voltage per phase fed to the stator if the voltage drop in the stator impedance, being small, is neglected, i.e., $V_s \approx E_s$. As shown before, the ratio of the induced emfs per phase in the stator and rotor winding can be taken as the ratio of the effective turns in the two windings, i.e.,

$$\frac{E_s}{E_r} = \frac{T'_s}{T'_r}$$

where $T'_s = k_{ws}.T_s$ and $T'_r = k_{ws}.T_r$. The winding factor for the stator winding is $k_{ws} = k_{ds}.k_{ps}$. The same formula is used for the above factor in the rotor winding, assuming it to be wound rotor machine.

The starting current in the stator winding can be shown to be $(I_s)_{st} = (I_r)_{st}.\frac{T'_r}{T'_s}$, neglecting the no-load current. This current is normally large, much greater than full load current. It is reduced by using starters in both types (squirrel cage and wound rotor) of the induction motor, which will be taken up in Section 13.6.

The starting torque in N.m is

$$(T_0)_{st} = 3.((I_2)_{st})^2.\frac{r_2}{\omega_s} = \frac{3.(E_r)^2.r_2}{\omega_s[(r_2)^2 + (x_2)^2]}$$

This expression is obtained by substituting $s = 1.0$ in the expression of T_0 derived earlier. If the starter is used, the starting torque is also reduced, as is the case with starting current.

The set of torque–slip characteristics with variation in input (stator) voltage is shown in Fig. 13.10(a). The point to note is that the torque at a given slip decreases with the decrease in input (stator) voltage, as $T_0 \propto V^2$. The characteristics shown are for decreasing stator voltage ($V_1 > V_2 > V_3$). The speed decreases or the slip increases with constant load torque, as the input (stator) voltage decreases. The region for stable operation with constant load torque remains the same ($0.0 < s < s_m$), as given earlier. Stable operation can be obtained in the region ($s_m < s < 1.0$), with fan type loads with the torque as ($T_L \propto (n_r)^2$). One disadvantage is that the maximum or pull-out torque decreases as ($(T_0)_m \propto (n_r)^2$), which is a drawback with constant load torque operation.

The set of torque–slip characteristics with variation in rotor circuit resistance is shown in Fig. 13.10(b). The characteristics shown are for increasing rotor circuit resistance ($r_2 > R_2 > R_3 > R_4$). The point to note is that the maximum torque remains the same for all the characteristics. It has been shown earlier that the maximum torque depends on rotor reactance only, not on rotor circuit resistance. Only the slip at maximum torque increases with the increase in rotor circuit resistance. Hence for constant load torque operation, the slip increases or the speed decreases with the increase in rotor circuit resistance. Motor efficiency decreases, as the rotor copper loss increases with the increase in slip. The load torque remains the same, but the output power decreases, as the speed decreases. Also, it may be observed that the starting torque increases with the increase in rotor circuit resistance, with the total rotor circuit resistance lower than the rotor reactance. The starting torque is equal to the maximum torque, when the total rotor circuit resistance is equal to the rotor reactance. If the rotor circuit resistance is more than the rotor reactance, the starting torque decreases.

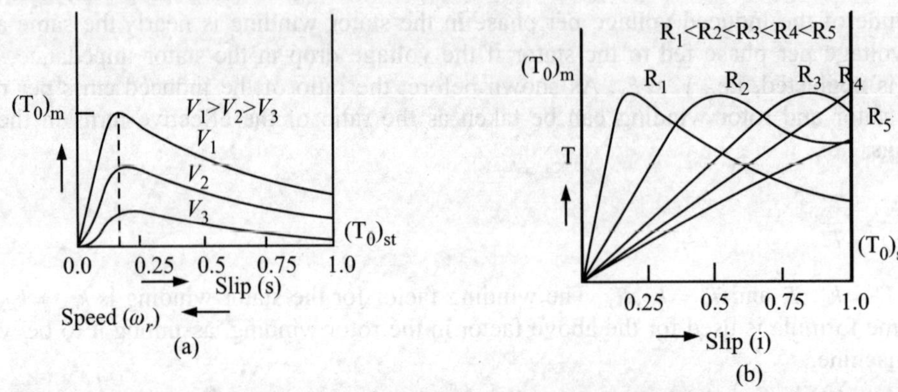

Fig. 13.10 *Torque–speed characteristics for variation in (a) input (stator) voltage, and (b) total rotor circuit resistance*

EXAMPLE 13.1 A 50 Hz, 8-pole, three-phase induction motor has a full load slip of 4%. The rotor resistance and standstill reactance are 0.01 ohm and 0.1 ohm, per phase respectively. Find the ratio of the maximum torque to the full load torque and the speed at which it occurs.

SOLUTION: Frequency, $f = 50$ Hz; number of poles, $p = 8$; $r_2 = 0.01\ \Omega$; $x_2 = 0.1\ \Omega$.

Synchronous speed is, $N_s = \dfrac{120 \times f}{p} = \dfrac{120 \times 50}{8} = 750$ rpm.

Slip at maximum torque is $s_m = \dfrac{r_2}{x_2} = \dfrac{0.01}{0.1} = 0.1 = 10\%$.

Speed under the above condition is $N_m = (1 - s_m).N_s = (1 - 0.1) \times 750 = 675$ rpm.

As given earlier, the two relationships are, $r_2 = s_m.x_2$, and $x_2 = \dfrac{r_2}{s_m}$.

Slip at full load is, $s_1 = s_{fl} = 0.04(4\%)$.

$$\frac{r_2}{s_1} = \frac{0.01}{0.04} = 0.25.$$

Maximum torque is, $T_m = \dfrac{K}{2x_2} = \dfrac{K}{2 \times 0.1} = 5.0 \times K$

Torque at full load is,

$$T_{fl} = K \frac{\left(\frac{r_2}{s_{fl}}\right)}{\left[\left(\frac{r_2}{s_{fl}}\right)^2 + (x_2)^2\right]}$$

or, $$T_{fl} = K.\frac{0.25}{[(0.25)^2 + (0.1)^2]} = (3.4483 \times K).$$

The ratio of maximum torque to full load torque is, $\frac{T_m}{T_{fl}} = \frac{5.0 \times K}{3.4483 \times K} = 1.45.$

Alternatively, it can be derived from the expressions given earlier that

$$\frac{T_m}{T_{fl}} = \frac{\frac{s_m}{s_{fl}} + \frac{s_{fl}}{s_m}}{2} = \frac{\left(\frac{0.1}{0.04}\right) + \left(\frac{0.04}{0.1}\right)}{2} = 1.45$$

EXAMPLE 13.2 A 4-pole, 50 Hz, three-phase, slip-ring (wound rotor) induction motor has a rotor resistance of 0.24 Ω per phase, and runs at 1,455 rpm at full load. The motor develops maximum torque at a speed of 1,275 rpm. Calculate the external resistance that must be added in series with each phase of the rotor to decrease the speed to 1,125 rpm, with torque being kept constant.

SOLUTION Frequency $= f = 50$ Hz ; No. of poles, $p = 4$; $r_2 = 0.24\,\Omega$

Synchronous speed is, $N_s = \frac{120 \times f}{p} = \frac{120 \times 50}{4} = 1{,}500$ rpm.

Speed at full load is, $N_{r_1} = N_{fl} = 1{,}455$ rpm.

Slip at full load is, $s_1 = s_{fl} = \left(\frac{N_s - N_{r1}}{N_s}\right) = \left(\frac{1{,}500 - 1{,}455}{1{,}500}\right) = 0.03(3\%).$

$$\frac{r_2}{s_1} = \frac{0.24}{0.03} = 8.0$$

Speed at maximum torque, is, $N_m = 1{,}275$ rpm.

Slip at maximum torque, is, $s_m = \left(\frac{N_s - N_m}{N_s}\right) = \left(\frac{1{,}500 - 1{,}275}{1{,}500}\right) = 0.15(15\%)$

As $s_m = \frac{r_2}{x_2}$,

Standstill reactance of the rotor per phase is, $x_2 = \frac{r_2}{s_m} = \frac{0.24}{0.15} = 1.6\,\Omega.$

Speed $= N_{r_2} = 1{,}125$ rpm

$$\text{Slip} = s_2 = \left(\frac{N_s - N_{r2}}{N_s}\right) = \left(\frac{1,500 - 1,125}{1,500}\right) = 0.25(25\%).$$

$$T_1 = T_{\text{fl}} = K.\frac{\left(\frac{r_2}{s_1}\right)}{\left[\left(\frac{r_2}{s_1}\right)^2 + (x_2^2)\right]} = K.\frac{8.0}{[(8.0)^2 + (1.6)^2]} = (0.1202 \times K).$$

Let R be the external resistance per phase to be added in series with each phase of the rotor.

$$T_2 = T_{\text{fl}} = K.\frac{\left[\frac{(r_2+R)}{s_2}\right]}{\left[\left\{\frac{(r_2+R)}{s_2}\right\}^2 + (x_2)^2\right]} = (0.1202 \times K)$$

Assuming, $y = (r_2 + R) = (0.24 + R)$, from the expression given earlier,

$$\frac{y}{[(y)^2 + (1.6)^2]} = 0.1202$$

or, $\quad (y^2 + 2.56) = \left(\frac{1.0}{0.1202} \times y\right) = (8.32 \times y)$

or, $\quad y^2 - (8.32 \times y) + 2.56 = 0$

or $\quad y = 4.16 \pm \sqrt{(4.16)^2 - 2.56} = 8.0 \quad \text{or} \quad 0.32.$

The second value is rejected as $y_2 = \dfrac{(0.24+R)}{0.25} = 0.32$, from which the value of R is found as $R = -\,0.16\ \Omega$ (negative).

As $y_1 = \dfrac{(0.24+R)}{0.25} = 8.0$, the external resistance to be added in the rotor circuit, R is obtained as

$$R = (2.0 - 0.24) = 1.76\ \Omega.$$

Alternatively, the developed torque (gross) can be rewritten as,

$$T_1 = T_{fl} = \frac{K(s_1.r_2)}{[(r_2)^2 + (s_1.x_2)^2]}$$

$$= \frac{K.\left(\frac{r_2}{s_1}\right)}{\left[\left(\frac{r_2}{s_1}\right)^2 + (x_2)^2\right]}$$

From this expression, it can be easily proved that if the developed torque remains constant, when external resistance, R is added in the rotor circuit, then,

$$\frac{r_2}{s_1} = \left[\frac{(r_2+R)}{s_2}\right]$$

Substituting the values given earlier, we get,

$$\frac{0.24}{0.03} = \frac{(0.24+R)}{0.25} = 8.0,$$

from which the external resistance to be added is found as $R = (2.0 - 0.24) = 1.76\ \Omega$.

POWER FLOW DIAGRAM

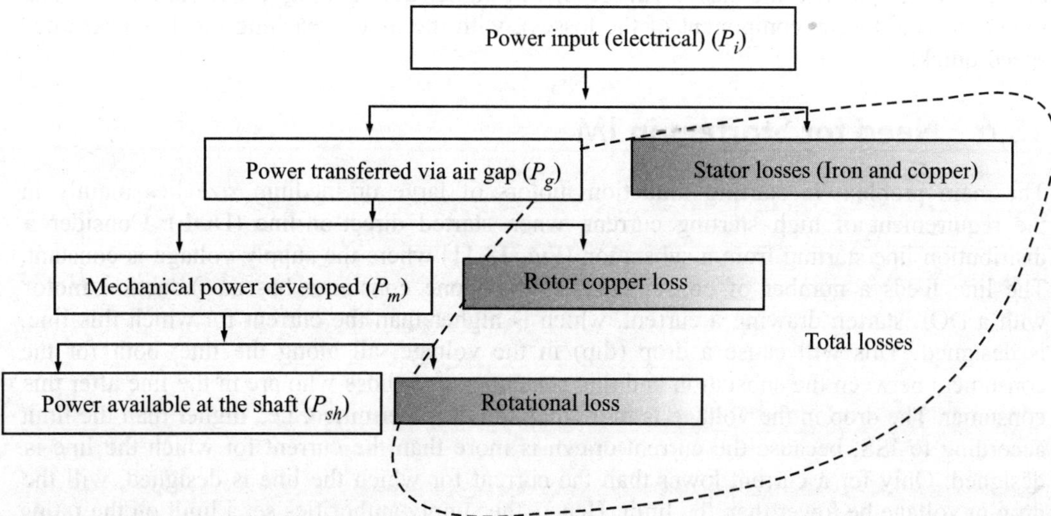

Direct-On-Line (DOL) starters

Induction motors can be started direct-on-line (DOL), which means that the rated voltage is supplied to the stator, with the rotor terminals short-circuited in a wound rotor (slip-ring) motor. For the squirrel cage rotor, the rotor bars are short-circuited via two end-rings. Neglecting stator impedance, the starting current in the stator windings is

$$(I_1)_{\rm st} = \frac{E_r'}{\sqrt{(r_2')^2 + (x_2')^2}}$$

where,

$$(I_1)_{st} = (I_2')_{st} = \frac{(I_2)_{st}}{a} = \text{Starting current in the motor (stator)}$$

$$a = \frac{T_s'}{T_r'} = \text{Effective turns ratio between stator and rotor windings}$$

$E_s = E'_r = aE_r$ = Input voltage per phase to the motor (stator)

E_r = Induced emf per phase in the rotor winding

$r'_2 = ar_2$ = Rotor resistance in terms of stator winding

$x'_2 = ax_2$ = Rotor reactance at standstill ($s = 1.0$, or $n_r = 0$) in terms of stator winding

The input voltage per phase to the stator is equal to the induced emf per phase in the stator winding, as the stator impedance is neglected (also shown in Section 13.5).

In the formula for starting current, no-load current is neglected. It may be noted that the starting current is quite high, about 4–6 times the current at full load. It may be higher, depending on the rating of IM, as compared to the no-load current.

The starting torque is $\{(T_0)_{st} \propto (I_1)^2_{st}\}$, which shows that as the starting current increases, the starting torque also increases. This results in higher accelerating torque (minus the load torque and the torque component of the losses), with the motor reaching rated or near rated speed quickly.

13.6 Need for Starters in IM

The main problem in starting induction motors of large or medium size lies mainly in the requirement of high starting current, when started direct-on-line (DOL). Consider a distribution line starting from a substation (Fig. 13.11) where the supply voltage is constant. The line feeds a number of consumers, of which one consumer has an induction motor with a DOL starter, drawing a current, which is higher than the current for which this line is designed. This will cause a drop (dip) in the voltage, all along the line, both for the consumers between the substation and this consumer, and those who are in the line after this consumer. The drop in the voltage is more than the drop permitted, i.e., higher than the limit according to ISS, because the current drawn is more than the current for which the line is designed. Only for a current lower than the current for which the line is designed, will the drop in voltage be lower than the limit. Hence, the supply authorities set a limit on the rating or size of IM which can be started DOL. Any motor exceeding the specified rating is not permitted to be started DOL—for these motors, a starter is to be used to reduce the current drawn at starting.

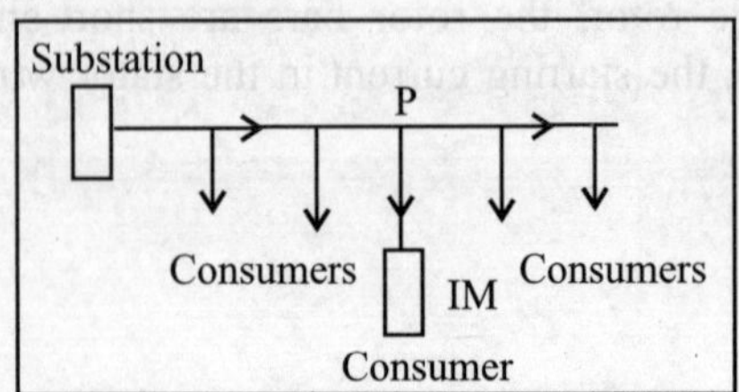

Fig. 13.11 *Distribution line fed from a substation for supply to various consumers*

13.6.1 Starters for Squirrel Cage IM

The starting current in IM is proportional to the input voltage per phase (V_s) to the motor (stator), i.e., $(I_1)_{st} \propto E_s$, where, $|V_s| \approx |E_s|$, as the voltage drop in the stator impedance is small compared to the input voltage, or $|V_s| = |E_s|$, if the stator impedance is neglected. This has been shown earlier. Hence, in a squirrel cage induction motor, the starter is used only to decrease the input voltage to the motor so as to decrease the starting current. As described later, this also results in decrease of starting torque.

STAR–DELTA STARTER

The voltage supplied to the stator winding, when it is connected in star, instead of delta (Fig. 13.12(a)), is reduced by ($\frac{1}{\sqrt{3}} = 0.577$) of rated voltage. This is a simple starter which can be easily reconfigured as shown in Fig. 13.12(b). As the voltage per phase in delta connection is V_s, the phase current in each stator winding is $\left(\frac{V_s}{Z_s}\right)$, where Z_s is the impedance of the motor per phase at standstill or start (stator impedance and rotor impedance referred to the stator at standstill). The line current or the input current to the motor is

$$\left[(I_1)_{st} = \frac{(\sqrt{3}.V_s)}{Z_s}\right],$$

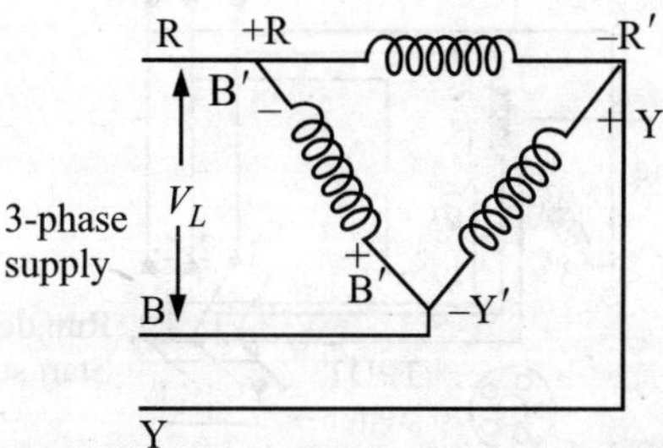

Fig. 13.12(a) *Delta connected stator winding of three-phase IM at run*

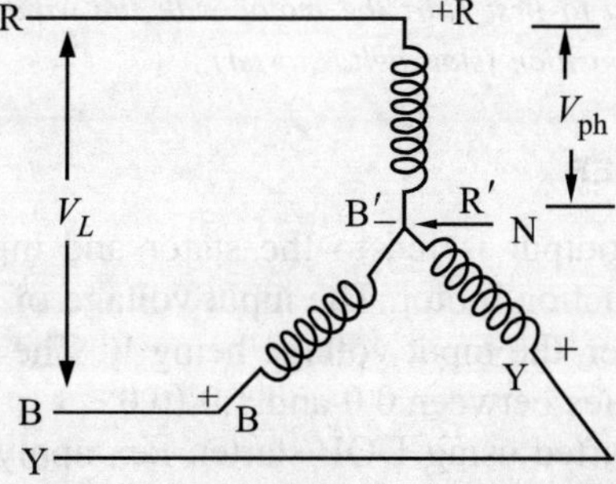

Fig. 13.12(b) *Same winding of three-phase IM, reconnected as star at start*

which is the current if the motor is started direct-on-line (DOL). Now, if the stator winding is connected as star, the phase or line current drawn from supply at start (standstill) is $\left[\frac{(V_s/Z_s)}{\sqrt{3}}\right]$, which is $\left[\frac{1}{3}=\left(\frac{1}{\sqrt{3}}\right)^2\right]$ of the starting current, if DOL starter is used. The voltage per phase in each stator winding is now $\left(\frac{V_s}{\sqrt{3}}\right)$. Hence, the starting current using a star–delta starter is reduced by 33.3%. As for starting torque, being proportional to the square of the current in each of the stator windings in two different connections as shown earlier, it is also reduced by $\left[\frac{1}{3}=\left(\frac{1}{\sqrt{3}}\right)^2\right]$, as the ratio of the two currents is $\left(\frac{1}{\sqrt{3}}\right)$, same as that (ratio) of the voltages applied to each winding as shown earlier. Hence, the starting torque is reduced by 33.33% which is a disadvantage for this starter. The load torque and the loss torque, must be lower than the starting torque if the motor is to be started using this starter. The advantage is that no extra component, except that shown in Fig.13.12(c), need be used, thus, making it simple. As shown later, this is an auto-transformer starter with voltage ratio as 57.7%. Alternatively, the starting current in the second case with the stator winding reconnected as star, can be found by using star–delta conversion, with the impedance per phase after converting to delta, found as $(3.Z_s)$ and the starting current now being reduced to $\left(\frac{1}{3}\right)$ of the starting current, obtained using DOL starter, with the stator winding connected in delta.

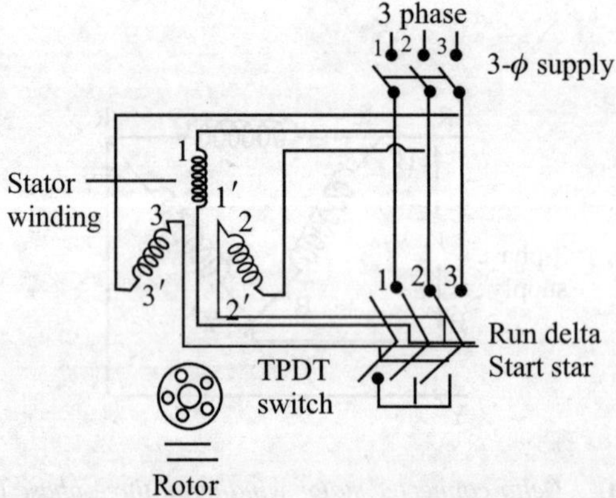

Fig. 13.12(c) *TPDT switch used to first start the motor with the windings connected in star and then switch to delta connection in run position (star–delta starter)*

AUTO-TRANSFORMER STARTER

An auto-transformer, whose output is fed to the stator and input is from the supply (Fig. 13.13), is used to start an induction motor. The input voltage of IM is V_s, which is the output voltage of the auto-transformer, the input voltage being V_s. The output voltage/input voltage ratio is x, the value of which lies between 0.0 and 1.0 $(0.0 < x < 1.0)$. Let $(I_1)_{st}$ be the starting current, when the motor is started using DOL starter, i.e., applying rated input voltage. The input current of IM, which is the output current of the auto-transformer is $x.(I_1)_{st}$, when this starter is used with input voltage as $x.V_s$. The input current of the auto-transformer, which is the starting current drawn from the supply is $x^2.(I_1)_{st}$, obtained by equating the input and output volt–amperes, neglecting losses and assuming nearly the same power factor on both sides. As discussed earlier, the starting torque, being proportional to the square of the input currents in the IM (with and without auto-transformer i.e., direct) is also reduced by x^2, as the ratio of the two currents is x, same as that (ratio) of the voltages applied to the motor. Hence, the starting torque is reduced by the same ratio as that of the starting current. If the ratio is $x = 0.8(80\%)$, both the starting current and torque are $x^2 = (0.8)^2 = 0.64(64\%)$ times the values of starting current and torque with DOL starting, which is nearly 2 times the values obtained using star–delta starter. Hence, the disadvantage is that the starting current is increased, with the result that the lower rated motor can now be started, as the current drawn from the supply is to be kept within limits, while the advantage is that the starting torque is now doubled, such that the motor can start against higher load torque. The star–delta starter can be considered equivalent to an auto-transformer starter with the ratio, $x = 0.577(57.7\%)$. If $x = 0.7(70\%)$, $x^2 = (0.7)^2 = 0.49 \approx 0.5(50\%)$, both starting current and torque are $x^2 = (0.7)^2 = 0.49 \approx 0.5(50\%)$ times the values of starting current and torque with DOL starting, which is nearly 1.5 times the values obtained using a star–delta starter. By varying the value of the voltage ratio x of the auto-transformer, the values of the starting current and torque can be changed. However, additional cost of the auto-transformer with intermittent rating is to be incurred for this purpose.

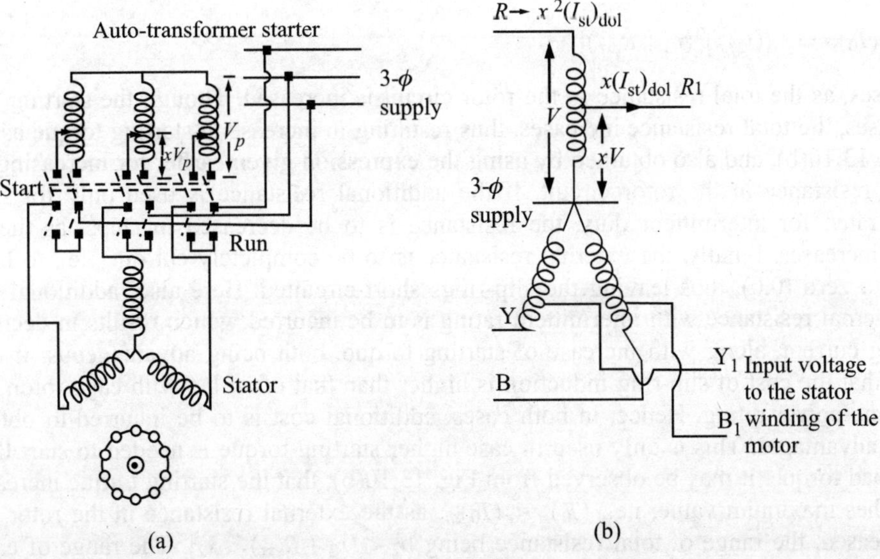

Fig. 13.13 *Auto-transformer starter for a three-phase IM*

13.6.2 Rotor Resistance Starters for Slip-ring (Wound Rotor) IM

In a slip-ring (wound rotor) induction motor, resistance can be inserted in the rotor circuit via slip-rings (Fig. 13.14), in order to increase the starting torque. The starting current in the rotor winding is

$$(I_2)_{st} = \frac{E_r}{\sqrt{(r_2 + R_{ext})^2 + (x_2)^2}},$$

where, R_{ext} = additional resistance per phase in the rotor circuit.

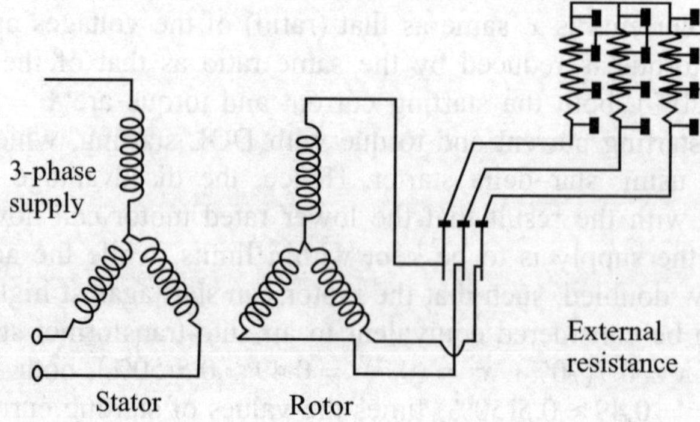

Fig. 13.14 *Rotor resistance starter for a three-phase wound rotor IM*

The input (stator) current is proportional to the rotor current as shown earlier. The starting current (input) reduces, as resistance is inserted in the rotor circuit. However, the starting torque,

$$[(T_0)_{st} = 3.[(I_2)_{st}]^2(r_2 + R_{ext})]$$

increases, as the total resistance in the rotor circuit is increased. Though the starting current decreases, the total resistance increases, thus resulting in increase in starting torque as shown in Fig. 13.10(b), and also obtained by using the expression given earlier for increasing values of the resistance in the rotor circuit. If the additional resistance is used only for starting, being rated for intermittent duty, the resistance is to be decreased in steps, as the motor speed increases. Finally, the external resistance is to be completely cut out, i.e., to be made equal to zero (0.0), thus leaving the slip-rings short-circuited. Here also, additional cost of the external resistance with intermittent rating is to be incurred, which results in decrease of starting current, along with increase of starting torque, both being advantageous. It may be noted that the cost of slip-ring induction is higher than that of an IM with cage rotor, having the same power rating. Hence, in both cases, additional cost is to be incurred to obtain the above advantages. This is only used in case higher starting torque is needed to start IM with high load torque. It may be observed from Fig. 13.10(b), that the starting torque increases till it reaches maximum value, i.e., $(T_0)_{st} < (T_0)_m$, as the external resistance in the rotor circuit is increased, the range of total resistance being $[r_2 < (r_2 + R_{ext}) < x_2]$. The range of external resistance is in between zero (0.0) and $(x_2 - r_2)$. The starting torque is equal to the maximum value, i.e., $(T_0)_{st} = (T_0)_m$, if the external resistance inserted is equal to $(x_2 - r_2)$. However, if the external resistance in the rotor circuit is increased further, i.e.,

$$[R_{ext} > (x_2 - r_2)],$$

the starting torque decreases $(T_0)_{st} < (T_0)_m$. This is because the starting current decreases at a faster rate, even if the total resistance in the rotor circuit is increased.

13.7 Methods of Speed Control for a Three-phase IM

The speed control methods used for IM are presented in brief in this section. The motor speed is,

$$n_r = \frac{120f}{p}.(1-s),$$

where, synchronous speed is $\left(n_s = \frac{120f}{p}\right)$.

The methods used are control of input voltage, rotor resistance, and frequency. To control input voltage, the input (stator) voltage (V_s) is decreased from the rated value, resulting in the decrease of output (developed) torque, as $T_0 \propto (V_s)^2$, given in Section 13.5. Hence, the motor speed decreases. If the load torque is assumed as constant, the load torque is now near the value of pull-out (maximum) torque as $(T_0)_m \propto (V_s)^2$, which may make the motor unstable. Also, the range of speed control is low under this condition. This is the main demerit. However, for fan-type loads, i.e., compressors, pumps etc, where the load torque varies as square of the speed, both output torque and load torque decrease, resulting in stable operation, with wider range of speed control.

To control rotor circuit resistance, additional resistance is inserted at the slip-rings of the wound rotor IM, thus increasing the rotor circuit resistance. The motor speed decreases at constant load torque. This results in the increase of rotor copper loss (see Section 13.5), thus decreasing the efficiency. Also, the cost of additional resistance increases, as it must be rated as continuous here, not intermittent for rotor resistance starters. These are the disadvantages of this method.

For frequency control, synchronous speed changes according to the formula given, with the number of poles being the same. For speeds below base (rated) speed, constant voltage/frequency control is used such that the flux $\left(\phi \propto \frac{V}{f}\right)$ remains constant in IM, as given in Section 13.4. For speeds higher than base (rated) speed, variable frequency higher than line frequency is used, but not $\left(\frac{V}{f}\right)$ control as used earlier. The input (stator) voltage is kept as constant, resulting in reduced flux as $\left(\phi \propto \frac{V}{f}\right)$. If the earlier type of control was used, the input voltage would have to be increased, which may cause damage to the motor. With the availability of solid state devices, such as transistors, and associated control circuitry, inverters using these devices are being developed, which are suitable for driving motors of medium size. As frequency can be changed, along with $\left(\frac{V}{f}\right)$ control in these inverters, they are now being increasingly used.

EXAMPLE 13.3 *A three-phase, 8-pole, squirrel cage induction motor, connected to 400 V, 50 Hz supply, rotates at 4% slip at full load. The total stator loss is 2 kW. If the motor takes 40 kW at full load, find the developed torque (gross) under that condition.*

SOLUTION $f = 50$ Hz, $p = 8$.

Synchronous speed $= N_s = \dfrac{120f}{p} = \dfrac{120 \times 50}{8} = 750$ rpm.

Angular speed $= \omega_s = \dfrac{2\pi}{60} . N_5 = \dfrac{2\pi \times 750}{60} = 78.54$ rad/s.

The input power to the rotor at full load = air gap power

= input power to the motor – total stator loss $= P_{ag} = 40 - 2 = 38.0$ kW.

The developed torque (gross) $= T_0 = \dfrac{P_{ag}}{\omega_s} = \dfrac{38{,}000}{78.54} = 484$ N.m.

EXAMPLE 13.4 *The rotor of a three-phase, 8-pole, slip-ring (wound rotor) induction motor has a resistance of 0.25 Ω per phase, and it runs at 720 rpm at full load. How much external resistance/phase must be added to the rotor circuit to reduce the speed to 600 rpm, the torque remaining constant?*

SOLUTION $f = 50$ Hz, $p = 8$.

Synchronous speed $= N_s = \dfrac{120f}{p} = \dfrac{120 \times 50}{8} = 750$ rpm.

Motor speed at full load $= N_r = 720$ rpm.

Slip at full load $= s_{fl} = \dfrac{(N_s - N_r)}{N_s} = \dfrac{(750 - 720)}{750} = 0.04 = 4\%$.

$r_2 = 0.25\ \Omega$.

Reduced motor speed $= N_{r1} = 600$ rpm.

Slip at this speed $= s_1 = \dfrac{N_s - N_{r1}}{N_s} = \dfrac{(750 - 600)}{750} = 0.2 = 20\%$.

If R is the additional resistance per phase in a rotor circuit, it can be shown that if the developed torque remains constant, the condition to be satisfied is given by

$$\frac{r_2}{s_{fl}} = \frac{(r_2 + R)}{s_1}.$$

Substituting the above values, we get, $\dfrac{0.25}{0.04} = \dfrac{(0.25 + R)}{0.2} = 6.25$, from which the external resistance to be added is found as $R = (1.25 - 0.25) = 1.0\ \Omega$.

EXAMPLE 13.5 *A 10-pole, 50 Hz, 3,300 V, three-phase, star-connected induction motor has a slip-ring (wound) rotor of resistance and standstill reactance of 0.025 Ω and 0.28 Ω, per phase, respectively. The motor runs at a speed of 582 rpm when full load torque is applied. Calculate (a) slip at maximum torque, and (b) the ratio of maximum torque to full load torque. Neglect stator impedance.*

SOLUTION $f = 50$ Hz, $p = 10$.

Synchronous speed $= N_s = \dfrac{120f}{p} = \dfrac{120 \times 50}{10} = 600$ rpm.

Motor speed at full load $= N_r = 576$ rpm.

Slip at full load $s_{fl} = \dfrac{N_s - N_r}{N_s} = \dfrac{600 - 582}{600} = 0.03 = 3\%$

$$r_2 = 0.025\,\Omega, x_2 = 0.28\,\Omega$$

(a) Slip at maximum torque $= s_m = \dfrac{r_2}{x_2} = \dfrac{0.025}{0.28} = 0.0893 = 8.93\%$.

(b) The ratio of full load torque to the maximum torque

$$= \frac{(T_0)_{fl}}{(T_0)_m} = \frac{2}{\left(\dfrac{s_{fl}}{s_m} + \dfrac{s_m}{s_{fl}}\right)} = \frac{2}{\left(\dfrac{0.03}{0.0893} + \dfrac{0.0893}{0.03}\right)} = 0.604 = 60.4\%.$$

EXAMPLE 13.6 *A 3-phase induction motor is operated by a 440 V, 50 Hz system and is taking 80 kW from the supply system. If the stator loss is 10 kW, estimate the following:*

(i) Rotor copper loss at 4% slip

(ii) Efficiency of the motor (neglecting the iron loss).

SOLUTION Power input to the stator = 80 kW, slip = 4% = 0.04.

Thus, input to rotor = power input – stator loss = 80 – 10 = 70 kW.

As we know, rotor copper loss = slip × input to rotor = $0.04 \times 70 = 2.8$ kW.

Total mechanical power developed = rotor output [Neglecting iron loss]

= 70 – 2.8 = 67.2 kW

Hence efficiency $= \dfrac{\text{Total mechanical power developed}}{\text{Input to the motor}} \times 100\% = \dfrac{67.2}{80} \times 100\% = 84\%$.

Conclusion

In this chapter, we studied the three-phase induction motor, a rotating one, which is run from an ac supply. We discussed the construction of two types of this motor, i.e., cage and slip-ring rotors. Then, the principle of operation, followed by the equivalent circuit, was presented. The torque–speed (slip) characteristics were described, along with various starters. Speed control methods are briefly given. In the next chapter—the fourth and last one in this series, synchronous machines, which can work both as a generator and a motor, as in the case of DC machines described in Chapter 12, are discussed, in detail.

Multiple Choice Type Questions

Tick the correct option(s) out of four choices provided in each question

13.1 For a three-phase induction motor, fed from a 400 V, 50 Hz supply, the rated speed is 1,440 rpm. The number of poles required in the stator is

(a) 10 (b) 8 (c) 6 (d) 4

13.2 For a 6-pole, three-phase induction motor, fed from a 400 V, 50 Hz supply, the rated speed is 970 rpm. If the supply is changed to 320 V, 40 Hz, find the rated speed in rpm (the slip is assumed to remain constant).

(a) 776 (b) 800 (c) 784 (d) 768

13.3 The fictitious resistance, representing the output power, in the rotor circuit of a three-phase induction motor, fed from a 50 Hz supply, is given by,

(a) $\frac{r_2'}{s}$ (b) $\frac{r_2'.(1-s)}{s}$

(c) $\frac{r_2'.(1+s)}{s}$ (d) none of (a)–(c)

13.4 For a three-phase slip-ring (wound rotor) induction motor, which is being fed from a rated supply, the maximum value of the torque obtained, using torque–speed characteristics, remains constant, when the total rotor resistance per phase, including that inserted in the rotor circuit at the slip-rings,

(a) is increased. (b) is increased

(c) remains same. (d) none of (a)–(c)

13.5 The power factor (lagging) of a three-phase induction motor at full load, fed from a 50 Hz supply is

(a) 0.9 (b) 0.83 (c) 0.75 (d) none of (a)–(c)

13.6 The starting current of a three-phase induction motor, using a direct-on-line (DOL) starter, is x times full load (fl) current. The value of x is given by,

(a) 5–6 (b) 2–3 (c) 3–4 (d) none of (a)–(c)

13.7 The starting current, using star–delta starter, for a delta-connected stator of a three-phase induction motor, is x times full load (fl) current. The value of x is given by,

(a) 3–4 (b) 1–2 (c) 1.7–2.0 (d) none of (a)–(c)

13.8 If the stator input voltage of a three-phase induction motor fed from a 50 Hz supply is varied, the speed changes to a large extent when the load torque

(a) is proportional to the speed
(b) is proportional to the square of the speed
(c) remains the same.
(d) none of (a)–(c)

13.9 The starting current of a three-phase induction motor, using an auto-transformer starter with the ratio of output to input voltage as 0.6, is x times the starting current, using a direct-on-line (DOL) starter. The value of x is,

(a) 0.6 (b) 0.78 (c) 0.36 (d) none of (a)–(c)

13.10 The efficiency of a three-phase induction motor at full load, fed from a 50 Hz supply, is in the range of,

(a) 97% (b) 70% (c) 90% (d) 85%

Answers

13.1	(d)	13.2	(a)	13.3	(b)	13.4	(a, b)
13.5	(b)	13.6	(a)	13.7	(c)	13.8	(b)
13.9	(c)	13.10	(d)				

Problems

13.1 A three-phase squirrel cage induction motor has a full load torque, which is one-third of the maximum torque. The rotor resistance and reactance per phase are 0.25 Ω and 3.0 Ω respectively. Calculate the ratio of starting torque to full load torque.

13.2 An 8-pole, 50 Hz, three-phase squirrel cage induction motor has a full load torque of 200 N.m, when the frequency of rotor emf is 2.5 Hz. If the mechanical loss is 15.0 N.m, determine the rotor copper loss, and efficiency of the motor. The total stator loss is 1,000 W.

13.3 A 12-pole, 50 Hz, three-phase induction motor develops maximum torque at a speed of 460 rpm. The star-connected rotor has a resistance of 0.4 Ω per phase. How much resistance must be inserted in series with each phase of the rotor to have a starting torque at one-third of its maximum value? [Hint: use the formula given in Example 13.1.]

13.4 A three-phase, squirrel cage, delta-connected induction motor draws a starting current of 80 A in each stator phase, when it is connected to a 400 V, 50 Hz supply. Find

(a) The line current for DOL starting,

(b) The line and phase currents, when a star–delta starter is used, and

(c) The line and phase currents, when an auto-transformer with 70% tapping is used for starting.

13.5 A 10 HP, 400 V, 50 Hz, three-phase, delta-connected induction motor has a full load power factor of 0.8 and efficiency of 0.9 (90%). The motor draws 7 A, when a voltage of 160 V is applied directly across the motor at stand-still. Determine the ratio of starting to full load current, when star–delta starting is used.

13.6 A three-phase, 8-pole, squirrel cage induction motor, connected to 400 V, 50 Hz supply, rotates at 4% slip at full load. The total stator loss is 2 kW. If the motor takes 40 kW at full load, find the developed torque (gross) under that condition.

13.7 The rotor of a three-phase, 8-pole, slip-ring (wound rotor) induction motor has a resistance of 0.25 Ω per phase, and it runs at 720 rpm at full load. How much external resistance/phase must be added to the rotor circuit to reduce the speed to 600 rpm, the torque remaining constant?

13.8 A 10-pole, 50 Hz, 3,300 V, three-phase, star-connected induction motor has a slip-ring (wound) rotor of resistance and standstill reactance of 0.025 Ω and 0.28 Ω, per phase, respectively. The motor runs at a speed of 576 rpm, when full load torque is applied. Calculate (a) slip at maximum torque, and (b) the ratio of maximum torque to full load torque. Neglect stator impedance.

CHAPTER 14

SYNCHRONOUS MACHINES

The synchronous machine is the last of the series of four electrical machines presented in this book. It is a rotating ac machine, like the dc machine discussed in Chapter 12, which can be used both as a generator and a motor. As a generator, when run by a prime mover at rated speed, the machine is widely used to provide power, be it small or large. All large generators are run mainly by steam turbines in thermal power stations. The synchronous machine can also be used as a motor because of certain features, as will be described later. We will first describe the construction of two types of synchronous machines—salient pole and non-salient pole. This is followed by the emf equation, along with distribution and pitch factors, for distributed winding in the stator, as derived for the three-phase induction motor in Chapter 13.

For the synchronous generator, we will first discuss the synchronous impedance, followed by the determination of regulation when different types of loads, i.e., resistive, inductive and capacitive, are connected across its terminals. Then, the operation of the synchronous machine as a motor is presented. The V-curve is discussed in detail. When the excitation of the motor is increased at a constant output, the input or armature current first decreases, and then increases. The latter case occurs when the field is overexcited, with the input current at leading power factor. This gives the synchronous motor an edge over the induction motor which always takes lagging current. It may be noted that the synchronous motor is costlier than an induction motor of the same rating.

14.1 Construction

Like all rotating machines, the synchronous generator consists of two parts—the stator and the rotor. In the stator, the winding used is a balanced three-phase one, which means that the number of turns in each phase is equal, as in a three-phase induction motor (explained in Chapter 13). The windings of the three phases are placed 120° (electrical) apart. The multi-turn conductors are placed in slots. The schematic of the stator winding is given in Fig. 14.1, where a single multi-turn coil per phase in one slot, is shown. The rotor is not included there, but presented later. As explained in the next section, in general, windings are distributed in number of slots per phase, giving rise to a distribution factor. The reader can refer to any standard textbook.

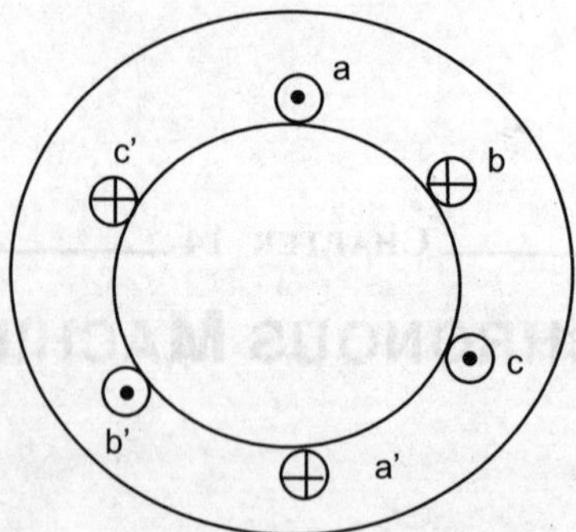

Fig. 14.1 *Stator winding for a three-phase synchronous machine*

There are two types of rotors used in synchronous machines, viz., salient pole and non-salient one. The salient pole machine is normally used (Fig. 14.2(a)) in hydropower stations where the water turbine runs at low speed (N in rpm). The number of poles (p) needed is high, so as to keep its output frequency, f at 50 Hz. The relationship, as derived earlier, is given by

$$f = \frac{p.N}{120}$$

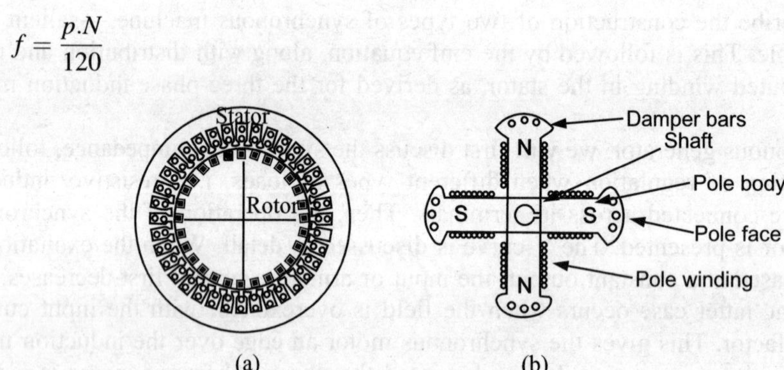

Fig. 14.2 *Two types of rotors for a synchronous machine: (a) Salient pole; (b) Non-salient pole*

Hence, salient pole is used here, as the diameter of this type of turbine is large, with linear speed being kept within limit. The field windings are wound on the salient poles. When current flows in the windings, magnetic flux is set up in the air gap.

The other type of rotor, i.e., a non-salient pole (cylindrical one) is used (Fig. 14.2(b)) in thermal power stations where the steam turbine runs at 3,000 rpm. There are two poles, thus giving an output frequency of 50 Hz, using the formula given earlier. The diameter, in this case, is much smaller than the salient pole type, keeping the linear speed within limit. As the rotating speed is high, there will be centrifugal force which a salient pole rotor cannot bear. To manage this speed, a cylindrical rotor with distributed winding is used in a non-salient pole machine. The air gap is uniform, and the field winding is suitably distributed along the rotor periphery. As one moves away from the field axis, the flux paths link progressively smaller number of field ampere turns, which can be distributed to produce a nearly sinusoidal flux in the air gap.

14.2 EMF Equation

EMF GENERATED PER PHASE, IN THE STATOR WINDING

Firstly, it is assumed that a full pitched coil with N turns is placed in the slot on the stator of a synchronous generator with salient poles (Fig. 14.3). There are only two poles. The machine is rotated at the rated speed to generate an output frequency (f) of 50 Hz. Initially, the axis of the field winding and the coil are coincident. The angle, $\theta = \omega t = 2\pi \ ft$, is measured from the coil axis (Fig. 14.3(a)), where the flux, ϕ is maximum in the forward direction. The angle, θ is $0°$ at that time. After the pole rotates, it goes into a position, $\theta = 90°$ (Fig. 14.3(b)), where the flux is zero. The next position is shown in Fig. 14.3(c)), when $\theta = 180°$, where the flux is maximum, but in the reverse direction. The flux linkage of the coil is given by $\psi = N.\phi.\cos\theta = N.\phi.\cos\omega t$.

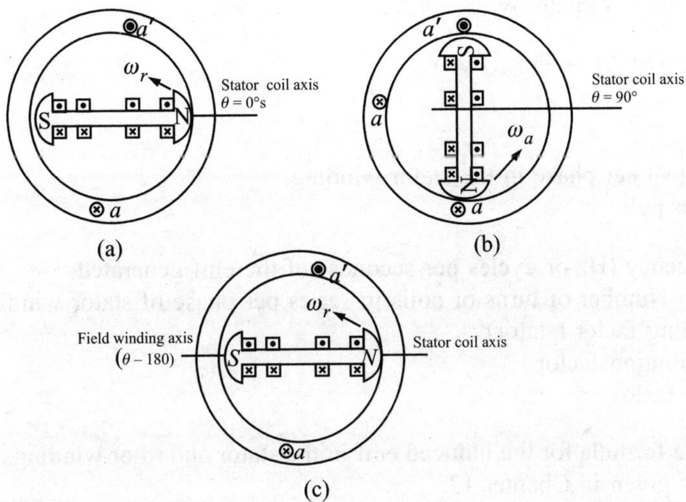

Fig. 14.3 (a)–(c) *Single winding in the armature of a synchronous machine at different positions with respect to the field winding*

The induced voltage in the coil is,

$$e = -\frac{d\phi}{dt} = N.\phi.\omega.\sin\theta,$$

where $\theta = \omega t$.

This can be written as $e = E_m.\sin\theta = N.\phi.\omega.\sin\theta$ where $E_m = N.\phi.\omega = N.\phi.2\pi.f$.

Note that E_m is the maximum value of the induced or generated voltage.

The rms value of the induced voltage is,

$$E = \frac{E_m}{\sqrt{2}} = \sqrt{2}.\pi.f.N.\phi = 4.44.f.N.\phi$$

It is similar to the formula for the induced emf in a transformer derived in Chapter 11. If the number of conductors is Z, then $Z = 2.N$, or $N = \frac{Z}{2}$. Substituting this in the equation, we get, $E = 2.22.f.Z.\phi$.

Normally, the stator winding is distributed in a number of slots, as there is a limit to the number of conductors placed in a slot as there is a chance that they may get overheated. This gives rise to a distribution factor, which is less than 1.0, as given in the next section. Sometimes, the coils are short chorded (explained later). In this case, the pitch factor is to be taken into account. All the factors are combined to form the winding factor in the stator, which is also less than 1.0. Taking into account all these factors, the induced emf per phase in the stator winding is given by,

$$E = 2.22.k_w.f.\phi.Z = 4.44.k_w.f.\phi.N$$

where,

E=Induced emf (V) per phase in the stator winding
ϕ=Flux (Wb) per pole

f =Frequency (Hz or cycles per seconds) of the emf generated
Z $=2.N$ = Number of turns or coils in series per phase of stator winding
$k_w = k_d.k_p$=Winding factor (stator)
k_d =Distribution factor
k_p =Pitch factor

It is similar to the formula for the induced emf in the stator and rotor windings of three-phase induction motors given in Chapter 13.

DISTRIBUTION FACTOR

In most cases, a large number of coil turns is needed to produce rated voltage (or induced emf) in the stator winding of the three-phase synchronous generator. It is not possible to accommodate all these in a single slot, as this would result in high temperature rise. If one coil/phase/pole pair is used, the number of slots is less, with the periphery of the stator not utilized fully, as the slot pitch is large. Hence, for a reasonable slot pitch, a large number of coils/pole/phase is used, say three or four, as the case may be. If the total number of slots is S, then the slot/pole/phase is given by $m = \frac{S}{q.p}$, where q is the number of phases ($q = 3$) in this case and p = number of poles. A two-pole, 3-phase machine is shown in Fig. 14.4(a). The phase angle between adjacent slots in electric radians is $\gamma = \frac{\pi p}{S}$. The total angle, $\sigma = m.\gamma$, occupied by the phase winding is termed as phase spread, which is assumed as 60°, in this case. This is called distributed winding. The induced voltage in the conductors in each slot is equal in magnitude, as the number of conductors in each slot is equal. However, the phase angle in electric radians between the induced emfs in adjacent slots is γ. The phasor diagram of the induced emfs in three coils ($m = 3$) is shown in Fig. 14.4(b). If the voltages (or emfs) in the three coils are AB, BC and CD, the phasor sum is AD. The phase angle between adjacent sides AB and BC is γ. The sides form a regular polygon. A circle is drawn touching the points, A, B, C and D, with its centre at O. The angle subtended at the centre, O by the side, AB is γ. Hence, the side, AB is $2.\text{OA}.\left[\sin\left(\frac{\gamma}{2}\right)\right]$, in terms of OA, the radius of the circle. The angle subtended at the centre, O by the resultant, AD is $\sigma = m.\gamma$. Hence, the resultant, AD is given by $2.\text{OA}.\left[\sin\left(\frac{m.y}{2}\right)\right]$. The distribution factor, for line frequency, which is the ratio between the phasor sum and the algebraic sum, is obtained as

$$k_{d1} = \frac{m.\text{AB}}{\text{AD}} = \frac{\sin\left(\frac{m.y}{2}\right)}{m.\sin\left(\frac{y}{2}\right)}$$

If the frequency of the nth harmonic is used, the distribution factor is

$$k_{dn} = \frac{\sin\left(\frac{m..n.y}{2}\right)}{m.\sin\left(\frac{n.y}{2}\right)}$$

If the winding is assumed to be distributed over a large number of slots, with a phase spread of 60°, the distribution factor is given by

$$k_{d1} = \frac{\sin\left(\frac{\pi}{6}\right)}{\frac{\pi}{6}} = 0.955$$

γ being small, it is taken in rad. k_{d1} is less than 1.0.

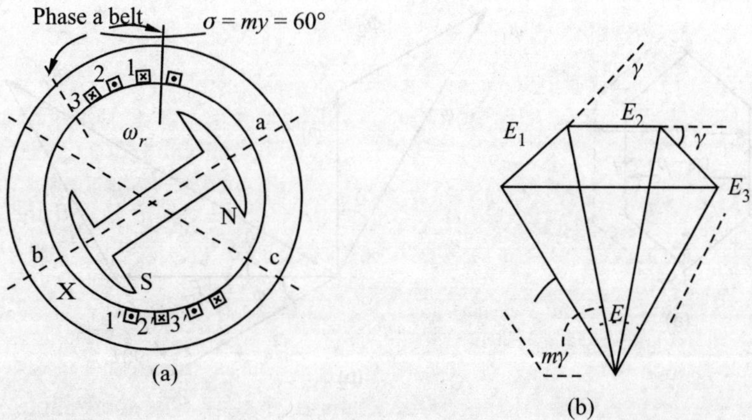

Fig. 14.4 *(a) Two-pole three-phase synchronous generator with distributed winding (b) Phasor diagram of coil emfs to determine the resultant voltage*

Pitch factor

Stator coils are normally of full pitch. The conductor in the forward direction is placed under the North Pole, while the other one in the return direction is under the South Pole. The induced emf in the conductor in the forward direction is positive, while that in the return conductor is negative. Taking their difference, the induced emf is doubled, with the coil span being π rad. (180°). In certain cases, the coils may have a span of less than full pitch. This is used in the stator winding of three-phase induction motors (Chapter 13). The coil span in radians is $(\pi - \rho)$, which is less than 180° (Fig. 14.5(a)). This results in lower overhang of the conductors, resulting in copper saving. Two induced emfs in the forward and return conductors have a phase difference of $(\pi - \rho)$ radians; their sum is given by $[\cos(\frac{\rho}{2})]$ times the voltages induced in a full pitch coil (Fig. 14.5(b)). Hence, the pitch factor for a fundamental (line frequency) is obtained as

$$k_{p1} = \cos\left(\frac{\rho}{2}\right)$$

If the frequency of the nth harmonic is used, the pitch factor is

$$k_{pn} = \cos\left(\frac{n\rho}{2}\right)$$

If the chording angle is assumed as 150°, ρ is 30° = 180° – 150°. The pitch factor in this case is

$$k_{p1} = \cos 15° = 0.966$$

being less than 1.0. The pitch factor for a full pitch coil is 1.0. The winding factor is taken as the product of distribution and pitch factors, found as $k_{w1} = k_{d1}.k_{p1}$. An example is given to compute the induced emf or voltage in a three-phase synchronous generator.

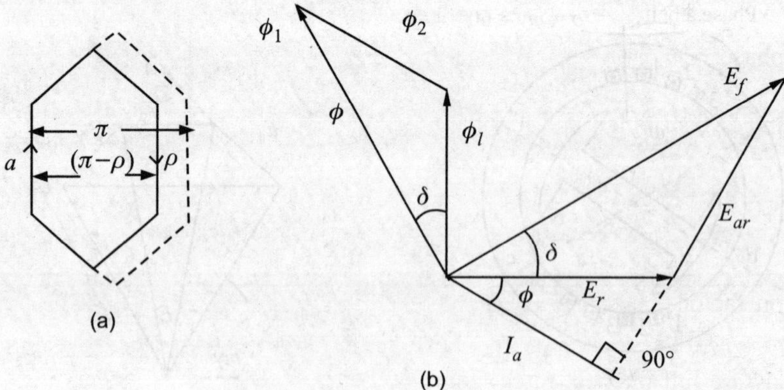

Fig. 14.5 *(a) A short-pitched coil (b) Phasor diagram of induced voltages in two conductors to find the resultant voltage*

EXAMPLE 14.1 *A three-phase, 50 Hz, salient pole, star-connected synchronous generator is driven by a hydro-electric turbine at a speed of 125 rpm. There are 576 slots, with two conductors per slot. The air gap diameter is 6.1 m, and the stator length is 1.2 m. The sinusoidally distributed flux density has a peak value of 1.1 Wb/m². Find the generated emf per phase. If the coil span is now changed to 10 slots, find the new generated emf per phase.*

SOLUTION Frequency, $f = 50$ Hz

Speed (synchronous) $= N_s = 125$ rpm

No. of poles $= p = \dfrac{120.f}{N_2} = \dfrac{120 \times 50}{125} = 48$ (even number)

As the flux density is sinusoidally distributed, its average value

$$B_{\text{av}} = \frac{2}{\pi} \times 1.1 \approx 0.7 \text{ Wb/m}^2$$

Length of stator $= D = 1.2$ m

Air gap diameter $= L = 6.1$ m

Pole pitch $= \tau = \dfrac{\pi.D}{p} = \dfrac{\pi \times 6.1}{48} \approx 0.4$ m

Area under pole $= a = \tau.L = 0.4 \times 1.2 = 0.48$ m^2

Flux per pole $= \phi = B_{\text{av}}.a = 0.7 \times 0.48 = 0.336$ Wb

Number of phases $= 3$

Number of slots $= S = 576$

Slot/pole/phase $= m = \dfrac{576}{3 \times 48} = 4$

The winding is distributed and its phase spread $= m.\gamma = 60$

Slot angle $= \gamma = \dfrac{60°}{4} = 15°$

Distribution factor $= k_{d1} = \dfrac{\sin\left(\frac{60°}{2}\right)}{4.\sin\left(\frac{15°}{2}\right)} = \dfrac{\sin 30°}{4 \times \sin 7.5°} = \dfrac{0.5}{4 \times 0.1305} = 0.958 > 1.0$

As the coils are full pitched, pitch factor $= k_{p1} = 1.0$

Winding factor $= k_{w1} = k_{d1}.k_{p1} = 0.958 \times 1.0 = 0.958$

Number of conductors $= Z = \dfrac{576}{3} \times \dfrac{4}{2} = 384$

Generated emf per phase $= E = 2.22.f.\phi.Z.k_{w1} = 2.22 \times 50 \times 0.336 \times 384 \times 0.958$

$$= 14.322 \times 0.958 = 13.72 \text{ kV}$$

For a short pitched coil, coil span = 10 slots

Span angle $= 10 \times 15° = 150°$

$$\rho = 180° - 150° = 30°$$

Pitch factor $= k_{p1} = \cos\left(\frac{30°}{2}\right) = \cos 15° = 0.966$

Winding factor (new) $= k_{w1} = k_{d1}.k_{p1} = 0.958 \times 0.966 = 0.9254$

Generated emf per phase (new) $= E = 2.22.f.\phi.Z.k_{w1} = 2.22 \times 50 \times 0.336 \times 384 \times 0.9254$

$$= 14.322 \times 0.9254 = 13.25 \text{ kV}$$

14.3 Synchronous Impedance

The equivalent circuit of a three-phase synchronous machine working as a generator is developed here. The output of the generator is connected to a balanced load (Fig. 14.6(a)). When the field is excited, an mmf, F_f is formed in the air gap which rotates at synchronous speed. The salient pole field winding is in the rotor. As the generator output terminals are connected to the load, a three-phase balanced current flows in the armature, creating a rotating magnetic field at synchronous speed, same as that of F_f. The armature mmf, F_{ar} is in the direction of armature current, which is assumed to be lagging the output voltage. The resultant mmf phasor, F_r is given by, $F_r = F_f + F_{ar}$.

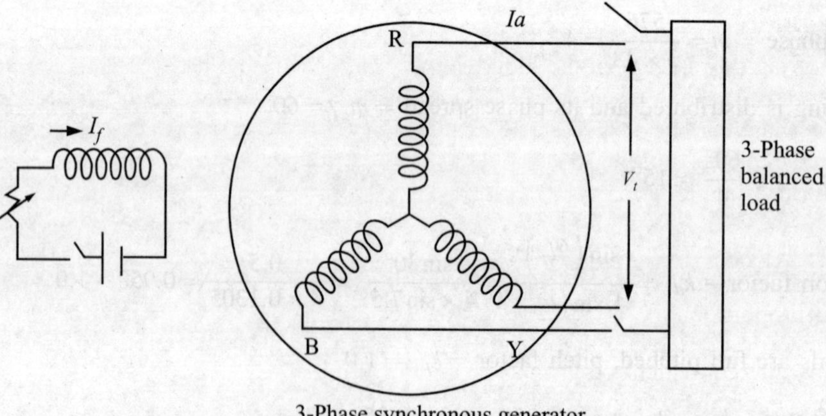

Fig. 14.6(a) *A three-phase synchronous generator is connected to a balanced load*

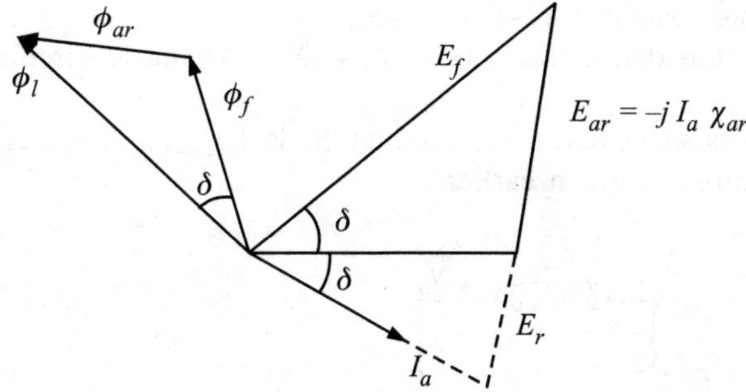

Fig. 14.6(b) *Phasor diagram of the flux and emf components in the machine*

If the magnetic circuit is assumed to be linear, the flux is taken as proportional to mmf. Hence, the resultant is also written as $\phi_r = \phi_f + \phi_{ar}$,

where,

ϕ_r = flux produced by F_r (resultant mmf)
ϕ_f = flux component produced by F_f acting alone
ϕ_{ar} = flux component produced by F_{ar} acting alone

Since, the induced emf is proportional to the flux/pole, and lags behind it by 90°, from the flux phasor, the phasor equation of the induced emf is, $E_r = E_f + E_{ar}$,

where,

E_r= emf induced by ϕ_r (resultant one)
E_f = emf induced by ϕ_f acting alone (excitation emf)
E_{ar} = emf induced by ϕ_{ar} acting alone

The angle of lag (90°) between the induced emf and the flux has been obtained earlier in the case of the transformer in Chapter 11. The emf phasors are proportional to the corresponding flux phasors, lagging the respective flux phasors by 90°. The phasor diagram is shown in Fig. 14.6(b), with the flux phasor triangle and the emf phasor triangle similar to each other; the emf phasor triangle being rotated anti-clockwise from the flux phasor triangle by 90°.

As ϕ_{ar} is in phase with I_a (generating machine), and is proportional to it, the induced emf, E_{ar} is also proportional to I_a and lags behind it by 90°, i.e., $E_r = -jI_a.x_{ar}$, where x_{ar}, the constant of proportionality, is, indeed an inductive reactance. Thus, we get, $E_r = E_f - jI_a x_{ar}$.

From this, the equivalent circuit per phase can be obtained as shown in Fig. 14.6(c). The reactance, x_{ar} equivalently replaces the effect of armature reaction. The effects of the armature resistance, r_a, and leakage reactance, x_l are taken as $V_t = E_r - I_a(r_a + jx_l)$. Substituting the earlier equation, we get,

$$V_t = E_f - I_a.[r_a + j(x_l + x_{ar})] = V_t = E_f - I_a.(r_a + jX_s),$$

where the synchronous reactance is, $X_s = x_l + x_{ar}$.

The synchronous impedance is $Z_s \angle \theta_s = (r_a + jX_s)$. All these circuits are given in Fig. 14.6(c).

The synchronous reactance takes into account the leakage reactance and also the reactance due to armature reaction, as given earlier.

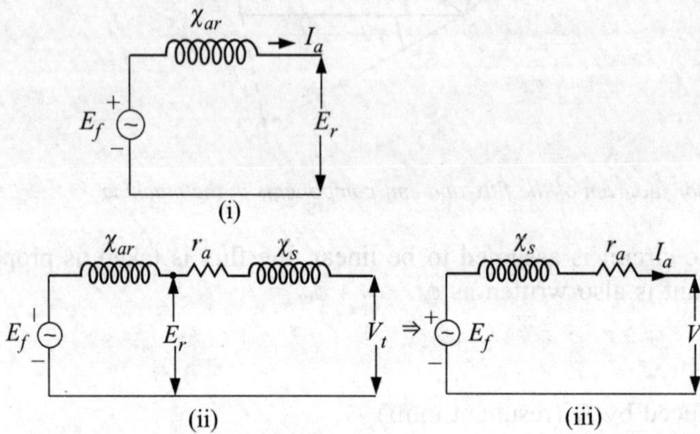

Fig. 14.6(c) *The equivalent circuit per phase*

14.4 Determination of Regulation of Synchronous Generator Using the Synchronous Impedance Method

The regulation of a synchronous generator is defined as the difference of the output voltage at load from the same voltage at no-load, in terms of the output voltage at load. The equivalent circuit of a synchronous generator (Fig. 14.6(a)) is given by the generated or induced emf in series with a synchronous impedance (i.e., resistance and synchronous reactance), as derived in the previous section. This is similar to the regulation of the transformer (Chapter 11), except that here, the synchronous reactance is quite large when compared to the total leakage reactance in the case of a transformer. Note that as the synchronous generator is put on the load, the load current is in the direction shown in the figure, which is used in deriving the relation between E_g and V_t, as given later.

If the output voltage at load and at no-load are V_t and E_g respectively, the per unit voltage regulation is given by $\dfrac{E_g - V_t}{V_t}$.

It can also be defined as the difference of the output voltage on load from the same voltage at load, in terms of output voltage at load, when the load is switched off, i.e., it is now on no-load. The output voltage at load is assumed as rated value. Assume that the load as inductive, i.e., the output current, I_L lags the output voltage, V_t by an angle, φ. The phasor diagram is shown in Fig. 14.7(a), with the output or terminal voltage as the reference phasor.

In Fig.14.7(a), the output voltage phasor at no-load, E_g, leads V_t by an angle, δ. The equation between E_g and V_t is written as,

$$E_g\angle\delta = [V_t\angle 0^\circ + I_L\angle -\phi \cdot z_s\angle\theta_s] = [V_t\angle 0^\circ + I_L\angle -\phi \cdot (r_a + jx_s)]$$

where z_s, r_a and x_s are the synchronous impedance, resistance and synchronous reactance of the synchronous generator respectively. The load impedance is inductive.

Resolving the equation into its X- and Y-axis components, we get,

$$E_g.\cos\delta = V_t + (I_L.Z_S).\cos(\theta_s - \phi) = V_t + I_L.(r_a\cos\phi + x_s\sin\phi)$$

$$E_g.\sin\delta = (I_L.Z_s).\sin(\theta_s - \phi) = I_L.(x_s\cos\phi - r_a\sin\phi)$$

Let us take the square of the above two equations, and add them,

$$(E_g)^2 = (V_t)^2 + (E_g)^2 + (I_L.z_s)^2 + [2V_t.(I_L.z_s).\cos(\theta_s - \phi)]$$

The regulation is positive, i.e., ($|E_g| > |V_t|$), as the magnitude of the output voltage at no-load is higher than that at load. This can be proved from the above expression.

An alternative method is given here, with the output or load current as reference phasor (Fig. 14.7(b)). The output voltage leads the load current by an angle, ϕ The equation now takes the form,

$$E_g\angle(\delta + \phi) = [V_t\angle\phi + I_L\angle 0^\circ.z_s\angle\theta_s] = [V_t\angle\phi + I_L\angle 0^\circ.(r_a + jx_s)]$$

Resolving this equation into its X- and Y-axis components,

$$E_g.\cos(\delta + \phi) = V_t.\cos\phi + I_L.r_a$$

$$E_g.\sin(\delta + \phi) = V_t.\sin\phi + I_L.x_s$$

Using same steps as given earlier and simplifying, we get the same equation, as derived earlier.

$$(E_g)^2 = (V_t)^2 + (E_g)^2 + (I_L.z_s)^2 + (2V_t.(I_L.z_s.\cos(\theta_s - \phi)]$$

This will be used in EXAMPLE 14.2 as given later.

If the load is taken as resistive, the phase angle between the output voltage and the load current is taken as $\phi = 0^\circ$. In this case, the output voltage and current are in phase. Substituting this value in the earlier expression, we get,

$$E_g\angle\delta = [V_t\angle 0^\circ + I_L\angle 0^\circ \cdot z_s\angle\theta_s] = [V_t\angle 0^\circ + I_L\angle 0^\circ \cdot (r_a + jx_s)]$$

Resolving the equation into its X- and Y-axis components, we get,

$$E_g.\cos\delta = V_t + (I_L.z_s).\cos\theta_s = V_t + I_L.r_a$$

$$E_g.\sin\delta = (I_L.z_s).\sin\theta_s = I_L.x_s$$

Using the same steps as given earlier,

$$(E_g)^2 = (V_t)^2 + (E_g)^2 + (I_L.z_s)^2 + [2V_t.(I_L.r_a)]$$

The regulation is positive in this case also, i.e., ($|E_g| > |V_t|$), as the magnitude of the output voltage at no-load is higher than that at load. This can be easily proved from the above expression.

If the load is taken as capacitive, the phase angle between the output voltage and the load current, is taken as $\phi = -$ve (Fig. 14.7(d)). Hence, the sign of the phase angle is to be reversed from the + ve to the –ve. Substituting this value in the earlier expression, we get,

$$E_g\angle\delta = [V_t\angle 0° + I_L\angle\phi \cdot z_s\angle\theta_s] = [V_t\angle 0° + I_L\angle\phi \cdot (r_a + jx_s)]$$

Resolving the equation into its X- and Y-axis components, we get,

$$E_g.\cos\delta = V_t + (I_L.z_s).\cos(\theta_s - \phi) = V_t + I_L.r_a\cos\phi - x_s\sin\phi$$

$$E_g.\sin\delta = (I_L.z_s).\sin(\theta_s + \phi) = I_L.x_s$$

Let us take the square of the above two equations and add them,

$$(E_g)^2 = (V_t)^2 + (E_g)^2 + (I_L.z_s)^2 + [2V_t.(I_L.z_s).\cos(\theta_s + \phi)]$$

The regulation is negative in this case also, i.e., ($|E_g| < |V_t|$), as the magnitude of the output voltage at no-load is lower than that at load, i.e., the output voltage at no-load increases from the output voltage at load. This can be proved using the above expression. A numerical example is given here to illustrate the determination of regulation.

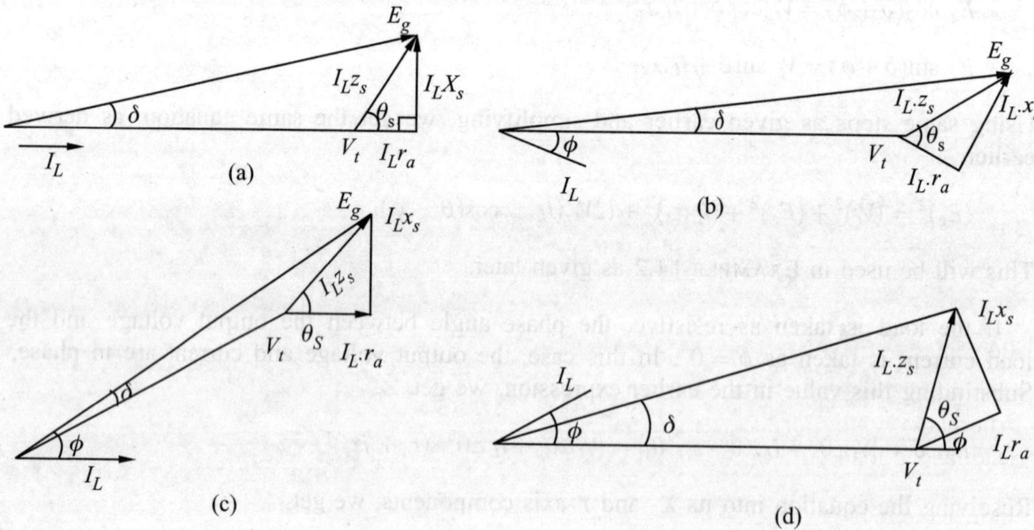

Fig. 14.7 *(a) Resistive (upf) load, determination of regulation of a synchronous generator, connected to different types of loads, such as (b) inductive, (c) inductive (with I_L as reference), and (d) capacitive*

EXAMPLE 14.2 *A 1.6 MVA, 13.5 kV, three-phase, 50 Hz, star-connected synchronous generator has an armature resistance and synchronous reactance (both in per phase) of 1.5 Ω and 30.0 Ω respectively. Find the regulation on full load at (a) unity power factor (pf), (b) 0.8 pf lagging, and (c) 0.8 pf leading.*

SOLUTION $S = 1.6$ MVA = 1,600 kVA, $V_L = 13.5$ kV = 13,500 V

$f = 50$ Hz, $r_a = 1.5\,\Omega$/ph, $x_s = 30.0\,\Omega$/ph.

Synchronous impedance per phase $= z_s = (r_a + jx_s) = (1.5 + j30.0) \approx 30.0375\angle 87.14°\Omega$

$$V_{ph} = \frac{V_L}{\sqrt{3}} = \frac{13,500}{\sqrt{3}} = 7,794.0 \text{ V}$$

$$\text{Full load current (A)} = I_a = \frac{S}{3 \cdot V_{ph}} = \frac{1,600 \times 10^3}{3.0 \times 7,794.0} = 68.4$$

In all three cases, the terminal voltage is taken as reference.

(a) pf = 1.0 (unity); $\phi = \cos^{-1}(1.0) = 0°$

Induced emf (Fig. 14.8(a)) $= E_g\angle\delta = [7,794.0\angle 0° + 68.4\angle 0° \times (1.5 + j30.0)]$

$$= [(7,794.0 + j0.0) + (102.6 + 6,718.4)] = (7,896.6 + 6.718.4) = 8,159.0\angle 14.57° \text{ V}$$

$$\text{Percentage voltage regulation} = \left(\frac{8,159.0 - 7,794.0}{7,794.0}\right) \times 100 = 4.68\%$$

(b) pf = 0.8 (lagging) $\phi = \cos^{-1}(0.8) = 36.87°$

Induced emf (Fig. 14.8(b)) $= E_g\angle\delta = [7,794.0\angle 0° + (68.4\angle -36.87° \times 30.0375\angle 87.14°)]$

$$= (7,794.0\angle 0° + 2,054.6\angle 50.27°) = [(7,794.0 + j0.0) + (1,313.2 + j1,580.1)]$$

$$= (9,107.2 + j1,580.1) = 9,243.2\angle 9.84° \text{ V}$$

$$\text{Percentage voltage regulation} = \left(\frac{9,243.2 - 7,794.0}{7,794.0}\right) \times 100 = 18.6\%$$

Alternatively, the load current is taken as reference, in this case, with the terminal voltage leading the load current by an angle, ϕ

Induced emf in V (Fig. 14.8(c)) $= E_g\angle(\delta + \varphi) = [V_t\angle\varphi + I_L\angle 0° \cdot (r_a + jx_s)]$

$$= [7,794.0\angle 36.87° + 68.4\angle 0° \times (1.5 + j30.0)]$$

$$= [(6,235.2 + j4,676.4) + (102.6 + j2,052.0)]$$

$$= (6,337.8.0 + j2,052.0) = 9,243.3\angle 46.71°$$

Load angle $= \delta = (46.71^\circ - 36.87^\circ) = 9.84^\circ$

The values of E_g and δ are the same as found earlier. The regulation can be obtained in the same way, as given earlier.

(c) pf = 0.8 (leading) $\phi = -\cos^{-1}(0.8) = -36.87^\circ$

Induced emf (Fig. 14.8(d)) $= E_g\angle\delta = [7,794.0\angle 0^\circ + (68.4\angle 36.87^\circ \times 30.0375\angle 87.14^\circ)]$

$= (7,794.\angle 0^\circ + 2,054.6\angle 124.0^\circ) = [(7,794.0 + j0.0) + (-1,148.9 + j1,703.3)]$

$= (6,645.1 + j1,703.3) = 6,860.0\angle 14.38^\circ$ V

Percentage voltage regulation $= \left(\dfrac{6,860.0 - 7,794.0}{7,794.0}\right) \times 100 = -11.98\%$

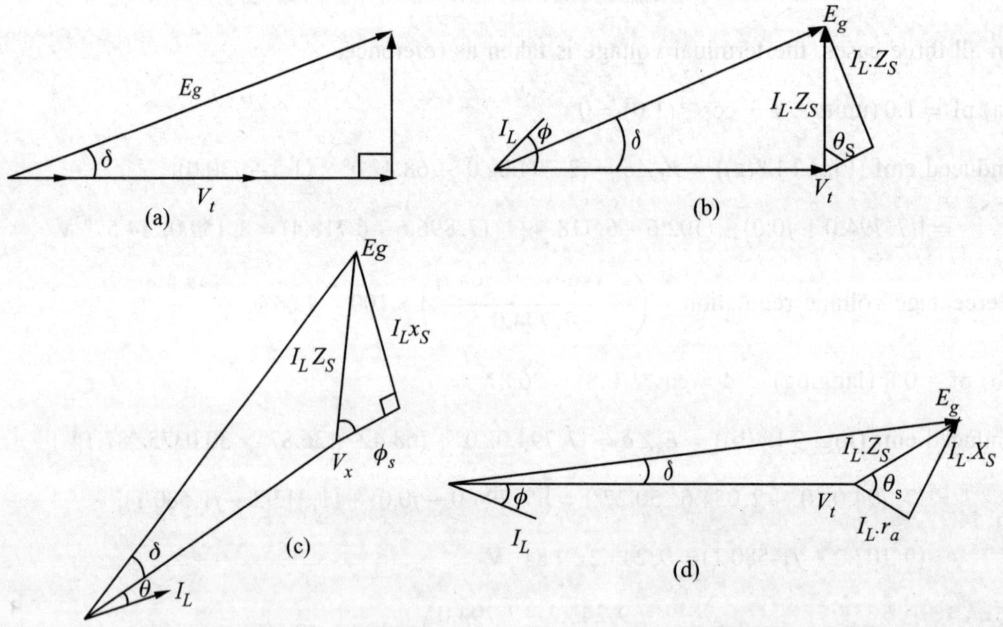

Fig. 14.8 *Determination of regulation of a synchronous generator, connected to different types of loads, such as (a) resistive, (b) inductive, (c) inductive (with I_L as reference) and (d) capacitive*

14.5 Motor Operation of Synchronous Machines

A synchronous machine with salient pole rotors is shown in Fig. 14.9. The rotor is initially stationary with fixed North–South Poles when the field winding, wound over the salient poles, is excited. The stator winding is fed from a three-phase supply at line frequency. As a result, a rotating magnetic field is created in the air gap—it rotates at a synchronous speed in rpm, given by

$$N_s = \frac{120.f}{p}.$$

This has been given in Chapter 13, where the three-phase induction motor is presented. It is assumed here that the magnetic field rotates in the clock-wise direction. Since the rotor is stationary, and cannot pick up speed instantaneously due to its inertia, average torque is not developed, as the two fields (stator and rotor) move relative to each other.

If the motor is run by auxiliary means, to bring its speed close to its synchronous speed in the direction of motion of the stator field, two fields now have the opportunity of locking into each other, or in other words, the rotor *pulls into step* with the stator field, and then runs at exactly the synchronous speed. Thus, as shown in Fig. 14.9, the electro-magnetic torque developed, acts on the rotor in the direction of the rotor which balances the load torque. The mechanical power therefore, flows to the load (motoring action) and, by the principle of conservation of energy, an equal amount of electrical power (plus losses) is drawn from the electric supply. It can be noted that, in this motor, mechanical torque is developed only at synchronous speed, not at any other speed—that is why the motor is termed as synchronous, unlike the induction motor (Chapter 13), which runs at any speed, but not at the synchronous speed.

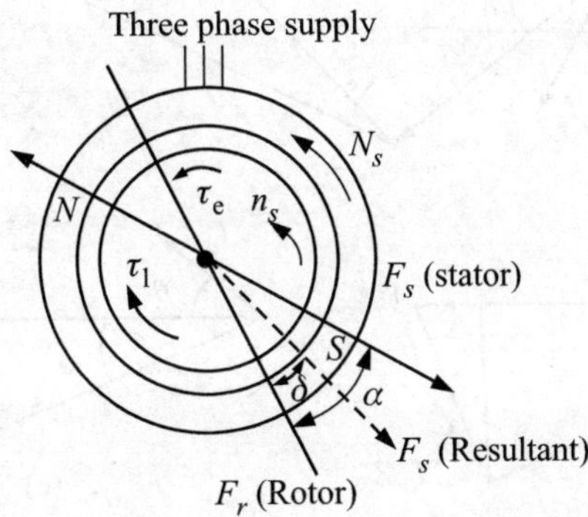

Fig. 14.9 *Synchronous machine operates as motor (torque production)*

RELATION BETWEEN THE INPUT VOLTAGE AND INDUCED EMF IN TERMS OF MOTOR PARAMETERS

The equivalent circuit of a synchronous motor (Fig. 14.10) is given by the induced emf in series with a synchronous impedance (i.e., resistance and synchronous reactance), as given in the previous section. When the shaft is loaded, the synchronous motor delivers an output load. The load current is in the direction shown in the figure, which is used in deriving the relation between E_b and V_t, given as,

$$E_b\angle-\delta = [V_t\angle 0° - I_a\angle-\phi\cdot z_s\angle\theta_s] = [V_t\angle 0° - I_a\angle-\phi\cdot(r_a + jx_s)]$$

where, z_s, r_a and x_s are the synchronous impedance, resistance and synchronous reactance of the synchronous generator respectively, as given earlier. It is assumed here that the input (or armature) current taken by the synchronous motor is lagging. In the phasor diagram (Fig. 14.11(a)) for this case, the induced emf phasor, E_b, is lagging V_t by an angle, δ. All steps are given here.

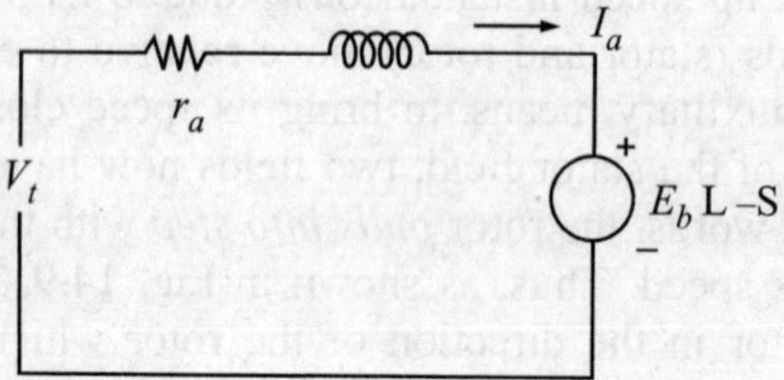

Fig. 14.10 *Equivalent circuit per phase, of a three-phase synchronous motor*

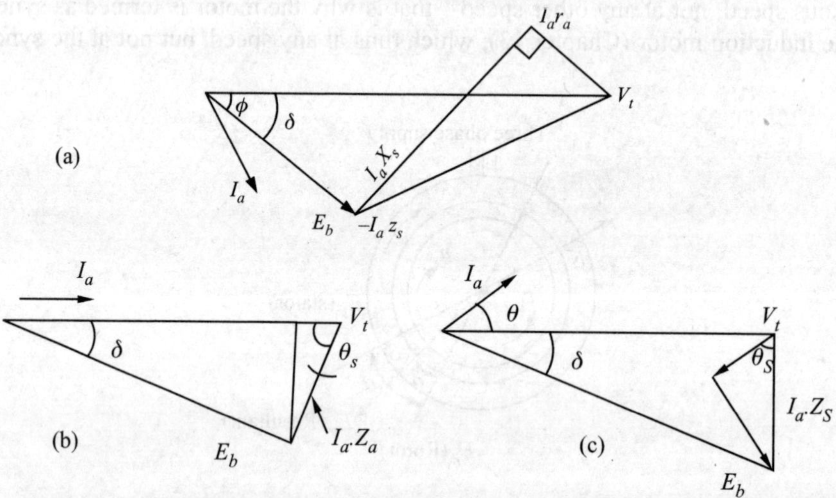

Fig. 14.11 *Determination of induced emf in a synchronous motor, when the power factor of the input current is (a) lagging, (b) unity and (c) leading*

Resolving the equation into its X- and Y-axis components, we get,

$$E_b.\cos\delta = V_t - (I_a.z_s).\cos(\theta_s - \phi) = V_t - I_a.(r_a \cos\phi + x_s \sin\phi)$$

$$E_b.\sin\delta = (I_a.z_s).\sin(\theta_s - \phi) = I_a.(x_s \cos\phi - r_a \sin\phi)$$

Let us take the square of the above two equations. Adding them,

$$(E_b)^2 = (V_t)^2 + (I_a.z_s)^2 - [2V_t.(I_a.r_a).\cos(\theta_s - \phi)]$$

It may be noted here that the magnitude of the input voltage is higher than the induced emf, as ($|V_t| > |E_b|$), which means that the motor is under-excited. This can be proved from the above expression. We can also use an alternative method as given earlier, while determining the regulation of a synchronous generator, with an inductive load. This has also been used in Example. 14.3

There may be two cases—when the input or armature current is (a) unity power factor, and (b) leading power factor, as shown earlier in the case of a generator. The derivation is given in brief, here.

(a) If the power factor of the input current is unity, the phase angle between the input or supply voltage and current is taken as $\phi = 0°$(Fig. 14.11(b)). In this case, the input voltage and current are in phase. Hence, the following expression can be obtained.

$$E_b\angle-\delta = [V_t\angle 0° - I_a\angle 0° \cdot z_s\angle\theta_s] = [V_t\angle 0° - I_a\angle 0° \cdot (r_a + jx_s)]$$

Using same steps as given earlier, we get,

$$(E_b)^2 = (V_t)^2 + (I_a.z_s)^2 - [2V_t.(I_a.r_a).\cos(\theta_s + \phi)]$$

It may be noted here that the magnitude of the input voltage is higher than the induced emf, as ($|V_t| < |E_b|$), which means that the motor is slightly under-excited. This can be easily proved from the above expression.

(b) If the power factor of the input current is taken as leading, the phase angle between the output voltage and load current is taken as ϕ = –ve (Fig. 14.11(c)). Hence, the sign of the phase angle is to be reversed from + ve to –ve. Substituting this value in the earlier expression, we get,

$$E_b\angle-\delta = [V_t\angle 0° - I_a\angle\phi \cdot z_s\angle\theta_s] = [V_t\angle 0° - I_a\angle\phi \cdot (r_a + jx_s)]$$

Using the same steps as given earlier,

$$(E_b)^2 = (V_t)^2 + (I_a.z_s)^2 - [2V_t.(I_a.r_a).\cos(\theta_s + \phi)]$$

It may be noted here that the magnitude of the input voltage is lower than the induced emf, as ($|V_t| < |E_b|$), which means that the motor is over-excited. This can be proved from the above expression.

From the above, it can be concluded that, as the field current in a synchronous motor is increased, the input or armature current first decreases, and then, reaches a minimum value and finally increases. The power factor of the armature current is lagging when the field is under-excited. However, if the field is over-excited, the input current is drawn from the input voltage at leading power factor. Thus, the synchronous motor has an edge over the induction motor, where the input current is always at lagging power factor. However, it may be noted that the synchronous motor costs more than the induction motor having the same rating.

14.6 V-curve

If a graph of the armature current in a synchronous motor at constant output, is drawn against the field current as shown in Fig. 14.12, it resembles the alphabet, V. If a constant load is applied at the motor shaft, with increase in field current, the armature current first decreases and then reaches a minimum value. Finally, the armature current increases again. As derived earlier, the current is first at lagging power factor. It then goes to unity, and finally, the power factor is leading. A set of V-curves are drawn in Fig. 14.12, each for a different output power, with the minimum current increasing, as the output power increases. This is because the minimum input current is nearly proportional to the output power. Another numerical example is given here to determine the induced emf in the motor, when the power factor of the input current changes.

EXAMPLE 14.3 *The input current of a 400 V, three-phase, 50 Hz, star-connected synchronous moor is 50 A. The synchronous impedance per phase is (0.25 +j3.2) Ω. Find the induced emf when the power factor of the input current is (a) unity power factor (pf), (b) 0.8 pf lagging, and (c) 0.8 pf leading.*

SOLUTION: $V_L = 400$ V, $I_a = 50.0$ A, $f = 50$ Hz

$$V_{ph} = \frac{V_L}{\sqrt{3}} = \frac{400.0}{\sqrt{3}} \approx 231.0 \text{ V}$$

Synchronous impedance per phase $= z_s = (r_a + jx_s) = (0.25 + j3.2) \approx 3.2\ 1 < 85.53°\Omega$

In all three cases, the terminal voltage is taken as reference.

(a) pf = 1.0 (unity), $\phi = \cos^{-1}(1.0) = 0°$

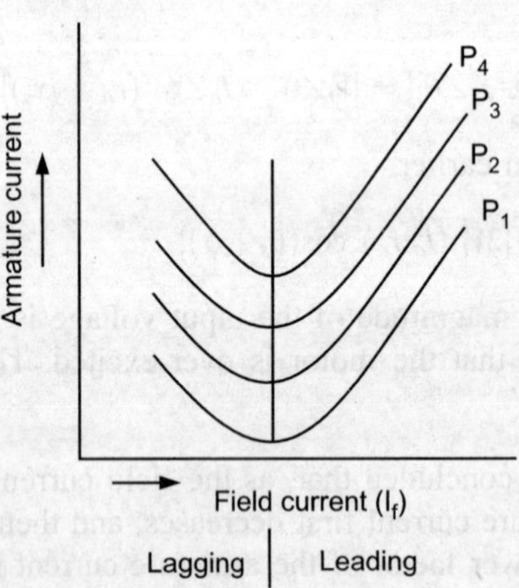

Fig. 14.12 *V-curves of a synchronous motor*

Induced emf per phase (Fig. 14.13(a)) $= E_b\angle-\delta = [231.0\angle 0° - 50.0\angle 0° \cdot (0.25 + j3.2)]$

$$= [(231.0 + j0.0) - (12.5 + j160.0)] = (218.5 - j160.0) = 270.8\angle -36.21° \text{ V}$$

Induced emf (line) $= \sqrt{3}.E_b = \sqrt{3} \times 270.8 = 469.0$ V

(b) pf = 0.8 (lagging), $\phi = \cos^{-1}(0.8) = 36.87°$

Induced emf per phase (Fig. 14.13(b))

$$= E_b\angle -\delta = [231.0\angle 0° - (50.0\angle -36.87° \cdot 3.21\angle 85.53°)] = (231.0\angle 0° - 160.5\angle 48.66°)$$

$$= [(231.0 + j0.0) - (106.0 + j120.5)] = (125.0 - j120.5) = 173.6\angle -43.95° \text{ V}$$

Induced emf (line) $=\sqrt{3}.E_b = \sqrt{3} \times 173.6 = 300.7$ V

Alternatively, the input or armature current is taken as reference. In this case, with the terminal voltage leading the load current by an angle, ϕ

Induced emf (Fig. 14.13(c)) $= E_b\angle(\phi - \delta) = [V_t\angle\phi - I_L\angle 0° \cdot (r_a + jx_s)]$

$$= [231.0\angle 36.87° - 50.0\angle 0° \cdot (0.25 + j3.21)] = [(184.8 + j138.6) - (12.5 + j160.0)]$$

$$= (172.3 - j21.4) = 173.6\angle -7.08° \text{ V}$$

Load angle $= \delta = (36.87° + 7.08°) = 43.95°$

The values of E_b (per phase) and δ are the same as found earlier. The induced emf (line) can be obtained in the same way as given earlier.

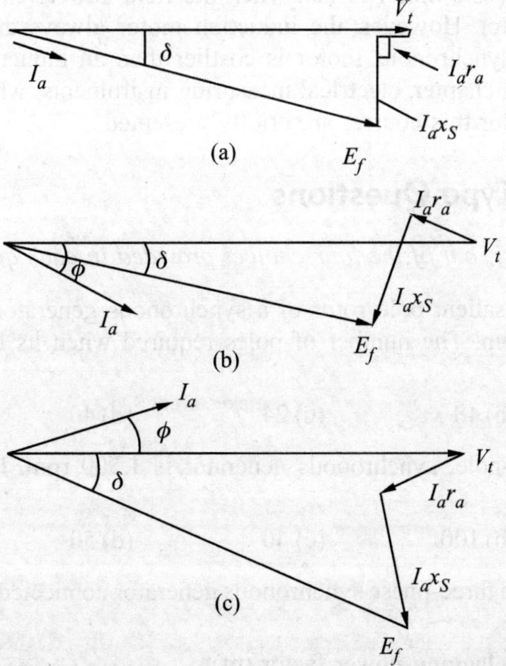

Fig. 14.13 *Determination of induced emf in a synchronous motor when the power factor of the input current is (a) unity, (b) lagging, (c) leading*

(c) pf = 0.8 (leading), $\phi = -\cos^{-1}(0.8) = -36.87°$

Induced emf per phase

$$= E_b\angle - \delta = [231.0\angle 0° - (50.0\angle 36.87° \cdot 3.21\angle 85.53°)] = (231.0\angle 0° - 160.5\angle 122.4°)$$

$$= [(231.0 + j0.0) - (-86.0 + j135.5)] = (317.0 - j135.5) = 344.75\angle - 23.14° \text{ V}$$

Induced emf (line) = $\sqrt{3}.E_b = \sqrt{3} \times 344.75 = 597.1$ V

Conclusion

This is the concluding chapter of the series where most of the commonly used types of electrical machines have been discussed. The synchronous machine, which is a rotating ac machine, is used mainly as a generator in both thermal and hydro power plants. This generator is run by a prime mover at rated speed, as the case may be. It is also used as a motor. The construction of two types of synchronous machines, non-salient pole and salient pole, is described. The emf equation was derived, along with distribution and pitch factors, for distributed winding in the stator. The synchronous impedance of the synchronous generator was discussed. Then, the regulation when different types of loads, such as, resistive, inductive and capacitive are connected across its terminals, was discussed. The operation of the synchronous machine running as a motor was presented. The V-curve was discussed in detail. The advantage of this motor is that when the field is over-excited, the input current is at leading power factor. However, the induction motor always takes lagging current. It may be noted that the synchronous motor is costlier than an induction motor of the same rating. In the concluding chapter, electrical measuring instruments, which are mainly used in laboratory experiments for this course, are briefly presented.

Multiple Choice Type Questions

Tick the correct option(s), out of the four choices provided in each question.

14.1 The speed of the salient pole rotor of a synchronous generator, used in hydro power stations, is 200 rpm. The number of poles required when its output frequency is 50 Hz is,

(a) 30 (b) 48 (c) 24 (d) 40

14.2 The speed of a 4-pole, synchronous generator is 1,500 rpm. Its output frequency in Hz, is

(a) 60 (b) 100 (c) 40 (d) 50

14.3 The regulation of a three-phase synchronous generator connected to a load whose power factor may change

(a) decreases with lagging power factor (pf).
(b) is constant at unity power factor.
(c) increases with lagging power factor.
(d) none of (a)–(c)

14.4 The regulation of a three-phase synchronous generator connected to a particular load is negative (–ve). What is the power factor of the load?

(a) lagging (b) unity (c) leading (d) none of the above

14.5 When the field current or excitation of a three-phase synchronous motor is not changed, but its load is increased, the load angle (δ),

(a) increases (b) remains constant
(c) decreases (d) none of (a)–(c)

14.6 The power factor of the armature current drawn by an over-excited three-phase synchronous motor, fed from a 50 Hz supply, is,

(a) lagging (b) unity (c) leading (d) none of (a)–(c)

14.7 At the minimum point on the V-curve giving the variation of armature current with change in the field current or excitation of a three-phase synchronous motor, fed from a 50 Hz supply, the power factor of the armature current is

(a) lagging (b) unity (c) leading (d) none of (a)–(c)

Answers

14.1 (a)	14.2 (d)	14.3 (c)	14.4 (c)
14.5 (a)	14.6 (c)	14.7 (b)	

CHAPTER 15

ELECTRICAL MEASURING INSTRUMENTS

In this last chapter, we will deal with electrical measuring instruments, mainly those used in laboratory experiments for a basic course in electrical engineering. We will first classify the measuring instruments. Then, the principle of operation of indicating instruments, along with the different components used there, has been taken up. Lastly, the different types of measuring instruments, such as ammeters and voltmeters, are discussed.

15.1 Classification of Measuring Instruments

Instruments may be broadly classified into two categories (Table 15.1). These are absolute and secondary type instruments.

Table 15.1

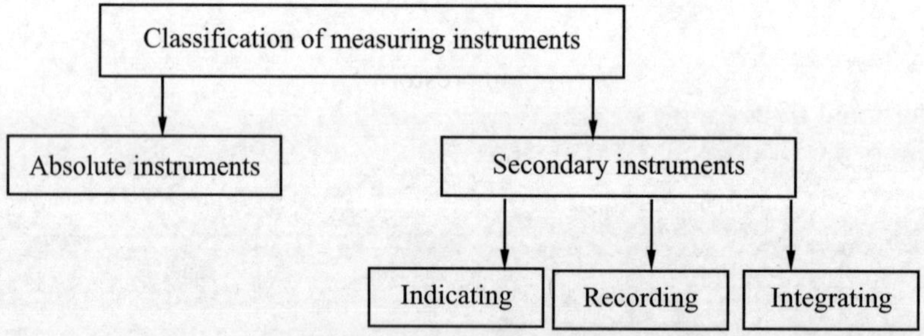

1. An *absolute instrument* gives the amplitude of the quantity which is measured in terms of physical constants of the instruments. A tangent galvanometer is an absolute type of instrument.

2. *Secondary instruments* can be classified into three types, viz., indicating, recording and integrating.

Indicating instruments indicates the magnitude of quantity, and, generally, makes use of a dial pointer. Ammeters, voltmeters and wattmeters are examples of indicating instruments.

Recording instruments gives a continuous record of quantity. The quantity is measured by an ink on paper record or a graph for a selected period of time.

Integrating instruments gives the total amount of energy or quantity of electricity over a period of time. Ampere-hour and watt-hour meters are examples of indicating instruments.

15.2 Principles of Operation of Indicating Instruments

All indicating instruments have the following three essential parts:

1. A deflecting system
2. A controlling system
3. A damping system

DEFLECTING SYSTEM

It converts electric quantity, like current or voltage, into mechanical force. This results in motion of the moving system (pointer).

CONTROLLING SYSTEM

It provides a force, which is equal and opposite to the deflecting torque. The pointer would swing over its maximum deflected position for a current of any amount without such a control.

The controlling torque for an indicating instrument is obtained by spring and gravity control. It is also referred to as restoring or balancing torque.

Spring control It is obtained by a hair spring (phosphor bronze) attached to the moving system (pointer) (Fig.15.1). The spring is twisted in the opposite direction for the movement of the deflection system (pointer). As a result, restoring torque is produced. The restoring torque is proportional to the angle of deflection, $T_c \propto \theta$, i.e., $T_c = k.\theta$, k being constant. This restoring torque is the controlling torque of the instrument.

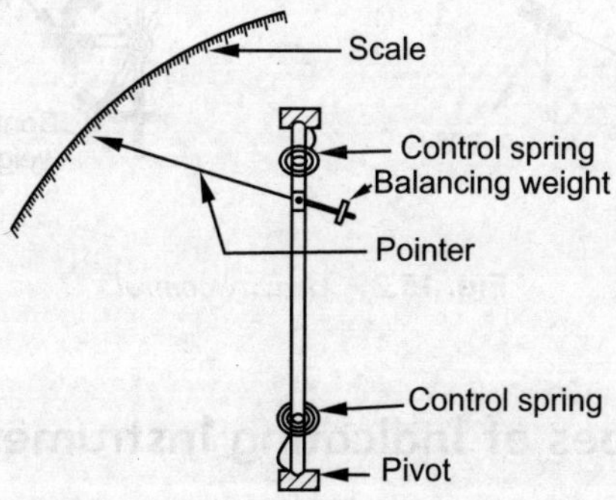

Fig. 15.1 *Spring control system*

The deflection of the pointer remains stationary, if deflecting torque (T_d) is equal to controlling torque (T_c).

Gravity control: It is done by attaching a small weight to the deflecting system (Fig. 15.2). The small weight produces a restoring or controlling torque, when the moving part of the system is at motion. The degree of control is made by adjusting the position of weight up and down from its position.

DAMPING SYSTEM

It acts as a brake on mechanical oscillations so that the moving system moves rapidly, but smoothly, to its deflected position.

Damping torque is provided by air friction, fluid friction and eddy current, which are described in brief below.

Air friction: To obtain air friction damping, normally an aluminum piston or vane is attached to the moving system.

Fluid friction: To obtain fluid friction damping, a light vane is attached with the moving system completely submerged in oil. The fluid will produce thc opposing force to the vane.

Eddy current damping: This type of damping is used in PMMC (permanent magnet moving coil) instruments, hot wire and induction type instruments. Magnetic flux linkage occurs in the conducting material during movement, resulting in a flow of eddy current. Due to this eddy current, a force is experienced opposite to the direction of motion.

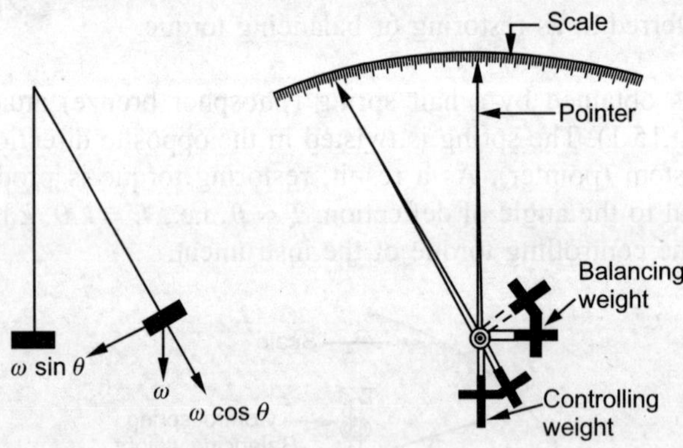

Fig. 15.2 *Gravity control*

15.3 Different Types of Indicating Instruments

15.3.1 Permanent Magnet Moving Coil (PMMC) Instrument

This type of instrument provides the most accurate measurement of quantities in DC circuits. It is made of a strong permanent magnet with soft iron pole pieces (Fig. 15.3). The moving or driving element of the deflecting system is a light rectangular coil (made of many fine wires) which lies in the narrow air gap between U-shaped pole pieces of the permanent magnet and the iron core.

Two phosphor bronze hair springs control the movement of the coil. The spring is connected at the two terminals of the coil. Hence, it solves the purpose of terminals for passing current through the coil. An aluminum former is used for damping the moving coil by introducing eddy current effect.

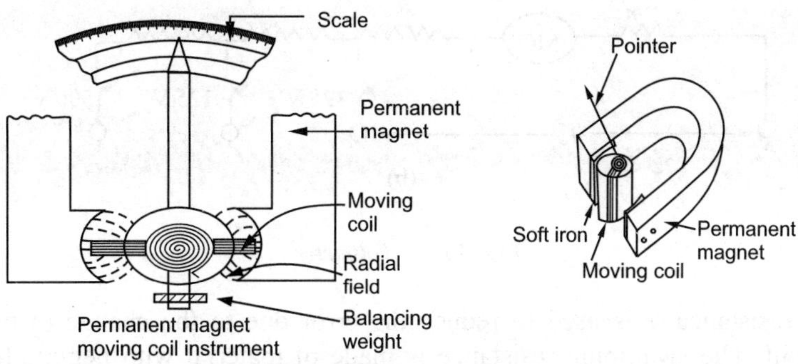

Fig. 15.3 *PMMC instruments*

PERMANENT MAGNET MOVING COIL USED AS AN AMMETER

Moving coil instruments are very sensitive. They are normally used to measure 10 μA to 50 mA of current, but may also be used to measure higher current in the range of A (ampere), by adding a shunt in parallel. The shunt forms a bypass and only a part of the actual current will be allowed to pass through the moving coil. Multi-range instruments may be designed as shown in Fig. 15.4 and Fig. 15.5. The ammeter is connected in series with the circuit, thus the resistance of the ammeter would be such that it will not affect the load current and voltage drop across the meter would be low.

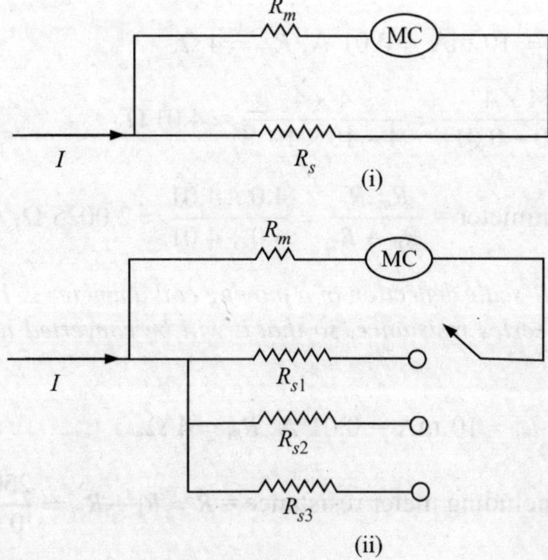

Fig. 15.4 *Ammeter*

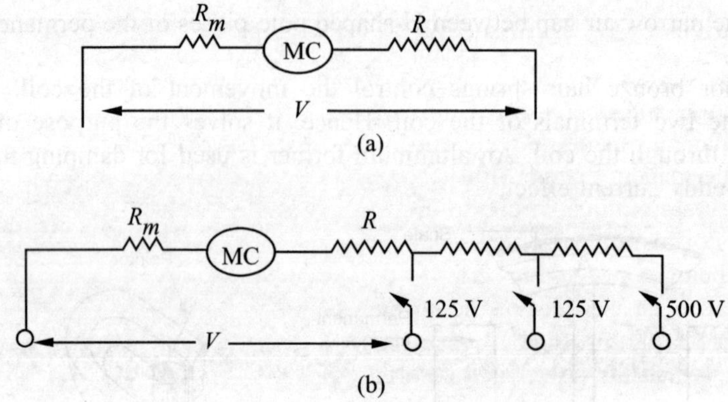

Fig. 15.5 *Voltmeter*

A swamping resistance, r is used to reduce the error due to the change of temperature of the moving coil. The swamping resistance is made of material with negligible temperature co-efficient. R_m is a resistance, which is the sum of the swamping resistance. The resistance of the moving coil is connected in series with the moving coil (MC) and the total series part is connected in parallel with the shunt, R_s.

PERMANENT MAGNET MOVING COIL USED AS A VOLTMETER

The moving coil instrument can be used as a voltmeter by adding high resistance, R in series with the coil. For multi-range operation, different tapping of high resistance R can be used.

EXAMPLE 15.1 *The full-scale deflection of a moving coil ammeter is 10 mA, and the resistance is 4 Ω. Find the value of the shunt resistance so that the ammeter will be converted into a 4 A range instrument. Also calculate the total resistance of the meter.*

SOLUTION $I = 4$ A, $I_m = 10$ mA $= 0.01$ A, $R_m = 4\ \Omega$.

$$R_s = \frac{I_m.R_m}{(I - I_m)} = \frac{4\times 4}{(4.0 - 0.01)} = \frac{4\times 4}{4 - 10\times 10^{-2}} \approx 4.01\ \Omega.$$

$$\text{Total resistance of the ammeter} = \frac{R_m.R_s}{R_m + R_s} = \frac{4.0\times 4.01}{4.0+4.01} \approx 2.0025\ \Omega, \text{ which is low.}$$

EXAMPLE 15.2 *The full-scale deflection of a moving coil ammeter is 10 mA, and the resistance is 4 Ω. Find the value of series resistance, so that it will be converted into a voltmeter of 250 V range.*

SOLUTION $V = 250$ V, $I_m = 10$ mA $= 0.01$ A, $R_m = 4\ \Omega$.

$$\text{Total circuit resistance, including meter resistance} = R = R_1 + R_m = \frac{250.0}{0.01} = 25,000\ \Omega = 25\ \Omega.$$

$$\text{Additional resistance} = R_1 = R - R_m = 25,000 - 4 = 24,996\ \Omega \approx 25\ \text{k}\Omega.$$

15.3.2 Moving Coil Dynamometer Instruments

The construction and operation of moving coil dynamometer instruments are the same as those of permanent magnet moving coil ones, except that two fixed air cored coils are used instead of a permanent magnet (Fig.15.6). Hysteresis and eddy current errors for measuring the ac quantity are eliminated due to the use of air cored coils. There are two types of coils in these types of instruments. One, a moving coil, is attached with a pointer, while the other one is fixed, being connected in series or parallel with the moving coil. The deflecting torque is proportional to the product of the current and voltage to be measured by the instrument. Hence, the scale of this type of instrument is non-uniform. They are widely used as a wattmeter, by which the power is measured in ac circuits.

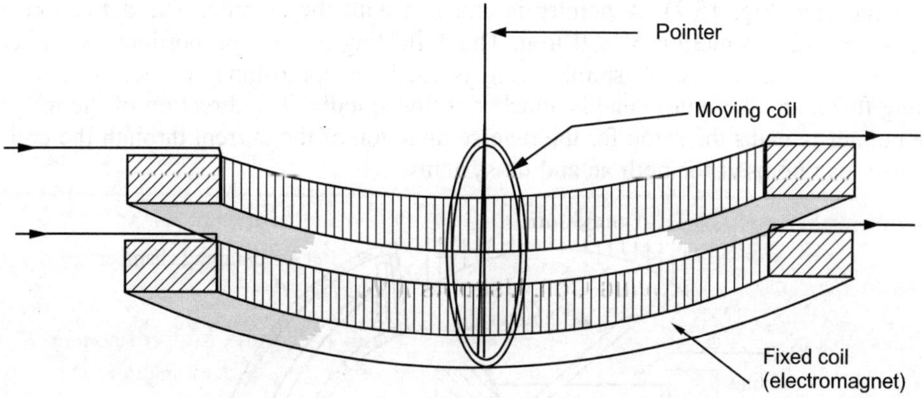

Fig. 15.6 *Electro dynamometer type instrument*

DYNAMOMETER INSTRUMENT USED BOTH AS AN AMMETER AND A VOLTMETER

For the ammeter, moving coils are connected in series as it is designed for low current. A shunt is connected across the moving coil. For the voltmeter, the moving coil is connected in series with high resistance and the combination is connected across the load whose voltage is to be measured.

15.3.3 Dynamometer Wattmeter

This type of wattmeter has two coils, viz. current coil and voltage or potential coil. The fixed coil or current coil is thick and has low resistance. It is connected in series with the load. On the other hand, the moving coil is connected in series with a high resistance across the load. The moving coil is known as the voltage coil or potential coil. The torque developed is proportional to the product of the voltage, current and power factor, which gives power consumed in ac circuits.

15.3.4 Moving Iron Instrument

The ammeters and voltmeters used in the laboratory and industry at power frequencies are moving iron instruments. This type of instrument is cheaper in cost than any other ac instrument of equal accuracy and ruggedness. There are two types of moving iron instruments, viz. attraction type and repulsion type. The instrument's operation is based on the magnetic field, which is set up by the measured current, or the current proportional to the voltage measured passing through certain number of coils.

Attraction Type

In the attraction type instrument, a small soft iron piece which is attached with a spindle moves toward the stronger field inside the coil from the weaker field while current is passing through the coil (Fig. 15.7). A pointer is attached with the spindle. The deflection of the pointer occurs due to motion of soft iron. The deflecting force is proportional to the current in the coil as given above. A spiral spring is used for controlling torque. To provide the damping force, an aluminum vane is attached to the spindle. The direction of the movement of the pointer remains the same for the reverse direction of the current through the coil. This instrument can be used for both ac and dc systems.

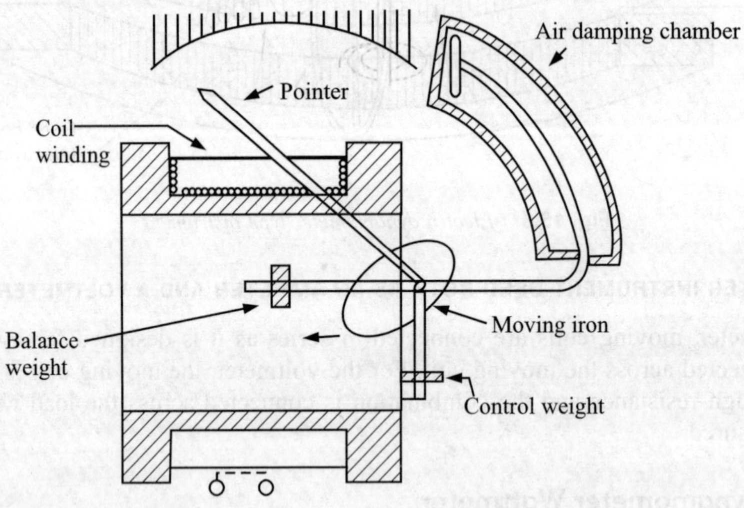

Fig. 15.7 *Attraction type moving iron instrument*

Repulsion Type

This type of instrument has two vanes or soft iron pieces within the coil, one stationary and the other movable (Fig. 15.8). The current to be measured is passed through the coil. The coil is wound with copper wire on a circular insulating former. For an ammeter, a few turns of large gauge wire are used. However, a large number of fine gauge wires are used for a voltmeter. The two soft iron pieces are magnetised due to the magnetic field produced by the current passing through the coil.

As a result, similar polarities are developed at the same end of the soft iron and repel each other. One of the two soft iron pieces is stationary, while the other one (movable) deflects, which causes the pointer to move. There is a spring which provides controlling torque. There is no movement of the pointer when deflecting torque is equal to the controlling torque. This type of instrument can also be used for both the ac and dc system, similar to the attraction type moving iron instrument.

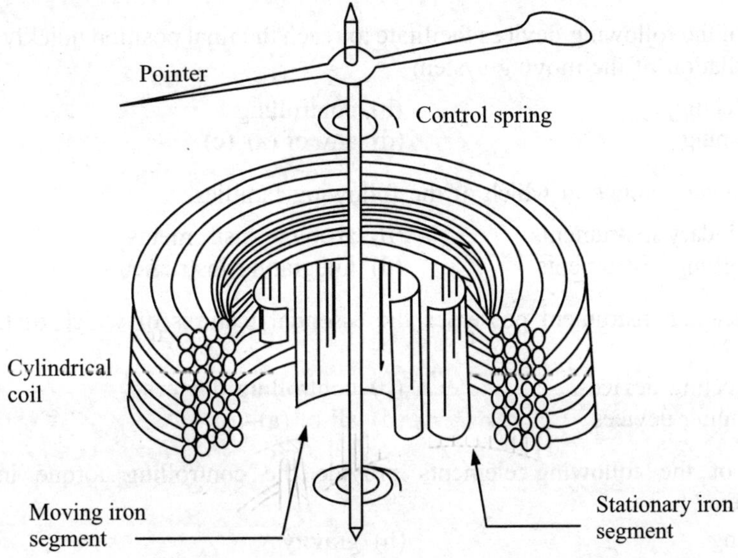

Fig. 15.8 *Repulsion type moving iron instrument*

Conclusion

In this chapter, we learnt about some electrical measuring instruments. Starting with the classification of measuring instruments, we studied the principle of operation of indicating instruments, along with the different components used there. Lastly, different types of measuring instruments, such as ammeters and voltmeters, are discussed.

Multiple Choice Type Questions

Tick the correct option(s), out of the four choices provided in each question.

15.1 Three moving iron type voltmeters are connected as shown in Fig. MCQ15.1. Voltmeter readings are V, V_1 and V_2 as indicated. The correct relation among the voltmeter reading is

(a) $V = \frac{V_1}{\sqrt{2}} + \frac{V_2}{\sqrt{2}}$ (b) $V = V_1 + V_2$

(c) $V = V_1 V_2$ (d) $V = V_2 - V_1$

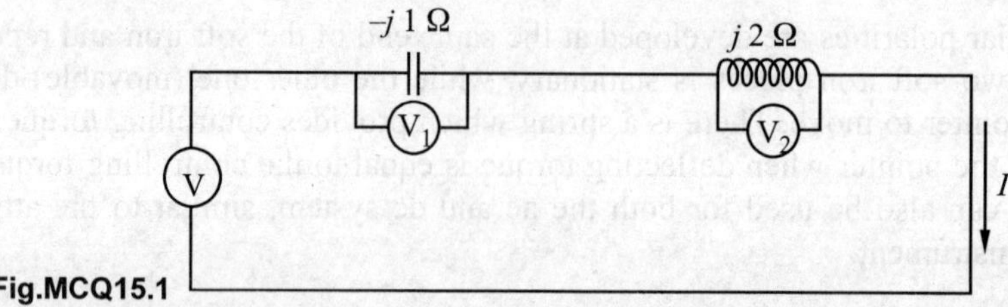

Fig.MCQ15.1

15.2 Which of the following devices facilitate to reach the final position quickly and prevents the oscillation of the moving system

(a) deflecting (b) controlling
(c) damping (d) any of (a)–(c)

15.3 An ammeter belongs in which of the following families?

(a) secondary instruments (b) absolute instruments
(c) recording instruments (d) integrating instruments

15.4 An indicating instrument possesses the essential features of which of the following families?

(a) deflecting devices (b) controlling devices
(c) damping devices (d) all of (a)–(c)

15.6 Which of the following elements provide the controlling torque in a portable instrument?

(a) spring (b) gravity
(c) eddy currents (d) all of (a)–(c)

Answers

15.1 (d) 15.2 (c) 15.3 (a) 15.4 (d) 15.5 (a) 15.6 (a)

Appendix

General Structure of an Electrical Power System

An electrical power system has three distinct components based on voltage level, and operation. They are (a) generation, (b) transmission, and (c) distribution. The layout of an electrical power system is shown in Fig. A1.

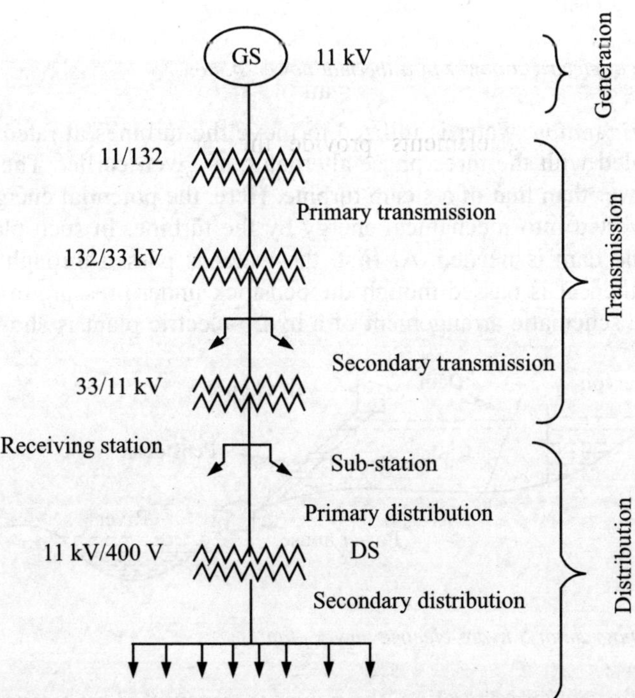

Fig. A1 *Basic layout of a simple power system*

Generation: The electric power is generated in generating stations, which are also called power stations. The generated voltage levels are 6.6 kV and 11 kV. Here, energy is converted from one form, mainly mechanical, as given in the examples, to another form, i.e., electrical. The main conventional power stations in our country are thermal and hydro-electric.

In a *thermal power station*, the source of energy is in the form of heat in the fuel, which may be coal, gas or oil. The heat is liberated by the combustion of the fuel, say coal, in the boiler furnace, and then, it is used to convert water to steam at high pressure and temperature, which, when passed through a steam turbine, gives off some of its internal energy, i.e., heat, as mechanical energy, driving it at rated speed. The steam turbine is mechanically coupled with the shaft of the three-phase alternator, which is termed as an ac generator – synchronous type, and connected to the grid. The entire operation of a thermal generating station is shown in Fig. A2.

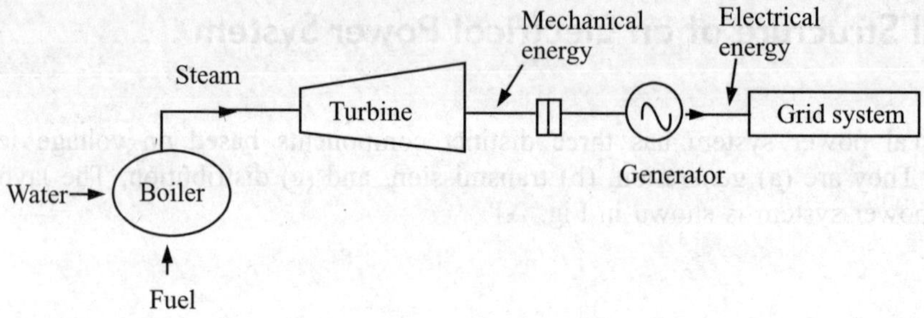

Fig. A2 *Production of electrical power in a thermal power-station*

In a *hydro-electric station*, water is utilized to move the turbines at rated speed, which are mechanically coupled with the three-phase alternator, as given earlier. The speed of a water turbine is much lower than that of a steam turbine. Here, the potential energy of water stored at a height is converted into mechanical energy by the turbine. In such plant, a river or the reservoir behind the dam is needed. At first, the water is passed through the intake works and fore-bay, and then, it is passed though the penstock under pressure to the turbine in the power station. The schematic arrangement of a hydro-electric plant is shown in Fig. A3.

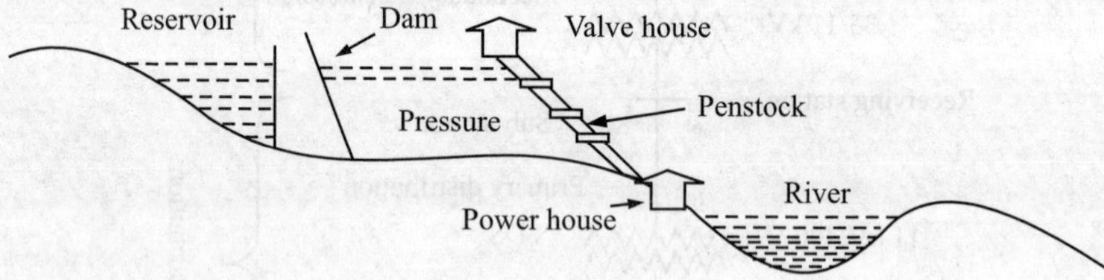

Fig.A3 *Schematic diagram of a hydro-electric power plant*

Transmission: After power generation, the power is transmitted by means of overhead lines over long distances near the city up to the distribution system at high voltages for economic reasons. The generating voltage is stepped up by a transformer. Usually materials, such as copper, aluminum, steel-cored aluminum, galvanized steel and cadmium copper are used as conductors, for overhead transmission. The voltage level used is determined on the basis of line length and power transmission. The different transmission voltage levels are 33 kV, 132 kV, 220 kV or even more. Sometimes, the transmission line may be divided into two parts for a big system. The voltage level of the first part of the transmission line is high (normally 132 kV, 220 kV, 400 kV), which is referred to as primary transmission line. The second part of the transmission line is known as secondary transmission. Here, the voltage is reduced by using a step-down transformer. It may be noted that from the receiving station at primary transmission, power is delivered to some bulk consumers through various substations.

Distribution: The distribution is further divided into two parts, primary and secondary distribution system. Primary distribution starts from the end of secondary transmission. Here, step-down transformers are used to convert higher voltage into lower level (33 kV/11 kV). Equipments such as circuit breaker, relay for protection, and various measuring instruments, are also used in the substation.

In secondary distribution, the voltage of 11 kV, available from the primary distribution, is delivered to distribution sub-stations near the consumer. This voltage is stepped down to 400 V and distributed by three-phase four-wire systems to the consumer. The voltage between any two phases is 400 V and between any of the phases to neutral is 230 V. For a three-phase motor operation, 400 V is used and for domestic purposes, 230 V is used.

Index

N

O

P

Q

R

501047 20072015